INTRODUCTION TO
AQUACULTURE

INTRODUCTION TO
AQUACULTURE

Matthew Landau
Stockton State College

JOHN WILEY & SONS, INC.

New York   *Chichester*   *Brisbane*   *Toronto*   *Singapore*

ACQUISITIONS EDITOR / Sally Cheney
COPY EDITOR / Marjorie Shustak
PRODUCTION MANAGER / Joe Ford
SENIOR PRODUCTION SUPERVISOR / Savoula Amanatidis
DESIGNER / Kevin Murphy
ILLUSTRATION COORDINATOR / Sigmund Malinowski
MANUFACTURING MANAGER / Lorraine Fumoso

Recognizing the importance of preserving what has been written, it is a policy of John Wiley & Sons, Inc. to have books of enduring value published in the United States printed on acid-free paper, and we exert our best efforts to that end.

Library of Congress Cataloging in Publication Data:

Landau, Matthew.
    Introduction to aquaculture / Matthew Landau.
        p.   cm.
    Includes bibliographical references and index.
    ISBN 0-471-61146-8
    1. Aquaculture.   I. Title.
    SH135.L36   1991
    639′ .8--dc20   91-4714
                                                                                       CIP

Printed in the United States of America

10 9 8 7 6 5 4 3 2 1

*This book is dedicated to
Jeane and Ely*

PREFACE

During the past three decades, the science of aquaculture has grown dramatically in importance, not only in a few places around the world, but almost universally. In areas with a long tradition of aquatic culture, like China, there has been recent expansion; and in other countries with little in the way of an aquaculture tradition, sudden interest has burgeoned.

When I was in college, few, if any, aquaculture courses were taught. Now, courses on the subject are offered at many of the major universities. In at least 34 of the 50 United States, one course on aquaculture is being taught at one school; most states offer more. Because the interest in aquaculture education is a rather recent phenomenon, a "teaching method" has not yet developed. And while there are many good aquaculture books available to the public, to the best of my knowledge, there has been no *general textbook* on this subject. *Introduction to Aquaculture* is an attempt to fill this void. It is hoped that, by filling the textbook void, a teaching method may gradually develop that will produce a crop of better-educated students in the field.

This text covers a wide spectrum of aquaculture-related topics. Many different fields play a part in aquaculture, and accordingly I have included sections on history, economics, law, engineering, chemistry, and biology. Not every aquaculture enthusiast will be interested in every aspect, and not every instructor will try to paint with such a broad brush. However, since this is an introductory book, all this information is at least touched upon.

Again, because this is an introductory text, no subject is covered in the great depth that the interested student might desire. At the end of the book is a bibliography containing a suggested reading list for each chapter; this is a starting place for seeking more information, but it is far from a complete bibliography. Most of the citations are from books and journals that are fairly easy to locate, and for the most part I stayed away from the "gray literature" of unpublished reports and preliminary talks presented at meetings. (However, a good deal of this unpublished information exists, and it may be enlightening for the more advanced student.) Instructors should be aware that some aquaculture videotapes currently available are excellent instructional aids and give a "living" picture of what is described here.

When a description of a process is presented here (for example, how to grow catfish), the student should understand that a generalized method is being described; many of the details and nuances have been left out because these vary with the particular conditions and traditions of the farm. There is no *one* way to grow catfish, to clean water entering a raceway, or to apply for permits to start a farm; it is clearly beyond the scope of an introductory book to explain everything in detail. Again, start with the suggested reading list.

I have included commercial culture procedures as well as experimental results that have been published in the scientific literature. In an introduction to aquaculture like this text, both should be available for consideration. Discussions of experimental results have been prefaced with such phrases as, "Biologists have recently shown . . ." or "Researchers have reported that . . . ."

Most important, I have tried to make this text "readable" for the person who is unfamiliar with the field. No assumptions are made about the background or education of the readers, other than that they are interested enough to spend some time with this text.

I welcome any suggestions from students and instructors on how to improve this book; any other comments on the material are also greatly appreciated.

Matthew Landau

ACKNOWLEDGMENTS

Many people have helped me with the publication of this textbook. The most helpful has been my wife, BJ, who painstakingly pored over the manuscript, pointing out what was clear and what wasn't, what contained too little or too much detail, and which topic flowed best into the next one.

Scientists in the United States and several other countries kindly sent me unpublished manuscripts and photgraphs, and answered countless questions during innumerable phone conversations and in a myriad of letters. None of them, however, bears any responsibility for any errors in this book. Thanks are offered to all these kind people, who are listed in alphabetical order here:

Dr. Bart Baca, Coastal Science Associates, Inc.; Dr. Kimon Bird and Mr. LeRoy Creswell, Harbor Branch Oceanic Institute; Mr. Tom Capo, The Howard Hughes Medical Foundation; Dr. Ernest Chang, University of California; Dr. Dudley Culley, Louisiana State University; Ms. Julie Delabbio, New Brunswick Community College; Dr. Eirik Duerr, Oceanic Institute; Dr. Earl Ebert, California Department of Fish and Game; Dr. Charles Epifanio, University of Delaware; Dr. Randy Hagood, Noraqua; Dr. Roger Hanlon, University of Texas Marine Biomedical Institute; Dr. Mike Hartman, Florida Institute of Technology; Mr. Liang Hongwu, Bureau of Aquatic Products (People's Republic of China); Mr. Roberto Hu, Nan Rong Fishing Machinery Company (Taiwan); Dr. Eugene Kaplan, Hofstra University; Drs. Hans Laufer and Shelagh Campbell, and Ms. Ellen Homola, University of Connecticut; Dr. Jasper Lee, Mississippi State University; Dr. David McKee, Corpus Christi State University; Ms. Shirley Metzger, Sort-Rite International, Inc.; Dr. David Moscatello, Stockton State College; Dr. Ferenc Muller, Fish Culture Research Institute (Hungary); Mr. Wayne Peterson, Peterson Fiberglass Laminates, Inc.; Ms. Laurie Pettigrew, New Jersey Division of Fish, Game, and Wildlife; Mr. R.J. Pierce, College of the Redwoods; Dr. Anthony Provenzano, Jr., Old Dominion University; Dr. Amir Sagi, The Hebrew University of Jerusalem; Dr. Evelyn Sawyer, Sea Run Incorporated; Mr. Allen Tom, University of Hawaii Sea Grant Extension Service; Dr. Fredrick Wheaton, University of Maryland; Dr. Bob Winfree, U.S. Fish and Wildlife Service.

M.L.

ix

CONTENTS

PART ONE

NONBIOLOGICAL METHODS

CHAPTER 1

INTRODUCTION

INTRODUCTION

Aquaculture (or sometimes, incorrectly, "aquiculture") has been defined in a number of ways. It has been called "the art of cultivating the natural produce of water; the raising or fattening of fish in enclosed ponds." Another author has said, "Aquaculture is the rearing of aquatic organisms under controlled or semicontrolled conditions. Thus aquaculture is underwater agriculture," and a lawyer writing on the subject has called aquaculture "the culture or husbandry of aquatic flora and fauna, but does not include the raising or breeding in tanks, nets, pens, or cages of aquatic flora and fauna (1) as aquarium specimens, (2) in laboratory experiments, or (3) by individuals on their own property as food for their own use." No single definition is universally accepted, but as the term is used in this text, we will consider aquaculture broadly and state it to be simply the large-scale husbandry or rearing of aquatic organisms for commercial purposes. (It differs, therefore, from **hydroponics,** which is the culture of terrestrial plants in an aquatic medium. **Mariculture** is aquaculture in brackish water or seawater.)

Does aquaculture have a role in our society? Each year the United States spends billions of dollars importing edible fish. In 1978, fishery products amounted to nearly 10% of the national trade deficit (that is, 28% of the deficit for nonpetroleum products). But aquaculture can be more than a potential means of reducing our need to import fisheries products; it can mean an increased number of jobs, enhanced sport and commercial fishing, and a reliable source of protein for the future. All of these are worthy goals and should be national and international concerns.

There are principles of aquaculture that must be addressed. In some ways, aquaculture is unique; in other ways, it is similar to fishing; and in still other ways, it most closely resembles farming. We will consider it to be a blend of fisheries biology, agriculture, limnology and oceanography, chemistry, animal and plant physiology, physics, engineering, and even law and business. Each aquaculture venture has its own special set of problems, be it salmon ranching, mussel rafting, kelp farming, or operating a shrimp hatchery. An aquaculture project should start with a preliminary economic analysis that asks, "Is this a reasonable business proposition?" Next, a suitable site is located for the business, then it is purchased or leased, and permits are applied for. Engineers should be consulted to design a facility that works efficiently. Ponds or raceways often must be built, pumps and filters purchased, and personnel hired. Some sort of water treatment may be needed. Stock animals must be bought, collected, or obtained by spawning reproductively active adults. The stock animals are normally grown out to a marketable size, harvested, and sold for the best possible price. During this process, water quality and health problems may arise. The "best" culture methods for the particular farm must be established. Aquaculture can be a complicated business, and those who enter it should be aware of all that it entails.

THE HISTORY OF AQUACULTURE

Early Farmers

Aquaculture has a substantial history, although its exact origin is still somewhat of a mystery. An Egyptian bas-relief on the tomb of Aktihetep (2500 B.C.) shows what appears to be men removing tilapia from a pond. In China, carp are known to have been spawned and reared about 2500 years ago, although many scholars believe the practice in that country may be twice that old. Wen Fang, founder of the Chou Dynasty, is often called the first fish farmer; it is said that when Wen Fang's predecessor, the emperor of the Shang Dynasty, had Fang confined to an estate in Hunan Province (1135 to 1122 B.C.), Fang built ponds and kept records on the growth and behavior of the fish. Fan Li, who wrote about aquaculture in 475 B.C., supported the idea of carp culture and described its practice in his *Yang Yu Ching* (*Treatise on Fish-breeding*). During the Tang Dynasty, 1400 to 1100 years ago, major breakthroughs in carp culture came with the initiation of polyculture. Before that period, the only fish cultured in China was the common carp; however, the family name of the emperors of the Tang Dynasty was Lee, which sounds like the Chinese word used for the traditional common carp. It therefore became a reli-

gious offense to grow or eat the common carp, so the early Chinese aquaculturists were forced to consider using other fish. It was found that when certain combinations of fish were used, greater yields were achieved, and this system of polyculture has been in use ever since. Aquaculture methods spread from China to Korea and then to Japan about 1700 years ago.

The Greek philosopher Aristotle also spoke of carp, suggesting that the Europeans were also interested in farming these fish. Carp culture was brought to England in the fifteenth century for the first time. Wild fish had been kept previously in England, in "stewponds," as security for times when red meat was difficult to obtain; this practice was not uncommon throughout Europe at that time. The death penalty could be imposed on anyone caught stealing a fish from another's pond.

Carp were not the only organisms of interest to early culturists. Mullet or other fish may have been cultivated off the Italian coast by the Romans; it is thought that this skill had been passed down from the Egyptians, by way of the Phoenicians and later the Etruscans. We know that Lucinus Murena, a Roman living in the first century B.C., dug fish ponds at Grotta Ferraia. During the Middle Ages, the lagoons and canals near Comacchio by the Adriatic Sea became a center for fish culture. The Comacchio waters were made significantly more productive later after a new canal system was built by Cardinal Palotta between 1631 and 1634. In France, during the mideighteenth century, salt pans were constructed in the Arcachon Basin, which soon also became great reservoirs to be fished by those granted permission. The most important species at Comacchio were eels and sandsmelt, while at Arcachon, the harvests were largely eels and mullets. Stephen Ludwig Jacobi built and operated a trout hatchery in Germany in 1741 to help support the increasingly popular recreational activity of sport fishing. Jacobi discussed his fertilization technique in the *Hannoverschen Magazin*, but it attracted little attention; the method was rediscovered and widely publicized in 1842 by Professor Coste of the College de France and two Vosges anglers, Gehin and Remy. In Czechoslovakia there is a nearly 900-year-old tradition of fish pond culture; the first ponds from Bohemia and Moravia date back to the tenth and eleventh centuries. Later, in the sixteenth century, extensive fish farms were built in those areas of Czechoslovakia, and it is estimated that they covered more than 120,000 hectares.

The practice of fish farming was probably brought from Polynesia in about 1000 A.D. to Hawaii, where, especially on Oahu, ponds were built to catch, store, and grow small brackish water organisms such as shrimp and fish, including mullet and milkfish (see Figure 1.1). These ponds were generally built of stone and could be quite large. Taro patches, which had to be kept under constantly flowing water, were also used as subsidiary fish ponds. Milkfish ponds were probably first built in Indonesia 600 years ago, although records have been kept only since 1821 when the Dutch government began to register the ponds.

Besides fish, bivalve mollusks have also been cultivated because of their great popularity as food. Perhaps no invertebrate or marine organism has enjoyed a culture history as long as the oyster, the rearing of which is believed to have started during the Roman Empire, perhaps by Sergius Aurata. According to Varro, other mollusks were also popular. It was either Fulvius Herpinus or Lippinus who first collected cockles and fed them in a vivarium for later consumption. Off the coast of France in 1235, a shipwrecked Irish sailor, Patrick Walton, made an interesting discovery: in order to catch birds for food, he suspended a large net on poles stuck in the mud flats of the area. Walton soon found that the poles became covered with mussels, and that the mussels on the poles grew more quickly than those in the mud. This was essentially the birth of the French mussel growing industry. Three hundred years ago, the Japanese began to develop a similar technology involving spat collection, the "hibitate" culture technique. In 1673, Gorohachi Koroshiya found that oyster larvae would settle and grow on upright bamboo stakes that were attached or driven into the ocean bottom; this laid the

Figure 1.1   Remains of an old fish pond enclosure on the coast of Oahu, Hawaii.

foundation for culture methods using floats and rafts.

Today, the production of nori in Japan is extremely important. This seaweed was cultivated in Hiroshima Bay from 1596 to 1614 and later was grown on poles in Tokyo Bay between 1675 and 1680.

Aquaculture During the Nineteenth and Early Twentieth Centuries

In Europe, the techniques of finfish culture were well established by the 1850s. In 1854, a major facility was built in Alsace by the French government, but in the United States the industry was just beginning. In 1853 an experimental fish farm was established just outside Cleveland, and another was built shortly thereafter in West Bloomfield, New York. In 1856 in Russia, Vrassky discovered the "dry method" for fertilization of fish eggs that calls for directly mixing eggs and milt, without the addition of water. The dry method greatly increased the chances of successful fertilization in many fish.

Richard Nettle, the first superintendent of fisheries for lower Canada, incubated and hatched the eggs of brook trout and Atlantic salmon in 1857. Pacific anadromous salmonids, in particular the chinook and coho salmon, were spawned soon after that, and the techniques soon spread south into the United States. In 1864, Seth Green started a highly successful trout breeding enterprise in Caledonia, New York, making use of his discovery of a new method for fertilization of eggs in a hatchery that increased fertility by about 50%. In 1877, culture methods from North America were brought to Japan to be applied to the chum salmon. Salmon culture then spread to mainland Asia, and the first hatcheries in the Soviet Union were built in 1927–28 at Lake Teplovka and Lake Ushkovskoye. However, failure was the rule for most salmon hatcheries until the 1940s, when much was learned about the physiology,

dietary requirements, and behavior of these fish. Today, salmonid hatcheries, with their freshwater nurseries, are important fisheries management tools.

The use of hatcheries has not fared as well for the Atlantic fisheries. During the second half of the nineteenth century, fishermen began to notice a decline in the catch per vessel. It was assumed that this was a result of overfishing and might be compensated for by the controlled spawning of traditionally economically important species, followed by the release of young fish from hatcheries. During the 1860s, this strategy became a possibility with the discovery by G. O. Sars that eggs from a marine fish, cod, could be artificially fertilized. Events followed quickly.

Fish culture became a recognized project of the federal government in 1871 when Spencer F. Baird became the first U.S. commissioner of fish and fisheries. At that time there was particular concern over the once common shad, which had been declining in catch during the previous 50 years. In 1873, approximately 35,000 shad were brought from the east coast for release in the Sacramento River, where the fish eventually established itself and became an important west coast species. Shortly thereafter, successful artificial fertilization experiments were carried out on the herring, haddock, and the American pollack. This paved the way for the construction of the first U.S. Fish Commission commercial marine fish hatchery in 1885 in Woods Hole, Massachusetts. Hatcheries were soon built at Gloucester Harbor and Boothbay Harbor that concentrated on the production of cod and similar species of commercial interest. By 1917, the three hatcheries produced 3 billion young fish per year, consisting of mostly pollack and flounder and lesser numbers of cod and haddock.

Similar events were taking place in Europe. In 1882, Captain G. M. Dannevig of Norway founded a commercial hatchery, which was established with the aid of both private and public funding, to supplement the cod fisheries in the fjords. That particular hatchery was taken over by the Norwegian government in 1916. Dannevig's techniques were different from those used in the United States in that he used large spawning ponds to obtain eggs, while in North America a stripping technique was used.

Dannevig also invented a rocking incubator that increased the hatching efficiency of the eggs. Progress made elsewhere stimulated the Scottish government to build a hatchery at Dunbar near the Marine Biological Station. The Dunbar hatchery was largely devoted to the production of plaice, and was completed in 1883. Very similar to the Norwegian facility, it operated under the direction of Harald Dannevig, who acted as supervisor. Dannevig left Dunbar in 1902 and moved to Australia, taking with him plaice and turbot from Europe in tanks located between the decks of a mail steamer. A hatchery was built for him at Gunnamatta Bay in New South Wales; by 1906 he had a stock of a few thousand adults, and released 20 million young fish that year. In England, W. A. Herdman, an enthusiastic scientist of the 1890s who did much to establish the Liverpool Marine Biology Committee at University College, set up a hatchery at Liverpool. Herdman's aggressive style also resulted in operations at Piel for plaice and flounder production, and in another hatchery at Port Erin, Isle of Man.

Yet some scientists were not convinced that the hatchery program actually helped the fishing industry. They believed that the increased landings of cod after their experimental release, which was heralded as justification for the continued building of hatcheries, was merely an artifact of the natural ups and downs seen in the catch statistics. The leading skeptics, Kurt Dahl and J. Hjort, pointed out that although the Norwegian catch had increased immediately after the hatchery release program began, so had the catch in other countries. In a joint study conducted by Dahl and Dannevig from 1903 to 1906, cod fry were sampled from the nearby fjords before and after the hatchery releases. The results were inconclusive but spurred a series of studies by each party that continued over a number of years. Although Dahl and Hjort's efforts to "prove" their case were not completely successful, there was a general decline in support for the fry release programs because hatchery supporters could not demonstrate clearly that their critics were wrong. This notwithstanding, hatchery production in the United States remained high for a number of years until the Woods Hole hatchery was taken over by the Navy Department in 1943; the hatcheries at Boothbay and Gloucester were closed in 1950 and 1952, respectively. The government apparently was not convinced that the stocking program was sufficiently beneficial to the Atlantic fisheries industry to justify the cost of the operation. The promise of large-scale propagation that would support the fisheries off the coasts of Europe and the United Kingdom was not realized either. These hatcheries began to lose support during the years just before and after World War I, and most were closed or converted into research stations.

Although many of the large propagation–release programs had run their course, the science of aquaculture was moving forward. Artificial spawning of mullet was achieved for the first time in Italy in 1930. In the 1920s, in a trout hatchery in New Jersey, G. C. Embody began to selectively breed animals for increased growth and resistance to disease with impressive results. The use of hormone injections for inducing breeding of fish began in Brazil in about 1932, and has been the key to the spawning of many difficult to culture species. In the 1930s it was reported that newly hatched brine shrimp, which were cheap and very easy to hatch from cysts in the laboratory, were an excellent food for fish larvae; these are now the most important live food for young fish and crustaceans.

Tilapia culture experiments were begun in 1924 in Kenya and later in the Congo in 1937. Naturally reproducing populations of this African fish were found in Java in 1939, although how they got there is still a mystery. The culture of tilapia became popular in Java during the World War II Japanese occupation of the island, when milkfish culture began to decline. The Japanese brought the tilapia to mainland Asia by way of Malaya, and it has spread throughout the continent, becoming an extremely important organism for aquaculturists.

For a long time the oyster was a staple of many coastal communities, especially in New England. However, as oysters were taken from their beds, there was a decrease in the setting of oyster larvae—not only because the breeding adults were being removed from the population, but probably more importantly because the shells that the larvae prefer to set on were not being returned to the water, so the larvae had less available substrate upon which to settle. Laws in the United States restricting the collection of this shellfish date back to colonial times, but by the early 1800s the oyster was clearly on the decline. Modern oyster culture research began in France in the 1850s with M. de Bon and M. Coste. In 1879, William Brooks of Johns Hopkins University first reared oysters to the larval stages in the laboratory. However, scientists were unable to sustain the oyster past this point in their development; the problem was that cultured or artificial food was not available, and when water with fresh food was introduced into a culture, the larvae were washed away. William Wells and Joseph Glancy, working for the New York State Conservation Commission in 1920, succeeded in carrying the larvae to the setting stage with the use of an aerating system and centrifugal apparatus. Victor Loosanoff, who became the first director of the National Marine Fisheries Service Laboratory at Milford, Connecticut, along with his coworkers, made a major contribution to the culture of this species in 1950 by describing the mass rearing of the oyster through its delicate larval stages. Loosanoff also added to our understanding of the induced spawning of other bivalves.

Other mollusks were also being studied. At the Fairport Fisheries Biological Station, beginning operation in 1910 on the Iowa banks of the Mississippi River, studies were carried out on the life cycle and culture of the pearly freshwater mussels (see Figure 1.2). In 1935, Saburo Murayama in Japan reared abalone larvae he had obtained through artificial fertilization. Collection of natural scallop spat was first tried in 1935 by T. Kinoshita in Lake Saroma, and he induced spawning in 1943 by increasing the temperature and the alkalinity of seawater during the reproductive season.

The culture of planktonic microalgae to be used as a food for young oysters and other filter feeders has also been a long-evolving process. P. Miquel grew a few diatoms in his laboratory in the early 1890s, making the observation that water from lakes, ponds, and the ocean were unable to support much growth unless they were enriched with mineral solutions. In 1934 Føyn developed the "Erd-Schreiber" medium, which contained a mineral solution and a soil extract that provided an excellent growth medium for many species of algae. During the 1940s, S. P. Chu made pioneering discoveries on the development of various media. Continuous cultures of algae, that is, cultures that are partially harvested while new medium is added, were first attempted in Czechoslovakia during the early twentieth century, but it was not until the 1940s that practical continuous culturing was developed by J. Monod, B. H. Ketchum, and A. C. Redfield.

Figure 1.2 The Fairport Fisheries Biological Station, where research was carried out on the culture of freshwater mussels. (From Coker, 1914.)

Lobsters, like the fish released to increase the Atlantic fisheries harvests, were cultured during the late nineteenth and early twentieth centuries to slow the decline of the natural populations. (By 1873, Massachusetts had already begun to restrict this crustacean's collection.) At Wickford, the Rhode Island Commission of Inland Fisheries developed a floating laboratory with screened compartments holding larval lobsters that were raised from eggs; the lobsters were reared to the fourth larval stage, when they begin their benthic existence, and then released (see Figure 1.3). Motosaku Fujinaga (sometime written as Hudinaga) first successfully spawned and partially reared a marine penaeid shrimp in Japan in 1934. During the 1950s and 1960s, Fujinaga's techniques spread to the United States, thanks in part to his stay in the National Marine Fisheries Laboratory in Galveston in 1963. Meanwhile, J. B. Panouse, working in France in the 1940s, made basic discoveries dealing with the hormone systems in crustaceans and the induction of maturation following the removal of the eyestalks that are the centers of crustacean endocrine activity.

Perhaps the most interesting story of all is that of the Malaysian prawn, *Macrobrachium rosenbergii*. Shao-wen Ling, a fisheries consultant to the Food and Agriculture Organization of the United Na-

Figure 1.3 A technician counting fourth-stage lobster larvae for release at the Wickford marine laboratory. (From Mead, 1908.)

tions (FAO), noticed this large crustacean in many markets throughout Asia. He began to experiment with the rearing of the animals and found that they bred very readily in the laboratory (see Figure 1.4). However, all the larvae of the first batch died after five or six days; his experiments continued for months, but always had the same results. The culture of the larvae frustrated him, and Ling began to try adding almost anything at hand, including tea or some fish his wife had made him for dinner to the larval cultures, but with no luck. Then one day he added soy sauce to the container with the larvae and they lived. (Ling's use of soy sauce is a popular piece of aquaculture folklore, but the validity of this event has been questioned.) It turned out that the prawns, like eels, are catadromous organisms that spend most of their life in freshwater but must begin their existence in seawater to complete their early development; therefore, the salt in the soy sauce was the key.

The Recent Status of Aquaculture

An International View   The study and practice of aquaculture has come a long way in the past three decades. This has been a worldwide phenomenon, extending into all the continents except Antarctica. It is estimated that global aquaculture production in 1987 reached 13.1

Figure 1.4 A gravid female *Macrobrachium rosenbergii*. Fisheries biologist Shao-wen Ling found that these prawns would breed easily, but salt must be added to the water for the larvae to complete development. (Photo courtesy of the University of Hawaii Sea Grant Extension Service.)

million metric tons, that is, about 12.3% of the total world fish and shellfish harvest. Furthermore, the International Aquaculture Foundation has predicted continued expansion with an annual growth of 5.5% between the present and the year 2010.

The culture of aquatic organisms is most important in Asia, where over 80% of the world aquaculture harvest is taken (see Table 1.1). External loans from development banks for aquaculture projects on that continent increased from $15.5 million in 1978 to $52.8 million in 1981. Along with seaweeds, mollusks, and a variety of crustaceans, the most important finfish produced in this area traditionally have been carp, tilapia, mullet, and milkfish. In 1982, a combined milkfish harvest of almost 30,000 metric tons was realized in Indonesia, Taiwan, and the Philippines. However, the production of milkfish may be on the decline because of changing consumer demographics; in Taiwan the total production of that species in 1982 was only 70% of that in 1975. (Note: Accurate and timely data on aquaculture production is often difficult to obtain. In many instances, the production values in this book may be from the late 1970s to the mid-1980s, as they are in the examples given above. While this is unfortunate, it is assumed that old data are better than no data.)

In Asia, because of the availability of extensive warm brackish water sites, farmer's produced 192,000 metric tons of marine shrimp in 1986, about 79% of the total world aquaculture harvest. Japan, where the value of shrimp is extremely high, has been the leader in production technology and has until recently also been the major producer (an honor that now goes to the People's Republic of China, which harvested about 79,000 metric tons in 1986 with production constantly growing since then). Taiwan produced 60,000 metric tons of shrimp in 1987 on less than 7000 hectares, more than 15 times the 3800 metric tons of 1979 (see Table 1.2.)

TABLE 1.1 Leading Aquaculture Producing Countries in 1985

	Finfish	Mollusks	Other
China	2,392,800*	1,120,000	1,689,400
Japan	283,900	359,000	540,600
Korea, Rep.	3700	369,000	417,500
Philippines	243,700	37,900	212,800
United States	195,200	128,000	30,000

* Values given as metric tons.

Seaweed aquaculture has also been on the rise in Asia. While all edible seaweeds have a market in Korea, China, and Japan, it is clearly the red algae, nori, that dominates trade. In Japan, the production of nori has increased more than twofold in the past two decades and is currently about 325,000 metric tons per year (making it the most cultured aquatic product in that country). Japan also grows more than 100,000 metric tons of wakame, a brown seaweed. Korea farms the seaweed *Hizikia*, much of which is exported to Japan. Besides being eaten, seaweeds are used for a number of industrial processes, such as the production of agar, carrageenan, and alginates; in 1980 it was estimated that 176,000 dry metric tons of seaweeds were used in the manufacturing of these colloids. The value of seaweeds varies greatly according to the demand. For example, in 1981 nori sold for $8300 per ton in Korea, while seaweeds for industry were rarely valued at over $1000 per ton and were more often sold for $200 to $400 per ton.

The People's Republic of China, with its long history of aquaculture, remains the world's largest producer of aquatic products. In the late 1970s, the government began to encourage aquaculture as a means to increase the country's protein consumption; in 1984, it was estimated that the Chinese harvested 2.5 million metric tons of freshwater and saltwater organisms. This total was doubled just one year later. In just the past few years, the number of communes and households engaged in culture of fish and shrimp has increased to over 2.3 million. Under the system of economic management before the Cultural Revolution, the rural populace had no say in the crops to be harvested, and much emphasis was placed on grain production. An incentive system called the "Responsibility System" was adopted in 1980, under which the farmers can contract with the government to raise fish in ponds (remember, all land and ponds belong to the government); the government sets an annual production minimum for each area, and the farmers are able to keep, as profit, all the fish that exceed this set production level. With the increased demand for fish in China's growing urban areas, many people who had been engaged in traditional farming have turned to aquaculture.

Much of China's pond fish culture is carried out on the Yangtze River Delta

TABLE 1.2 Production of Crustanceans and Seaweeds in 1985

	Crustaceans	Seaweeds
Africa	100*	0
North America	33,800	200
South America	32,900	4900
Asia	198,500	2,767,500
Europe	300	4500
Oceania	100	100

* Values given in metric tons.

and in the Pearl River Delta area (which contains about 28,000 hectares of ponds, about 10% of the delta). Aquaculture products represent 87% of the aquatic harvest of the delta. In particular, Shunde County has about 16,500 hectares of ponds, although most ponds are less than $\frac{1}{3}$ hectare in size (ponds are usually measured in "mu" which are $\frac{1}{15}$ hectare) (see Figures 1.5 and 1.6). The fish raised here are tilapia and various species of carp; these are primarily reared in integrated polyculture systems. In addition, the areas around the ponds are often used for cultivating plant crops; grass cuttings, weeds, and animal wastes are used to fertilize the ponds, and pond soil is used in the fields of vegetables.

India has a potential 67 million hectares of land for freshwater aquaculture and another 2 million for marine culture. Like China, India has a strong polyculture program centered on several species of carp. Marine shrimp and freshwater prawns are also cultured.

Salmon production is also very important in Asia. In 1983, Japan released al-

Figure 1.5 Net cage culture in the Xing'an River in Zhejiang Province. (Photo courtesy of Liang Hongwu, Bureau of Aquatic Products, People's Republic of China.)

most 2 billion salmon fry from hatcheries. Ocean cage culture of salmon in Japan began in the 1970s and yielded less than 100 metric tons in 1978, but this increased to about 4400 metric tons by 1984. Eel, too, is a widely eaten fish in Korea and Japan, where it is cultured and considered to be a gourmet food that commands premium prices. Ayu or sweetfish is also popular, along with carp, crucian carp, and trout. The yellowtail is the most commonly cultured marine fish in Japan; in 1978 about 122,000 metric tons of yellowtail were produced while the wild catch was less than 30 metric tons. In Ja-

Figure 1.6 Collecting fish for later growout on a Chinese fish farm in Shunde County, Guangdong Province. (Photo courtesy of Liang Hongwu, Bureau of Aquatic Products, People's Republic of China.)

pan and Korea the relative value of the cultured fish, as compared to the total harvest of fish, has increased sharply in recent years.

While the Japanese are interested almost exclusively in chum salmon, in the Soviet Union significant attention is also paid to the pink salmon. About 860 million pink and chum salmon were released by the USSR in 1978. New Zealand also has over a dozen active salmon farms that use both cage culture and open ocean ranching techniques; the resulting output is probably about 500 metric tons per year, but the government of New Zealand expects that this will increase dramatically before the turn of the century. Although the aquaculture effort in Australia is modest at the moment, research on the culture of a number of species, along with warm waters and an extensive coastline, suggest that Australian aquaculture has a good deal of potential. At present there are about 10,000 metric tons of oysters produced in the New South Wales area of Australia, and there are trout farms throughout much of the continent. Microalgae is being mass cultured for industrial purposes in Western Australia. In addition, experimental culture of native crayfish is underway, with government facilities producing about 300 metric tons each year.

Latin America accounts for less than 1% of the world aquaculture yield. However, in recent years, culture in that area—particularly in Ecuador and Panama—has been dominated by the successful production of saltwater shrimp for export. In 1987, Ecuador harvested an estimated 70,000 metric tons of cultured shrimp, an impressive increase over the 3900 metric tons harvested eight years before. In the late 1970s and early 1980s, shrimp production was highly variable, the reason being largely a function of the availability of seed animals. Shrimp culturists had to depend on wild, seasonally available postlarval shrimp, but there

have been a significant number of shrimp hatcheries built since 1983, and the availability of the postlarvae has begun to stabilize.

The average annual shrimp production in 1984 was about 605 kg/ha/year in Latin America as compared to less than 400 kg/ha/year in Asia. There were approximately 8200 hectares used for farming shrimp in 1979, but there was a fivefold increase in this by 1983. The success of Ecuador and Panama has encouraged other countries to try shrimp farming. New methods of culture are being developed in Brazil, and projects are under way on the Caribbean Islands (including Antigua, the Bahamas, and the Dominican Republic), Colombia, Costa Rica, French Guiana, Guatemala, Honduras, Mexico, and El Salvador. Peru has a small shrimp industry, as well as an expanding scallop industry, and although it is just now finally recovering from the damage caused in 1983 by the El Niño phenomenon, aquaculture is expected to grow quickly in the future. In Chile, a rapidly growing salmon industry exists that concentrates largely on coho; salmonid production grew from 1720 metric tons in 1987 to 15,410 metric tons in 1989 (over 29% of the worldwide Pacific salmon aquaculture harvest). There is also a large hatchery for the production of the Pacific oyster as well as the Peruvian scallop.

Mexico, with its extensive coastlines and semitropical climate, also has a great aquaculture promise. Currently catfish are being produced, as well as oysters, prawns, and shrimp. Culture of marine organisms such as oyster and shrimp must, by law, be done by the country's aquacultural cooperatives, which receive permission to farm from the Federal Department of Fisheries. If the government approves of the project, the cooperative is eligible for assistance from BANESCA, a special bank that provides low-interest loans to fishing and aquaculture co-ops. It is estimated that 400 tons of freshwater prawns and 300 tons of marine shrimp were produced in Mexico in 1984; a harvest of about 4000 metric tons of shrimp was projected for 1990.

It is generally felt that the aquaculture potential of Latin America is substantial, but increased production is hampered because fishery products are not in high demand locally (in fact, despite protein deficiencies in a number of countries, these products are often exported). Other problems include high spoilage because of the warm temperatures, and poor transportation and marketing networks in many of the Latin American nations. However, several native freshwater fish show considerable potential and may impact the market in the future.

In Europe, one of the most rapidly developing mariculture industries is that of Norway's salmon farming (see Figure 1.7). Norwegian salmon culture, which has grown significantly since 1977, was responsible for nearly 63% of the world's cultured Atlantic salmon in 1989. A large number of these salmon are exported to France and the United States; about 40% of the 33,700 metric tons that were

exported in 1986 went to these two countries. Currently, over 300 hatcheries in Norway sell trout and salmon fry for growout, and this has led many observers to estimate that Norway soon will be producing 80,000 metric tons of salmon each year. Production in 1989 was 74,000 metric tons. Salmon farming has spread to several other countries in northern Europe, most notably Scotland.

Trout, especially the rainbow trout, are important culture items in many European countries. About 150,000 metric tons were grown in 1985 throughout the continent. Denmark's production of 24,000 metric tons plus France's production of 27,000 metric tons accounted for over one-third of the total. Italy, Finland, Germany, Spain, and the United Kingdom all harvested significant numbers of cultured trout, too. Even more often grown in Europe than the trout, carp accounted for over 400,000 metric tons of harvested fish (of which 250,000 metric tons were grown in the USSR). Common carp is the warm water fish most often grown in France, although the greater portion of these fish are exported to Ger-

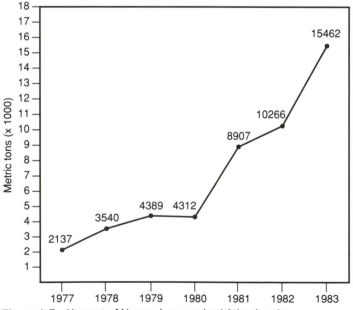

Figure 1.7 Harvest of Norway's pen-raised Atlantic salmon.

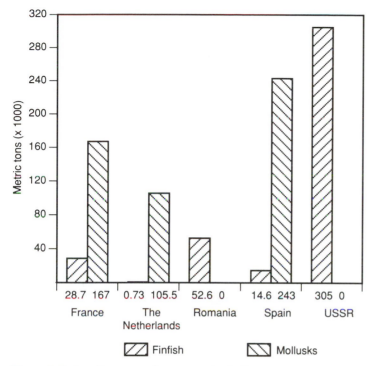

Figure 1.8 Leading aquaculture countries in Europe.

many. Carp are the most frequently reared fish in the approximately 20,000 hectares of ponds in Hungary, which has vigorously supported government research programs on the culture and biology of these animals (see Figure 1.9). In Germany, there are aquaculture operations dedicated to the production of carp, as well as several species of trout, the European eel, and tench. Trout, eels, and to a lesser extent, carp and catfish, are grown in Italy. In Portugal, with an annual per capita fish consumption of 36 kg (which is second only to Japan), trout, salmon, and carp are being explored as culturable species. In 1985, Poland produced about 19,000 metric tons of aquaculture products, Romania about 52,000 metric tons, and Czechoslovakia about 16,000 metric tons.

The most commonly cultured shellfish in Europe is the mussel. It is extremely popular in Spain, where, on the Galicia coast alone, there are about 3500 mussel culture operations that each year produce about 250,000 metric tons. The Netherlands has also become a center of mussel culture in recent years. France leads Europe in the production of oysters with over 100,000 metric tons being grown annually. Crayfish are demanding premium prices, so some small farms for their production have been started in France, al-

Figure 1.9 The Aquaculture Research and Development Center at Szarvas, Hungary. A modern laboratory that includes 55 hectares of experimental ponds, a fish hatchery, and a warm water system heated by geothermal energy. (Photo courtesy of Dr. Ferenc Muller, laboratory director.)

though the leading European producers of these crustaceans are still Sweden and Finland.

Perhaps no country relies on fish production more than Israel, where over half of the fish consumed are grown in ponds. Common carp and tilapia are the most important products, but shrimp, prawns, mullet, silver carp, and several other species are receiving attention. The amount of pond acreage in Israel has remained in the neighborhood of 4500 to 5100 hectares since the mid-1960s because of a limited water supply, but production (fish harvested per hectare) over that period has risen by about 40% from 3490 kg/ha in 1979 to 4910 kg/ha in 1987. Israel's government, by way of the Ministry of Agriculture, has long encouraged experimental culture techniques; this has resulted in Israel's status as one of the most advanced aquaculture countries in the world.

Elsewhere in the Middle East, aquaculture has been limited. In Jordan the first commercial facility was established in 1983 in the northern Jordan Valley. Several private fish farms have begun operations since then, producing limited amounts of carp, trout, and tilapia. There is every reason to believe aquaculture will become increasingly more important in this area.

Despite its history of tilapia research, African aquaculture is still in many ways in its early developmental stages, but this may be changing. New research facilities have been established, while existing facilities have been enlarged with the understanding that aquaculture may be a key to solving the problem of protein deficiency in the diet of many of Africa's people. The most popular fish to culture in Africa are still tilapia, and new methods of rearing these fish, including cage culture, are being explored in Egypt, Niger, and the Ivory Coast. Finfish culture in Africa in 1986 resulted in a 11,616 metric ton harvest, but the shellfish accounted for only about 278 metric tons.

Canadian aquaculture has been evolving for over 100 years. Salmonids are the most important fish cultured, especially trout, salmon, and char. The salmon harvest in 1988 was 10,000 metric tons, but production is expected to increase rapidly in the near future. In British Columbia, there were roughly 10 licensed salmon farms in 1982, 82 farms using 283 hectares of crown land by the end of 1986, and by 1988 this had grown to 178 farms covering 940 hectares; this phenomenal growth rate shows no signs of slowing in the near future. The major cultured shellfish in Canada is the oyster, which is grown on both the east and west coasts.

Aquaculture in the United States   The annual per capita seafood consumption in the United States has been growing steadily; in 1987 it reached a record 7.0 kg, up 0.32 kg from 1986. The greater demand for fish and shellfish suggests that aquaculture will become increasingly important in the United States. Not surprisingly, commercial culture of aquatic organisms in the United States has significantly increased over the last 20 years, as has aquaculture research supported by the government, private companies, and universities. Most of the oysters, catfish, crawfish, and rainbow trout that are consumed are cultured, but little attention is paid to "low-cost" fish such as carp, tilapia, and mullet.

Despite new increased levels of production, most experts feel that aquaculture has not expanded as quickly as it could have. The reasons are to some extent technological, but political, social, and economic obstacles seem to be more serious. When these limitations are addressed, it is possible that there will be a substantial increase in aquaculture production in the United States.

Although catfish farming in the United States has not always been a profitable and smoothly running industry, it is currently enjoying good times. At the moment, more catfish are produced than any other species, and by some estimates there are more pounds of catfish grown in the United States than all other species of aquatic organisms combined. In 1982 there were 987 operating catfish farms, but this increased to 2003 in 1988, covering 57,730 hectares and is still growing. The warm water channel catfish is the most frequently farmed species (see Figure 1.10), with the blue catfish and the white catfish running a distant second and third. The center of the catfish farming community is Mississippi, where 18% of the country's catfish farms produce 80% of the catfish. There has been a 600% rise in production in the United States since 1980. In 1987, processing plants purchased over 127,000 metric tons of farm-reared catfish.

The culture of salmon is carried out by commercial and government hatcheries that release fry for later harvests. In 1982, 620 million salmon and steelhead trout were released on the Pacific coast. Government-bred fish probably represent about 40% to 50% of the chinook and coho salmon taken by sport and commercial fisheries. Studies conducted on re-leases in the Columbia River have shown that there was a $4.20 return on each dollar spent on culturing the fish, and coho had a benefit-to-cost ratio of 7 to 1. Most salmon culture takes place in the Pacific Northwest, but there has been recent interest in salmon farming in New England. The trout industry, based in Idaho because of its clean water and suitable temperatures, follows catfish and salmon in finfish production. Despite its long history, which began over 100 years ago, trout farming has remained stagnant in recent years, with production staying in the area of 20,000 to 26,000 metric tons per year. Besides private trout growout facilities, there are also many government trout hatcheries, largely financed by money collected from the sale of fishing licenses, that produce small trout to be released into streams to support the sport fishing industry (see Figure 1.11).

Baitfish have long been a popular item for culturists. There are probably over 14,000 baitfish farms in the United States producing over $50 million worth of fish each year. In 1987 there were about 12,200 metric tons harvested. Baitfish

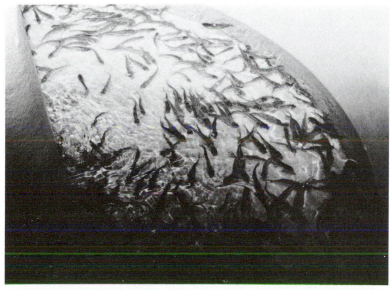

Figure 1.10 Channel catfish are the most commonly cultured fish in the United States. (Photo courtesy of the New Jersey Division of Fish, Game, and Wildlife.)

Figure 1.11 Trout are reared both for direct consumption and (as shown here) for restocking streams fished by the public. (Photo courtesy of the New Jersey Division of Fish, Game, and Wildlife.)

trail only salmonids, crawfish, and catfish in annual production.

Tropical fish and invertebrate culture for the aquarium hobbyist is also important; on the retail level, over $200 million is spent each year. This is a Florida-based operation that consists of at least 220 trop-

ical fish farms in that state alone. Because of the unpredictable nature of the wild supply and import problems, aquarium fish culture is seen by some as a potentially large field. At the moment, most of the culture work concerns the more easily bred freshwater species, although salt-water fish traditionally have a greater retail value.

Other important fish include large-mouth bass, which are largely produced in government hatcheries for the sport fishing industry; over 6 million were reared in 1982. Striped bass culture is also carried out in government hatcheries as well as some private facilities. Sun-fish, smallmouth bass, pike, and walleye, among others, are also currently in culture.

The crawfish is the most commonly cultured invertebrate and is second only to catfish in private production. (In the United States the term "crawfish" is more commonly used by culturists than "cray-fish," but either is correct.) Essentially all of the farms are located in the Southeast, most in Louisiana where the growth of the industry has been a very important part of that state's economy (see Figure 1.12). For

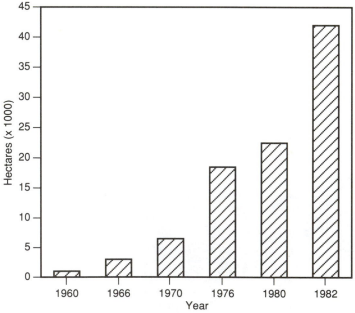

Figure 1.12 The change in Louisiana acreage dedicated to the production of the crawfish.

the past several years the annual production of crawfish has been greater than 40,000 metric tons. The 1987 harvest was valued at approximately $53 million. Commercial culture of other species of crustaceans is limited, although some operations are successfully producing brine shrimp in California, and hatcheries for marine shrimp and freshwater prawns exist in Hawaii and the Southeast, along with a few growout farms.

Many people feel that the culture of abalone on the west coast is very near at hand, and mussels and hard clams (see Figure 1.13) are cultured by companies largely on the east coast. However, the most commonly grown bivalve mollusks are the various species of oyster. In 1987, the oyster harvest in the United States was about 11,800 metric tons (valued at $46 million). Like the rest of the bivalve, shrimp, and prawn culture industries, oyster farming is divided into those operations that are self-sufficient, those that are only growout operations, and those that specialize in the production of seed. The oyster industry covers almost the entire coastal United States, although the species used varies with the location.

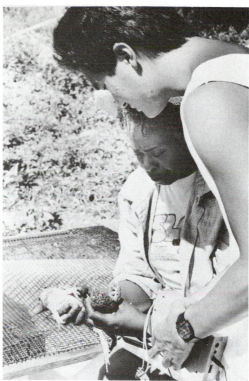

Figure 1.14 Frog culture is carried out both for the production of laboratory specimens for scientists and for food. Here Mr. Allen Tom, an aquacultural extension agent, is examining animals on a frog farm on the north shore of Oahu. (Photo courtesy of the University of Hawaii Sea Grant Extension Service.)

Figure 1.13 The culture of clams is a growing industry in the United States, especially on the East Coast. (Photo courtesy of LeRoy Creswell, Harbor Branch Oceanographic Institution.)

Besides finfish, crustaceans, and mollusks, other aquatic animals have been or are being cultured in the United States. Frogs are grown for laboratory scientists as well as for food (see Figure 1.14). Sea turtles have been reared to protect dwindling natural populations. Alligators were originally grown for the same purpose as the sea turtles, although now the wild stocks are so strong that the cultured animals can be sold for meat and skins.

SUMMARY

Aquaculture is the large-scale production of aquatic organisms. It is related to other applied sciences such as farming and fisheries, but has elements that are unique. Some of the disciplines that the aquaculturist may need to be familiar with include law, economics, chemistry, engineering, and biology.

The origins of aquaculture are not clear, but it was probably first practiced by either the Egyptians, who may have reared tilapia, or the Chinese, who grew carp. It then spread through Asia and Europe. During the nineteenth century, fish cultured for release to support Atlantic commercial fisheries became very important. Fisheries' release programs became unpopular during the twentieth century, but fish, mollusk, crustacean, and algal culture technologies made significant advances during this period.

Today, aquaculture accounts for over 13 million metric tons of aquatic products harvested each year, and the industry is growing rapidly. It is extremely important in Asia, where carp, tilapia, yellowtail, salmon, shrimp, and seaweeds are grown. In Central America, aquaculture is dominated by a very productive shrimp industry. In Europe, the Atlantic salmon, eels, trout, carp, oysters, and mussels are cultured in large numbers. Israel's freshwater and marine culture systems are among the best in the world. In Canada, salmonids are the most cultured species. In the United States, catfish, salmonids, baitfish, crawfish, and several species of mollusks also generate significant amounts of income.

CHAPTER 2

WATER QUALITY

Water quality is a critical concept for the aquaculturist. If the water is "bad," plants and animals won't grow or reproduce. Animals stressed because of poor water quality are also prime targets for pathogens and parasites. Just as people who work in factories or offices that are stuffy and have smoke or chemical fumes in the air are more apt to be sick, so it is with aquatic organisms grown in poorly oxygenated water in which toxins can accumulate.

To fully appreciate the rearing of an aquatic organism, one must have some idea about the medium it inhabits. Water chemistry can be an extremely complex subject, but even the most basic information about the properties of water is valuable to the culturist.

THE CHEMISTRY OF WATER

Properties of Pure Water

A review of pure water will be helpful to the culturist because typical freshwater has properties that are nearly identical to it and seawater's behavior can be better understood.

Water is composed of two hydrogen atoms and one oxygen atom in a configuration resulting in a "four-pronged" molecule. This simple structure is responsible for the unique properties of water that were critical for establishing life on earth. The oxygen atom is large, with an atomic weight of 16, compared to the hydrogen with weights of about 1 per atom. The hydrogen atoms form an angle of about 105° and compose two prongs of the molecule. The result is a water molecule that has a positive charge on one side where the hydrogen atoms are found and a negative charge on the other side where the electrons dominate the structure. The two unshared electron pairs can be pictured as the other two prongs of water. Such a molecule is said to be **polar** because there are strong charge differences at its two ends (see Figure 2.1).

Water molecules are attracted to each other because of these negatively and positively charged areas. The unshared negative electron prongs form weak bonds with the positively charged hydrogen prongs of other nearby water molecules. These attractions between water molecules are called **hydrogen bonds.**

Water can be found in three forms: as a solid (ice), liquid, and gas (steam or water vapor). When liquid water is heated, steam forms. As steam, the molecules of water are independent, and there is no hydrogen bonding. The energy that has been added to the liquid water by heating it has made the molecules vibrate and

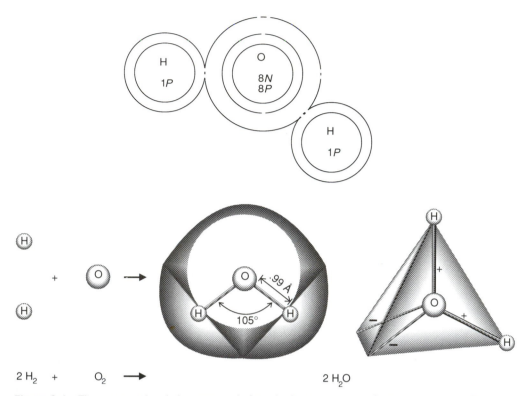

$$2\,H_2 \;\; + \;\;\; O_2 \;\; \longrightarrow \;\;\; 2\,H_2O$$

Figure 2.1   The water molecule is composed of two hydrogen atoms and one oxygen atom. The outer electron shell of the oxygen is completed by electrons shared with the hydrogens. Thus, two electron pairs are associated with the hydrogen atoms, and two electron pairs are not. The unshared pairs are the negative prongs of the water molecule while the hydrogen atoms are the positive prongs. (From Nelson, Robinson, and Boolootian, 1970, *Fundamental Concepts in Biology,* and Nason and DeHaan, 1973, *The Biological World,* John Wiley & Sons, Inc.)

move so quickly that the existing bonds are broken. Since there are no bonds holding the molecules together, they separate and move apart; water in the form of steam no longer has a particular volume, rather its volume is defined only by the structure containing it. Just the opposite is true for ice; not only is the volume defined, but so is the shape. In this case, there is little energy available, so water molecules can form relatively permanent bonds with their neighbors. The molecules do not easily rotate or move, and there is minimal molecular vibration; the lower the temperature of the ice, the less the vibration. Ice is considered a crystal, held together by hydrogen bonds in a very regular manner. This crystal results in a relatively open structure, making ice

less dense than liquid water (this is why ice cubes float).

Liquid water with a temperature greater than 4°C behaves like most known liquids because as it cools, it becomes more dense. This increase in density occurs because there is a decrease in molecular vibration, allowing the molecules to be packed more closely together. However, when it reaches about 4°C, water becomes atypical because unlike most other compounds the structure starts to become more crystal-like and therefore less dense; its density continues to decline as it cools from 4°C to 0°C when it freezes. Most liquids simply continue to get denser as they get colder. This property of water is important for the culturist since it means that in cold weather, ice forms on the top of a pond, acting as a barrier between the cold air and the water below. If water behaved like other substances, there would be a danger of ponds freezing solid from the bottom up.

Of course, aquaculturists are most concerned with water as a liquid, when it is neither a gas nor a solid. The liquid's properties are a result of the rapid formation and breaking of hydrogen bonds that is constantly taking place. (This is why liquid water doesn't hold a shape as ice does.) As the temperature increases, the rate at which bonds break and form also increases.

The density of water in ponds is a function of temperature. When ice melts, it becomes denser as it warms to 4°C. Once 4°C is reached, further warming causes the density to decrease. In some deep bodies of water, mixing between the upper layer of water and the lower layers is inhibited because a density-stratified system has developed. This stratification is not very pronounced in shallow ponds, but becomes more significant as pond depth increases. The upper layer of water gets so warm that it floats on the colder deep water that may eventually become depleted of oxygen. The area where the warm water changes quickly to cold water in a vertical profile of the pond or lake is called the **thermocline.**

In some lakes and ponds there is a phenomenon called **turnover.** This refers to those periods during fall and spring when the surface water increases in density and vertical mixing can take place throughout the water column (see Figure 2.2). From an ecological standpoint this is critical since it allows nutrient-rich deep water to come to the surface where it can be used by phytoplankton that must also have enough sunlight to carry on photosynthesis. For the culturist, depending on how a pumping or aeration system is set up, winter and summer stratifications can result in drastic changes in a culture system receiving lake water.

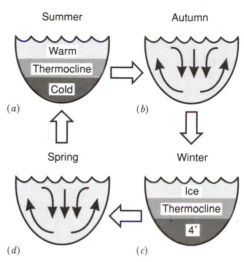

Figure 2.2 Turnover in a deep pond or lake. (*a*) In summer, the upper waters are warmed by solar radiation and therefore, decrease in density. A thermocline develops between the upper water and the denser deep water. There is little water exchange between the surface and the water below the thermocline because of the difference in densities. (*b*) In autumn, the surface water begins to chill and gradually reaches the same density as the lower waters. Since there is no difference in the densities, vertical mixing can take place. (*c*) In winter, the surface waters continue to chill. As the temperature of the surface water drops below 4°C, the density begins to decrease, and again the less dense water floats on top of the denser water. Unlike the situation in the summer, now colder water is floating on denser warm water. (*d*) In spring, the surface water begins to warm. When it reaches the density of the lower water, there again may be a vertical mixing.

Water has a very high **specific heat value** of 1.0, which means that large amounts of energy must be used to increase its temperature. Other compounds, such as alcohol, for example, have smaller specific heat values, so the amount of energy (measured in calories or joules) that must be added to a liter of alcohol to raise its temperature 1°C is much less than the amount of energy added to a liter of water to raise it by 1°C. The reason that water has this high specific heat value is that much of the added energy is used to break the hydrogen bonding as opposed to just increasing the motion and the speed of the vibrations of the molecules (which is what we measure as heat). Extra energy is also needed to go from one phase of water to another. While it takes about one calorie to raise a gram of liquid water 1°C, it will take approximately 540 calories to change 1 gram of liquid water at 100°C to steam at 100°C. To change a gram of ice at 0°C to liquid water at 0°C, 80 calories must be added. The high specific heat of water has made the evolution of life on earth possible, since the temperature of the earth's surface is stabilized by the ocean's ability to buffer temperatures, thus preventing large changes from day to night or season to season. Likewise, deep aquaculture ponds are more temperature stable than are shallower ponds.

Properties of Seawater

Seawater has many characteristics that are similar to pure water, but there are also certain differences that merit discussion. Seawater is 96.5% water and 3.5% salts, although this may change in estuaries and the coastal oceans, where significant evaporation can take place raising the salinity, or more commonly, where there is a large amount of river runoff diluting the seawater. Of the salts dissolved in seawater, sodium (Na^+) and chlorine (Cl^-) make up about 86% of the ions, and sulfates (SO_4^{-2}), magnesium (Mg^{+2}), calcium (Ca^{+2}), and potassium (K^+) compose about 13%. The rest of the ions identified in seawater combine to make only 1% (see Table 2.1).

Most of the major and minor elements and ions in seawater are said to be **conservative,** meaning that their ratios to each other are constant despite any changes in the total salinity; they are not involved to any great extent in biological activities. **Nonconservative** elements and ions, such as nitrogen (N), phosphorus (P), and sometimes under culture conditions silicon (Si) and carbon dioxide (CO_2), are taken up by phytoplankton at significant rates; therefore, their ratios to each other and to the conservative elements change

TABLE 2.1   Composition of Seawater

Constituent	Concentration (ppm)
Chloride	18,980
Sodium	10,560
Sulfate	2560
Magnesium	1272
Calcium	400
Potassium	380
Bicarbonate	142
Bromide	65
Strontium	13
Boron	4.6
Fluoride	1.4
Rubidium	0.2
Aluminum	0.16–1.9
Lithium	0.1
Barium	0.05
Iodide	0.05
Silicate	0.04–8.6
Nitrogen	0.03–0.9
Zinc	0.005–0.014
Lead	0.004–0.005
Selenium	0.004
Arsenic	0.003–0.024
Copper	0.001–0.09
Tin	0.003
Iron	0.002–0.02
Cesium	~0.002
Manganese	0.001
Phosphorous	0.001–0.10
Thorium	≤ 0.0005
Mercury	0.0003
Uranium	0.0015–0.0016
Cobalt	0.0001
Nickel	0.0001–0.0005
Radium	8×10^{-11}
Beryllium	.
Cadmium	.
Chromium	.
Titanium	Trace

Source: Spotte, 1973.

with the biological activities in the system. Nutrients will be discussed in greater detail in appendices 1 and 5.

Salts in water, naturally, will alter some of the properties of water. The degree of the effect is a function of the amount of salt in the water. For example, the more salt that is dissolved in water, the denser the saltwater becomes. (Density is also affected by temperature, just as it is in the case of freshwater.) Viscosity, the resistance to flow, also increases as salt is added to water, but vapor pressure will decrease as salinity increases. The refractive index will change, so that light entering seawater is bent to a greater extent than light entering freshwater (that is, the speed of light slows down more in the more dense medium). The freezing point and temperature of maximum density of water are lowered as salt is added.

SOURCES OF WATER

Freshwater

There are two important sources of freshwater: ground water, taken from below the earth's surface, and surface water, which comes from lakes, streams, and rivers.

About 97% of the planet's unfrozen freshwater is ground water. Normally, ground water originates as rain or snow that eventually percolates through the soil. The water in the soil is stored or flows, according to gravity, through fractures in rocks and areas of unconsolidated sand and gravel. The rate at which water can travel through soils is variable; it will move relatively quickly in a coarse sand that has large spaces between the grains, but shales that have a very limited void volume are much less efficient at conducting water. Soils that are more uniform in grain size (these are called **well-sorted** soils) are generally better sources of ground water than are soils that are poorly sorted.

Any geological formation that is capable of holding or transmitting water is called an **aquifer.** Unlike water in a river, there is considerable resistance to flow in an aquifer, so a velocity measured in terms of centimeters per day is not uncommon. The velocity through the aquifer can be expressed in terms of Darcy's equation,

$$v = kh/x$$

where v = velocity measured in meters per second, k = the permeability of the soil measured also in meters per second, x = the distance (measured in meters) through the aquifer that the water has to travel, and h = is the head loss due to resistance, also measured in meters. (**Head** can be thought of as the pressure at the bottom of a column of water, the higher the column the greater the pressure. The force that the head pressure exerts is reduced by the aquifer material.)

Aquifers are either contained, meaning that the water-bearing material is confined between two relatively impermeable layers of soil or rock, or uncontained, meaning that they are subject to contamination from the surface. The uppermost or surface aquifer can be either contained or uncontained, and is called the **water table.** Below this there may be one or more confined aquifers (see Figure 2.3).

The depth of the uncontained water table will vary somewhat from season to season because the amount of water that percolates down varies with the amount of rain or melting snow. This affects the height of the water above the confining lower boundary of the aquifer. The surface aquifer's depth can also change from one location to the next as the height of soil above the confining bottom changes. Areas where a stream has eroded into the surface aquifer and there is an exchange of water are called **seeps.** When water enters the aquifer from the surface and the height of the water table is raised, it is termed **recharge.** The surface aquifer

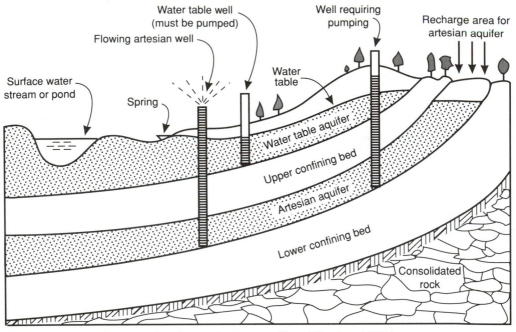

Figure 2.3 Ground water and surface water sources. (From Wheaton, 1977.)

that communicates with streams and rivers that drop in level during dry seasons will experience a lowering of the water table. Such changes in the height of the water table will affect the culturist that is using the surface aquifer as a source of water. Deeper aquifers are not affected by dry and rainy seasons and are therefore normally used for drinking water supplies.

The composition of ground water is different from the original rain or snow. Rainwater is naturally acidic, having a pH of roughly 5.7, although this can vary with the location. The acidic character of the rainwater comes from the CO_2 in the air that mixes with the water to form carbonic acid (**acid rain** refers to rains containing stronger acids that form when fossil-fuel fumes dissolve in the rainwater resulting in pH values of 3 to 4). The slightly acidic rain landing on the ground is able to dissolve some of the minerals in the soil, so these minerals will be found in the ground water but were not present in the original rainwater.

Ground water is obtained by drilling a **well** into an aquifer. If the recharge area for an aquifer is below the top of the well, then the water must be pumped to the surface. If the recharge area is above the top of the well, which is going into a confined aquifer, then the water will simply flow out of the well; this is called an **artesian well.** Artesian wells may be pumped to increase the speed of the water flow.

Freshwater can be taken from surface sources as well as from the ground. These surface sources can be slow moving or stagnant (lakes and ponds) or quickly moving (rivers and streams). While surface waters may be obtained without drilling and are rich in oxygen, unlike ground waters, they are also subject to pollution to a much greater extent than even the surface aquifers and have more variable temperatures than ground waters, especially in nontropical environments.

Rivers and streams tend to be particularly rich in oxygen, but there are problems associated with their water. The amount of water flowing in smaller streams and creeks will often vary a great deal from season to season so, during a dry period, there may not be enough water for the culturist to use. Smaller and slower streams and creeks tend to change temperature quickly, so they may become too warm or freeze. Rapidly moving streams tend to carry a large amount of suspended material like silts and small sand grains; when the quickly moving water is diverted into the culturist's ponds, the suspended materials will begin to settle, resulting in a sediment layer forming over the bottom of the pond (and over any benthic organisms). The silt can reduce the amount of light available for phytoplankton or clog the gill mechanisms of aquatic animals.

Water from lakes and ponds tends to differ from water taken from streams and rivers. Lakes and ponds usually have fewer particulates suspended in the water, but the oxygen content is often not as great as in fast-moving streams. The temperatures, water chemistry, and nutrient levels of lakes will not change as dramatically over a short period of time as a river's will because there is a smaller surface area: volume ratio. However, any changes that do occur, such as contamination by a particular pollutant, may last longer. Slower-moving water from lakes may also mean a greater chance of disease organisms being present.

Saltwater

Although the use of seawater wells and artificial sea salts are clearly on the rise, most land-based aquaculture facilities that cultivate marine organisms either pump water from intakes suspended in bays or inlets, or use tidal changes (as in the case of Asian milkfish and shrimp culture operations). As stated earlier, such coastal waters can be very variable in their chemical and biological compositions. The number of particulates tends to be

high, not only because waves suspend bottom sediments but also because seawater is more dense than freshwater; therefore, suspended materials will resist sinking to a greater extent than in freshwater.

Seawater wells are built in a number of ways. Generally, a portion of the sea floor is excavated and then partially filled with gravel. A pipe is laid on the gravel bed that runs to a pump at the culture facility; the portion of the pipe that is on the gravel has holes or slits cut into it. After the pipe is laid, more gravel is placed over it, and the gravel bed is finally covered with the sand that was originally present. Water that penetrates the sand can flow into the gravel and pipe. The gravel functions to keep sand from clogging the holes in the pipe. Water from seawater wells tends to be much more uniform in quality and has fewer particles than does water from suspended intake pipes, although the dissolved oxygen will be lower. The site of the well must be an area with sand that is neither too fine nor has too high an organic content.

Seawater is subject to contamination by pollutants just as freshwater is. In addition, seawater itself is highly corrosive, and therefore special materials must often be used when a marine culture facility is being built. These include tanks, pipes, pumps, and anything else that will be in contact with the water or be in the "spray zone." Most plastics can be used without being affected, but metals generally should be avoided if possible, even stainless steel. When concretes that are reinforced with steel are used, care must be taken that salts are not present in the concrete (that is, the builder should not use seawater or unwashed sand when mixing the concrete) since salts will corrode the metal; a dense, impermeable type of concrete must be used when reinforced with steel.

Fouling of saltwater pipes by barnacles, mussels, tunicates, bryozoans, and other invertebrates and algae can be troublesome. The fouling organisms effectively reduce the internal diameters of the pipes and increase frictional loss of pressure. Antifouling paints cannot be used to prevent settling in pipes since these paints tend to work by slowly leaching a toxin (often copper) into the water and therefore would contaminate the animals being grown, or at least slow down their growth and reduce their reproductive capacity. Several methods have been devised to keep fouling to a minimum; these can be classified as

1. those that prevent the buildup of fouling organisms (e.g., keep water speeds in the pipes greater than 4 meters per second thus preventing attachment)
2. those that remove fouling organisms that have already settled (e.g., a plastic "snake" or "pig" that is forced through the pipe and scrapes away the attached organisms).

Regardless of the source of the water, there may be legal problems associated with either taking it or discharging it after it is used. These will be discussed later, in Chapter 12. In addition, all seawater and surface freshwater sources may contain parasites or larval predators that could damage the stock (this is not a problem with ground water); the elimination of these will be the subject of Chapter 5.

WATER QUALITY PARAMETERS

There are several commonly defined water quality parameters with which the aquaculturist should be familiar: pH and alkalinity, salinity and hardness, temperature, dissolved oxygen, and nutrients (including nitrogen, phosphorus, and silicon, among others). These parameters are considered critical because they affect the health and productivity of the culture

system, and in some cases the effectiveness of chemical filtering systems or drugs added to treat diseases. A general understanding of water quality can be a powerful tool. Unlike a broken pump or a diseased fish, bad water cannot be seen; water must be tested, and the results of the tests must be understood in the context of the aquaculturist's system. Methods for measuring several of these parameters are covered in Appendix 1.

pH and Carbonate Alkalinity

Scientists express the concentration of hydrogen ions $[H^+]$ as **pH.** (Note: The use of brackets [] denotes molar concentration, that is, the number of moles of a substance per liter. Therefore, $[H^+]$ refers to the number of moles of hydrogen ions in a liter solution.) By definition,

$$pH = -\log[H^+] \quad \text{or} \quad [H^+] = 10^{-pH}$$

Pure water is a very weak electrolyte and will ionize

$$H_2O \leftrightarrow OH^- + H^+$$

We can define a general dissociation term, K_{eq}, which for water is

$$K_{eq} = \frac{[H^+][OH^-]}{[H_2O]}$$

and another term, K_w, specifically for the water, can be generated as

$$K_{eq}[H_2O] = K_w = [H^+][OH^-]$$

It has been shown experimentally that for pure water, at a temperature of 25°C, $K_w = 1 \times 10^{-14}$. And since the concentration of H^+ must be equal to the concentration of OH^-,

$$[H^+][OH^-] = [H^+]^2 = 10^{-14}$$
$$[H^+] = 10^{-7}$$
$$pH = 7$$

That is, the concentration of the H^+ (or the HO^-) is 10^{-7} moles/liter (M), a very small amount because water is about 55.4 M. To put it another way, 1 H^+ forms in every 5.54×10^8 water molecules. If an acid, a substance that gives up hydrogen ions, is added to water, the concentration of the H^+ will increase. If, for example, the H^+ concentration is raised 100 times so that it is 10^{-5}, then the pH is 5. If a base, which will take up H^+, is added to the water, the concentration of the H^+ will be lowered and the pH will rise.

Many water sources are acidic. Extremely low pH values (<4) have been recorded; these are unable to support plants, fish, or invertebrates. These low pH values are caused by the presence of strong acids, such as sulfuric acid, in the environment. Many water sources are slightly acidic (pHs of 4.5 to 6.5), which may be caused by the presence of weak acids leaching in from the soil. The pH of a pond can also be affected by the activity of plants and phytoplankton (see Figure 2.4); this will be discussed in the next section.

Acidic waters in ponds can be neutralized, but such water treatment is almost never a one-time operation, and the water should be constantly monitored for a declining pH. The most common treatment for acidic water is the addition of agricultural lime (calcium carbonate, calcite, $CaCO_3$), quicklime (calcium oxide, CaO), dolomite ($CaMg(CO_3)_2$), or builder's lime (calcium hydroxide, $Ca(OH)_2$) to the water. The addition of lime to an acid pond will result in water molecules and calcium salts of the acid, so for example, sulfuric acid in the water becomes calcium sulfate.

Water can also become too alkaline, although this is less frequently seen than acidic waters. Fish cannot live at pHs above 11. A major fish kill was recorded in the late 1800s when a large amount of alkali water was dumped into the Colorado River from the Arizona Gila River. Ammonium sulfate has been sug-

Figure 2.4 The pH of a hatchery in China is measured. (Photo courtesy of Liang Hongwu, Bureau of Aquatic Products, People's Republic of China.)

gested as a treatment for high pH, but caution should be used since alkaline conditions favor the formation of toxic ammonia (NH_3) rather than ammonium (NH_4^+).

There have been many studies on the effects of pH in aquatic culture systems. Northern pike can be conditioned to survive in alkaline lakes with pH values of 9.5 to 9.8, but bullfrogs cultured in waters with pHs of 7.2 and 8.3 showed mortality rates of 10% and 97%, respectively; this may have been due to the direct effects of the alkaline water, or it may be related to a general weakening of the organism that allowed bacterial pathogens to gain a foothold. In acidic water fish increase their production of slime, become hypersensitive to bacterial infections, and will sometimes show a discoloration of the gill tissue.

It is clearly advantageous to be able to stabilize the pH of water. **Alkalinity** is an expression of the anions in the water that will act to neutralize H^+ when an acid is added. Carbonate and bicarbonate ions are the most important of these anions, so the term **carbonate alkalinity** has been developed to express the amount of acid that the CO_3^{-2} and HCO_3^- in the water can react with. Alkalinity affects the behavior of some compounds in the water; for example, copper sulfate, used to control algal growth, snails, and many protozoan ectoparasites, will vary in its toxicity, both to the target and the nontarget organisms, according to water alkalinity (it is more toxic in less alkaline water).

Unlike most gasses, when CO_2 molecules enter the water, they react with the water molecules. The result is a buffering system like the one that most animals have in their blood. First, there is a production of carbonic acid, which partially dissociates to bicarbonate ions, and the bicarbonate ion itself will partially dissociate to a carbonate ion:

$$CO_2 + H_2O \leftrightarrow H_2CO_3 \text{ (carbonic acid)}$$
$$H_2CO_3 \leftrightarrow H^+ + HCO_3^- \text{ (bicarbonate)}$$
$$HCO_3^- \leftrightarrow H^+ + CO_3^{-2} \text{ (carbonate)}$$

The dominant ion in water systems with a pH of 6.5 to 10.5 is the bicarbonate ion; at lower pH values carbonic acid becomes the major form, and in basic waters the bicarbonate ions are replaced by carbonate ions. This system acts to buffer the water, preventing large changes in the H^+ concentration. If an alkaline substance is added to water having a pH of 7, bicarbonate ions begin dissociating into H^+ and carbonate ions to maintain the pH. If an acid is added, the equation shifts and the bicarbonate combines with the H^+ and becomes carbonic acid. Under experimental conditions, when eggs of Atlantic salmon, arctic char, and brook trout were allowed to develop in water with a low pH, the mortality rates to the alevin stage were 23%, 40%, and 25%, respectively; when the water was first passed through a shell–sand filter, the pH rose and mortalities of 1%, 6%, and 9% were recorded.

A pond with a lot of plant or phytoplankton activity will have a high pH during the day while photosynthesis is taking place because CO_2 is being removed from the system. During the evenings, or when it is overcast, there is more CO_2 being produced than is being used so the pH will drop and the pond will become more acidic.

Salinity and Hardness

While natural waters vary in their purity, none are "just water." Other compounds are always found, to some extent, dissolved in the water. Salinity and hardness are closely related terms that refer to the nature of these dissolved substances.

Hardness is normally used when we are discussing freshwater. Hardness originally referred to the ability of water to precipitate soap. The precipitation is largely a function of the concentration of calcium and magnesium in the water, but may also be caused by ions of several metals as well as by hydrogen ions. Hardness has now come to mean the total concentration of calcium and magnesium ions expressed in terms of parts per million (ppm) of calcium carbonate. Soft water is 0–55 ppm, slightly hard water is 56–100 ppm, moderately hard is 101–200 ppm, and very hard water is 201–500 ppm.

Calcium is important in water systems since it is taken up and used in the bones of fish and in the shells of crustaceans and mollusks. Soft water cannot be used in the culture of crawfish because the exoskeletons will be too thin to offer the sort of protection that the animals need. Calcium may also be important in the hatching of some fish eggs. The eggs of the marine dolphin fish, *Coryphaena*, will not hatch in calcium-free seawater, and magnesium seems to be important for its development, especially immediately before and after hatching.

Salinity is usually discussed when speaking of saltwater, and is defined as the number of grams of dissolved inorganics in 1 kg of seawater after the bromides and chlorides have been replaced with an equal amount of chlorine. Typical open ocean seawater has a salinity of about 3.5% (often written 35‰ or 35 ppt, both meaning "35 parts per thousand"). Most ocean culture takes place in the coastal ocean, often near the surface, where the salinity is below that of open ocean seawater; the salinity of the water in the coastal ocean is apt to be in the range of 28–33 ppt.

If salinity has to be adjusted in a culture system, it can be lowered by adding freshwater or raised by adding sea salts or hypersaline water that has been formed during evaporation. If a **euryhaline** species of organism (one that can withstand large changes in salinity without being greatly affected) is being grown, the culturist may wish to adjust the salinity to prevent other species or disease organisms from becoming established. For example, if the brine shrimp *Artemia* is reared in a high-saline water (>60 ppt), marine competitors or pathogens will probably be unable to survive in the culture environment, so the crop would be relatively pest-free (see Figure 2.5).

There have been a vast number of experiments carried out on the effects of salinity on growth and reproduction of cultured aquatic species. In one study, microalgae (*Isochrysis* and *Tetraselmis*) were grown in a series of enriched artificial seawaters; both species appeared to grow better at about 25 ppt than at higher or lower salinities. Several studies have also demonstrated that salinity changes result in an alteration of the organism's physiology. For example, the brine shrimp has different nutritional requirements at different salinities; in the bream *Sparus*, fewer abnormal fish developed in experiments when the eggs were incubated at 30–40 ppt at 18°C, or 35–45 ppt at 23°C, than when other combinations of temperature and salinity were used.

Figure 2.5 A synthetic salt mix that can be used to increase the salinity of water. (From Spotte, 1973.)

Dissolved Oxygen

All commonly cultured organisms need oxygen to survive. Higher green plants and phytoplankton can make their own oxygen as long as there is enough sunlight and CO_2 to carry on photosynthesis, but animals must extract oxygen from the water (or air). The oxygen in the air, just like that in a pond, is the product of plants containing chlorophyll. During daylight plants are both making *and using* oxygen, and in the dark they are still using it even though photosynthesis is no longer taking place. It naturally follows that the ponds showing the greatest range in oxygen content over a 24-hour period are the ponds with large phytoplankton or benthic plant populations. The oxygen levels will be lowest near dawn, while in the midafternoon the water may be supersaturated with oxygen.

Aquatic animals are generally very good at removing oxygen from the water. Fish and higher invertebrates do this with gills (or similar structures), which are thin tissues with relatively large surface areas. As water passes over the gill tissue, oxygen diffuses from the water into the blood or hemolymph and is carried away by a pigment molecule, such as hemoglobin, that has a strong affinity for the oxygen. While a complete discussion of countercurrent gill flow and oxyhemoglobin dissociation curves is beyond the scope of this text (but fascinating reading for those interested in the evolution of physiological adaptation), it should be made clear that the gills are wonderfully designed tools, and parasites or pathogens that attack the gill structures are very serious indeed.

Our atmosphere contains over 20% oxygen, but water will hold only a very small fraction of that. The actual amount of oxygen that can be found in a sample of water is a function of the chemical characteristics of that sample, in particular the salinity and temperature. Heating the water will reduce its capacity to hold oxygen, as will increasing the salinity.

Dissolved oxygen (DO) in aquatic culture is one of the most critical parameters. This is the reason that so much time and effort have been devoted to developing and testing aeration systems. Recent advances in electronics and computer microprocessors have spurred interest in automatic monitoring of assorted parameters in aquaculture ponds (see Figure 2.6). Automatic sampling of DO at various depths in ponds is perhaps the best use that these sampling devices can be put to, since DO can change so quickly and can vary so drastically with depth when the pond is stratified.

In the natural environment there is normally enough oxygen in water to support the animals that are living there. But under aquaculture conditions, especially in intensive culture, we often find ourselves with very unnatural conditions where the demand for oxygen by the fish or other organisms is extremely high because

1. The cultured organisms are stocked at high densities.
2. The microbial breakdown of excess wastes in the ponds (a result of feeding for quick growth) also requires large amounts of oxygen.
3. Phytoplankton, macroalgae, and higher plants also require O_2, but

when there is insufficient light, they do not "put it back" into the water by way of photosynthesis.

We call this oxygen that is required by the cultured organisms, and by bacteria and plants, the **biological oxygen demand** or **BOD.**

Because most aquaculture systems have a high BOD, culturists sometimes use aeration devices (which bring air and water together so that there can be a transfer of oxygen into the culture medium) (see Figure 2.7). Aeration systems will be discussed in greater detail in Chapter 5. The rate of transfer is a function of a number of factors. Not only will it depend on the ability of the water to hold a given amount of oxygen, which as we have stated is related to temperature and salinity, but it is also related to at least three other criteria:

1. *The amount of air that comes in contact with the water.* It should be obvious that even if air very rich in oxygen is used, if the volume of air in question is small, then it will not contain the amount of oxygen required. That is, it will not meet the BOD, and there will be an oxygen shortage.
2. *A large enough contact area to allow exchange.* A liter of air can be broken into 10,000 little bubbles or into 10 big bubbles. Because the 10,000 bubbles have a greater combined surface area than do the 10 big bubbles, it is easier for oxygen to leave the small bubbles. Remember, oxygen transfer takes place at the bubble surface.
3. *The oxygen gradient between the water and the air will dictate the rate of oxygen transfer.* If there is very little oxygen in the water, then the *rate* that the gas moves into the water will be quicker than if the water is near oxygen saturation.

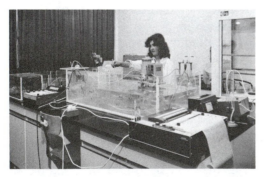

Figure 2.6 Oxygen consumption by fish and feed organisms is measured by a flow-through respirometer in a research laboratory. (Photo courtesy of Dr. Ferenc Muller, Fish Culture Research Institute, Szarvas, Hungary.)

Figure 2.7 In high-density stocking situations aeration must be used to increase the DO levels because of the high BOD. (Photo courtesy of Liang Hongwu, Bureau of Aquatic Products, People's Republic of China.)

Although aeration is the normal means used to increase the amount of oxygen in a pond, farmers sometimes add calcium hydroxide or potassium permanganate when the threat of a fish kill is near, because these will oxidize the organic matter and lower the demand for the dissolved oxygen. These techniques have been questioned by many researchers.

It is unlikely that oxygen will be evenly distributed in a pond. This is especially true in systems that are not well mixed. Surface water will probably have a higher DO level than the bottom water because

1. There is a greater chance for water to mix with the air at the surface.
2. Phytoplankton are more active near the surface.
3. There is probably more bacterial consumption of oxygen on the bottom.

Often in ponds the organisms may crowd into a small area rather than be distributed evenly in the water; this results in a drop in DO at that particular spot.

As mentioned earlier, ponds and lakes may become stratified and a thermocline may become established, restricting the downward movement of oxygen-rich surface water. In oceans and some lakes there is an oxygen-minimum zone associated with the thermocline. It is thought that organic material from the surface water sinks at a fixed rate until it reaches the denser water in the thermocline where the material is held up and used (oxidized) by bacteria. This layer of organic material and active bacteria in the thermocline results in an oxygen minimum at that horizon.

Many culturists will try to maintain a DO of at least 5 parts per million (ppm). Some species, for example salmon, are very sensitive to drops in DO, while others such as anabantid fish like the gouramis, which can breathe air, can live in water with almost no dissolved oxygen. The actual levels of oxygen that are needed by particular species tend to vary greatly with the size of the animal, the temperature, and stress. For example, in the crawfish *Procambarus*, the LC_{50} (lethal concentration when 50% of the animals die) for juveniles 9–12 mm long is 0.75–1.1 ppm of oxygen, but for larger 31–35 mm long juveniles, it is about 0.5 ppm. Animals that are under stress also use more oxygen. In the crab, *Carcinus*, the rate of oxygen consumption is greater in low-saline water than in normal seawater, and this higher rate of oxygen consumption is most pronounced immediately after a change in salinity takes place.

Mathematical models have been developed to predict changes in DO overnight, when the biggest declines in oxygen content take place. The nature of the pond will dictate the model used; factors to consider are temperature, biomass, bacterial and plankton activity, water exchange, composition of the water and the sediment, and oxygen transfer from the air.

Temperature

All of the commonly reared aquatic organisms have traditionally been considered cold-blooded (or "poikilothermous"). However, this is not exactly true; many poikilotherms may not be able to regulate their body temperatures in the same way or to the same extent that birds and mammals can, yet are able to maintain internal temperatures within certain limits by physiological and behavioral mechanisms. These adaptive mechanisms may include light-seeking or migrational adaptations, reduced circulation to the surface of the body, or countercurrent heat exchange between arteries and veins. The line between warm-blooded and cold-blooded has grown very indistinct during the past 30 years.

Although some animals can partially regulate their body temperature, the culturist may still try to maintain an optimal growth temperature so that all the organism's energy can go toward producing more tissue rather than toward staying warm so it can remain alive. Optimal temperature may be considered the temperature at which total weight gain is greatest, although there may be certain exceptions to this rule. This optimal temperature, as mentioned before, may be related to other environmental factors, so what is an ideal temperature at one salinity or DO level may be slightly different from what is ideal at other DO levels or salinities. In practice, culturists may raise animals at temperatures below what is optimal because of disease problems in warm water.

The actual optimal temperature is based on the sum of the internal chemical (mostly enzymatic) reactions that are taking place. While different enzymes may have efficiencies that are maximized at slightly different temperatures, the optimal temperature for the organism is the one that allows the most reactions to go on at efficiencies closest to maximum. For example, good growth over a 12-week period for the tubificid worm *Branchiura* occurs at 21°–29°C, but growth at 33°C or 17°C is marginal. Growth of many cultured animals will take an extended amount of time to measure, but can usually be translated into some more easily (and quickly) recognizable behavior; this is demonstrated in the mussel *Mytilus*, which will filter algal cells out of seawater very quickly as long as water temperatures are 15°–25°C, but filtration will slow

considerably when the temperature either increases or decreases. This change in filtration rate will eventually result in changes in the growth rate of the mussel.

As temperatures are reduced below optimal, growth slows down because the rates of the metabolic reactions decrease. Quick decreases in temperature are often fatal to aquatic organisms, although many can adapt to slow reductions, especially organisms from coastal temperate environments where natural seasonal changes in the temperature may be drastic. There are times when a reduced metabolism may be beneficial, such as during shipping. Culture of nori in Japan is currently very successful in part because spores of the seaweed can be frozen, stopping the metabolism, for several months and remain viable to be used when other spores are not available.

Significant increases in heat are also often fatal, partially because they mean a quick decrease in DO while at the same time raising the BOD by increasing the metabolism of the organisms in the pond. Higher temperatures can also lead to **thermal death** because the structures of the life-supporting enzymes have subtly changed and they are no longer able to regulate the needed reactions with sufficient speed (if at all). Increases in diseases are also common at high temperatures since cultured animals are weakened, and bacteria, which tend to be able to withstand relatively large shifts in temperature, increase in number. But raising the temperature of a pond may also be extremely beneficial if done correctly. The lobster, *Homarus*, is normally found in cold waters, but researchers have found that not only can it withstand higher temperatures, but it can grow more quickly at these higher temperatures, suggesting the use of thermal effluent in commercial lobster culture.

Temperature can also affect the chemical composition of an organism. The saturation of fatty acids is known to vary with the temperature, and although this may not be a great concern in most cases, it can be critical when raising feed organisms going to support a primary crop.

Nutrients

In the context of water quality, **nutrients** refers to those molecules in the water that can be used by aquatic plants, in particular the nonconservative elements and ions discussed earlier in this chapter. If these elements are in the wrong molecular form, they cannot be used by the plants, and if they are present in concentrations that are too high, they can even damage the plants and animals rather than stimulate primary productivity. In many marine systems, the nutrient that controls plant growth is nitrogen (see Figure 2.8), while in freshwater it is more often phosphorus. These are the nutrients that plants most often "run out of"

first, and are therefore called **limiting nutrients.**

Nitrogen is used by organisms in a large number of chemical reactions and is an essential component of amino acids, the building blocks of proteins. Nitrogen can occur in two oxidation states, +5 and +3, and is most commonly found as nitrogen gas (N_2). From the standpoint of the culturist, nitrogen is most important as a nutrient in the forms of

1. ammonia or ammonium (NH_3 or NH_4^+, respectively)
2. nitrite (NO_2^-)
3. nitrate (NO_3^-)
4. organic molecules

While N_2 is the most common form, it is only important for certain nitrogen-fixing organisms (such as some of the cyanobac-

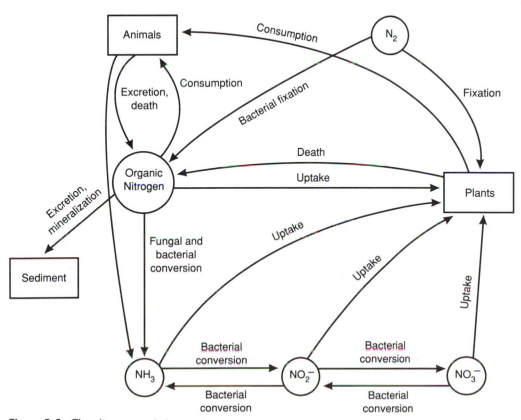

Figure 2.8   The nitrogen cycle in a pond.

teria) and otherwise is essentially unimportant in terms of being a nutrient.

Although there are plants that can use organic nitrogen or nitrite as sources of nitrogen, most require, or at least prefer, that nitrogen be either in the form of nitrate or ammonia. In most established ponds there is a natural **nitrogen cycle** that will convert the organic nitrogen to a form that plants can assimilate. Such a cycle can look very complex, partly because of the changes in oxidation states, and will be dominated by plants (producers), bacteria, and to a lesser extent fungi (reducers), which carry out many of the conversions. The specific bacteria and nature of the chemical reactions will be explained in greater detail when biological filters are discussed in Chapter 5.

Nitrogen compounds can also be toxic to plants and especially to animals, when these compounds are present at high concentrations. Besides certain organic nitrogen molecules that are rare in aquatic systems, the most toxic of the nitrogen compounds is ammonia. Ammonia and the less toxic ammonium are in an equilibrium that depends on pH; the more acidic the water is, that is, the more free hydrogen ions that are present, the more the equilibrium shifts from NH_3 to NH_4^+. As temperature rises, the ammonia/ammonium ratio rises, but as the salinity increases this ratio decreases; in general, temperature and salinity are less important than pH in determining the form of the molecule. Just how toxic ammonia is may be a function of a number of factors. Obviously, different species of organisms have different sensitivities, but different developmental stages of each organism may also differ in sensitivity. For example, rainbow trout sac fry and older fish are both more susceptible to ammonia intoxication than are juveniles. Fluctuating exposures of different concentrations of ammonia, and other forms of environmental stress, are also factors in determining ammonia toxicity; again returning to the trout fingerlings, the tolerance for

ammonia in the water is 30% less at a DO level of 5 mg O_2/liter than it is at 8.5 mg O_2/liter.

The toxicity of ammonia is usually reduced somewhat when it is oxidized to nitrite, and more dramatically when it is oxidized again to nitrate. Nitrite may also be formed from the reduction of nitrate. Since nitrite and especially nitrate are less toxic than ammonia, and because they are found in lower concentrations in culture systems, they are generally of less concern to culturists than is ammonia. This is not to say that they are harmless. They can cause unwanted algal blooms, and nitrite is known to oxidize hemoglobin in fish blood to methemoglobin, thereby reducing the blood's ability to pick up oxygen molecules as water passes over the gills. Catfish exposed to high levels of nitrite are more apt to be attacked by pathogens, such as *Flexibacter*, than are those kept in water without nitrite.

Phosphorus, usually in the form of orthophosphate (PO_4^{-3}), is also a critical nutrient for plants. It is usually found in concentrations lower than nitrogen, but is needed in lesser quantities. As in the case of nitrogen, cold and deep waters tend to have relatively high levels of phosphates, while productive waters that are warm and well lit have little free phosphate, most of it being incorporated in organic molecules in plants and animals. And again, as in the case of nitrogen, there is a "phosphorus cycle": plants take up phosphates and fix them into organic molecules, and bacteria and fungi release the phosphates from these organic compounds.

Phosphates in general do not often pose a direct threat to the fish farmer, although there are a number of extremely toxic pesticides that include phosphorus in their structure. The greatest problem associated with this nutrient relates to cyanobacteria blooms caused by high levels of phosphorus entering a pond where nitrogen is limited; under these conditions cyanobacteria that can use N_2 proliferate and exclude algae. The cyanobacteria often cannot support the normal phytoplankton grazers in the pond, and sometimes actually release toxins or die, causing a sudden depletion of the DO.

At first glance it would seem that high levels of N and P, when properly balanced, would not be a problem for the culturist, since this would only result in an increase in the number of higher plants and algae. However, too much floral production can also be a problem. At night, the plants are using up O_2 rather than producing it, so the DO in a fish pond can fall to lethal levels. During daylight the plants will be removing CO_2 from the water, allowing the pH in the pond to rise. The less acid pond means that NH_4^+ is converted to the more toxic NH_3; this problem can be minimized by liming the ponds since this will increase the carbonate alkalinity, thus preventing dramatic shifts in the pH of the water.

We have stated that under natural conditions either N or P will probably be the limiting nutrient in water, but this is not always true in aquaculture. Farmers may add large quantities of these nutrients to the water but get only a small increase in the phytoplankton density. This often happens because after fertilization some other factor has become limiting, it may be potassium which can be added in the form of potash (K_2O), or more commonly it is CO_2, which can be added in the form of lime. Silicon, probably in the form of orthosilicic acid, $Si(OH)_4$, is normally found in high enough concentrations to sustain all life in the water. However, there are some instances, for example when diatoms are being cultured, when Si can be limiting. Theoretically, any of several dozen inorganic or organic molecules, such as vitamins, can also limit the productivity of the water.

Other Water Parameters

Turbidity refers to the amount of suspended material in the water, so it is a measurement of the inhibition of light passing through a water sample. Suspended materials are often clay particles that stay in suspension because they are small and negatively charged; these suspensions are called **colloids.** Any addition of positively charged ions into the water will reduce the turbidity. Therefore, builder's lime, limestone, or gypsum are sometimes added to the water to reduce the turbidity. Many fish farmers wish to have the water in a nonturbid condition since turbidity

1. Makes it difficult to observe the fish directly.
2. The plant/phytoplankton productivity is reduced, thus making a build up of nitrogen products possible.
3. The suspended material can interfere with oxygen transport across the gills

Heavy metals are a problem in some areas. The effects may be dramatic resulting in fish kills, deformities, or simply reduced growth or reproduction. Brine shrimp eggs, for example, can be stopped from hatching by several heavy metals. The effects of the metals are a function of

1. the species of organisms in the pond (some plants will remove metals so that little damage is done to the animals in culture)
2. pH
3. salinity
4. temperature of the water

Metals may interact with other metals or organic molecules to increase or decrease their toxicity. Metals may be found in different concentrations in different tissues of the same animals; mercury, for example, is much more likely to be found in the lipid-rich head of a shrimp than it is to be found in the shrimp's edible muscle.

Organic compounds in the water may also change the flavor of the product. An off-taste may make an entire crop of otherwise healthy organisms worthless. Oils can give a bad taste and smell to crustaceans. Cyanobacteria in a pond can sometimes produce geosmin ($C_{12}H_{22}O$), which gives the animals a musty flavor. Actinomycetes living in pond mud may produce geosmin-like compounds that have a similar effect. The amount of off-flavor that is associated with a piece of tissue may be related to the amount of lipid that is present.

SUMMARY

If aquaculture is the rearing of aquatic organisms, it is important to understand the aquatic medium: water. Water is a pronged polar molecule. Hydrogen bonds that are responsible for many of water's properties, such as density and specific heat, form because of the attraction between the water molecule's differently charged poles. Seawater is about 3.5% salt and 96.5% water; the ions that are most important in seawater are Cl^-, Na^+, Mg^{+2}, SO_4^{-2}, Ca^{+2}, and K^+. The amount of salt in water will affect properties of seawater such as density, freezing point, vapor pressure, viscosity, and refractive index. Most elements and ions in seawater remain in constant ratios to each other, with the exception of those involved in biological reactions.

Water may be obtained from surface sources (lakes, ponds, streams, rivers) or from the ground (aquifers) by the use of wells. The source of the water will partially determine its properties. Seawater is generally obtained from a surface source, although seawater wells are sometimes used. Seawater must be treated differently than freshwater because of its corrosive properties.

There are several important water quality parameters that the culturist should be aware of. pH is a measure of the H^+ concentration and should be maintained near neutral (7 ± 1). Carbonate alkalinity is a measure of the water's ability to resist rapid pH shifts. Salinity is the amount of salt dissolved in the seawater. Hardness is the amount of Ca^{+2} and Mg^{+2} dissolved in freshwater. Dissolved oxygen (DO) will affect the health of the pond environment; the DO may change quickly during a 24-hour period. The optimal temperature for each species is the one that allows the most efficient growth of the culture organism. Nutrients are taken up by plants in culture and include N, P, and some other elements of lesser importance. These elements must be in the proper form to be used. Other parameters to consider are turbidity, metal content, and compounds that give an off-flavor.

CHAPTER 3

CULTURE SYSTEMS

It would be fair to say that no two aquaculture facilities operate the same way. Many are based on traditional ideas that have been used for years, but some encompass new and sometimes radical concepts that make them unique. Yet all can be divided into essentially three major system groups: open, semiclosed, and closed. Each has its special characteristics, advantages, and disadvantages. The choice of system is largely a function of the organisms to be grown and the resources and ideas of the fish farmer. (The reader should be aware that the meanings of these systems vary, especially "open" and "semiclosed," and are not identical in all aquaculture literature.)

OPEN SYSTEMS

These are the oldest of the aquaculture systems. **Open system** farming is the use of the environment as the fish farm. An open system does not require water to be pumped out of a sea or a lake; rather, the organisms to be cultured are kept *in* the sea or lake. Clam beds, oyster rafts, and fish cages are all examples of open system techniques.

There are a number of advantages to the open system. It usually requires that land be leased from some government agency rather than purchased outright. In fact, capital expenses in general are low for the open system farmer. Also, there is less management than in the other systems; the more natural and uncrowded the culture environment, the less time must be spent in monitoring the condition of the culture organisms.

Open systems also have some disadvantages. Disease in general is not common, but predation and poaching can be a problem. The grower has less control over environmental conditions, so the rate of growth and the uniformity of the product are variable compared to other systems (this is less true of cage/pen systems in which food is supplied by the culturist).

A discussion of protection against poachers is beyond the scope of this book; however, there are prescribed ways to reduce predation and, to a lesser extent, ways to reduce the environmental variability that contributes to irregular growth rates.

Protection from Predators

One of the most successful strategies for open system aquaculturists is the use of **off-bottom** culture. While there are many reasons why this is advantageous (which will be covered later in this chapter), one of the principal reasons is that the predators of bottom organisms also live on the bottom, so by moving the culture organism they are no longer available to their benthic predators. For example, oysters living in natural oyster reefs are eaten by starfish (also called "sea stars") and a snail, *Urosalpinx,* known as the **oyster drill.** When the oyster is no longer on the bottom, these predators, which cannot swim, are unable to reach the oysters suspended in the water.

An alternative technique, if off-bottom culture is not possible, is the use of **traps.** Cement blocks or tiles are placed nearby and barnacles are allowed to settle. Oyster drills prefer barnacles to oysters, so if the barnacle traps are checked daily and the drills and their egg cases are removed, there will be a marked decrease in predation on the oyster crop. This practice can be adapted to animals other than oysters.

Clam beds may also be protected from many predators by a short chicken wire fence. This is especially effective against predators such as the horseshoe crab, green crab, and large moonsnails and whelks. There are other kinds of **fences.** Simply spreading gravel around the clam

bed will prevent many crabs from entering the beds, and **bubble walls** made using an air source and some type of diffuser will inhibit the entrance of several invertebrates as well as fish.

Starfish are a substantial problem for shellfish farmers using open culture systems. In shallow water, they can be picked up by hand and killed. Starfish usually are killed by drying or dropping them in hot water; it should be remembered that because these animals can regenerate body parts very efficiently, just cutting a starfish in half with a knife and throwing the parts back in the water will probably result in a three- or four-legged starfish that is also destructive to the crop. In deeper waters a **mop** is attached to an iron frame that can be 2 or more feet wide. The mop is dragged through the water and the spine-covered starfish are tangled in it. The mop and starfish are then brought aboard and dunked in a saturated brine solution or hot water (see Figure 3.1). Broadcasting **quicklime** over a bed of oysters being attacked by starfish has also been used to halt predation. Quicklime (calcium oxide) produces lesions on the starfish's surface that quickly deepen.

Figure 3.1   Starfish caught using a mop in a 4-hour period in Buzzard's Bay near Cape Cod. (Photo from Galtsoff and Loosanoff, 1939.)

Only a very small amount of quicklime landing on a starfish will cause death in a few days. Some understanding of the local water currents, both above and below the surface, is required, since the quicklime can easily be carried away from the beds.

A few predators have some value themselves. The blue crab, *Callinectes*, considered a delicacy by many people, is a pest for shellfish farmers. Traps can be set for this crab, and in shallow water they can be caught at night with a flashlight and a dip net. The crabs can be sold, if possible, or used for personal consumption. Some large whelks are also eaten and can be harvested in the same way.

Chemicals have been used to protect organisms from predators, but only in a few isolated cases have these been successful. Lampreys, which had been reducing the populations of commercial fish in the Great Lakes, have been successfully controlled by a specific pesticide. But in most cases these chemicals are expensive, so adding them to open systems such as bays and lakes is impractical. Also, they may do more harm than good. A pesticide called polistream-sevin was supposed to rid the open system farmer of oyster drills, but it was later shown that the drills were not greatly affected although most other benthic invertebrates died on exposure.

Site Selection

Perhaps the best way to begin the process of site selection for an open system is to have a few organisms placed in several available locations and then select the site that gives the best production as the one of choice; however, this is often not possible. Alternatively, the farmer can compare the growth of wild specimens at each location. For example, if bivalves are to be grown, the farmer can estimate growth at different locales by examining the annual growth rings of bivalves collected at those sites.

Another major consideration is that the site should be accessible. How close

can a truck be brought to the area? Is a boat needed? If so, what type? Is there a chance that poachers will remove some of the crop? If this is a possibility, how much time should the culturist dedicate to surveillance? The proximity to markets must also be noted since this will determine the cost of moving the product.

In the case of bottom culture, the sediment type should be examined because what is good for one species of shellfish may not be good for another. For example, soft-shell clams prefer muddy bottoms, while oysters must have a hard substrate on which to settle.

Tides are another important consideration. If bottom culture is attempted, how much of the time will the crop be covered? If rafts or some other off-bottom technique is used, it is important that at low spring tides the organisms never touch the bottom. The tides or currents should be strong enough so that wastes are removed from the area, phytoplankton is brought in for filter feeders, and the water remains well oxygenated. The tides should not be so strong that they damage the equipment, make working in the area dangerous, or prevent feeding caged fish.

Naturally, the farmer will avoid areas where there are signs of pollution since this may not only inhibit growth, but may also make the crop unfit for consumption. If possible, something should be learned of the disease and parasite problems in the area as they pertain to the species of interest.

Cage Culture

Originally, **cages** were retaining structures, poles or stakes driven into the sediment of shallow water lakes and bays with nets stretched around them. These are still used in parts of the world for rearing certain types of fish, but perhaps more importantly, they have inspired the modern floating cages (see Figure 3.2) that are increasing in popularity today. There have been many recent developments in

Figure 3.2 A net cage used to grow catfish. (Conrad, 1988. Copyright © Archill River Corporation. Reprinted by permission.)

cage technology, especially in marine (coastal and offshore) designs.

Cages can be used in invertebrate culture, but are especially important for rearing fish. Large cages, some over 200 m³, are sometimes called **pens.** (For some farmers, the word "pen" means only those cages with no top netting.) They are most important when the open system's farmer has leased only a part of a large body of water.

Cages are currently used in many salmon rearing programs in Europe and the northwestern United States and have enjoyed great success with catfish farmers as well. Although cages are expensive and have a limited life (normally not over 5 years), require constant attention, and are mended frequently, they are essentially the only way to raise highly mobile species and harvest 100% of the crop. Besides the ease of harvesting, fish this close together do not waste energy swimming or hunting for food.

Cages may be stocked at densities far greater than ponds; among the most important of the factors in determining the stocking density in a cage is the circulation, that is, the exchange of water through the mesh walls of the cage. If the water exchange is sufficient, then the cage may be densely stocked, but if there is poor water circulation, then fewer organisms must be used; this may make the use

of cages uneconomical. Yellowtail are raised in floating net pens under very crowded conditions, up to 20 kg/m³, on the Japanese coast.

Most cages are rectangular or cylindrical in shape; large cages are cheaper to build than several smaller cages, but are difficult to handle out of water and may buckle or collapse under their own weight if not constructed correctly. Large cages that are damaged, allowing animals to escape, are more likely to mean a substantial loss for the farmer. Cages have three parts:

1. The cage itself, having some sort of mesh sides
2. Floats that keep the cage in the water column
3. Anchors that prevent the cage from drifting away

The mesh material is usually a sturdy fabric netting such as nylon, a plastic (for example, polypropylene or polyethylene) (see Figure 3.3), or metal wire coated with plastic. Most uncoated metals will corrode, especially in saltwater. Mesh width will vary with the species and age of the organisms being cultured. Some cages are made of a single layer of mesh material; others have a second outer wall that is made of a stronger material acting as ex-

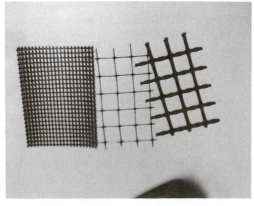

Figure 3.3 A variety of plastic mesh materials can be used to build cages.

tra protection from predators and possible damage to the inner walls. The tops of the cages are made of either a solid material like wood or the same mesh as the sides are made of, and may contain a feeding ring. The feeding ring extends a few centimeters below the top of the cage, so floating pellets are retained inside the ring (see Figure 3.4). Naturally, sinking pellets are lost through the cage bottom unless they are eaten very quickly. An alternative to the feeding ring is surrounding the upper few centimeters of the cage walls with a fine material so the pellets do not float out the sides.

The most common material-related problems in cage culture involve

1. Biofouling in marine systems, which leads to increases in drag and weight, and decreases in flow through the nets (resulting in a reduced water quality)
2. Corrosion of metallic components
3. UV light deterioration of plastic materials
4. Damage by waves or ice

There have been several attempts to solve these problems. A fiberglass framed cage with copper–nickel mesh has been found to be more satisfactory than typical nylon net cages in many respects because it costs less to maintain and replace, is less easily damaged, and has reduced levels of fouling. For these reasons, despite the higher capital costs, such rigid cages may be considerably more economical than nylon cages. Another suggested alternative involves the use of neoprene bladders that can be inflated and deflated; these allow the cage to rotate in the water with different sides being periodically exposed to the surface to be cleaned of fouling organisms.

The most common flotation material used is styrofoam because it is cheap, light, durable, and does not corrode in saltwater. For small cages, plastic milk jugs are sometimes sealed shut and used, and for very large cages air-filled steel drums may be best. Some of the very large cages have floating walkways to make feeding and harvesting easier.

The bottoms of the cages must be high enough off the sediment to prevent wastes from building up near the fish; therefore, water that is too shallow is to be avoided. Only the very top of the cage should be above the water surface; if more than a few centimeters of the cage

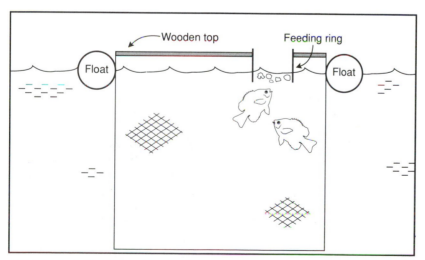

Figure 3.4 A fish cage with a feeding ring. The ring extends down into the water several centimeters. Floating feeds are placed in the ring; the food pellets cannot be swept from the cage by water currents so there is little waste.

are out of the water, the floats should be readjusted because valuable growing space is being wasted.

Cages can be anchored easily to almost any stationary object on the bottom. Wave action on a large cage may exert considerably more force than might be suspected, and therefore anchors that may appear to be sufficiently heavy may not be. Sometimes cages are not anchored individually, but rather are strung together with a rope or a cable that is fixed in the sediment or to a dock. Cages are often arranged in one or two parallel rows, which makes working the cages with a small boat easier. If more than two rows are used, there is a danger that the currents that bring in oxygen and take out wastes will be inhibited, resulting in a decline in water quality in the cages. Water currents in freshwater lakes and brackish lagoons are almost always weaker than in the marine environment; it has been shown that under these conditions, the swimming of the fish in the cages adds significantly to the water exchange rate.

Cages used to grow marine fish are generally kept in bays and protected waterways. These coastal environments are used because it is easy to get to the cages when it is time to feed the animals and because the cages can be easily anchored. It has been suggested, however, that the culture of marine fish should be done offshore where there is less variability in the salinity and temperature, and pollution is less. When trout were grown by Swedish researchers in cages that were either onshore or offshore, it was demonstrated that the offshore trout were in better condition and had a much lower mortality rate. (Beside water quality stress, it was suggested that the superior growth and survival of offshore trout was due to continuous automated feeding, which was probably superior to feeding large amounts of food four times each day, as was done for the onshore fish.)

Predation is usually not a great problem in cage culture, but occasionally a farmer may experience problems with diving birds and seals. The use of sound-producing devices to scare away seals has become more common in recent years. Jellyfish also sometimes pose a threat to caged fish: they may clog the mesh as well as kill fish that come in contact with the stinging cells on the jellyfish's tentacles.

A variation on the theme of cages is the **floating vertical raceway.** This is essentially a vertical frustum-shaped plastic tube that extends 75 cm to 80 cm above the water surface and several meters below. There is a screened outlet at the bottom and a flotation collar with a walkway around the top. Water flows into the top of the raceway, mixes with the water already in the raceway, and exits through the bottom.

Floats, Rafts, and Trays

Cages are used primarily for fish, while **floats, trays,** and **rafts** are most often used for filter-feeding sessile invertebrates, especially bivalves. These organisms are naturally found on, or in, the sediments; they are therefore often some considerable distance from the greatest concentration of **phytoplankton,** which is their food source. Unicellular algae are normally most concentrated in the first few meters below the surface. (Phytoplankton are often not as common at the water surface because the light is too strong, resulting in a **photoinhibition**). By using rafts, the filter-feeding animals are closer to the phytoplankton maximum zone. Even more important is that in culture systems using rafts and long lines, more animals can be grown because a greater volume of water is being used as culture changes from a two-dimensional operation (bottom area only) to one of three dimensions (area × height of the water column). In addition, floats keep the animals away from their natural predators and make harvesting easier.

Rafts are usually made of wood and some sort of float, such as styrofoam or steel barrels (see Figure 3.5). Long-line

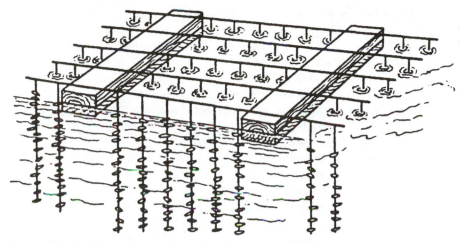

Figure 3.5 Raft for shellfish culture. (From Wheaton, 1977.)

floats may be used in place of rafts when culture takes place in rough water (see Figure 3.6). Anchors are important to keep rafts from drifting away, and must be particularly strong for long lines because as the weight of the crops and fouling organisms increases, the floats tend to be pulled together, allowing the crop to sink lower in the water.

Suspended from the long lines or rafts are strings or bags of shells that attract the planktonic larvae of the animals to be raised. Rafts may be left in one area to collect larvae, but can later be towed to another site if there are better water conditions for growout to be found. Determining the number of strings or bags, as well as their size, is a function of several parameters, including

1. depth of the water at low tide
2. speed of the currents
3. the weight that the float can support

A heavy set of bivalves or fouling organisms can eventually sink a raft or cause the rope to which they are attached to break.

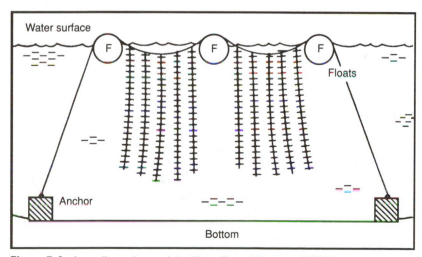

Figure 3.6 Long-line culture of shellfish. (From Wheaton, 1977.)

Rope with bivalves growing on it may become abraded and eventually break, resulting in loss of the crop.

Tarred sticks supported horizontally by posts are a form of rack (see Figure 3.7). Larvae settle on the sticks, and either the organisms are allowed to grow out there or the sticks are transferred to trays. Trays may be attached to rafts or floats or may be supported by stakes driven into the bottom (see Figure 3.8). Trays may be either single or stacked with sufficient space between them to allow unrestricted circulation of water. Trays may be constructed simply of some sort of wide mesh material that allows the animals to be relatively exposed to the environment. Alternatively, the trays can be constructed of fine meshed or solid material and have a layer of sand in which the animals can bury themselves. If there is a danger of predation, another layer of mesh may be placed above the animals.

Management

Management of open systems is different from management of other culture systems. Although there is generally less manipulation associated with open systems, the successful farmer cannot just "let nature take its course."

An example of how management can improve production may be found in Old Saybrook, Connecticut. The Oyster River in Old Saybrook was at one time a very productive area, but was closed in 1971 to shellfishermen because of high bacterial counts. It was left for nine years and surveyed in 1980. The beds showed characteristics of being unmanaged (that is, uncultivated) such as overcrowding, poor oyster seed setting, and high mortality. In 1981, six sites in the Oyster River were examined for seed oysters (spat that have grown to 1–2 cm in diameter) and only one specimen was found. At that point 2200 bushels of adult oysters were removed ($66,000 value) and replaced with 2000 bushels of clean shells. Thus, overcrowding was reduced while good setting sites were added. Within one year the number of seed oysters found at the originally sampled sites increased from 1 to 996. The population has become very productive again with high-quality oysters being grown for depuration at another location (the concept of depuration is discussed in Chapter 8).

SEMICLOSED SYSTEMS

For many types of organisms, the **semiclosed** system is the most popular method of culture. Water is taken from a lake, bay, well, or other natural source and is directed into a specially designed facility. The water either makes a single pass and

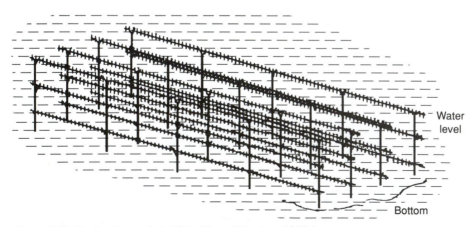

Figure 3.7 Rack culture of shellfish. (From Wheaton, 1977.)

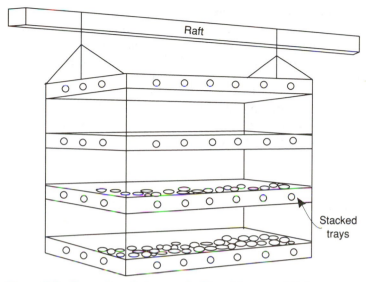

Figure 3.8   Tray culture below a raft. (From Wheaton, 1977.)

is discharged, or it may be partially replaced so that a certain percentage of the total water in a system is retained and recirculated.

Semiclosed systems offer an advantage over open systems in that they allow greater control over the growing conditions. A greater production per unit area is possible in addition to the crop being more uniform. This is because

1. Temperature, in some systems, can be regulated.
2. Prepared feeds can be used easily with much less waste.
3. Water volume and current speeds can be controlled.
4. Aeration can be increased simply.
5. Poaching is severely restricted to all but the boldest offenders.

Water can be filtered to remove predators or even disease organisms. Disease or other health problems, if they arise, can be observed and treated more easily than in open systems. All these things have made semiclosed systems popular.

There are also disadvantages to these systems. First, they are sometimes more expensive to develop and operate than open systems, although this may be offset by an increase in the value of the crop. Second, they require a more complex management scheme. Because organisms are crowded together and higher levels of feed or nutrients are being added, there is a greater chance of disease and stress.

Fertilization

Fertilization refers to the addition of nutrients (discussed in Chapter 2) to the water to increase productivity. This means that for filter-feeding organisms such as clams, oysters, and mussels, there will be more food resulting in faster growth. Fish ponds are often fertilized to increase the pond's productivity, so even if the crop organism will not graze the phytoplankton, it will benefit by feeding on zooplankton and/or benthos which are supported either directly or indirectly by the phytoplankton. Fertilization has been carried out in open systems, but there is always the possibility that the nutrients or resultant phytoplankton will be swept away by water currents. Semiclosed systems such as ponds are often fertilized, but raceways, because of their rapid water exchange, are not fertilized.

When fertilization is used, there is a choice of nutrient sources. The most common type of fertilizer is organic in nature, including sewage effluent, manures, vegetable wastes, grass clippings, and seaweeds. Ideally, this represents the disposal of unwanted waste materials in exchange for good nutrients. Realistically, there are major problems associated with this:

1. Often, much material has to be used since the nutrients represent a relatively small portion of the waste matter.
2. Nutrients must be released from the material by bacteria and fungi, and therefore are not immediately available.
3. These reducing microbes use oxygen while releasing nutrients, so the BOD may increase to dangerous levels.
4. If animal (and especially human) wastes are used, there is a chance that viral or bacterial pathogens may be taken up by the crop organism.
5. The organic materials may give the products an off-flavor.

The use of special organic wastes, such as sewage, is discussed elsewhere in this chapter.

Alternatively, inorganic fertilizers can be used. These are popular in the United States. Inorganic fertilizers have none of the disadvantages of the organic wastes, but do have two of their own. First, because the nutrients are available immediately to the phytoplankton, they must be added slowly over time or the blooms will become so large that BOD levels at night may be dangerous. Second, even when these nutrients are added in the most economic forms—agricultural fertilizers and industrial-grade salts—thay may be more costly than the organic wastes.

Water Replacement

In semiclosed systems the water replacement rate can be readily calculated if a portion of the water is removed and then is replaced by an equal volume. However, most culture systems are run on a continuous flow basis, where new water is flowing into the system, mixing with the existing water, and flowing out. Thus, all the clean water coming into a full system is not retained.

The rate that water is exchanged in a fully mixed system can be expressed as

$$T = -\ln(1 - F) \times (V/R)$$

where T = the time that water is flowing, F = the fraction of new water that is actually replaced, V = volume of the system, and R = rate of the water input (volume/unit time). This is used to calculate the time needed to replace a certain fraction of the volume of the culture system. This can be rearranged to

$$R = -\ln(1 - F) \times (V/T)$$

which indicates the flow rate that is required to replace a fixed volume of water in a given time.

Site Selection

The ideal site for a semiclosed system is determined by how the animals are to be grown. **Embankment ponds, excavated ponds,** and **raceways** all require different types of sites. Other factors also come into play.

The source of water must be good, and it must be available in quantities sufficient for the farmers' needs. Water quality should always be established before a site is selected. If water does not come from a constant source such as a deep aquifer, the amount of water will fluctuate, and this must be planned for. Legally, there may be a question concerning how much surface water may be taken for an aquaculture project without affecting the rights of others. There may also be a problem of too much water; there may be flooding during the spring when ice melts or during periods when there are heavy rains. Besides getting water, the discharge of the water after it has gone through the semiclosed system must be considered. Many municipalities have special regulations about the amount and quality of water that can be discharged into the environment. If there are regulations on discharge quality, then the cost of postculture treatment must be examined. A significant amount of paperwork and number of permit applications may be necessary for using and discharging water, and this can slow down construction and therefore production. These regulations may vary from site to site.

Absolute location has to be considered just as in the case of the open system. Is the site convenient to a market? Where must those who work at the farm live? How easy will it be to get supplies for both building and running the farm? What is the tax (income, sales, property, etc.) status of the farm in that particular area? What will be the value of the farm if the farmer decides to sell it? Is poaching a problem, and if so, can a fence be constructed without difficulty?

The soil, in the case of earthen ponds, is also critical. While some measures can be taken to seal a pond if it does not hold water, this is troublesome and expensive. The slope of the land will also affect the cost of construction. The soil should be tested by an experienced soil engineer before a large pond project is begun to determine its ability to hold water.

Although it is not an insurmountable obstacle, the predators and pests in an area can cause problems. Raccoons, snakes, and even some kinds of insects (see Figure 3.9) can raid ponds. Moles, gophers, and muskrats may dig into the soil and damage the pond walls. In some cases, birds can be major consumers of

Figure 3.9 The water beetle, *Hydrous triangularis*, the larvae of which prey on small fish. (Photo from Wilson, 1924.)

cultured fish (especially bait fish and ornamentals). Birds will congregate around densely populated ponds once they have learned that it is a good source of food; the most important bird pests in the United States are cormorants, herons, egrets, ducks, coots, gulls, and common grackles. Frogs and turtles may enter a pond from the local environment and compete for space and food with the cultured species; this can be a major obstruction to efficient use of the pond.

Raceways

Raceways are long, relatively narrow rectangular tanks. They are generally shallow, and, as the name suggests, water moves quickly through the structure. As the rate at which the water is moved through the raceways increases, the production rate increases, until the organisms begin to have difficulty moving easily in the raceway and the amount of energy the organisms use to maintain position rather than for growth becomes excessive. Raceways are usually associated with trout farming, but they have also been used with success for other types of fish, such as catfish, as well as for shrimp. The rate that water flows through a raceway depends on the type of organism that is being raised, the temperature of the water (since this relates to the amount of oxygen that can be held by the water and used by the organisms), the stocking density, and the feeding rate. Ideally, the flow should be such that the raceways are essentially self-cleaning, but not so fast that pumping costs are excessive and the organisms in culture are stressed.

The water flow rate through the raceway should be adjusted so that the water quality will not deteriorate appreciably from one end of the raceway to the other; therefore, fish can be stocked more densely in raceways than ponds. Because the fish are held so closely together, raceways have several distinct advantages over ponds:

1. Fish can be grown in high densities, so a greater yield per unit area is realized.
2. Feeding and harvesting take less time since the animals are held in a small area.
3. Because the raceways are shallow and the water is moving quickly, water tends to be rather clear com-

pared to ponds, so problems arising from predators that have been accidentally introduced or diseases are more quickly seen.

Treatment of diseases is also facilitated because the volume of water that receives the treatment is restricted. Raceways do have one major disadvantage: there is a significant pumping expense because so much water is moved so quickly.

One way to reduce the cost of pumping is to arrange a cascade of raceways in which the water from one raceway flows into the next. When using this **series design,** the raceways are typically built in steps so water is gravity fed down the line of raceways (see Figures 3.10 and 3.11). The difference in the heights of the raceways is related to the rate of flow that is required and the oxygen that is used by the fish. There are several problems with series raceways:

1. Metabolites build up in the water so the fish in the last raceway get water that is lower in quality than water in the first raceway.
2. The oxygen content may drop as the water is passed along.
3. Disease organisms in a raceway may move to all raceways "downstream" of the original infection.

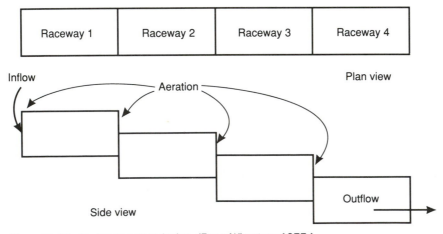

Figure 3.10   Series raceway design. (From Wheaton, 1977.)

Figure 3.11 Series raceways used to culture trout near Beijing, China. (Photo courtesy of Liang Hongwu, Bureau of Aquatic Products, People's Republic of China.)

Parallel raceway systems refer to a design in which all raceways receive water of equal quality from a single water source (see Figure 3.12). Water flows through the individual raceways and out to a common outflow. The amount of water that is pumped is substantial, but the greatest problems of the series design are minimized.

There are some **circular "raceways."** These differ from what are normally considered to be raceways because the water does not leave the system after a single pass, but rather is retained for a much longer period of time. The distinction be-

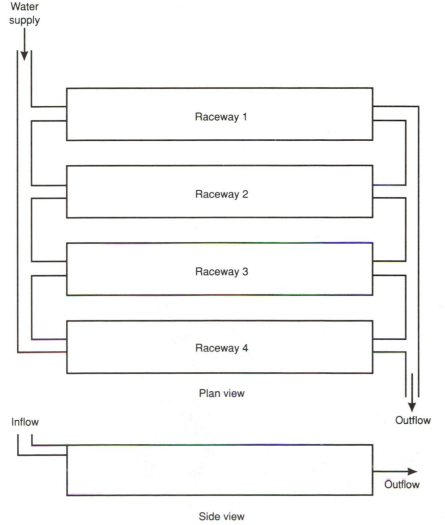

Figure 3.12 Parallel raceway design. (From Wheaton, 1977.)

tween a circular tank and a circular raceway, although this is fairly arbitrary, is that raceways tend to be shallower and more quickly moving. The circular raceway is used in the culture of algae when water must flow rapidly so that the cells will stay in suspension, and the depth of the water is limited because of the amount of light that can penetrate to the bottom. Such raceways are often painted white to increase the reflection of light off the bottom back into the culture (see Figure 3.13).

Raceways in most commercial farms are constructed of concrete. Soil is sometimes used, but because of the swiftly flowing water, there may be some damage to sides that are too steep. Earthen raceways should be lined. Wood, plastics, and fiberglass are also used to construct raceways, but generally for smaller operations (see Figures 3.14 and 3.15).

Deep circular tanks are known as **silo tanks,** or sometimes **vertical raceways** (not to be confused with the *floating* vertical raceways discussed earlier in this chapter). In vertical raceways the water is pumped down a pipe in the center of the raceway and the water exits this pipe

Figure 3.14   Redwood raceways used to culture *Tilapia*. (Photo from Uchida and King, 1962.)

near the bottom through a screen. The water flows upward and over the sides of the raceway around which there is a trough used to collect the water (see Figure 3.16).

Figure 3.13   A pair of circular raceways. (Photo courtesy of The Oceanic Institute, Hawaii.)

Figure 3.15 Fiberglass raceways. (Photo courtesy of The Oceanic Institute, Hawaii.)

Ponds

Ponds are the most common structures for culturing fish in semiclosed systems. Water exchange, if employed at all, is much less rapid than in raceways, and circulation is also reduced. Unlike the fast-moving and oxygen-rich raceways, ponds are sometimes stratified and require aeration to maintain acceptable DO levels. Aeration will be discussed in Chapter 5.

Some fish ponds are simply holes in the ground, but much more common is the above-ground **embankment pond.** These may be any size or shape, but most are rectangular because they are easy to build and little space is wasted between ponds. Some large ponds may be built in valleys because only one or two walls need be constructed between the hills. Ideally, so costs are minimized, ponds should receive their water by a gravity feed, and discharge of water from the ponds should also be gravity driven; in practice, pumping water either in or out is often required.

Ponds are generally at least 1 m in depth, with soil on the banks sloping at 2 : 1 to 4 : 1, depending on the stability of the soil. If rock, cement, or blocks are used for the sides, pond walls may be vertical. The bottom of the pond may be sloped to make it easier to drain; however, if the slope is too steep, the water flowing out to the drain may wash away some of the soil on the pond bottom. In some ponds there may be a depression around the drain pipe or monk where the fish will collect as the pond water is leaving the system; if the fish collect in this small area, it will make harvesting easier, but the depression should not be so small that the fish are damaged or starved for oxygen. Some ponds have a series of interconnecting ditches on the bottom to make the final steps in draining go quickly to completion.

The walls of the ponds, called **dams** or **dikes (dykes),** are usually less than 8 m in height. The walls must be constructed so that they can withstand the pressure that the water is exerting; watery soils or soils with a high organic content that might decompose in time are rarely used. The soil should have a low water permeability; those with a relatively high clay content should be used, while sandy or rocky soils should be avoided. The culturist may wish to consult a soil engineer before actual construction takes place. If the ground is covered with a layer of permeable soil, the wall must be constructed of an impermeable soil, and a portion of it should penetrate through the permeable layer into a deeper impermeable layer to prevent seepage. (If the depth of the permeable layer is too great, then a liner must be used.) If impermeable soil to be used is in short supply, only the middle of the wall, called the **key,** needs to be made of this material (see Figure 3.17). The soil for the

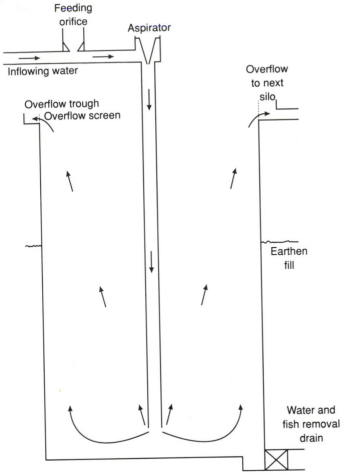

Figure 3.16 A cross section of a silo tank or vertical raceway. (Redrawn from Moody and McCleskey, 1978. Courtesy of the New Mexico Department of Game and Fish.)

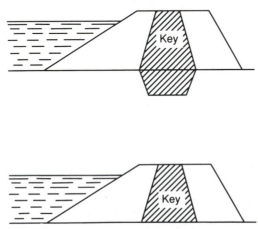

Figure 3.17 Dam construction using an impermeable soil key for seepage reduction. (From Wheaton, 1977.)

wall should be added in layers of 15 cm to 20 cm; each layer is compacted before the next layer is placed above it. In cases where there is no available impermeable soil for the wall, a key can be constructed from steel or concrete. The outsides of the pond walls should be covered with grass to stabilize the wall and prevent erosion from rain or in the event of water washing over the sides; trees should not be grown on or by the wall because their roots will increase the flow of water into the wall thereby destabilizing the compacted material.

In unlined ponds, vegetation present before building is usually not a problem unless it is extensive. In *Macrobrachium* cultures, it has been shown that by shortening the interior bank grass (planted to prevent erosion), the harvesting efficiency is increased; in one experiment, ponds in which short bank grass was maintained yielded 2300 kg/ha, while those with long bank grasses produced only 1916 kg/ha. Some grass may be useful for offering protection, but clearly, too much can hinder crop removal by the farmer because protection is too good! Some plants may actually grow through plastic liners and puncture them, so when liners are used, the plants must be removed, either by hand or with an herbicide, before the liner is put in place.

The tops of the walls should be wide enough so that there is enough space for the farmer to work comfortably. It may be a good idea to have the top of one wall wide enough so that a car, truck, or tractor can be driven on it if needed for feeding or harvesting. The width of the top of the wall is a function of the wall height. If the walls are less than 3 m high, a width of 2.5 m may be acceptable, but if the wall is 5 m tall, the top width should be about 3.5 m.

There are several variables that determine the height of the pond walls.

1. **Water depth** is chosen as a function of the organism to be grown; other considerations may include using a depth that inhibits extensive plant growth on the bottom or prevents wading birds from standing in the water to hunt.

2. **Wave action** may result in water washing over a wall that is too short. In small areas like a pond, the maximum wave height can be estimated by the equation

$$h_w = 0.014(F)^{1/2}$$

where F = fetch (in meters), the greatest distance a straight line can be drawn from one wall to another wall (or to another structure that can break the wave). Waves can erode the side of a pond, so the inside of the walls can be protected with **booms** made up of several logs chained together and anchored in front of the wall, or **rip-rap,** which are stones or blocks placed on the wall so that erosion is prevented (see Figure 3.18).

3. The extra safety height that is added to be sure that water will not

Figure 3.18   Rip-rap along the side of a pond.

overlap the walls is termed **free-board.** The height of the freeboard is related to the length of the wall; a wall 200 m long should have a freeboard of 0.3 m, while if 500 m long there should be a freeboard of 0.6 m.

4. Even if the soil is well packed, there will be some **soil settlement** that will reduce the wall height; the more traffic on top of a wall, the quicker the settlement.

5. Water in the soil is subject to freezing and expanding, then thawing. **Frost action** tends to destabilize the pond wall, causing the height to drop.

The minimum height of a wall can be calculated using the formula

$$H = h + h_w + h_f + h_s + h_{fr}$$

where H = final height of the wall, h = depth of the water, h_w = dam height needed because of waves, h_f = freeboard height, h_s = height to compensate for settlement, and h_{fr} = height needed to compensate for frost action. A soil engineer should be consulted concerning an estimate of h_s and h_{fr}.

Sometimes there is no available impermeable soil to make the ponds, so the soil that is available must somehow be sealed to prevent excessive water loss by seepage. There are several ways that this can be done.

1. A mixture of **clay** and coarser soil is layered upon the bottom and inside walls of the pond. Since clay is very fine, it will block the spaces between the coarse particles through which the water can flow. However, because it is very fine, it can also easily be washed away, so the clay layer is covered by a layer of gravel 0.3 m to 0.5 m thick for protection.

2. **Bentonite,** which is also mixed with the coarser soil, is a special clay that expands up to 20 times its original volume when it is wet; thus it also plugs up spaces through which water can leave. Bentonite will shrink when dried, so if water levels change quickly, it may not be satisfactory.

3. Clays and silts may bind together so that they make aggregates that increase the size of the pores in the soil. These can sometimes be broken up with **chemical additives** such as sodium polyphosphates, sodium chloride, or technical-grade soda ash (99% sodium carbonate). Before adding these materials, it must be established that aggregates are the cause of the problem.

4. **Waterproof liners** may also be used (see Figure 3.19). These liners are made of materials like PVC, butyl rubber, polyethylene, and chlorinated polyethylene. Some liners have a reinforcing material added

Figure 3.19 Small ponds with plastic liners.

to them. Each of these liner types has advantages and disadvantages relating to cost, damage done by UV light, and resistance to impact. For protection, a layer of soil is sometimes placed on top of a liner that is subject to mechanical damage. Pond liners eliminate all seepage but are expensive. They are available in rolls that are joined by solvents and cut so they conform to the curves of the pond. The liners must be anchored in place when the ponds are initially filled; this is done by digging a trench around the top of the pond wall, placing the outer edges of the liner in the trench, and then backfilling the trench with the soil that was removed.

Liners come in different thicknesses that are measured in units called **mils** (1 mil = 0.025 mm). Many liners are made for purposes other than aquaculture and may have protective chemicals associated with them that are dangerous to culture organisms. Lined ponds may therefore have to be washed several times. In addition, it may be worthwhile to wait until a layer of algae has grown over the liner before adding the culture organisms.

Ponds should have at least two **outlets,** or **spillways,** for the water. One allows water to leave under regular operating conditions, and another is needed when the primary spillway is clogged with debris or the volume of water suddenly increases so that the water level rises before the excess water can exit in the normal manner. The **riser and drain** or **standpipe** design (and the related **monk** design) and the **drop outlet** design are two principal types of spillways.

Although not commonly used, the drop inlet design is simple, cheap to build, and essentially cannot be blocked by normal debris. It is a modified section of the upper portion of one of the walls from which soil has removed and has been replaced by steel or concrete so that erosion does not undermine the wall soil (see Figure 3.20). Unless a drain of some sort is added, water must be pumped out to empty the ponds (a major disadvantage of this type of spillway).

The standpipe design consists of two pipes, one that runs parallel to the pond bottom and goes through the wall. A valve is placed in this pipe on the section inside the pond; this valve is normally kept shut. The end of the pipe may be covered with a screen to keep the animals from leaving when the valve is opened, and should be placed in the deepest part of the pond. A second pipe is joined perpendicularly to the first, between the valve and the pond wall, forming a "T" structure. This second pipe rises straight to the surface of the water and has a small hood that extends a few centimeters below the water surface. When the water level reaches the standpipe top, it begins to flow out, but the hood keeps the pipe from becoming clogged with material floating on the water surface. When the pond must be drained, the valve on the bottom pipe is opened. Outside of the wall where the water exits there is some sort of erosion control structure, and **antiseep collars,** that increase the distance that water

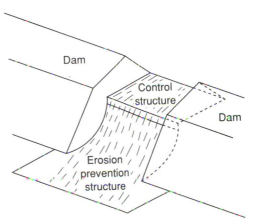

Figure 3.20 Drop design outlet. (From Wheaton, 1977.)

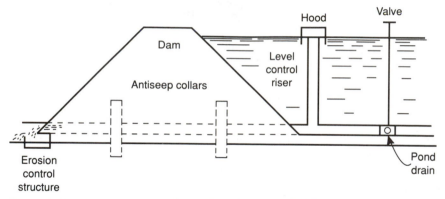

Figure 3.21 Riser and drain spillway. (From Wheaton, 1977.)

(which is seeping along the side of the pipe and washing away material) will have to travel, may be placed along the pipe inside the pond wall. The greater the distance that the water must travel, which is increased by the addition of the collars, the greater the resistance and therefore the less the flow (see Figure 3.21). In ponds where clogging by floating material is severe, a **trickle tube** may be used (see Figures 3.22 and 3.23).

Sometimes a **monk** is substituted for a system of pipes. The monk is a solid three-sided structure made of wood, bricks, or concrete and is connected to a pipe that goes out of the pond. On the fourth side of the monk is a series of wooden boards that fits into a groove on the inside of the structure (see Figure 3.24). As each board is removed, the level of the water drops. Clay or sawdust can be added to the space between the boards so they are watertight. Screens may be used above the top board to prevent the fish from escaping.

Water inlets should be above the level of the water inside the ponds so that fish cannot escape. The falling water is also an additional source of aeration for the system. Inlet pipes are also sometimes covered with a screen to prevent the entrance of undesired wild fish. There should be some device or valve on the inlet pipe that will allow the amount of water that is entering the pond to be regulated.

CLOSED SYSTEMS

Closed systems are defined in this text as those in which little or no water is exchanged and the water is subject to extensive treatment. These have made little impact on the commercial aquaculture industry, although some laboratory reports are very promising. Extremely high densities of organisms may be raised under these conditions if they are managed

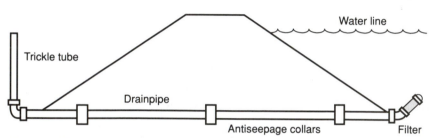

Figure 3.22 A trickle tube variation on the riser. (Redrawn from Inman, 1980. Courtesy of the Texas Parks and Wildlife Department.)

Figure 3.23 Pond outlet with plastic netting to prevent floating debris from entering and clogging the drain pipe.

properly. The major advantage to closed systems is that they allow the farmer to have complete control over growing conditions. The temperature may be carefully regulated, which is not economically possible in semiclosed systems where much of the water is replaced on a regular basis. No parasites or predators are introduced from the environment, and microbial diseases are less often introduced. Weather conditions are never a problem, and harvesting is simple. Food and drugs can be added efficiently into the system. All this allows the organism to grow quickly and uniformly. Because there is a reduced need for a large water supply, the choice of the site for the farm is expanded; a farmer could raise marine organisms in artificial seawater hundreds of kilometers from the sea. Another future consideration is that with our ever-dwindling water availability, and the possibility of space travel/colonization, closed aquaculture systems and hydroponic systems for terrestrial plants may be the only viable culture methods in many instances; other types of systems require much more water.

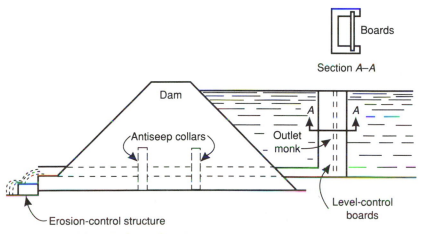

Figure 3.24 Monk design outlet. (From Wheaton, 1977.)

There are also problems with the closed system. Because water is reused and the density of the animals is so great, the filtration/treatment systems must be very good, and the water must generally be pumped through these systems at a fairly high speed. This, in turn, requires that there be exceptionally good managers operating these facilities since there is no natural safety buffer in the form of significant amounts of clean water brought in from outside the system. Because the system can be controlled with respect to optimal living conditions for the animals in question, disease control should not be a great problem. However, if a disease organism does find its way into the system, it can spread quickly because of the high stocking densities that are used and because the water is often warm (allowing rapid replication of the disease organisms). If diseases are undetected by the manager for even a short time, they can decimate a stock. There is also the problem of the capital equipment and operating costs, which may be substantial, since there is a greater dependence on machinery to treat the water.

At Ahrensburg in Germany, a closed system has been developed for testing conditions needed to grow carp. The system is 50 m³, of which 6 m³ is used for fish tanks and 44 m³ is used for water treatment; water flows through the system at 25 m³/hr. The system has a capacity to support 1.5 metric tons of fish. Fish are stocked in the system semi-monthly to make better use of the space; 10 g of juvenile carp can be harvested in less than six months at 500 g, giving an annual production of 8 to 9 metric tons. Elsewhere, a small closed system that has been used experimentally to maintain the northern squawfish has water flowing at 150 liters/min; at this rate 95% of the water is reused and 5% of the water is replaced. This may be considered somewhat of a semiclosed–closed hybrid system (see Figure 3.25). Fish have been held with no significant disease problems at densities of up to 0.97 kg/liter/min.

Tanks

The closed system is normally run in a **tank,** although semiclosed systems also make extensive use of tanks. These structures are often made of concrete, plastics, or wood. Each of these materials has its own set of advantages:

1. Concrete is strong and easy to use; it can be made into any size or shape with little trouble. The surface of the concrete can (and should) be smoothed.

2. Wood (see Figure 3.26) is easier to work with than concrete, and wooden tanks are more movable. Tanks made of plywood are not as strong as concrete tanks and require support bracings on the sides to resist the static pressure of the water. Wood will rot, so it must be coated with some type of sealer. It is useful to protect both the inside and outside of tanks since they are normally in moist environments and their life will be greatly reduced if they start to rot from the outside. **Epoxies** work well but may be brittle, so if the wood bends under pressure, the epoxy may crack and leak. **Fiberglass** resins are relatively strong and protect wood well, but are more difficult to work with; fiberglass cloth material must be completely covered by the resin while the tank is under construction because the small pieces of glass in the cloth could be very dangerous to the organisms in the tank that may rub up against it.

3. Plastic tanks are made out of any of a number of synthetic polymers, each with its own good and bad points. The greatest advantage of the plastic tanks is that they are lightweight and easily movable, and repairs are generally simple. Some of the more commonly used forms are described as follows.

   a. **Fiberglass,** one of the strongest plastics, is the most commonly

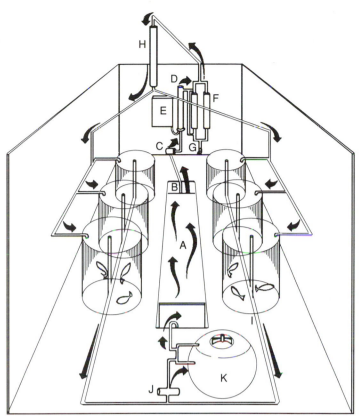

Figure 3.25 Diagrammatic view of a closed system used experimentally to rear squawfish. Overflow water from biofilter (*a*) is drawn from sump (*b*) by pump (*c*) and delivered to circulation heater (*d*) regulated by control panel (*e*). The water is then treated with UV lights (*f*) and released as effluent (*g*) or directed into packed columns for degassing and oxygenation (*f*) and distributed by gravity to the fish tanks. A pump (*j*) distributes overflow from tanks (*i*) through a sand filter (*k*) and the biofilter (*a*). Arrows show the direction of flow. (Redrawn from Lucchetti and Gray, 1988.)

Figure 3.26 Wooden tanks holding seawater. (Photo from Gordon and Boolootian, 1964.)

used; tanks can be made by the farmer, or they can be purchased from a company that has precast molds (see Figure 3.27).

b. **Plexiglass acrylics** are also rather strong, but are expensive and have limited flexibility; therefore, they are most often used for smaller tanks. Plexiglass can be machine worked like wood and is inert.

c. **Vinyl** and **polyethylene** are very flexible, so they are used as liners for tanks made of other materials.

d. **Polypropylene** is rigid and may be substituted for fiberglass. Its major drawback is that it is more difficult to repair than most other types of plastics.

Tanks are often circular in shape, having either a flat or conical bottom. Flat bottoms are sometimes preferred because they can rest on a flat floor, while conical bottoms require a support stand. The advantage of the conical bottom is that sunken food and feces will collect in a small area near the bottom and can be

easily removed. Any tank that is raised on a stand (see Figure 3.28) can be emptied by removing a standpipe attached to the bottom, but this cannot be done with tanks that sit flush against a floor.

Most circular tanks have a standpipe in the middle to accept overflowing water. It may be beneficial to enclose the standpipe in a perforated sleeve; this allows better mixing and exchange of the tank water rather than simply surface skimming. External standpipes with internal screen systems are also popular and may allow superior circulation and self-cleaning. Water is often added to circular tanks by nozzles that are placed so that there is water circulation around the tank (see Figures 3.29 and 3.30). This helps mixing and, if properly designed, is also a self-cleaning action.

There are several advantages to circular tanks. The velocity, circulation, and mixing of the water in circular tanks are greater than in rectangular tanks. Fish, especially wild fish that have just been captured and are placed in a tank, sometimes have a tendency to "get stuck" in corners of rectangular tanks, which may result in injury or a localized depletion of oxygen. The only advantages to the use of

Figure 3.27   Plastic tanks used to rear milkfish in a semiclosed system. (Photo courtesy of The Oceanic Institute, Hawaii.)

Figure 3.28 Tanks used for fish culture. The tanks are raised off the hatchery floor and can be easily drained into troughs built into the floor. (Photo courtesy of the New Jersey Division of Fish, Game, and Wildlife.)

Figure 3.29 Water entering a concrete tank by nozzles parallel to the tank surface. Water entering this way increases the circulation.

rectangular tanks is that they are easy to construct and little space is wasted when they are kept side by side.

A compromise between circular and rectangular tanks is the oval tank (see Figure 3.31). These are typically constructed with a dividing partition placed in the center that is parallel to the straight side walls. Water can be moved fairly efficiently around the tank by using nozzles that spray water out parallel to the straight portion of the wall or by using a paddlewheel that is placed between the straight

Figure 3.30 The central standpipe in these tanks controls the water level. (Photo courtesy of the New Jersey Division of Fish, Game, and Wildlife.)

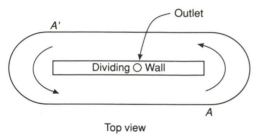

Top view

Figure 3.31 Basic configuration of an oval tank. (From Wheaton, 1977.)

outer wall and the inner partition. It is easier to change the speed of the water flow by using a variable speed paddle-wheel than by trying to adjust nozzles. Oval tanks have an outlet standpipe in the center of the internal partition.

NONTRADITIONAL SYSTEMS

The previous sections outlined some of the most commonly used culture systems. Sometimes, these systems are modified in special ways that make them work efficiently in a particular locale. Some of these "new" methods have attracted a good deal of attention and have been, to lesser or greater degrees, adapted by some commercial growers.

Greenhouses

Like other shelters, greenhouses offer protection from the weather. They differ from other structures in that they are generally less expensive to build, and less expensive to operate because they allow natural light to enter, which reduces lighting and heating costs. In aquaculture, greenhouses are used for production of larvae and algae, but not generally for growout, although this may change in the future (see Figure 3.32). Pilot studies have indicated there may be some commercial potential for greenhouses in cold regions, although this is far from established. In small solar greenhouses in Vermont, fish were grown with some success, and vegetables (tomatoes, peppers, and watercress) were planted in the biological filter

to help remove nitrogen and as a secondary crop.

A single greenhouse in colder climates, especially above 40° latitude, should be built running east to west so that the light in the winter, when the sun is lower in the horizon, can enter the side of the greenhouse rather than the end. When several greenhouses are connected in a **ridge and furrow** design, they should be built running north to south to compensate for the shadows cast by the adjacent greenhouse roofs and gutters; the shadows will move across the floors rather than being fixed in one spot (see Figure 3.33).

Greenhouses that are built alongside another structure, the **attached** greenhouses, have their own set of rules. In the northern hemisphere they should face south to collect as much light as possible. The rear wall, depending on the particular requirements, may be painted either a dark color to help retain heat or white to reflect light back to the tanks in the greenhouse. The angle of the walls that transmit the light will determine when transmission is greatest; since light best penetrates the greenhouse when it enters perpendicularly to the wall, a greenhouse with a very steep wall will be best for getting sunlight in the winter when the sun is

Figure 3.32 A small quonset-type greenhouse with a flexible plastic covering. This greenhouse is used for the production of soft crawfish. (Photo courtesy of Dr. D. D. Culley, Louisiana State University.)

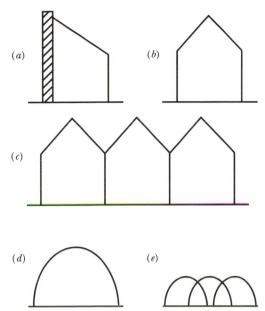

Figure 3.33 Common greenhouse designs. (*a*) attached lean-to, (*b*) single even-span, (*c*) ridge-and-furrow, (*d*) quonset, and (*e*) ridge-and-furrow quonset.

low in the horizon, while walls with a smaller pitch will transmit light better when the sun is higher in the sky.

Glass greenhouses were the most common agricultural types built 30 years ago, but they are impractical for aquaculture and have been replaced in most forms of agriculture by plastic film and rigid panel greenhouses. Flexible plastic covers for greenhouses may be mylar, vinyl, or polyethylene. Of these, polyethylene is clearly the most popular. Its disadvantages are that it will break down after a few years' exposure to the sun's UV light (which darkens it and makes it brittle) and that it requires ongoing repair and attention. However, it is very inexpensive and lightweight, making it easy to repair and replace.

To save heating costs and to add strength to the structure, a flexible film greenhouse can be built with two layers of plastic; these are almost always air-inflated systems. A low-pressure air pump is used to produce a cushion of air that pushes apart the inner and outer layers of plastic. The outer layer of film should not be in contact with any wood, metal, or the inside layer of film, or its insulating ability will be reduced and its life shortened. The airspace between the layers should be about 1.5 cm to 7 cm; if it is less, the insulation value is lost because the layers are too close, and it is also lost if the space is more than 7 cm since air currents become established between the layers.

The most popular design for flexible plastic greenhouses is the quonset. Pipes are used as ribs for the structure that has a dome of 180° and may have straight sides. A plank of wood is run along the base of the pipes that serves as the site of attachment for the plastic. These may be built singularly, or joined together in a ridge and furrow structure, which reduces the costs for building materials and heating.

Rigid-panel greenhouses are an alternative to flexible plastic structures. They may be constructed out of acrylic, PVC, or polycarbonate, but the most popular material is fiberglass-reinforced plastic (FRP). FRP is often corrugated (adding strength to the material) (see Figure 3.34). It is flexible enough to be used for quonset design greenhouses, or it can be used in frames like panes of glass. It is more resistant to shock damage than either plastic films or glass, and is also effective in scattering light. FRP is more resistant to UV damage than films, lasting 5 to 20 years, but is subject to etching and can become rough, which leads to dust collecting on the outer surface and algae

Figure 3.34 Modular type of fish hatchery that relies on solar energy. (Photo courtesy of Dr. Ferenc Muller, Fish Culture Research Institute, Szarvas, Hungary.)

growing on the inner surface, thus reducing light transmission.

Many greenhouses used in aquaculture are simply heated by solar energy, and the large mass of water inside the structure acts as a heat storage system. However, in other cases a greenhouse may have to be heated. This can be done with a central boiler that circulates hot water in pipes that radiate heat; for very large greenhouses, steam pipes are used. Other heating systems include the use of the following:

1. Forced air heating units blow hot air directly out into the greenhouse or conduct the air around the greenhouse through large plastic tubes with a series of open flaps along the sides.

2. Convection heaters are cheap but are efficient only for small greenhouses. These burn a fuel in a contained compartment, and the hot exhaust is directed through pipes that lead out the opposite side of the greenhouse. While the hot exhaust is traveling through the pipes, its heat is lost to the cooler greenhouse by convection.

3. Radiant heaters operate at very high temperatures but are fuel efficient because their radiant energy does not heat the air it travels through, only the water or structures at which it is directed.

There are several methods of cooling a greenhouse when it gets too hot. For aquaculture, probably the best cooling system is simply the use of thermostat-controlled vents and fans that allow cool air to enter at the base of the greenhouse and leave near the roof. Shades inside the greenhouse can also be used to reduce the amount of incoming solar energy (see Figure 3.35).

Although a greenhouse is designed to allow light to enter, there may be times

Figure 3.35   Large greenhouse housing a shrimp spawning facility in Zhongshan County, China. Note the shades that reduce incoming light. (Photo courtesy of Liang Hongwu, Bureau of Aquatic Products, People's Republic of China.)

when extra light is needed, such as overcast days, during the winter when the light is weakest, or at night. There are several types of lights that might be considered:

1. **Incandescent** (tungsten-filament) lights give off much red and far-red light, and much of the energy they receive goes to the production of heat rather than light.

2. **Fluorescent** lamps convert a larger amount of their supplied power to light. Cool white varieties are most commonly used; much of their light comes in the blue range, although some are available with extra emissions in the red and far-red ranges. The major disadvantage to these is that they are low-power lamps, so for large greenhouse areas, there must be many lights and fixtures, increasing the capital cost.

3. **High-intensity discharge (HID)** lights are popular for many agricultural operations. There are four varieties (see Table 3.1).

Thermal Waste Energy

Plants and animals have optimal temperatures (as discussed in Chapter 2) at

TABLE 3.1 Types of HID Lights

Type	Emissions	Life
1. High pressure	Similar to fluorescent, but much of the UV is converted to red.	1000-watt bulb retains up to 70% of output after 10,000 hours
2. High-pressure metal halide	20% of light at 400–700 nm	Shorter than simple high-pressure light
3. High-pressure sodium (HPS)	High in yellow and extends past visible to the 700–850 nm range; 25% energy in 400–700 nm range	Up to 24,000 hours
4. Low-pressure sodium (LPS)	About 27% energy goes to visible light, but most of this is in a narrow band of yellow	About 18,000 hours

which they can best maintain themselves and grow. This is especially true of poikilotherms, which cannot regulate their body temperature to any great extent. At temperatures significantly below the optimal, the animals are sluggish and grow slowly, and above the optimal, they quickly stop much of their activity, including feeding. If the environment of the cultured organisms can be maintained at or near their optimal temperature, organisms can be grown more quickly and efficiently.

The problem with growing animals in warm water is that water's high specific heat means high costs for the aquaculturist; to reach the temperature that might be needed for a large-scale growout operation, a substantial amount of heat energy may be expended.

There are some industries that produce **thermal waste** as a by-product, and this heat can be transferred to cool water when it is pumped past some sort of heat exchanger. Almost all of this "waste" heat is generated by power production; it has been estimated that about 3% of the total energy consumed by the United States ends up as thermal effluent from such power stations. Geothermal springs are another source of hot water. Some of this waste and geothermal heat can be used to extend the growing season or geographical boundaries of cultured aquatic organisms.

There are problems with the use of thermal waste. The heat that is generated will vary, often with the most heat being produced in the summer when it is least needed by the cultured animals. Also, within a 24-hour period, there may be significant shifts in temperature, and a highly variable temperature is worse for growing animals than a cooler constant temperature.

Thermal aquaculture generally takes one of two forms:

1. Cages or rafts are placed directly in the path of the warm water that is pumped from the cooling facility. In Long Island Sound, the maturation of oysters has been accelerated by 1.5 to 2.5 years using the effluent from a power plant that raises the temperature of the water about 11.1°C above ambient.

2. Warm water is pumped into ponds, raceways, or tanks at a specified rate to raise the temperature of the water already present. During the winter, the effluent from a coal-fired generating station on the Delaware River was pumped into raceways containing rainbow trout fingerlings (about 40 grams each); there

was about an 80% survival after 6.5 months, and the fish grew to an average weight of about 295 grams. Similar experiments have been carried out with shrimp (see Table 3.2).

Thermal waste heating may become more important in areas where solar heat is a problem. In Finland, for example, there are long winter periods when solar energy is too weak to do much heating. At the nuclear power plant station in Ol-kiluoto, waste heat from the reactor is used to speed up the production of Atlantic salmon smolts from fingerlings. Smolt production is accelerated so that the animals are released as 1+-year-olds rather than the typical 2+-year-old fish. Since the fish return in three to four years, saving one year during the culture period allows for a substantial increase in the annual catch because of the decrease in recruitment time.

Upwelling

In tropical ocean regions a strong thermocline exists that blocks the transport of nutrient-rich deep water to the surface where it can be used by phytoplankton to expand the base of the food chain. Tropical waters, except in areas where upwelling takes place naturally, are generally unproductive compared to temperate waters. If nutrients can be easily brought to the surface where they can be taken up by phytoplankton, they can be used for the production of feed organisms.

In St. Croix in the U.S. Virgin Islands, in experiments conducted in the 1970s, water was pumped from a depth of 870 m into 50,000 liter pools on the surface. These ponds supported a culture of the diatom *Chaetoceros;* diatom production was extrapolated, and it was estimated that large ponds could produce 8.1 tons protein/ha/year. In turn, the algae were used to support brine shrimp, which were hatched from cysts; researchers estimated that in 1 m^3, 15 grams of cysts could be converted to 8700 grams of adult *Artemia* in 14 days by feeding them the *Chaetoceros.*

While these results sound exciting, much work has yet to be done on these systems. The principal problem is the cost of pumping; even if windmills could be used to bring deep water to the surface, the capital costs would be very high. Theoretically, deep water can be brought to the surface using mechanical energy derived from the difference in temperature between warm surfaces and colder deep waters, the **ocean thermal energy conversion (OTEC).** This temperature difference is a relatively stable 15° to 20°C and could be used to power a heat exchange engine. Again, capital costs are a problem, but initial reports on the use of the OTEC system for abalone production in Hawaii are encouraging.

TABLE 3.2 The Effect of Heated Water on the Growth of the Shrimp *Penaeus vannamei* During the Winter*

Flow Rate	Mean Temperature Above Ambient	Mean Growth Rate	Percent Survival
High	6.9°C	0.04 g/day	47.25
Medium	4.8°C	0.015 g/day	7.45
Low	0 °C	0 g/day	0

* Culture ponds of 0.1 ha received low (35 liters/minute), medium (1500 liters/minute), and high (3000 liters/minute) flow rates from waters pumped from a cooling pond of a fossil-fuel power station.

Source: Chamberlain and colleagues, 1980 Copyright © World Aquaculture Society. Reprinted by permission.

Use of Plant, Animal, and Human Waste

The use of wastes to encourage production, both in agriculture and aquaculture, is very old. Waste is defined as "that which is unwanted"; organic wastes, including plant cuttings, manures, and sewage products, generally contain high levels of nitrogen and phosphorus, as well as trace nutrients like vitamins. In urban India alone, 4.4×10^4 metric tons of refuse are disposed of each day which, if recycled, would annually yield 8.4×10^4 metric tons of nitrogen and 3.5×10^4 metric tons of phosphorus. As discussed earlier in this chapter, when wastes are broken down by fungi and bacteria, and to a lesser extent some other organisms, the nutrients are returned to the environment and are available to plants. In some cases, the wastes can be fed directly to the cultured animals.

If wastes could be converted to plants, which in turn could either be harvested directly or fed to animals, the aquaculturist would be reducing the cost of production as well as helping to improve the environment. Perhaps the most promising scenario includes the use of liquid wastes, like wastewater effluent, for culturing phytoplankton that are then fed to filter-feeding organisms such as oysters or clams. In experiments with three types of filter-feeding bivalves, the animals were presented with phytoplankton grown in an artificial medium and in a medium containing effluent from a sewage treatment plant; the mollusks grew equally well, regardless of the type of nutrient solution used for algal culture. This sounds simple, but there are several scientific, economic, and sociological problems involved. Small-culture systems, especially in China, make regular use of wastes for aquaculture, but widespread use of sewage products in industrialized Western countries has not evolved as it was once hoped it would. However, research still progresses, and results are promising (see Figure 3.36).

Figure 3.36 An experimental fish farm that uses waste from ducks to fertilize fish ponds. (Photo courtesy Dr. Ferenc Muller, Fish Culture Research Institute, Szarvas, Hungary.)

The major objection to using wastes, especially human wastes, is that they contain more than just nutrients; wastes may also contain organic and inorganic pollutants, and possibly disease organisms, which may affect the cultured organisms directly or the consumers of those organisms. When trout were fed a pelletized diet that contained 30% activated sludge, there were significant increases in the concentrations of some metals (but not others) in whole body analysis of the fish. Preliminary tests suggest that many bacterial pollutants are not effectively transferred through the aquaculture food chain (although some viruses survived well in experimental systems). Most of the potentially dangerous organic and inorganic chemicals that are found in wastewater are probably either taken up in relatively small amounts or are concentrated in the fatty organs and blood that are usually removed before the organism is eaten. However, *caution should certainly be exercised* until much more research in this field is done.

Aquaculture–waste systems can be used for purposes other than the ultimate production of food for people, and this may be where their immediate future leads. The simple removal of nutrients from secondarily treated sewage effluent (i.e., an aquaculture system for tertiary water treatment) could be important in preventing unwanted eutrophication of coastal and lake waters. Experimentally, wastewater effluent was run through a serial–polyculture system at the Woods Hole Oceanographic Institution to see if the water could be effectively stripped of its nutrients. Effluent was mixed with seawater and used to grow phytoplankton that was harvested and fed to oysters, and the water from the oysters was shunted into a culture of the seaweed *Chondrus*. About 95% of the inorganic nitrogen was taken up by the phytoplankton. The oysters consumed about 85% of the algae, although they regenerated some of the nitrogen and returned it to the system in the form of ammonia. The regenerated ammonia was completely taken up by the seaweeds. That is, the system was about 95% effective at removing nitrogen, and was 45% to 60% efficient at removing phosphorus. In a similar experiment at the Harbor Branch Oceanographic Institution in Florida, the final step included a pass through tanks of either of two commercially important seaweeds: *Gracilaria*, which is used to make agar, or *Hypnea*, which has a high carrageenan content. Very high yields of 12 to 17 grams dry weight/m^2/day were realized for both species of the seaweeds.

SUMMARY

Open aquaculture systems are the oldest and perhaps still the most popular. In an open system, organisms are placed in a bay or lake, and the natural water currents bring in oxygen, remove waste, and in some cases also bring in food. Production may sometimes be enhanced by protecting bottom crops (such as clams and oysters) from predators. Crops may also be suspended in the water using cages for fish, while bivalve mollusks (such as oysters and mussels) can be grown using floats, rafts, and trays. Suspended cultures make better use of space, and crops are also protected from most natural predators.

In semiclosed systems, water is diverted into ponds and raceways and then discharged. Raceways are usually constructed of concrete, are generally characterized as being long and shallow, and have a rapid water flow. Ponds are usually earthen, often are above ground level, and have a much slower water exchange rate; since they must hold water within their soil embankments and may be periodically drained, special construction methods have evolved for ponds. Higher stocking densities are used more often in semiclosed than in open systems, there is some control over the environmental conditions, and a more uniform crop is produced. However, greater management skills are required to maintain semiclosed systems. Production in ponds can be increased in some cases by fertilizing the water.

In closed systems, water is replaced very infrequently. There is almost absolute control of water conditions in the tanks, and very high stocking densities can be used. These systems are costly to build and must be managed very carefully. Closed systems are not currently important in commercial aquaculture.

Nontraditional modifications of aquaculture systems have been used to make operations more efficient. Greenhouses make use of solar energy to supply light and heat to hatcheries. Thermal effluent has been used to increase the temperature of growout facilities, thus speeding the metabolism and growth of the culture organisms. Nutrients traditionally considered unavailable can be used to increase production; this includes the use of deep ocean waters and wastes such as sewage and manures.

CHAPTER 4

PUMPS AND THE MEASUREMENT OF FLOW

Pumps may be the most commonly used industrial machines. They are important to the aquaculture industry not only because they move water to and from the culture facility, but also because they can be used to drain ponds, force water through filter systems, increase water flow so that circulation is greater, and so on. When possible, the culturist will use gravity to move water, but since this often is not a viable option, pumps are used extensively.

The theory behind the workings of most pumps is rather simple: it is a machine that transfers energy (which originates, for example, from an electric generator) to a fluid. There are a number of types of pumps, each operating a bit differently, but all based on this principle. Fluid that has passed through a pump then has available energy in the form of either (1) **kinetic energy** (e.g., if the velocity of the water is increased with pumping, so is its kinetic energy) or (2) **potential energy** (e.g., pumping water up to a holding tower, so that it can be gravity fed to tanks or ponds later, increases its potential energy).

When a pump is selected for an aquaculture system, the needs of the system must be carefully considered. The farmer must be sure that the performance characteristics of the pump are those that are closest to the job requirements. These performance characteristics are normally expressed in a graphical form by the manufacturer (see Figure 4.1). There are several ways that these graphs can be drawn; normally they relate the amount of liquid discharged, the head generated, the power requirements (called **brake power**), and the **efficiency** (which is the ratio of the energy that is imparted to the liquid to the energy going to the pump from the driving device). These characteristics differ from pump to pump and change as the pump ages because parts begin to wear. Pumps of approximately the same size and similar design tend to have similar performance characteristics.

RECIPROCATING PUMPS

Reciprocating pumps include the old hand pumps used to get water from wells before electricity was available in rural districts. Many types are presently used, some of which are very sophisticated, and unlike the original reciprocating pumps, they require a significant amount of energy that is supplied mechanically or

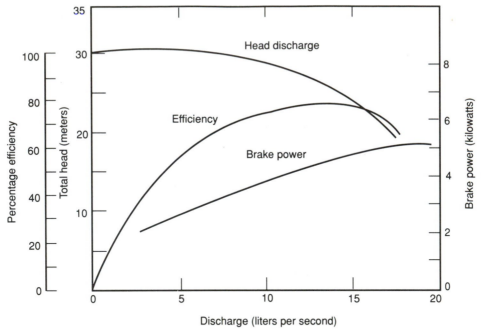

Figure 4.1 Graphical representation of the performance characteristics of a centrifugal pump. For this particular pump, the maximum efficiency is about 77%, which occurs when the flow is 13 to 14 liters per second and there is a total head of about 25 meters; the required power is about 4.6 kilowatts. (From Wheaton, 1977.)

derived from air, steam, or hydraulic pressure. Large motor-driven reciprocating pumps are simplex, duplex, triplex, or quintuplex in design according to the number of cylinders. Each of these pumps can be further classified in a number of other ways, including piston and plunger types, single- and double-acting types, inside- and outside packed types, and so on. Steam-powered pumps can be used economically in industries where the heat for the steam is a by-product and/or the steam energy can be used in some process down the line, but when these steam pumps exhaust directly into the atmosphere their steam/horsepower-of-pump-output ratio is high and therefore not economical. Steam lines usually are not available to the aquatic farmer, although they may be found in some industrial plants that process aquaculture products.

Hand pumps work because there is a plunger moving up and down inside a cylindrical chamber as the pump handle is raised and lowered. These are often called **piston pumps.** When the handle is raised, the plunger is forced down; this closes the lower check valve, but the upper check valve opens, allowing the plunger to descend freely. When the pump handle is brought back up, the upper check valve closes and forces water up and out of the pump; at the same time the pressure in the lower section of the cylinder is reduced, causing a partial vacuum, thus lifting the lower check valve and pulling water up into the cylinder. The amount of water that is pumped is controlled by the number of strokes per unit time and by the length of the stroke. Each time the plunger moves, it displaces a particular amount of fluid, and the speed that the fluid is pumped is linearly related to the speed at which the plunger is moved; therefore, these, along with rotary pumps discussed next, are called **positive displacement** pumps (see Figure 4.2).

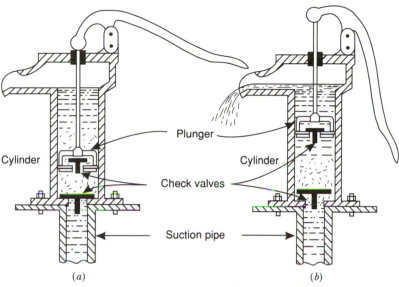

Figure 4.2 Operation of a piston pump. (*a*) Downstroke. (*b*) Upstroke. (From Wheaton, 1977.)

Reciprocating pumps are self-priming as long as the valves do not stick or leak, and as long as the plunger seal is good. The seals will need periodic replacement, but these pumps are generally durable machines with few maintenance needs. Piston pumps can handle water with significant amounts of suspended materials. Water is discharged from these pumps in pulses, but the pulsing can be reduced by using multipiston pumps in which each piston is in a different phase of the cycle at any moment.

ROTARY PUMPS

Rotary pumps consist of rotating cams, vanes, pistons, screws, gears, or other devices, inside a fixed housing. Generally, they operate with very little clearance within the housing, so that each rotation causes a positive displacement of the fluid. The energy imparted from the rotating part of the pump moves the fluid from the low-pressure side of the pump to the high-pressure side. Again, there are many varieties of rotary pumps, such as the spur-gear types, herringbone-gear

types, screw pumps, flexible vane pumps, and peristaltic pumps.

Flexible Vane Pumps

The **flexible vane pump** consists of a rotor placed eccentrically (i.e., the rotor is not centered) inside a housing. Attached to the rotor are vanes made of a flexible material. Vanes farther from the wall of the housing are bent only slightly, and it is here that the fluid enters the housing, while on another side of the rotor, where the vanes come very close to the housing wall, the vanes are bent much more, and the water is forced out of a discharge port. The bending of the vanes decreases the area that the fluid occupies and therefore increases the pressure on the fluid forcing it to exit (see Figure 4.3). These are generally small pumps with a maximum output of about 500 liters/minute, self-priming, and relatively simple to repair.

Peristaltic Pumps

Peristaltic pumps are not used in aquaculture to move large amounts of liquid, but can be valuable because they can

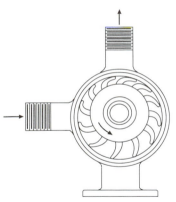

Figure 4.3 Flexible vane pump. (From Wheaton, 1977.)

very accurately deliver small volumes of liquid at a constant rate. Such a device could therefore be used, for example, to slowly add nutrients to an algal culture. The pump is primed by filling the flexible plastic tube with the liquid; a rotating member then squeezes the tube by pushing it against the wall of the housing, forcing the liquid through in the same direction as the member is moving (see Figure 4.4). Besides their accuracy, peristaltic pumps have an advantage over other types of pumps: the fluids never come in contact with the machinery, so a tube to be used to move algal media may be sterilized before it is put in the pump. These pumps can also handle highly cor-

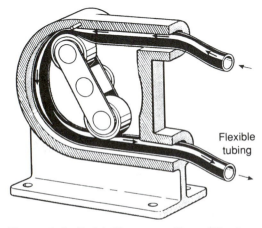

Flexible tubing

Figure 4.4 Peristaltic pump. (From Wheaton, 1977.)

rosive materials such as acids or chlorines without being damaged. The flexible tubes are fairly inexpensive. These pumps are highly reliable.

REGENERATIVE PUMPS

Regenerative pumps are sometimes called **turbine pumps, vortex pumps,** and **peripheral pumps.** They are sometimes considered to be rotary pumps. Regenerative pumps contain a small multibladed impeller that rotates inside a race chamber. As liquid enters the housing of the pump from a central suction passage, it flows to both sides of the impeller. Centrifugal force throws the liquid out to the periphery, and the rotating vanes hitting the liquid increase its velocity. (The head is not developed by centrifugal force so these are not considered centrifugal pumps.) As the liquid enters the casing's passage, there is a reduction in the velocity and an increase in the potential energy in the form of pressure. This pressure represents the **generation of static head.** The liquid continues to move inside the channel passage in the same direction as the impeller, but as the fluid loses velocity, its orbital becomes smaller, and it again comes in contact with the impeller vanes, which again hit, and therefore impart more energy, to the liquid. This cycle is repeated as the fluid moves through the channel, being hit a number of times by the impeller, and each time there being a "regeneration" of the head, from which the pump gets its name. The greater the head that the pump must work against, the greater the regeneration; thus the head that the pump can generate is limited by the diameter and the speed of the impeller. Eventually the liquid is guided out of a discharge nozzle (see Figure 4.5). Regenerative pumps develop higher heads per stage than do centrifugal pumps. They are most often used for capacities of 0.5 to 400 liters/minute and can develop head to 150 m or more.

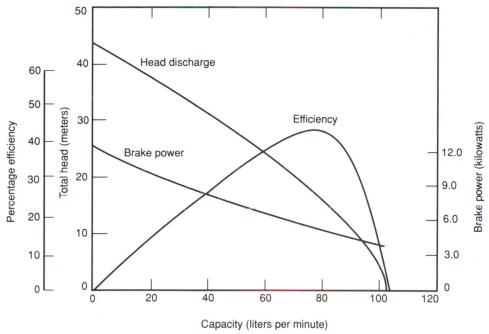

Figure 4.5   Performance characteristics of a regenerative pump. Note the difference in this graph and that for a centrifugal pump in Figure 4.1. (From Wheaton, 1977.)

These pumps work best with fairly low levels of suspended materials.

CENTRIFUGAL PUMPS

Centrifugal pumps are, by far, the most commonly used types of pumps in aquaculture. There are many types of centrifugal pumps; all have a power source that turns a shaft causing an **impeller** to rotate. They use centrifugal force to lift the fluid being pumped. The most important benefit of the centrifugal pump is its ability to be choked (i.e., restricting the outline and thereby reducing the flow).

The fluid enters the housing from the center or **eye** of the impeller. The rotating impeller forces the fluid to move outward as a result of centrifugal force. As in the case of the regenerative pump, the fluid that leaves the impeller to circulate in the housing rapidly has the energy that is associated with the velocity (dynamic head) converted to energy in the form of

pressure (static head). As the fluid leaves the center of the impeller, there is a drop in pressure in that area of the pump, so more fluid is sucked in to prevent a vacuum from forming. There are three major types of centrifugal pumps: the volute, diffuser, and mixed flow; of these, the volute and diffuser are strictly centrifugal.

Volute and Diffuser Centrifugal Pumps

In the **volute** type of pump, the impeller is placed in a housing that spirals out from the eye. As the volume of the housing increases, the velocity of the fluid that has left the impeller decreases and the dynamic head is converted to static head (see Figure 4.6).

Diffuser-type centrifugal pumps have an impeller that causes the fluid to accelerate outward into a set of diffuser vanes. This diffuser is attached to the housing (see Figure 4.7); it is able to convert much of the dynamic head to static head before

Discharge

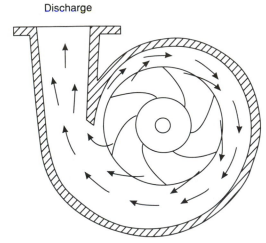

Figure 4.6 Volute centrifugal pump. (From Wheaton, 1977.)

regenerative "turbine," although it does not work on the same principle.

Impeller Types and Performance Characteristics

Both the volute and diffuser pumps may have one or more **stages.** A single-stage pump is one that develops all the head from a single impeller. If the total

the water reaches the outer walls of the housing. Thus, the diffuser pump is able to convert more energy to pressure, and therefore generally has a slightly higher efficiency than does the volute type, although the diffuser pump is more expensive and does not handle suspended particles as well. The diffuser pump is sometimes known as the "turbine" pump because it is similar in structure to the

Discharge

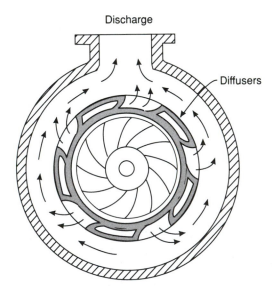

Figure 4.7 Diffuser centrifugal pump. (From Wheaton, 1977.)

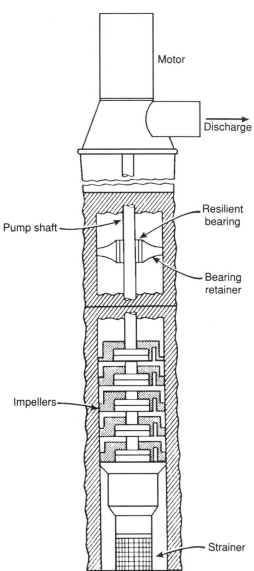

Figure 4.8 A multistage centrifugal pump used to bring water up from a well. The motor as shown here is at the top but may also be submerged. (From Wheaton, 1977.)

head that is required is greater than the head that can be produced by one impeller, a multistage pump is used (see Figure 4.8). Two or more impellers are used in a multistage pump, with the second impeller taking in fluid from the discharge of the first.

Volute and diffuser pumps can have several different types of impellers. Each of the different impeller designs will impart a different set of characteristics to the pump:

1. An **open impeller** is actually a set of vanes attached to a central hub. The vanes are given support by **ribs** (see Figure 4.9). These are generally not very efficient, but they are good for moving liquids with high levels of suspended materials.

2. A **semiopen impeller** has a wall or **face plate** on only one side of the vanes (see Figure 4.10).

3. The **closed impeller** is used when handling clear fluids. It has face plates on both sides of the vanes so the pathway of the fluid from the eye to the housing wall is enclosed, preventing **slippage** observed in the open and semiopen impellers (see Figure 4.11).

Centrifugal pumps are not positive displacement pumps. The relationships be-

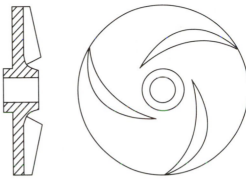

Figure 4.10 Semiopen impeller. (From Schwab et al., 1981.)

tween the size or speed of the impeller, the amount of fluid that is moved, and the head produced are not linear. However, a series of set relationships does exist. These pertain to pumps with variable impeller rotation speeds and to impeller pumps of the same design but with impellers of different sizes. These relationships can be stated as

$$\left.\begin{array}{l} \dfrac{Q_1}{Q_2} = \dfrac{N_1}{N_2} \\[2mm] \dfrac{h_1}{h_2} = \left(\dfrac{N_1}{N_2}\right)^2 \\[2mm] \dfrac{P_1}{P_2} = \left(\dfrac{N_1}{N_2}\right)^3 \end{array}\right\} \begin{array}{l}\text{for pumps of}\\\text{the same size}\end{array}$$

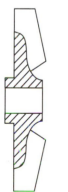 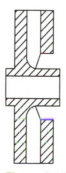 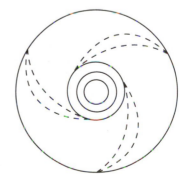

Figure 4.9 Open-type impeller. (From Schwab et al., 1981.)

Figure 4.11 Closed impeller. (From Schwab et al., 1981.)

$$\left.\begin{array}{l} \dfrac{Q_1}{Q_2} = \left(\dfrac{D_1}{D_2}\right)^3 \\[2mm] \dfrac{h_1}{h_2} = \left(\dfrac{D_1}{D_2}\right)^2 \\[2mm] \dfrac{P_1}{P_2} = \left(\dfrac{D_1}{D_2}\right)^5 \end{array}\right\} \begin{array}{l} \text{for pumps which} \\ \text{are similar but} \\ \text{of different sizes.} \end{array}$$

where Q = discharge, N = pump speed, D = impeller diameter, h = head, and P = power requirements.

For example, suppose a pump with an impeller diameter of 0.2 m is discharging at 20 liters/sec while it was rotating at 1000 rpm. Under these conditions the energy requirement is 4000 watts and the head generated is 36 m. What would be the new discharge rate, head, and power requirement if the impeller speed were increased to 1200 rpm?

$$\dfrac{20 \text{ liters/sec}}{Q_2} = \dfrac{1000 \text{ rpm}}{1200 \text{ rpm}}$$
$$Q_2 = 24 \text{ liters/sec}$$

$$\dfrac{36 \text{ m}}{h_2} = \left(\dfrac{1000 \text{ rpm}}{1200 \text{ rpm}}\right)^2$$
$$h_2 = 51.8 \text{ m}$$

$$\dfrac{4000 \text{ watts}}{P_2} = \left(\dfrac{1000 \text{ rpm}}{1200 \text{ rpm}}\right)^3$$
$$P_2 = 6912 \text{ watts}$$

That is, to increase the flow by 20%, the power requirement for this pump would have to be increased by 73%!

If we were to use a pump similar in design but larger than the one first described, and it had an impeller diameter of 0.4 m, what would be the resultant flow and the power requirement?

$$\dfrac{20 \text{ liters/sec}}{Q_2} = \left(\dfrac{0.2\text{m}}{0.4\text{m}}\right)^3$$
$$Q_2 = 160 \text{ liters/sec}$$

$$\dfrac{4000 \text{ watts}}{P_2} = \left(\dfrac{0.2\text{m}}{0.4\text{m}}\right)^5$$
$$P_2 = 128,000 \text{ watts}$$

PROPELLER AND MIXED FLOW PUMPS

Propeller pumps have open impellers that look like propellers of boats or airplanes and develop most of their head by a lifting action (see Figure 4.12). Propeller pumps have been marketed under names like **screw, spiral, axial flow,** and **straight flow pumps.** They are sometimes considered to be rotary pumps or centrifugal pumps, although since centrifugal force does little to generate the head and the flow of the fluid past the impeller is axial, there seems to be little reason for calling these centrifugal. Mixed flow pumps have impellers that develop head partly by centrifugal force and partly by the lift that the vanes give to the liquid, and therefore may be considered centrifugal–propeller hybrids.

Propeller pumps in particular are used to move large amounts of water, but they are not efficient at moving fluids against a large energy gradient. These pumps may be able to raise the water only several meters above the original surface (i.e., they can operate only under conditions of low head). The design of the propeller pumps is rather simple, consisting of a motor turning a shaft with an impeller at the end. The impeller is in a pipelike housing.

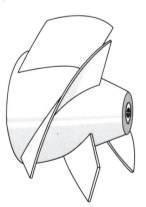

Figure 4.12 Mixed flow impeller. (From Wheaton, 1977.)

The impeller lifts the fluid, and the thrust bearing absorbs the accelerating force that is generated by the impeller. Guide vanes are sometimes used to eliminate the rotational component of the water.

FISH PUMPS

There are times when large numbers of fish must be transferred between two bodies of water. This must be done quickly and with limited stress, but nets and handling reduce survival of many species. Three types of **fish pumps** are available for this purpose:

1. **Centrifugal** pumps can be modified so that they can move live fish as well as water. These pumps are characterized by a large space between the ends of the impeller vanes and the walls of the housing, and all surfaces are polished to prevent damage to the fish being moved (see Figure 4.13).

2. **Liquid ring** pumps, with rapidly flowing water that produces a vacuum, have been modified into fish pumps (see Figure 4.14). These have no moving vanes that may damage the fish in transport, so the mortality rate is lower than with the centrifugal pumps, and larger fish may be moved.

3. The **auger** or **Archimedes screw** type of fish pump is basically a spiraling axis encased in a cylinder. Since this is not based on suction, it can be designed to carry large fish.

All these fish pumps are used with flexible hoses and may be attached to a sorting device if needed. Depending on the design, the original water the fish were in can be transferred with the organ-

Figure 4.14 A Transvac® liquid ring fish pump. (Photo courtesy of Innovac Technology, Inc., Washington.)

Figure 4.13 Workers at a trout farm transferring fish from a raceway to a truck. Men in the lower portion of the picture are slowly moving a screen toward the pump intake forcing the trout to that end of the raceway. (Photo courtesy of the New Jersey Division of Fish, Game, and Wildlife.)

isms, or much of the water can be split off. The flow rate through the pump is critical; if the flow is low, the fish may avoid the intake hose and time spent pumping may be excessive, but pumping rates that are too high may increase fish mortality.

AIRLIFT PUMPS

Airlift pumps are not mechanical pumps in the same sense that the others discussed are. They are often associated with undergravel filters seen in small saltwater aquariums.

The airlift pump consists of an open-ended tube that is upright and partially submerged in the water. The top of the tube is bent to be parallel with the surface of the water. Air is released at the bottom of the tube through an airstone or some device that will break the stream of air into small bubbles. Essentially an air–water mixture that is lighter than water is being produced inside the tube, and that therefore has a tendency to "float" above the surface level of water in the tank. The mixture would eventually stop rising, but before it can do so, it is directed out the upper end of the tube. As the mixture rises inside the tube, more water is drawn into the bottom of the tube to replace that which has spilled out of the top (see Figure 4.15).

Airlift pumps can move moderate amounts of water, but very little head is generated. Their chief advantage is the fact that they add oxygen to the water, and that there are no moving mechanical parts in the water. The efficiency of the flow is a function of the volume of air in the tube, the depth of air injection, the height of the outlet above the surface of the water, and the size of the bubbles. Smaller bubbles work best, and the volume of air should be about 10% at the bottom of the tube (since the bubbles will expand as they rise, the volume will increase near the surface), but if more than

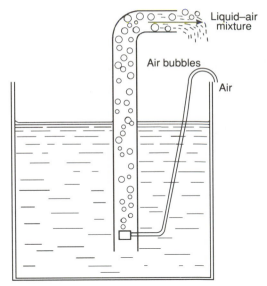

Figure 4.15 An airlift pump. (From Wheaton, 1977.)

10% is used, the little bubbles start combining to make the less efficient big bubbles.

These pumps are used in tanks more often than in ponds or raceways. Most trials in ponds have been unsatisfactory because of the fluctuating water levels; slight differences in submergence of these pumps will alter the flow rate significantly. The problem of variable water levels can be overcome by using a floating airlift pump that is attached to a fixed post by an arm that will rise and fall with the water level.

AIR PUMPS

Air compressors are used largely for small aquariums. They deliver small volumes of air at high pressures, but are not economical for large-culture operations. **Air blowers** deliver large volumes of air at low pressures and are more efficient for large farming operations (see Figure 4.16). Blowers have designs similar to water pumps in several respects; they can be

Figure 4.16 Air blowers in series. (Photo courtesy of Aquatic Eco-Systems, Inc., Florida.)

positive displacement machines with rotary or lobe vanes, or they may be non-positive displacement in design, using regenerative or fan mechanisms.

MEASUREMENT OF FLOW

While water will flow out of a new pump at fixed and determined speed, this flow rate will change as the pump ages. In addition, it is almost always true in aquaculture that the pump directs water into a pipe, and the pipe will reduce the flow rate because of internal friction or turbulence (see Appendix 2). The culturist will often need to know the flow rate inside the pipe and/or at the pipe exit.

In aquaculture, the most common method for measuring the flow rate can be characterized as the **bucket and stop-watch method.** A bucket, which must have a known volume, is placed under the pipe exit so that it catches the water flowing out. Simply by measuring the amount of time it takes to fill up the bucket, a flow rate can be established. Thus, if a 7 liter bucket takes 12 seconds to fill, the flow

rate must be 35 liters/min or 2.1 m^3/hr. The larger the bucket and the more times this procedure is repeated, the greater the accuracy of the measurement.

More sophisticated flow measurement techniques are sometimes employed. The devices used may be divided into three major groups:

1. **Head flowmeters,** which are designed to cause a pressure difference in the pipe that can be measured and related to the flow.
2. **Mechanical flowmeters,** which have moving parts.
3. **Other kinds,** such as electromagnetic flowmeters and weirs (many of these are rather complex and are unlikely to be used by the fish farmer, and therefore will not be considered here).

Head Flowmeters

The **orifice meter** consists of a flat plate with a hole drilled in the center that is perpendicular to the flow inside the pipe. Water going through the hole has an increased speed (see Appendix 2 for the reasons for the increased velocity). This increase in dynamic head results in a decrease in the pressure (see Figure 4.17).

This change in pressure can be measured by a device called a **differential manometer,** which relates the differences in the pressure before and after the restriction. The manometer is essentially a U-shaped tube partially filled with a liquid. If the pressures on both sides of the tube are the same, the levels of the liquid on both sides of the tube are the same, but if the pressure on one side is greater than the pressure on the other side, the fluid is pushed down on the side of the tube where the pressure is greater, so the levels of the liquid on the two sides of the U-tube are unequal. The difference in the levels of the fluid is directly related to the

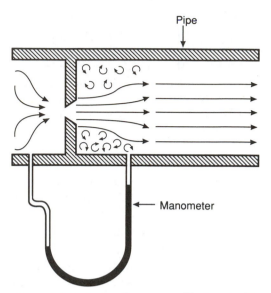

Figure 4.17 Orifice flowmeter with manometer. Since the pressure at the manometer opening is greater before the orifice than after, the heights of the liquids in the arms are unequal.

difference in the pressures exerted on the fluids in the pipe.

The **Venturi meter** operates on the same principle as the orifice meter, but is designed so that there is less eddy formation (see Figure 4.18). The reduced turbulence occurs because the approach to and exit from the area of restriction are gradual, so the loss of energy is reduced. Venturi meters can therefore handle up to 60% more fluid than the orifice meters for the same pipe diameter. Like the orifice meters, the Venturi meters can handle fluids that contain suspended solids.

Mechanical Flowmeters

The **rotameter** is one of the most common of the mechanical flowmeters for pipes with small diameters. It consists of a plastic or glass tube, held in a vertical position. Water enters the bottom of the flowmeter; the tube is a frustum shape, with the entrance below being smaller than the exit above. Because of the increase in the diameter of the tube, the water flowing through the lower part of the flowmeter travels at a greater velocity than the water near the exit.

Inside the tube is a bullet-shaped float that is heavier than the water. This float is being forced down by gravity, but also lifted up by the flowing water. At some point the float will come to an equilibrium, and at that point the lifting force of the water is balanced by the effect of gravity pulling it down. The float can rise no farther since the reduced velocity in the upper part of the tube will not support it. If there is an increase or decrease in flow, the float will move. Rotameters are normally marked with an arbitrary scale that must be calibrated (usually with the bucket and stopwatch method) (see Figure 4.19).

Turbine flowmeters are particularly well suited for measuring the flow through open channels as well as in pipes. These flowmeters consist of a propeller or turbine that turns as water goes past the blades (see Figure 4.20). The number of revolutions per unit time increases proportionally with the flow of the water. The turbine is connected to some sort of de-

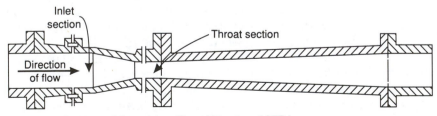

Figure 4.18 Venturi flowmeter. (From Wheaton, 1977.)

vice that counts the number of revolutions per minute.

SUMMARY

Pumps, machines that impart energy to fluids, are critical for aquaculture. There are many different types of pumps, each with advantages and disadvantages, so the fish farmer must be able to recognize the pump with the characteristics best suited for a particular job. Reciprocating pumps are positive displacement machines with pistons or plungers to move the fluid in and out of the pump housing. Rotary pumps, also positive displacement devices, feature a rotating member inside a housing; the fluids are pushed from a low-pressure to a high-pressure environment and are then discharged. Regenerative pumps have impellers with vanes on both sides of the face plate that impart energy to the fluid as it moves around the pump housing. True centrifugal pumps, like the volute and diffuser types, are based on the principle that dynamic head given to a fluid by a spinning impeller can be converted to static head by reducing the velocity of the fluid as it is flung by centrifugal force out from the pump eye. Propeller and mixed flow pumps use lift-

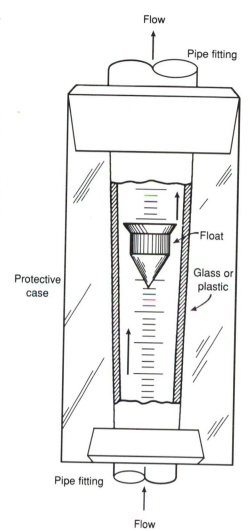

Figure 4.19 Rotameter. (From Wheaton, 1977.)

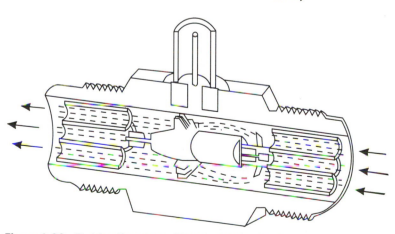

Figure 4.20 Turbine flowmeter. (From Wheaton, 1977.)

ing action to partially or completely move a fluid. Several types of pumps can be modified to transport live fish. Airlift pumps function by making a water–air mixture that is lighter than the water that will float out the pump tube through an exit above the surface of the water.

Knowing the velocity of a fluid flowing inside or exiting a pipe is often important. The pipe exit flow can be most easily measured with a bucket and stopwatch, but flow through a pipe is determined in most aquaculture operations by head flowmeters, such as the orifice or Venturi types, or mechanical flowmeters, such as the rotameter or turbine types.

CHAPTER 5

FILTRATION AND WATER TREATMENT

In open systems, the farmer rarely becomes involved in water treatment, but for those culturists using closed or semi-closed systems, it is likely that the water will be filtered, disinfected, degassed, aerated, or otherwise manipulated so as to improve the culture conditions. It is worth calling attention to the fact that there is no "best" way to treat the water; filtration may be beneficial in many cases, but in some cases the expense is not justified and may even hinder the growth of the organisms. The same may be true of aeration, chlorination, UV treatment, and so on. Each aquaculture facility must be looked at individually, and the benefits derived from the treatments should be weighed against their costs when the economic model is being developed (see Chapter 13). In many instances, the quality of the water can (and should) be improved, and in this chapter we will explore some of the most common ways of upgrading the aquatic medium.

We can improve the water quality in several standard ways. Filtration removes both particulate and dissolved materials from the water, including unwanted nutrients, pollutants, living organisms, and debris. Ozone, UV radiation, and chlorination are used to kill potential parasites, competitors, or predators and may alter some of the compounds dissolved in the water. Aeration is commonly employed in crowded culture conditions since O_2 levels often decline to suboptimal levels. Degassing removes excess N_2 from supersaturated water, thus preventing gas bubble disease.

MECHANICAL FILTERS

Screens and Bags

The simplest type of filter is the **mechanical filter,** sometimes called a **clarifier.** It is used to remove particulates from the liquid medium. In its most basic form, it is a screen placed in front of an inlet line to prevent logs, leaves, fish, and other objects from entering the culture system. This prevents potential damage to the pumping system, as well as halting the introduction of unwanted materials or organisms. There may be a single screen or a series of screens, progressing from those with a coarse wide mesh to those of a finer grade. A culturist may try a few different mesh sizes until the ideal size(s) are determined. The culturist will select the mesh that will remove most of the floating material but does not require constant cleaning.

If clogging of the screen becomes a consistent problem, there are special mechanisms that can be used to reduce time-consuming maintenance. The farmer can opt for the installation of **rotary screens,** so called because they rotate while they filter. Part of the rotating screen is submerged to filter the water, while the other part of the screen is being cleaned by a backwash; the backwash consists of a stream of water that is forced against the screen in a direction opposite to the flow of the water being filtered. The backwashed water goes into a trough that carries away the concentrated particles removed from the inflow. Slowly, the clean portion of the screen rotates into the water, replacing the clogged portion. The flow of the water may be along the axis of the screen (**axial flow**), or the water may flow outward from the axis of the turning screen (**radial flow**) (see Figures 5.1 and 5.2). In either case, the rate that the screen must turn is related to the size of the mesh, the flow of the water, and the character and amount of material suspended in the water.

One of the most common variations on the screen is the **bag filter.** Water that is discharged from a pipe flows into a bag made of nylon or some other synthetic material before entering the culture system. Particles are caught inside the bag and the water flows out. The bags are cleaned periodically; they are turned inside out, and a stream of clean water is forced through.

Particle Filters

Particle filters, sometimes also called **sand filters,** are housings containing beds of sand, gravel, clay, or some other unconsolidated material. Water passes through the beds, flowing by either gravity or being forced by a pump (see Figure 5.3). The particles in the bed physically filter out the suspended material while the water flows through the spaces between the particles (the **void volume**).

The size of the particles and their ability to compact will affect the filtering efficiency. Large filter particles allow water to flow quickly because there is little resis-

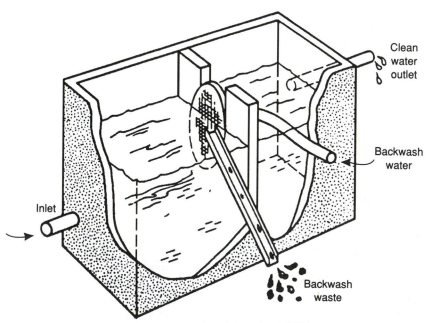

Figure 5.1 Axial flow rotary screen. (From Wheaton, 1977.)

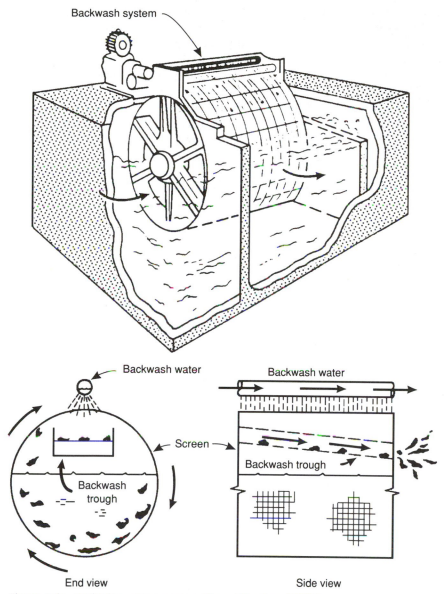

Figure 5.2   Radial flow rotary screen. (From Wheaton, 1977.)

tance; the filter therefore does not clog easily. However, the larger and less compact particle beds remove only the largest of suspended materials, allowing smaller materials to pass. If very fine particles are used in the filter bed, almost all the suspended material will be taken out, but the flow rate will be slow, and the filter will require frequent cleaning. These problems can be partially overcome if a series of two to four filters are used with progressively smaller particles in the beds (but also progressively increasing the total bed volumes).

Particle filters must be cleaned periodically by **backwashing.** During backwashing, the direction of the water flow is reversed and the flow rate is increased. Sometimes air is forced into the filter housing to add to the turbulence. This

Figure 5.3 Pressurized sand filters.

backwashing process increases the bed volume by suspending all the particles in the water–air mixture in the housing; at this point the bed is said to be **fluidized.** When the bed is fluidized, the particles are moving quickly, bumping into each other, and thereby cleaning themselves of the adhering materials that were filtered from the culture water. The backwash water containing the filtered material is then drained from the filter.

If a series of filters is unavailable a single filter can be used, although the efficiency of the system is greatly reduced. In such single housing systems, there may be a series of beds inside the housing. The beds are arranged with the finest particles on top and the coarsest at the bottom (see Figure 5.4). There are two reasons why this arrangement is used:

1. The coarser particles hold the finer ones off the bottom so that the receiving drain or plate at the bottom of the filter does not get clogged with the finer particles.
2. After backwashing, the larger particles sink first.

Because the water flows from the finer to the coarser beds, it is only the first bed that does the filtering, while the rest of the particles in the lower beds are used for support; thus the efficiency of this filter (per unit volume of housing) is not very good.

Sometimes, when very small particles and bacteria must be removed, **diatomaceous earth** is used as the filtering medium. Diatomaceous earth can be used in a simple particle filter, or may be used in a special filter designed for this particular medium. Diatomaceous earth filter elements consist of rigid cores, covered by removable sleeves made of a clothlike material such as polypropylene, inside a closed vessel. Water can flow through these filter elements. The filtering operation has three phases:

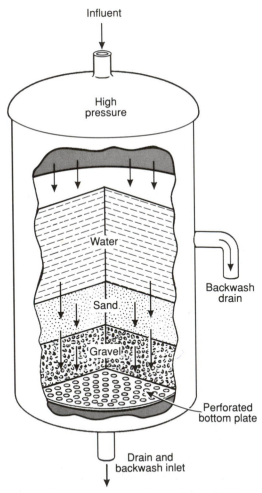

Figure 5.4 Sand filter showing fine sand layered above coarse gravel. The gravel prevents the sand from clogging the bottom plate or flowing out through the drain. (From Wheaton, 1977.)

1. Clean water and diatomaceous earth are pumped through the system, and the diatomaceous earth coats the filter elements.

2. The water to be filtered is mixed with more diatomaceous earth (usually of a different grade than the original coating material), and this mixture is pumped into the housing. The new diatomaceous earth and the suspended materials to be removed are caught on the coated filter element, while the filtered water passes into the core of the element and out of the system. Thus, the thickness of the layer of material on the filter element is constantly increasing.

3. When the layer of material on the element is so thick that water does not easily flow through, the system is backwashed, and the old diatomaceous earth and filtered materials are drained away (see Figure 5.5).

GRAVITATIONAL FILTERS

Gravity can be used to separate culture water and suspended particles that are denser than the water. The greater the difference in density, the greater the ease of separation.

Sedimentation

Simple sedimentation refers to the practice of holding water and allowing the suspended materials to settle. Small particles may be maintained in moving and turbulent water, but settlement takes place in the sedimentation facility, where

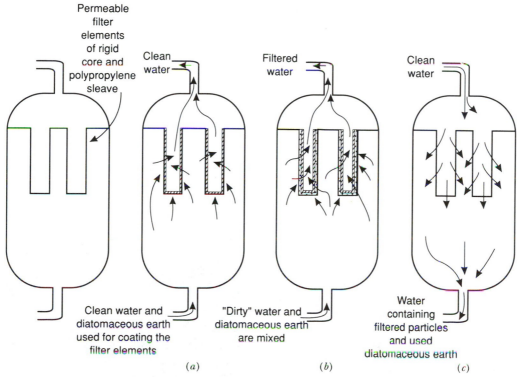

Figure 5.5 Diatomaceous earth filter. (*a*) The filter element being coated. (*b*) Water being filtered. The water contains suspended particles plus new diatomaceous earth that is being added; the layers thicken over the filter element. (*c*) The filter is backwashed.

the water is still or moving very slowly. In many cases, sedimentation is a simple, cheap, and fairly effective way of removing much unwanted material suspended in the culture water.

The sedimentation rate may be increased by the addition of a **coagulant,** which results in the formation of a cloudy **floc** in which suspended materials, plankton, microbes, and colloidal matter are caught. The floc will sink more quickly than the individual suspended materials. Common coagulants used in wastewater treatment are aluminum sulfate (alum), ferrous (copperas) and ferric sulfates, ferric chloride, lime, and clays. Sometimes the pH must be adjusted for good coagulation.

In addition to removing the unwanted materials in the water, coagulation may also be used to harvest microalgae. **Chitosan,** a partially deacetylated form of chitin (the organic base of the crustacean exoskeleton), has been used to bring several species of microalgae out of the water. Chitosan's advantage is that it is probably safer for consumers of the algae than other coagulants. In seawater, flocculation is inhibited by the water's ionic strength; combinations of different flocculants or pretreatment with ozone is required to produce an effective floc in seawater.

Centrifugation

Sedimentation works because gravity pulls dense particles through the less dense water. This process can be speeded up if the gravitational force on the particles can be increased—this is what **centrifugation** does. Most science students are familiar with the batch centrifuge in which centrifuge tubes or jars of a fixed volume are placed in the rotor head, spun, and decanted. It is impractical to remove the particles from large volumes of water in this manner, so the **continuous flow centrifuge,** which removes suspended particles from water that flows

into the instrument, is used in place of the batch centrifuge. When the fluid is discharged (spun) from the centrifuge, it is clean, and the suspended particles are retained in the centrifuge head (see Figure 5.6). Moderately large volumes of water can be treated this way, and continuous flow centrifuges can be used both for filtering culture water and, more commonly, for harvesting microalgae or a similar product.

There are a number of designs for continuous flow centrifuges; one common design is the **solid basket design** (see Figure 5.7). In these, the centrifugal force generated by the spinning rotor throws the fluid with the suspended material against an interior wall of the centrifuge head. As the water continues to flow, the heavy particles are trapped by the wall, but the clean water is able to flow over the wall and leave the centrifuge. The efficiency of the removal varies with the head that is used, the speed at which the head is spinning, the rate of the water flow into the centrifuge, the viscosity of the water, and the size and density of the materials that are being removed. Obviously, if the culturist decreases the flow rate, increases the speed of the centrifuge, decreases the water viscosity, or increases the size and density of the particle, the efficiency will improve.

The **hydroclone** also makes use of gravity to filter water. Hydroclones are composed of tapered cylinderlike tubes.

Figure 5.6 Continuous flow centrifuge.

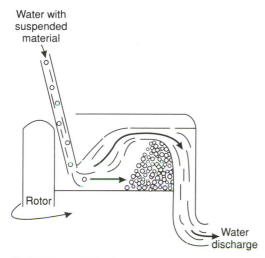

Figure 5.7 Half of the rotor of a basket-type continuous flow centrifuge. Water with suspended materials enters and is forced to the outer wall of the basket by centrifugal force. The water, being less dense than the particles, is able to flow over the wall.

Unfiltered water is pumped into the hydroclone along the top wall of the cylinder. The water flows down along the interior wall of the hydroclone. As the diameter of the cylinder decreases, the water rushing along the wall increases in velocity. The water circulating rapidly around the hydroclone walls causes a pressure drop in its center. The particles in the water continue to move down the hydroclone, but the lighter water is sucked upward, through the center of the hydroclone, because of the partially formed vacuum (see Figure 5.8). Experimentally, it was shown that a hydroclone removed most (87%) of the heavy particles and a majority (56%) of the total solids from a trout culture system, thereby extending the life of the biological filter that received water from the hydroclone.

BIOLOGICAL FILTERS

In almost all aquaculture systems, there is some degree of **biological filtration** taking place, planned or not. Some systems, especially closed saltwater sys-

tems (like home marine aquariums), rely on biofilters almost exclusively. Unlike the mechanical or gravitational filters, biological filters are not primarily designed to remove solid materials; rather, they remove dissolved nutrients from the water. Most important, they convert the harmful nutrient forms such as ammonia to less dangerous forms such as nitrate.

The primary biological agents in these filters are autotrophic (chemotrophic) bacteria, although algae, yeasts, protozoans, and some tiny animals may be present and aid in the removal of the "wastes" from the water. Bacteria are not found on the surfaces of filter particles as individuals; they form colonies and secrete a coat of slime for protection. Again, there are a number of filter designs to choose from: rotating disks and drums, submerged beds, fluidized beds, and

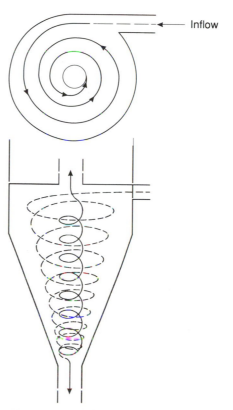

Figure 5.8 A hydroclone. (From Wheaton, 1977.)

moist beds over which water is sprayed and quickly drained (such as trickling filters used at wastewater treatment plants). All work well, and each has its advantages and disadvantages.

One of the most common biological filters is the **submerged bed,** or **undergravel filter.** Gravel, crushed coral, or broken shells are often used as the support medium for the growth of communities of bacteria and associated organisms; shells and coral have the advantage of being composed of calcium carbonate and therefore help to buffer the pH of the water. Water is directed through the bed of material containing the bacteria and then is returned to the culture (see Figure 5.9). As the water passes through the bed, the bacteria take up some of the wastes, but more important, they oxidize the ammonia and nitrite to the less dangerous nitrate. These oxidation reactions are an energy source for the bacteria. The major drawback of this submerged design is that the reactions are oxygen limited, and if there is not enough oxygen in the water, the conversions will be slow or not take place at all. It is important to remember that if a submerged filter is also used as

the tank bottom, organisms that are constantly disturbing the submerged filter bed by burrowing or digging (such as flounder or lobster) may prevent the filter from reaching its maximum efficiency.

Trickling filters are never oxygen limited and are therefore often more efficient than submerged filters. However, if for some reason the water flow over the gravel stops, the bacterial colonies and associated life will dry and the system can be difficult to revive.

Rotating disks and **drums** are also never oxygen limited. These turn slowly, part of the filter being submerged and the other part passing through the air. Disks are made of some rough material that encourages bacterial growth. Drums are made with some sort of mesh covering, and are usually filled with lightweight plastic particles designed to have a large growing area for microbes (see Figure 5.10).

Fluidized bed biofilters are composed of some lightweight substrate, like sand, plastic, or granular carbon, which is kept suspended in the water by an upflow current. These filters do not clog (which causes a restriction of the flow and re-

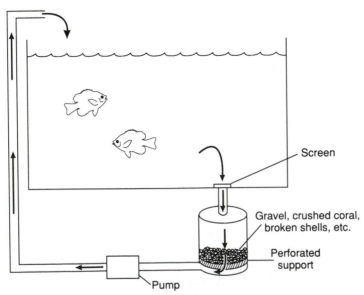

Figure 5.9 A submerged biological filter.

Figure 5.10 A rotating biodrum filter. (Photo courtesy of R. W. Hagood, NORAQUA, Florida.)

duced oxygen penetration) as can submerged filters. Experimentally, it has been shown that fluidization of a bed tripled the ammonia-removing capacity of a biofilter.

Nitrification

Animals, especially when they are fed the high-protein diets that are often used in culture, excrete most of their nitrogen in the form of ammonia or urea. Urea breaks down into two molecules of ammonia (or ammonium; remember, ammonia is in the pH- and temperature-dependent equilibrium with ammonium, the ionized form of the molecule) (see Figure 5.11). Ammonia is harmful to animals and must be removed from the system. In the biofilter bed is a bacterium, *Nitrosomonas*, that is able to convert the ammonia to a less toxic form of nitrogen, nitrite

$$NH_4^+ + 1.5 \, O_2 \rightarrow NO_2^- + 2 \, H^+ + H_2O$$

Although nitrite is usually less toxic than ammonia, it will still have an adverse effect on the culture. Another bacterium in the bed, *Nitrobacter,* is able to convert the nitrite to nitrate

$$NO_2^- + 0.5 \, O_2 \rightarrow NO_3^-$$

The conversions of ammonium to nitrite to nitrate are called the **nitrification** reactions. Note that both these steps are oxidations and may be restricted by the amount of oxygen in the water. Also note that the valence charge of the nitrogen changes from +3 in ammonium and nitrite to +5 in nitrate.

These bacteria are found in most natural bodies of water and will eventually develop in the culture system. Establishment of bacterial populations can be accelerated by bringing some bed material from an existing system to the new one that is being started. There are also saltwater and freshwater cultures of bacteria for biofilters; these bacterial cultures are marketed as **water conditioners.**

The culture system may be initially spiked with ammonia before the animals are added. If the bacteria are present, a chemical monitoring of the system will show the ammonia being converted first to nitrite and then to nitrate (monitoring is best done in the dark to prevent uptake of the nitrogen by plants and algae).

The culturist must remember that a new biological filter cannot support a

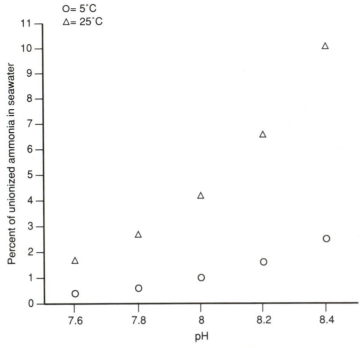

Figure 5.11 The effect of temperature and pH on the ionization of ammonia.

large population of animals; the bacterial colonies must be allowed to grow before the system is stocked to capacity (see Figure 5.12). Naturally, if this is not done, the bacteria will be unable to handle the waste load, and the culture organisms will show symptoms of nitrogen toxicity.

The efficiency of biological filters will vary with several parameters. Obviously, the environmental conditions—such as temperature, light, the concentration of ammonia in the water, and the presence of other dissolved nutrients or pollutants in the system—will affect the metabolic rates of the bacteria.

Among the more important design features affecting biological filter efficiency are the size of the particles in the bed, the volume of the bed relative to the volume of water, and the rate that water is passed through the bed. For example, by decreasing the size of the particles in the bed, the surface area for the bacteria is increased, and therefore, initially, the efficiency of the filter increases; however,

fine bed materials may lead to clogging of the bed and a restriction of the amount of water that can pass, both lowering the flow rate and introducing the possibility of having anaerobic pockets in the filter bed where the oxidation reactions cannot take place. (For these reasons, water should be gravity or mechanically filtered before it reaches the biological filter, since the removal of particulates will extend the life of the biofilter by decreasing the rate of clogging. In addition, organic particles such as fecal pellets are broken down by heterotrophic bacteria; if these particles enter the biological filter, a population of heterotrophic bacteria may develop that will compete for space and O_2 with the autotrophic bacteria carrying out nitrification.)

Increasing the rate of flow through a biofilter, even though this decreases the time that the water is in the filter, will increase the daily NH_3 removal if there is an increase in the biomass of the filter bacteria. This has been observed when a

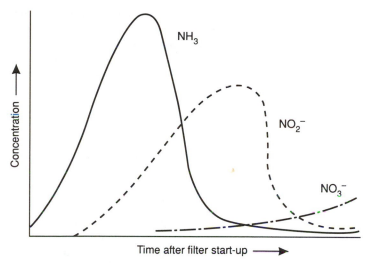

Figure 5.12 Oxidation of nitrogen over time in a new biological filter. (From Wheaton, 1977.)

material that will support very large bacterial populations, such as polyurethane, is used as the substrate.

All biological filters eventually do clog with particulate material and thick colonies of bacteria. They may be cleaned by gently disturbing the bed and siphoning off the material that suspends itself in the water. While this reduces the bacterial population, it increases the flow through the bed. Ideally, a biological filter should be designed to support very large populations of bacteria without reducing the flow of the water through the filter. Such filters have been suggested in the scientific literature.

Other Biological Filters

The nitrification biological filter is by far the most common type; there are other types, although they are infrequently used. Higher plants and algae are sometimes also employed since in some cases they can remove nitrogen from the water very efficiently.

The physiology of plants is complex and beyond the scope of this introductory text. The form of nitrogen, and the rate of uptake of the nitrogen, will change with the mixture of nitrogen sources the plant has to "choose" from and with the environmental conditions of the system. The advantage that some plants and algae have over bacteria is that they themselves may be a harvestable product. In addition, they offer increased protection for fish and crustaceans from birds or other predators. Their greatest disadvantage is that they may make harvesting of the animals difficult or impossible. Of course, if protection is not a problem, then the plant and animal cultures may be joined in series and harvest difficulties avoided (see Figure 5.13).

Denitrification filter systems can also be used. Although nitrate is less toxic than ammonia or nitrite, it may still have an effect on the organisms in culture if the concentrations are high enough. Denitrification, which takes place when there is limited oxygen or under anaerobic conditions, is the conversion of nitrate to nitrogen gas. Several bacteria can complete this reaction, including some species of *Pseudomonas*. Denitrification filters, which are more difficult to maintain than other biological filters, are unpopular in aqua-

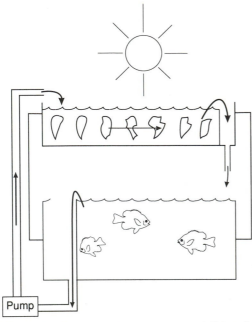

Figure 5.13 Segregated seaweed and fish culture system. Plants are used to remove ammonia from the water. The organisms are separated so that they may be harvested more easily.

culture. There are three major reasons for this:

1. Most culture systems require high levels of oxygen to be most productive, but for denitrification, oxygen must be removed from the water.

2. A carbon donor, such as methyl alcohol, is required since CO_2 is one of the end products of the reaction.

3. If there is too much oxygen or too little of the carbon donor, the reaction will not go to completion, resulting in the production of the nitrogen intermediate, nitrite, which is more toxic than the nitrate that was present at the start.

CHEMICAL FILTERS

Chemical filters, like biological filters, are used primarily to remove dissolved materials (the **solutes**) from water (the **solvent**). This may include nutrients such as ammonia, but these filters may also be used for ridding the system of other dissolved chemicals that bacteria and other biofilter microbes might be unable to remove efficiently.

Foam Fractionation

The basic theory of **foam fractionation** is simple. Air is pumped through a diffuser and into the water to be filtered. As the air bubbles rise through the water, the hydrophobic solute(s) will concentrate on the surface of the bubbles. At the top of the fractionator, a foam with a high concentration of the solute(s) will form. The foam is then collected, and the "clean" water can be used in the culture system (see Figure 5.14). If for some reason foam does not form, the water at the surface of the filter will still be richer in the solute than the water at the bottom, and this solute-rich water may be skimmed or otherwise removed.

Many factors contribute to efficient foam fractionation units. The chemistry of the water (pH, temperature, salt concentration), the chemistry of the dissolved solutes (solubility, equilibrium, interac-

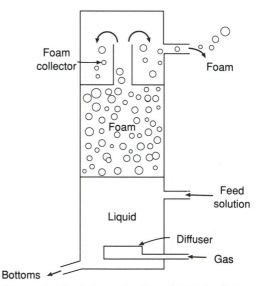

Figure 5.14 A foam fractionation unit. (From Wheaton, 1977.)

tions of solutes, concentrations, and so on), as well as the design of the fractionator (shape and depth of the filter) should all be considered. Operating features, such as the flow rate of the air, will affect performance.

Carbon Filtration

One of the most familiar methods of filtration involves the use of **activated carbon.** Commonly in aquaculture, a bed of activated carbon granules is enclosed in some sort of vessel (drum, column, plastic or metal housing), with water entering one side of the vessel, passing through the bed of carbon, and then exiting. There are many variations on this theme. Alternatively, it is possible to add powdered carbon to the water and then remove it by gravitational or mechanical filtration methods. The use of powdered carbon in aquaculture, however, is very limited; activated carbon is used primarily to remove

low levels of nonpolar organic compounds and may also separate some metals from the water, especially copper. Acid-treated carbon may also remove ammonia from water, although this should never be relied upon as the only ammonia removal step in the culture facility.

As water enters one end of a new filter, solute molecules are quickly taken up by the carbon at that end of the filter vessel. Soon the carbon particles near the vessel entrance are unable to adsorb any more solute, so the solute will be adsorbed by the carbon a little farther toward the exit. Slowly, the area of the vessel that is doing the adsorption moves backward, until the adsorption area is reduced in size at the back wall of the filter vessel. When the adsorption area begins to shrink, the efficiency of the filter begins to quickly change; this is called the **breakpoint** (see Figure 5.15).

Carbon filters are often placed downstream of biological filters and serve as a

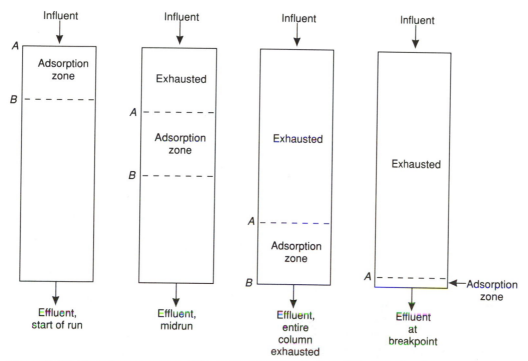

Figure 5.15 The adsorption zone of the carbon filter moves down. When the size of the zone begins to shrink after most of the bed is exhausted, it has reached the breakpoint. (From Wheaton, 1977.)

polishing step. However, if the water is rich in ammonia or nitrite, because the biofilter was unable to remove those nutrients prior to their reaching the granular carbon, then bacteria will begin to grow on the carbon's surface. These bacteria will block the carbon's pores, thereby inhibiting adsorption.

Activated carbon can be produced from a number of substances (wood, bone, nut shells, peat, saw dust, and others). There are two steps in the manufacture of the activated carbon from these bases:

1. The first step is the **slow heating,** in the absence of air, of the base material; this is sometimes done in the presence of dehydrating substances like phosphoric acid. Water is driven off, gasses are removed, and the organic carbon is converted to primary carbon (ash, tars, and other forms).

2. **Activation,** the second step, may be chemical or physical (heating). Heating burns off tar that may be blocking the pores of the carbon granules and enlarges those pores. Chemicals can be used to mimic the heating process. After this, the base material is crushed to the desired size particle.

Activated carbon works by **adsorption,** the attraction of the solute to the surface of the carbon. Therefore, the greater the surface area of the carbon, the more efficiently it will remove the solute. A single gram of activated carbon will have 500 to 1400 m^2 of surface area because of the highly irregular and fissured surface of the particles, along which are the pores formed during the activation (see Figure 5.16). Some of the pores may be too small to admit some of the solute molecules; molecules bound to the carbon most strongly are those that just barely fit inside the pores.

The speed and degree of adsorption of

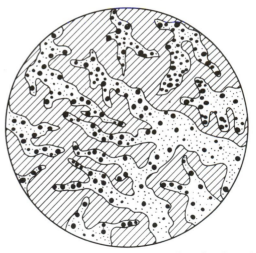

Figure 5.16   Magnified cross section of activated carbon. (From Wheaton, 1977.)

the solute is a function of the nature of that solute. In particular,

1. the solubility of the solute in the water; the more hydrophobic it is, the more readily it will be taken up by the filter.

2. the **affinity** (chemical, electrical, or van der Waals) of the particular solute for the activated carbon.

The removal of the solute is also related to

1. the **concentration gradient,** that is, the relative amounts of solute in the water and bound to the carbon. Newly activated carbon, with "empty pores," will begin to remove the solute quickly because the concentration gradient is steep.

2. the pH of the water, which can affect the adsorption process since it may affect the ionic charge of the solute.

3. the temperature of the water, which will also act to regulate the rate of uptake by the carbon. Since raising the temperature will increase the motion of the molecules, there is an increased rate of molecular escape from the carbon at high temperatures.

Ion Exchange

Ions are electrically charged molecules; they may have a positive charge, such as NH_4^+ and Mg^{++}, in which case they are called **cations,** or they may have a negative charge, such as F^- or SO_4^{--}, and are called **anions.**

Generally, a granular or porous **ion exchange** material, the **resin,** is contained in a vessel, just as activated carbon is. Particular ions are initially bound to the resin material when it is manufactured; as water containing some unwanted ions is passed through the resin bed, the ions that were bound to the resin are released to the water, and the unwanted ions are then bound to the resin in their places. That is, the ions are "exchanged." Resins are designed to exchange either cations or anions. Because of the high natural ionic strength of seawater, ion exchange is practical only in freshwater culture systems.

There are a great number of available ion exchange materials, many of them synthetic. In aquaculture, the most commonly used exchange resins are natural **zeolites,** which are hydrated silicates of aluminum and either sodium or calcium or both (see Figure 5.17). Synthetic zeo-

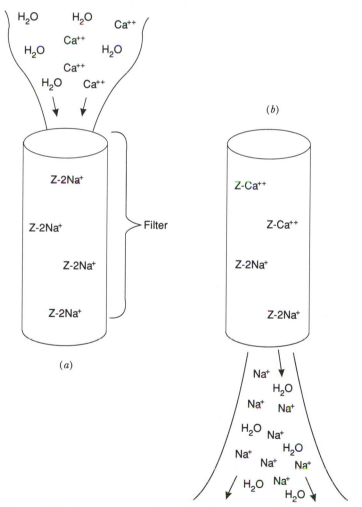

Figure 5.17 Water softening by zeolite ion exchange. Hard water with Ca^{++} passes through the resin bed; the Ca^{++} is exchanged for Na^+. Because calcium is divalent and sodium is monovalent, it takes two Na^+ to replace one Ca^{++}.

lites are cross-linked polystyrene materials. Zeolites are commonly used in wastewater treatment for water softening, and one, **clinoptilolite,** is used for the selective removal of ammonia. All resins have their own selectivity, picking up certain ions easily and others less so; for example, clinoptilolite has affinities for ions in the following order: $K^+ > NH_4^+ > Na^+ > Ca^{++} > Mg^{++}$. The affinity of a resin may be altered with changes in the pH.

If significant levels of dissolved organic substances are present in the water, it may be wise to install an activated carbon or foam fractionation filter upstream of the exchange resin, since organics tend to collect on the surfaces of the resin, thus blocking the ion exchange sites. A few methods have been developed to remove organic molecules from the resin, such as running NaOH or hypochlorite through the filter bed. Cleaning the resins, in some cases, may damage their ability to exchange ions, so the manufacturer should be consulted as to the best methods.

After a period of time, the sites on the resin become saturated with ions removed from the water. At this point, the resin should be **regenerated.** Again, specific methods are used with different resins. For clinoptilolite, one method that has been suggested is to pass a 2% NaCl solution with a pH of 11–12 through the zeolite to strip the ammonium ions from the resin and replace them with sodium. There are similar sorts of treatments for all exchange resins; again, the manufacturer will probably have the best information on how to regenerate the material.

DISINFECTION

The term **disinfection** refers to the killing of most of the small and microscopic organisms that may be entering the culture facility with the water. This is done to prevent disease organisms, predators, and competitors from getting a foothold. It is not the equivalent of **steril-** **ization,** which is the elimination of all life in the water, and is neither practical, in terms of economics, nor necessary. Ultraviolet radiation, ozone, and chlorination are the best agents for the disinfection of culture water.

Ultraviolet Radiation

Ultraviolet (UV) radiation is a term used to refer to that part of the electromagnetic radiation spectrum from about 10 to 390 nm, between the longest X rays and the shortest waves of visible light.

Ultraviolet waves are effective at killing microorganisms in the water. To do so, the UV waves must strike the organism and be absorbed. The germicidal properties of UV energy are well established, although just *how* UV energy works is still the subject of some investigation. It is thought that the UV radiation is absorbed by molecules in the cell's nucleus, leading to the disruption of unsaturated bonds; purine and pyridimine are believed to be the prime targets. The most effective wavelengths are from 250 to 260 nm. Since only the UV radiation that strikes the microbes is effectively removing the contaminants from the water, it is very important that the water be as free of shielding particles as possible.

Different organisms have different sensitivities to UV radiation. The length of exposure (i.e., the flow rate of the water) may have to be adjusted in the UV system depending on the microbes in question. UV energy is most commonly used to kill bacteria, microalgae, and the larvae of invertebrates, but it has also been shown to be effective against viruses, including Coxsackie and polio viruses. It is generally felt that UV radiation is ineffective against cysts and "large" organisms (a good rule of thumb is that if you can see it with the naked eye, it will not be killed by a standard UV system).

Along with the types of microbes in the water, there are a number of other factors that must be considered because they can interfere with the effectiveness of UV treatment. Temperature and pH probably have little direct effect on the germicidal efficiency of the UV radiation, while radiation intensity and length of exposure are probably the most critical factors.

Pure water, at the wavelengths used for disinfection, absorbs almost no UV energy, so the radiation is completely available for disinfection. The intensity of the radiation should relate to the size and number of particulates in the water, as well as what is dissolved in the water. Dissolved materials that reduce the transmittance of UV energy will naturally reduce the effectiveness of the treatment. It has been demonstrated that both ammonia and organic nitrogen compounds hinder the transmittance of UV energy at the commonly used wavelengths.

Since the intensity of UV energy is critical, the depth of the water as it passes the UV source is generally not very great. The more suspended and dissolved materials in the water, the shorter the path of the UV radiation must be. And as mentioned, the longer the water is exposed to the UV waves, the more effective the treatment is. The result of reducing the volume *and* slowing the flow rate of water past the UV source is that not much water can be disinfected per unit time. It is generally held that UV treatment is not practical for culture systems that require very large amounts of water. For smaller systems with fewer suspended and dissolved materials present in the water, UV radiation is a cheap and effective disinfection method. The chief advantages to UV radiation are that nothing is introduced to the water, the water's physical and chemical properties are not significantly affected, and overdosing produces no ill effects.

Several companies manufacture UV systems for use in aquaculture facilities. The focus of these systems is the UV source, a mercury vapor lamp. Electric currents passing through the lamp excite the mercury, and as the mercury atoms return to the original lower energy state, they emit UV radiation. There are two types of UV systems used by the culturist: the suspended and submerged systems.

Suspended systems consist of a bank of lamps and reflectors that are hung 10–20 cm over water flowing through a trough; the trough may have baffles to break up the flow (see Figure 5.18). The height, number, and spacing of the lamps and reflectors must be geared toward directing the UV light at the water. Radiation that does not enter the water is wasted. Submerged UV systems are designed so that the water flows around a quartz tube that houses the UV lamp (see Figure 5.19). In either system, the mercury vapor lamps must be kept clean; the suspended lamps must be attended to if water splashes on them and leaves a residue, and a tube cleaner around the tube holding the submerged lamp is needed to remove adhering particles.

Ozone

Ozone, an allotropic form of oxygen (O_3), has been used to treat wastewater, especially in Europe, for more than 70 years because of its ability to reduce color, taste, and smell. Recently, it has been incorporated into modern aquaculture technology.

Ozone is an unstable blue gas with a readily identifiable odor. The rate that O_3 degrades back to O_2 increases with rising temperatures. Because O_3 is unstable, it is made when and where it is needed, not shipped to the aquaculture facility. Ozone generators produce O_3 by passing O_2 or dry air through a high-voltage (4000 to 30,000 volts) corona between two electrodes.

Ozone works as a disinfectant because it is a powerful **oxidizing agent;** only fluorine has a greater electronegative oxidation potential. It is extremely corrosive and dangerous. Ozone may react with some plastics, but not with glass and porcelain. Ozone appears to be an excellent

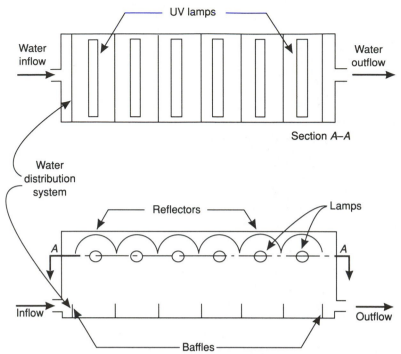

Figure 5.18   A suspended UV system. Water flows under the bank of lamps in reflectors; baffles are used to force water near the surface so it may receive the maximum radiation. (From Wheaton, 1977.)

virucide and works well as a bactericide by disintegrating bacterial cell walls.

Although O_3 is more soluble in water than O_2, it is still much less soluble than chlorine. Saturation at 20°C is about 570 mg O_3 per liter of water. It reacts with water

$$O_3 + H_2O \rightarrow HO_3^+ + OH^- \rightarrow 2\ HO_2$$
$$HO_2 + O_3 \rightarrow HO + 2\ O_2$$
$$HO + HO_2 \rightarrow H_2O + O_2$$

The free radicals, HO_2 and HO, are strong oxidizers and have the advantage of being quickly converted to oxygen. The HO_2 and HO oxidation reactions may be

1. Inorganic, including conversion of sulfides and sulfites to sulfates, nitrites to nitrates, chlorides to chlorine, and ferrous and manganous ions to their insoluble ionic forms resulting in precipitates.

2. Organic, rupturing compounds at unsaturated bonds and destroying humic acids, pesticides, phenols, and a host of other types of compounds.

It has been shown that ozonated water, despite the fact that O_3 breaks down quickly to O_2, can be dangerous to some delicate cultured organisms. However, if the ozonated water is passed through a bed of activated carbon, it loses its toxicity.

The effectiveness of ozonation is directly related to the amount of contact between the gas and microbes. Poor O_3 transfer into water therefore results in a low killing efficiency. There are several methods used to mix water and ozone, the most common being simple injection of the gas into the liquid using a diffuser of some sort, or ozone and water can be mixed in a **packed bed** of some sort of granular material. Packed beds increase the mixing surface area.

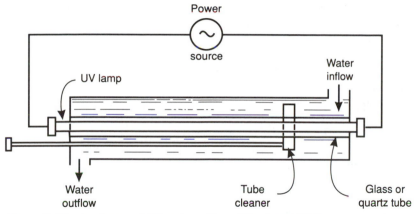

Figure 5.19 A submerged UV system. (From Wheaton, 1977.)

Chlorination

Chlorine is widely used in wastewater treatment in the United States. It is produced commercially by electrolysis of sodium chloride (common table salt). Chlorine gas has a green-yellow color and a strong odor; it is usually sold under pressure as a liquefied gas, as a dry powder (calcium hypochlorite, $Ca(OCl)_2$), or as liquid sodium hypochlorite (NaOCl). The gas mixes easily with water, forming a 0.7% solution at 20°C. Like the other halogen elements, fluorine, bromine, and iodine, chlorine is a potent bactericide as well as being a strong oxidizing agent.

When chlorine mixes with water, it rapidly hydrolyzes to form **hypochlorous acid:**

$$Cl_2 + H_2O \rightarrow HOCl + H^+ + Cl^-$$

Hypochlorous acid is a weak acid, meaning that it undergoes a partial dissociation:

$$HOCl \leftrightarrow H^+ + OCl^-$$

This reaction is clearly pH dependent; as the pH decreases, HOCl becomes more important. At pH 4, all the chlorine is in the form of HOCl, while at pH 11, only 0.03% of the chlorine is in the form of the acid and 99.97% exists as OCl^-, the **hypochlorite ion.** At the pHs normally found in aquaculture systems, both the acid and hypochlorite ion are present. HOCl and OCl^- are commonly referred to as the **free chlorine.** The free chlorine, rather than the chlorine molecule itself, is the active oxidizing agent.

The exact mechanism that chlorine employs to kill microorganisms is still unclear. It is believed that chlorine enters the cell and reacts with some enzyme(s). This assumption is based on the fact that chlorine reacts with nitrogenous compounds (enzymes are proteins and therefore composed of a string of nitrogen-containing amino acids). The more easily the free chlorine molecule enters the cell membrane, the quicker the organism is killed. It has been shown that hydrochlorous acid will diffuse into the cell more quickly than the hypochlorite ion; this is probably the reason that chlorine disinfection works more effectively at low pHs.

Chlorine chemistry has a number of specialized terms. The amount of uncombined HOCl and OCl^- in the water is called the **free residual chlorine;** in pure water this is the same as the amount of chlorine added to the water. The amount of chlorine needed to react with all the organic and inorganic dissolved compounds, plus all the microbes in the water, is called the **chlorine demand.** When chlorine reacts with ammonia (equations follow), the resultant compounds are **chloramines.** These chloramines have disinfecting properties and are therefore important themselves; the amount of chloramines in a water sample is the **combined available chlorine.** The chloramines react more slowly than does the free chlorine but are more effective at high pHs.

To understand how chlorine works to disinfect water, we must discuss the **breakpoint reaction.** There is still some disagreement as to the steps in this reaction, but the equations that follow have some strong support. In brief, the breakpoint reaction is reached when enough chlorine is added to the water sample to cause the oxidation of ammonia to nitrogen gas (N_2). The reactions are

$$NH_4^+ + HOCl \rightarrow NH_2Cl + H_2O + H^+$$
$$NH_2Cl + HOCl \rightarrow NHCl_2 + H_2O$$
$$0.5\ NHCl_2 + 0.5\ H_2O \rightarrow 0.5\ NOH + H^+ + Cl^-$$
$$0.5\ NHCl_2 + 0.5\ NOH \rightarrow 0.5\ N_2 + 0.5\ HOCl + 0.5\ H^+ + 0.5\ Cl^-$$

giving an overall equation of

$$NH_4^+ + 1.5\ HOCl \rightarrow 0.5\ N_2 + 1.5\ H_2O + 2.5\ H^+ + 1.5\ Cl^-$$

Note that one of the final products is Cl^-, which may recombine with water to form HOCl.

Chlorinated water is unsuitable for aquatic culture, so the chlorine must be removed before it reaches the organisms being raised. Several methods can be used to accomplish this. The most common dechlorination method for heavily chlorinated potable water involves the addition of sulfur dioxide to the water, which results in the conversion of chlorine to chloride and sulfite ions to sulfate ions. It is doubtful that this method will be widely used by the aquaculture community in the near future. Other methods that can be used include the use of ion exchange res-

ins, aeration and storage, and activated carbon. Activated carbon removes chlorine from the water by **contact catalysis;** the chlorine is absorbed by the carbon and then reacts:

$$2\ Cl_2 + 2\ H_2O \leftrightarrow 4\ HCl + O_2$$

Removal of chlorine is more efficient at low pHs than in basic solutions. Chloramines are not removed by carbon as well as chlorine.

Besides being used to treat culture water, chlorine can be sprayed in pipes and on the sides of tanks and raceways that have been drained. The chlorine solution will disinfect the surface, eliminating bacterial/algal communities that have developed.

AERATION

One of the most critical factors the culturist must consider is the amount of oxygen in the water, usually referred to as the **DO (dissolved oxygen).** Oxygen is required for all animals, and although plants can make their own O_2 while there is sufficient sunlight, they can be considered an oxygen sink at night or on cloudy days. Therefore, the culturist cannot rely on plants to produce enough oxygen to keep animals alive when they are stocked at high densities. DO requirements vary as a function of species, water temperature, stocking density, and water quality.

There are several commonly used ways to get oxygen into the water. In most cases this equates to mixing air and water, thereby transferring the oxygen from the air into the water across the gas/liquid boundary. Air is about 21% O_2 (by volume). The rate of oxygen transfer into the water is related to the following:

1. *Differential gradient.* If the water is low in O_2, then the transfer rate

from the air is greater than it would be if the water was near O_2 saturation.

2. *Temperature.* The absolute ability of water to hold oxygen increases as the temperature of the water decreases. However, theoretically the *rate* that transfer can occur is initially increased in low-DO water as temperatures rise.

3. *Impurities in the water.* Dissolved substances may affect the surface tension of the water or the solubility of oxygen.

4. *Surface area across which diffusion can take place.* Surface area is probably the most important aspect to consider in most aeration models. Oxygen molecules diffuse from the air into the water where they interface. By increasing the area of that water–air interface, the culturist can increase the rate of oxygen transfer. Most of the standard aeration techniques are designed to increase this water–air boundary surface area.

Another related consideration is the vertical motion of the water. In a stagnant pond, the water in the upper layer may have a high DO, but because O_2 diffuses slowly downward through the water, the water below the surface may have a low DO. It is therefore important that there be vertical motion in a culture situation, allowing bottom water to rise to the surface and O_2-rich surface water to be directed toward the bottom to keep the sediments from becoming anaerobic. This is especially a problem in the summer when a thermocline is likely to form.

In *Aquaculture Engineering,* F. W. Wheaton (1977) has divided aeration systems into four basic groups: gravity aerators, surface aerators, diffuser aerators, and turbine aerators. This method of classification will also be used in this text.

Gravity Aerators

Gravity aerators are among the most common of aerators because they are simple to construct and reliable. The principal requirement is that the water be raised above the level of the pond, tank, or raceway; as the water falls, the potential energy is converted to kinetic energy that serves to break apart the falling water. When the water is broken into droplets or a mist, the area over which diffusion can take place is increased, thereby increasing the DO of the water.

In the simplest form, this means water falling and splashing into a pond, but there are an assortment of ways to increase the efficiency of this method:

1. The water falls through a **screen** or **perforated plate,** or a series of screens or plates, that break up the falling stream of water (see Figure 5.20).

2. A **gravity-driven paddlewheel** is also effective. The falling water turns the paddlewheel; the paddlewheel throws more water into the air and, in some cases, even traps air below the surface of the water (see Figure 5.21).

3. A recent and somewhat popular variation on the traditional gravity aeration systems includes the use of specially designed **plastic rings, wheels,** or **balls.** These are commonly a few centimeters in diameter and have fringed interior walls or spokes. As water passes through a column or drum filled with these, the water is effectively broken up (see Figure 5.22).

Surface Aerators

Surface aerators increase the surface area of the culture medium by agitating it with some sort of mechanical device. The DO rises when the surface water is thrown

Figure 5.20 A gravity aerator that contains a series of perforated trays. (Photo from Clark and Clark, 1964.)

up and mixes with the air above the water, and then falls back down into the pond or tank. There are three commonly used types of surface aerators: nozzle aerators, spray aerators, and motorized paddle-wheels.

Nozzles are often used with round tanks. Water is pumped downward through a nozzle toward the surface of the water. When the jet of water hits the tank surface water, there is considerable turbulence; that is, the pump imparts energy to the water, which is forced out the nozzle, and the water leaving the nozzle transfers that energy to the water it hits. Besides increasing the DO in the

round tanks, nozzle aerators also set up a circular water flow (see Fig. 3.29).

Spray aerators are generally built as floating propellers. The propeller is beneath the water's surface, and as it turns, it brings subsurface water up and into the air (see Figure 5.23). The rate of oxygen transfer depends on the size and depth of the propeller and the speed at which the propeller turns.

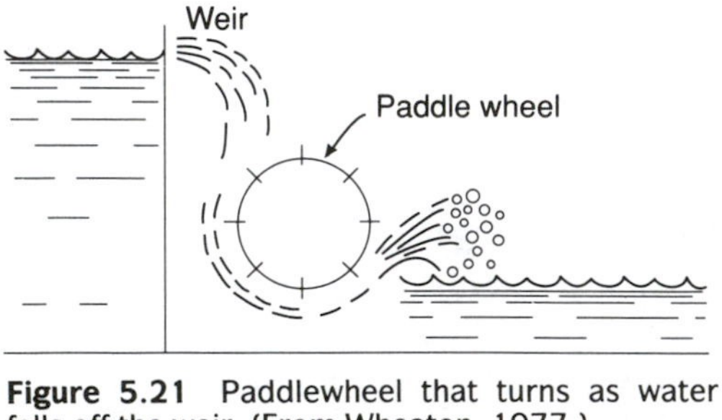

Figure 5.22 Water falls into a drum of plastic rings that break up the flow of the water. (Photo courtesy of the New Jersey Division of Fish, Game, and Wildlife.)

Figure 5.21 Paddlewheel that turns as water falls off the weir. (From Wheaton, 1977.)

(a)

(b)

Figure 5.23 (a) A spray aerator. (b) Water being thrown into the air by a spray aerator and splashing back into a pond. (Photo courtesy of The Power House, Inc., Maryland.)

Floating paddlewheels are very popular (see Figure 5.24). They are considered one of the most energy-efficient devices for increasing the DO. Besides increasing the DO as the wheel hits the water, they also increase both vertical and horizontal water movement. A number of paddlewheel designs can be used, but recent research shows that the most efficient paddles are triangular in cross section. The amount of oxygen that is transferred to the water increases with the size and speed of the paddlewheel, although energy efficiency (O_2 transferred per kilowatt) may decrease at high speeds and depths.

Diffuser Aerators

Surface aerators work to disturb the water surface, thereby putting water in the air, but **diffuser aerators** work to put air in the water. Oxygen in air bubbles will diffuse out into the water through the surface of the bubbles. The longer the bubbles stay below the water surface, the more time there is for oxygen to pass into the water. In some cases, diffuser aerators use pure oxygen in place of air.

The use of oxygen rather than air increases the efficiency of the transfer since the gradient is much greater, and the problems of nitrogen supersaturation are avoided. Since oxygen is more expensive than compressed air, some efforts have been directed toward designing systems that reuse O_2 that is not taken up by the water.

There are many types of diffuser aerators, including the simple diffuser, the Venturi diffuser, the U-tube diffuser, and the downward bubble contact aerator.

The aquarium airstone is an example of a **simple diffuser.** If an air line were simply dropped into the aquarium, the result would be large bubbles that rise quickly to the surface. The diffuser forces the air through small holes. This results in a greater bubble surface area per unit volume of air, yielding a better transfer of O_2

Figure 5.24 A floating paddlewheel. (Photo courtesy of the Nan Rong Machinery Factory, Taiwan.)

Figure 5.25 A simple diffuser used to bubble pure oxygen into the water can support a large fish biomass. (Photo courtesy of the Xorbox Corporation, New York.)

to the water (see Figure 5.25). In addition, the smaller bubbles have less slip and therefore lift some of the water to the surface with them as they rise; as stated, this improved vertical motion helps to distribute the water.

The **Venturi diffuser,** an aspirator, operates by forcing water at a high speed through a restriction of some sort. At the restriction site, there is an opening to the atmosphere. The quickly moving water causes a drop in pressure, which results in air being drawn into the water line (see Figure 5.26). Unlike other types of diffusers, a source of compressed air is not required.

The **U-tube** is a simple device designed to increase the time that bubbles are in the water. At the front of the U-tube, air bubbles are injected into the water as it flows down. The downward velocity of the water must be greater than the velocity at which the bubbles rise. This being the case, the bubbles are first forced to the bottom of the U-tube before they may rise (see Figure 5.27). The amount of O_2 transferred is related to the flow and composition of the air or oxygen, the depth of the diffuser, the velocity of the water, and the depth of the U-tube.

The **downflow bubble contact aerator** operates on the same principle as the U-tube aerator. A propeller forces water downward; a diffuser is placed in the stream of downward-flowing water. First, the bubbles are forced down and then they rise up, thus increasing the contact time with the water and therefore the amount of oxygen passed from the bubbles (see Figure 5.28).

Turbine Aerators

A **turbine aerator** is essentially a submerged propeller. Rather than injecting air into water or water into air, it is simply

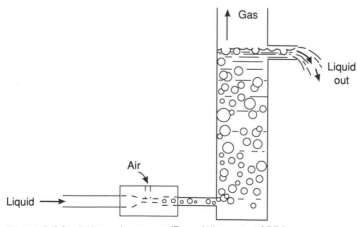

Figure 5.26 A Venturi aerator. (From Wheaton, 1977.)

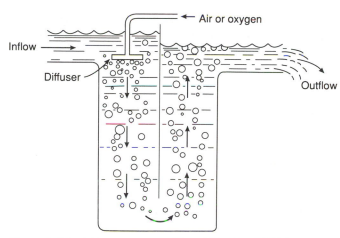

Figure 5.27 A U-tube aerator. (From Wheaton, 1977.)

a device that increases the circulation in the pond or tank that results in greater surface aeration. The size and speed at which the propeller rotates, along with the physical proportions of the tank or pond, and the oxygen gradient, govern the efficiency of the transfer. There are variations on this basic theme that may make the turbine a more popular way to oxygenate water than it is at present. One variation involves a rapidly spinning propeller mounted on a hollow drive shaft; the drive shaft has an opening at the water surface and by the submerged propeller. As the spinning propeller forces the water at a high velocity past the submerged hole in the drive shaft (see Figure 5.29), there is a drop in pressure in the shaft, which results in air being sucked through the surface hole and out the submerged hole near the propeller.

DEGASSING

Aeration is the addition of a gas, O_2, to the water. **Degassing** is the removal of gas, usually inert N_2, from the water. Since N_2 is the most abundant gas in the atmosphere (about 78% of air is N_2), it is not surprising that it is also the most common gas in water. High levels of N_2, reaching **supersaturation,** lead to gas bubble disease (the formation of bubbles

Figure 5.28 A downflow bubble contact aerator. (Photo courtesy of Aquatic Eco-Systems, Inc., Florida.)

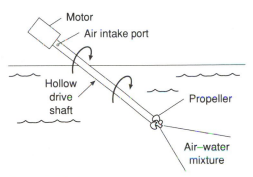

Figure 5.29 A turbine aerator with an air intake port to draw air down. The basic turbine aerator simply moves the water to promote mixing and disturb the thermocline.

in the blood of fish and shellfish, see Chapter 6). Supersaturation is an unstable condition. It is usually reached when gas-saturated waters are exposed to changing physical conditions in the pond, such as temperature or pressure. If the amount of gas that can be normally held under a particular set of conditions is exceeded because of that physical change, the water will temporarily contain levels of gas that are greater than the saturation limits. For example, suppose that the water in an aquifer is saturated with N_2 and then there is a rise in the height of a lake that feeds the aquifer; this will result in an increased pressure on the aquifer, which may result in supersaturation of well water taken from that aquifer for a culture facility. There are three commonly used ways to remove excess N_2 from culture water: vacuum degassers, oxygen injection, and the use of packed columns.

Vacuum degassers are sealed vessels that contain packing material, such as plastic rings. A vacuum is maintained in the vessel using a vacuum pump. Water is distributed over the packing material by a diffuser. The diffuser breaks up the stream of entering water and distributes it over the packing material, which results in more of the water's surface being exposed to the vacuum. The gasses flow from the liquid and are exhausted by the vacuum pump, while the degassed water flows out the bottom of the vessel through a pipe (see Figure 5.30). The weight of the water in the exit pipe must be great enough to overcome the vacuum in the vessel. These units are efficient and will decrease the levels of all gasses in the wa-

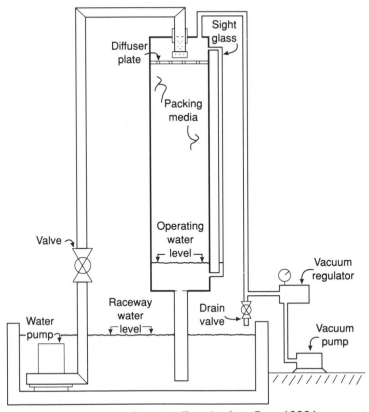

Figure 5.30 A vacuum degasser. (Drawing from Fuss, 1986.)

ter below saturation; the greatest problem is that they will slightly reduce the DO level.

Oxygen injection degassers are also packed columns. As water flows through the bed, oxygen is bubbled or injected into the system (see Figure 5.31). The surfaces of the packing particles again increase the surface area where mixing can take place. The oxygen scrubs the nitrogen from the water because there is a competition for "space" in the fluid. Oxygen injection often results in supersaturation of O_2; oxygen supersaturation is not a problem in most cases, allowing an increased stocking density and helping to meet the **chemical oxygen demand (COD)** of the water. While this is probably the most expensive way to remove high levels of N_2 in culture water, the usefulness of additional O_2 in the water (supplementing the aeration system) may make oxygen injection practical.

A **packed column degasser** is similar to the vacuum degasser without the vacuum. When levels of saturation are initially high, and the organism in culture can survive and grow in water with a slight supersaturation (101–103%), this is a simple and cost-effective method of dealing with high levels of N_2. The packed column degasser works because the packing material again breaks up the water passing through the bed, allowing the supersaturated gas to leave the water. The greater the saturation, the greater the efficiency of the bed; the major problem with these degassers is that they become very inefficient when the saturation level is approached, so the water always stays slightly supersaturated.

SUMMARY

Fish and shellfish must be raised in "good" water. The quality of the water can often be improved in a number of ways, depending on the animals and plants in culture, the physical culture system, and the condition of the water before it enters that culture system.

The removal of particles results in the elimination of shade from algal cultures, sediment that can clog gills or ruin machinery, and predators and competitors that will lower productivity. Particles are removed by mechanical filters that prevent large particles from passing through small holes and gravitational filters that decrease the settlement time of suspended particles that are heavier than water. Dissolved nutrients, which may lead to unwanted blooms of algae or bacteria, or may be directly toxic to culture organisms, can be removed by biological filters composed primarily of bacteria that convert NH_3 to NO_3^-. Nutrients and other solutes can be removed with chemical filters like activated carbon, which depend on the solute's hydrophobic properties, and the ion exchange resins, which replace unwanted ions with those ions of a similar charge that are less harmful.

Water may also be disinfected; microbes and small larvae are killed before they can establish themselves in the culture system. Common ways to disinfect water include the use of UV radiation,

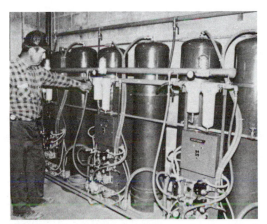

Figure 5.31 Oxygen generators and pressurized tanks, the basic elements of the oxygen injector degassers. (Photo from Marking, 1987.)

which is generated by a mercury vapor lamp; ozone, which is formed by passing dry air or oxygen between two electrodes; and chlorine, which dissociates in water to form effective oxidizing agents, hydrochlorous acid and the hyperchlorous ion.

The aim of aeration is the addition of oxygen to the water system. Low oxygen levels lead to poor growth and food conversion, so O_2 levels must be carefully monitored. The addition of oxygen can take place by using (1) gravity aerators that increase the surface water turbulence, and therefore oxygen diffusion, as the water falls, (2) surface aerators that inject water into the air, (3) diffuser aerators that inject air or oxygen into the water, and (4) turbine aerators that increase vertical mixing and therefore diffusion.

Supersaturation of N_2 can result in gas bubbles forming in the fluids of animals being reared, which in turn leads to tissue damage and death. The N_2 can be removed by several degassing methods, including the use of a vacuum, the injection of oxygen, and simple mixing in a packed column.

PART TWO

BIOLOGY AND CULTURE METHODS

CHAPTER 6

GENERAL BIOLOGICAL CONCEPTS

Aquaculture traditionally has been viewed as a biological science because it concerns living plants and animals. Certainly, an understanding of biological principles and ideas will help anyone involved in the business to better appreciate the dynamics of the culture operation. If a culturist is lucky, he may be able to proceed from year to year viewing his ponds essentially as a system of "black boxes" into which are added seed stock and some feed; sometime later, by processes he does not understand, a large number of animals of a marketable size can be harvested from those black boxes. But most people, in any business, do not want to count on being "lucky" to make a profit. The factors governing survival, reproduction, and growth of the species that are being cultured are important pieces of information for the aquatic farmer.

Knowledge of specific organisms in culture should be detailed and include an understanding of behavior, life cycles, and physiology; these will be covered in the chapters on plants and animals that follow this introductory section. This chapter deals primarily with some of the basic principles that govern life in the aquaculture facility as they pertain to the culturist: disease vectors, ecology, genetics, and nutrition.

PATHOLOGY AND PREDATION

Plants and animals, under most circumstances, must interact with other organisms in the culture facility. The number of these interactions is, to a large extent, a function of the design of the culture system and its management. Many types of interactions have insignificant effects on the cultured organism; some may have a beneficial effect; others may somehow result in damage to the stock. The cultured species may be exposed to disease organisms, which can include certain types of bacteria, fungi, protozoans, and viruses (although viruses are technically not considered living). Plants and animals are also **hosts** to **parasites**, organisms that extract their nutrients from the tissues of other organisms. Another type of interaction is predation, which will result in the death or injury of the cultured species. Finally, there are interactions of competition between organisms that have similar ecological niches, in which the cultured species may be prevented from getting the food, space, light, or oxygen that it needs to survive or grow well.

Pathology

All fish and invertebrates host a wide variety of parasites and bacteria, but these normally do not significantly inhibit the animal's ability to live, grow, and reproduce. Diseased animals, those with obvious symptoms, tend to be rare in nature. However, in the aquaculture environment, where animals are stressed by high densities and other unnatural conditions, diseases can spread very quickly and have serious effects. The key to preventing constant disease problems in an aquaculture operation is to manage the system carefully and to be alert to signs that something is wrong.

Good management means seeing that the animals are not excessively stressed, since a weakened animal is more likely to become sick than a strong one. Several basic practices are of key importance in preventing the outbreak of serious diseases. These include

1. Monitoring the quality of the water
2. Using feeds that supply all the essential nutrients
3. Avoiding overfeeding
4. Minimizing handling
5. Using stocking densities that are reasonable
6. Isolating and inspecting animals suspected of being sick

This means that the farm manager should examine the animals on a regular basis and immediately investigate any unexpected decreases in population densities, or unfamiliar behavior or appearances. For example, a reduction in feeding, gasping for air, animals resting on their side or upside down, and frantic or sluggish movements are signs of a problem, as are spots or lesions on the animal's surface.

If an infection is diagnosed, the condition can sometimes be treated by the culturist. In the United States, the Environmental Protection Agency (EPA) is responsible for the clearance of such substances as pesticides and algacides, that is, chemicals to be added to a pond to control a disease problem. The Food and Drug Administration (FDA) regulates chemicals used as food additives and therapeutants, that is, those chemicals that are given directly to the organism. *Of the compounds listed below, only a few antibiotics and other substances can legally be used for the treatment of fish intended for human consumption. Those that are legal for edible fish may not be approved for crustaceans, mollusks, or higher vertebrates such as frogs.* Some of the substances listed can be used to treat ornamentals or bait fish, but not edible fish. *The culturist should consult the local extension agent or other authority to determine which drugs may be currently used for which organisms.* Some treatment forms include

1. **Antibiotics,** substances produced by specific microorganisms that destroy or inhibit the growth of bacteria, such as tetracycline, neomycin, terramycin, erythromycin, and kanamycin.
2. **Nitrofurans,** including nitrofurazone, furazolidone, and furanace, which are used to control bacteria.
3. **Sulfonamides,** such as sulfamerazine and sulfamethazine, which have antibacterial properties.
4. **Acriflavine,** used to treat external parasites, some fungi, and protozoans.
5. **Copper sulfate** and **potassium permanganate,** used to control external parasites.
6. **Stains,** including crystal violet, methylene blue, and malachite green, which control fungal and protozoan infections.

7. **Formalin/formaldehyde,** for treatment of certain protozoans and fungi.
8. **Oxolinic acid,** a feed additive used to control certain bacteria.
9. **Iodine solutions,** for the treatment of eggs to prevent growth of fungi and bacteria.
10. **Di-n-butyl tin oxide,** given orally to fish to eliminate tapeworms, nematodes, and acanthocephalids.

It is impossible to discuss all the diseases that affect all the different animals currently being cultured. Only a few examples will be presented in this chapter's text and tables.

Viruses A **virus** is not a living organism, but rather a large **nucleoprotein particle** (a nucleoprotein is a conjugate, consisting of a protein combined with a nucleic acid; in normal cells nucleoproteins are found in the nuclei and cytoplasm, being concentrated in the chromosomes). Each virus is a bit of either DNA or RNA enclosed within a coat of protein that allows it to pass from cell to cell. All viruses are extremely small, ranging from about 10 to 300 nanometers, too small to be seen using a light microscope. A virus outside a living cell is not metabolically active, but will undergo replication, after entering the host cell, by using the host's enzymatic machinery to produce new virus particles.

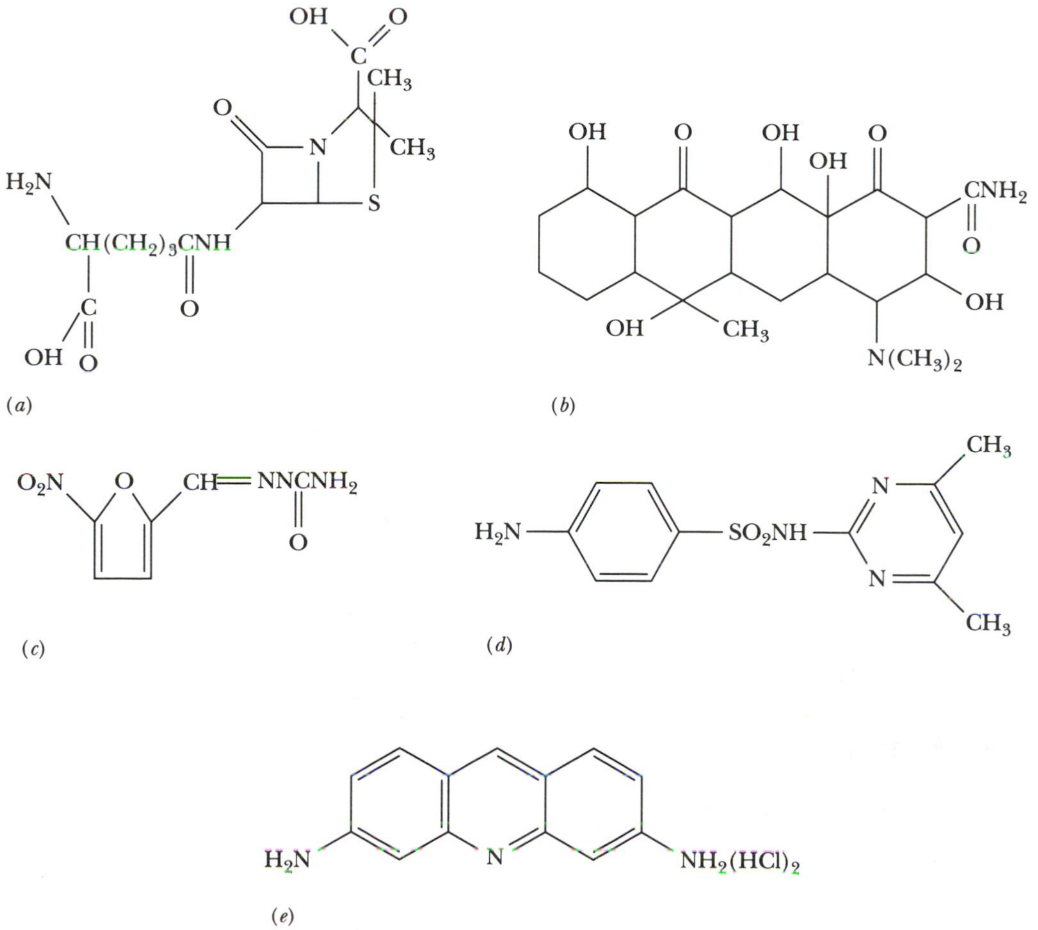

Figure 6.1 Some antibacterial compounds: (*a*) penicillin N, (*b*) tetracycline, (*c*) nitrofurazone, (*d*) sulfamethazine, and (*e*) acriflavine hydrochloride.

Figure 6.2 Catfish with bacterial diseases. (Collins, 1988. Copyright © Archill River Corporation. Reprinted by permission.)

Each type of virus usually attacks some specific part of the host and can only use certain types of cells for its replication. Some viruses, called **bacteriophages,** can only reproduce in bacterial cells.

Organisms can create an immunity to many virus infections by the production of **antibodies.** A virus infection can also stimulate the host cell's synthesis of **interferon,** a protein that "interferes" with the virus. It is probable that interferon is an animal's first line of defense and is important in halting the growth of the virus until the specific antibodies against that virus are produced.

Some shrimp, especially when stressed, are susceptible to "virus disease" that attacks the host's hepatopancrease (liver) cells. *Penaeus monodon,* a commonly cultured species of shrimp, can be infected with a baculovirus, **MBV,** which is associ-

Figure 6.3 Catfish with columnaris disease. (Griffin, 1987. Copyright © Archill River Corporation. Reprinted by permission.)

ated with high mortalities in some culture operations. MBV will infect adult and larval shrimp. A **herpeslike virus (HLV)** is hosted by the blue crab, being particularly dangerous for juvenile crabs. Blue crabs also can be infected by a **reoviruslike virus (RLV)** that inhibits molting and blood clotting, and can result in partial or complete paralysis. Probably the most important of the invertebrate viral diseases is known to strike several species of popularly reared shrimp; it is called **infectious hypodermal and hematopoietic necrosis virus** or **IHHNV,** and it spreads quickly through stocks of animals affecting many tissue types.

There is a herpes-type virus that is known to attack the American oyster when it is placed in thermal effluent. This virus results in significant mortalities, and there is no known cure, but fortunately it is extremely rare. Another viral disease of the oyster, **ovacystis disease,** which attacks the epithelial reproductive tissues, seems to be salinity dependent.

Many fish are also susceptible to viruses. One of the better known viral diseases of both fresh- and saltwater fish is **lymphocystis,** caused by an **irridovirus,** which results in whitish nodules on the body and fins. Many fish tumors are also believed to be caused by viruses, the most stunning of which is **cauliflower disease,** or **Blumenkohlkrankeit.** Cauliflower disease occurs principally in the European eel and results in the tumor overgrowing the mouth and parts of the head, eventually leading to loss of weight and death. At least four viral isolates have been made from tumors of the infected eels, all of which are probably **rhabdoviruses.** Other viral diseases of fish include **viral haemorrhagic septicaemia (VHS)** and **infectious pancreatic necrosis (IPN),** which infects freshwater trout. **Channel catfish virus disease (CCVD)** causes massive mortalities in young catfish, especially in warm water.

Bacteria **Bacteria** are some of the simplest forms of life. They, along with the blue-green algae **(cyanobacteria),** are **prokaryotes,** organisms without an organized nucleus. In any environment where we find "higher life," we also find bacteria; they can be found 5 m deep in soil or in the ice of a glacier. Bacteria typically range from 1 to 10 μm (micrometers) in length and are normally rod shaped (**bacilli**), spherical (**cocci**), or corkscrew shaped (**spirilli**). Many are covered with a protective capsule. Bacteria have a wide variety of environmental needs and types of metabolisms; most are either helpful or have no effect on the aquaculturist's system, but a significant few are associated with diseases (see Table 6.1).

Many bacteria (e.g., *Vibrio* and **chitinolytic** microbes) are a normal part of a culture ecosystem, but are also able to act as opportunistic pathogens when animals are environmentally stressed or are first infected by another pathogen. The best way to avoid bacterial diseases is to never let the bacteria gain a foothold; this is done by avoiding stress and maintaining water quality.

The cyanobacteria may also be a problem in culture. Blooms of some species may cause an off-taste and others may be toxic. Deaths of freshwater crustaceans have been linked to *Aphanizomenon* and *Microcystis*. *Spirulina subsalsa* infestations of shrimp raceways have resulted in the necrosis of the epithelial lining of the midgut, dorsal cecum, and hindgut gland.

Fungi The **fungi** are an extremely diverse group of heterotrophic (since they lack chlorophyll) organisms. The true fungi, the Eumycota, usually are multicellular and are characterized by branching tubular filaments called **hyphae.** A mass of hyphae is called a **mycelium** (such as the cobweblike spots of bread mold). Like bacteria, many species of fungi are normally in culture systems and become pathogenic only when the animals are stressed (see Table 6.2). Eggs and larval animals are often extremely vulnerable to fungal infection, as are weakened adults with exposed tissues (wounds or lesions).

The fungi can release **zoospores** into the water that settle on the surfaces of plants and animals and begin to grow into mycelium.

Recently, a small, spherical organism termed the **rosette agent** has been described in penned chinook salmon. Kidney and spleen are attacked, leaving the fish anemic and with lymph-filled cysts. It is suspected that this organism is a fungus.

Protozoans Many "higher" forms of one-celled and colonial organisms are **protozoans,** which include the familiar amoeba and *Paramecium*. Some are pathogenic (see Table 6.3). There are four major groups of protozoans:

1. The **Sarcodina,** which move by pseudopods.
2. The **Flagellata,** including the **dinoflagellates,** which move by one or more whiplike structures (dinoflagellates are one of the groups that has always resisted the traditional "animal versus plant" classification scheme, and are sometimes called algae rather than protozoans).
3. The **Ciliata,** which move by many short hairlike projections.
4. The **sporozoans,** a parasitic group with sporelike infective stages, often having complex life cycles.

Many of the sporozoan spores are resistant to treatment by the culturist. Most transmission of sporozoan diseases is probably via ingestion of the spores, intermediate hosts of the spores, or infected animals. Some of the protozoans, such as *Zoothamnium*, have free-swimming stages, called **telotrochs,** which can also spread the infection. When reaching maturity or when the host dies, each individual (**trophont**) of the parasitic dinoflagellate *Amyloodinium* divides into 256 freeswimming **dinospores** that are viable for nearly two weeks.

One of the most interesting of the protozoan pathogens is *Myxosoma cerebralis*; it

TABLE 6.1 Some Common Bacterial Diseases Affecting Cultured Organisms

Disease Name	Organisms Affected	Bacteria	Signs of Disease
Bacillary necrosis	Mollusk larvae	*Vibrio*	Decrease or halt of movement
Focal necrosis	Mollusks	Unknown gram-positive bacterium	Pale digestive gland
Septicemic bacterial disease	Crustacean larvae and post larvae	*Vibrio* and possibly *Aeromonas*	Animals become moribund
Brown spot or shell disease	Crustaceans	Chitin-destroying bacteria, including *Vibrio* and *Aeromonas*, and fungi	Spots or lesion on the shell
Filamentous bacterial disease	Crustaceans	*Leucothrix*	Filaments appear on appendages and gills
—	Crayfish	*Flavobacterium*	Appear to be affected with a neurotoxin
Red Tail	Lobster	*Aerococcus* (formerly called *Gaffkya*)	Pink color on dorsal abdomen and reduced clotting of the blood
Vibriosis	Fish	*Vibrio*	Reddish lesions and hemorrhaging of the intestinal tract*
Hitra disease	Salmon	*Vibrio*	Internal and (sometimes) external hemorrhaging
Edwardsiellosis	Catfish	*Edwardsiella*	Gas-filled lesions in muscles
Furunculosis	Fish	*Aeromonas*	Similar to vibriosis†

Kidney disease	Fish	*Corynebacterium* and *Renibacteri*	Variable, but may include abnormal behavior, swelling because of accumulated liquid, or nodules
Fin rot	Fish	Several types of bacteria including vibrios, pseudomonads, and aeromonads	Damaged fins
Pasteurella disease	Fish	*Pasteurella*	White nodules on some of the internal organs
Abdominal dropsy or bacterial hemorrhagic septicema	Fish	*Aeromonas* (there may also be a viral form of this disease)	Abdominal swelling and sometimes ulcers
Piscine tuberculosis	Fish	Unknown bacilli	Loss of color, ulcers, loss of weight, listless swimming
Redmouth	Fish	*Yersinia*	Inflammation around the mouth*
Columnaris disease	Fish	*Flexibacter*	Gray-white spots on the head and fins

* A vaccine is commercially available.
† A vaccine is commercially available, although its effectiveness is rather low.

TABLE 6.2 Some Common Fungal Diseases of Cultured Animals

Disease Name	Organisms Infected	Fungus	Signs of Disease
Molluskan larval mycosis	Mollusk larvae	*Sirolpidium*	Growth ceases and fungus is visible inside the larvae
Dermo	Mollusks	*Labyrinthomyxa* (=*Dermocystidium*)	Gaping, body emaciated and dark
Shell disease	Mollusks	*Ostracoblabe*	White spots inside shell
Crustacean larval mycosis or crab egg fungus disease	Crustacean eggs and larvae	*Lagenidium*	Fungus visible inside the eggs or larvae
Fungus disease	Crustaceans	*Fusarium*	Lesions
Lobster mycosis	Crustacean larvae	*Haliphthoros*	Red-brown discoloration and lesions
Ichthyophonosis	Fish	*Ichthyophonosis*	Roughness of the scales and ulcers

TABLE 6.3 Some Common Protozoan Aquaculture Diseases

Name of Disease	Organisms Affected	Disease Organisms	Signs of Disease
MSX or Delaware Bay disease	Mollusks	*Haplosporidium nelsoni*	Little or no growth and weak shell closure
SSO or Seaside disease	Mollusks	*Minchinia costalis*	Growth inhibited
Haplosporidan disease	Crustaceans	*Minchinia* sp.	Sluggishness
Microsporidios of reproductive organs	Crustaceans	*Thelohania*	Opaque white areas
Ciliate disease	Crustaceans	*Lagenophrys, Anophrys, Zoothamnium,* and *Epistylis*	Sometimes a fuzzy appearance on appendages and gills
Cotton crab	Crustaceans	*Nosema*	Muscles appear chalky
Paramoebiasis	Crustaceans	*Paramoebiasis*	Sluggishness and ventral surface may appear grayish
Hematodinium disease	Crustaceans	Dinoflagellate *Hematodinium*	Sluggishness
White spot disease	Fish	*Cryptocaryon*	Small white cysts over body
Cardiac myxosporidiosis	Fish	*Trachinotus*	Poor growth, white spots on and in heart
Myxosporidan disease	Fish	*Kudoa*	Cysts visible around dissected brain
Velvet disease	Fish	Dinoflagellate *Amyloodinium*	Erratic behavior
Costiasis	Fish	*Costia*	Sluggishness, blue-gray film on skin
Whirling disease or blacktail	Fish	*Myxosoma*	Whirling, deformation of spine or jaw
Ichthyophthiriasis or Ich	Fish	*Ichthyophthiriasis*	White spots on skin
Octomitiasis	Fish	*Octomitus*	Erratic swimming and sometimes weight loss

is well tolerated by the European brown trout, but in rainbow trout leads to **whirling disease,** a loss of equilibrium and tail-chasing behavior. Recent research on the complex life cycle of *Myxosoma* has demonstrated the mode of infection that has been a mystery to pathologists for over 80 years. The spores of *Myxosoma* actually infect tubificid worms; the spores are converted to a different form that had previously been identified as a separate genus, *Triactinomyxon*, which is either released into the water or ingested with the worm by the young trout (see Figure 6.4). In the trout the protozoan changes back to the *Myxosoma* form whose spores will infect the tubificid worms.

A protozoan that has done a significant amount of damage in recent years in the United States is *Haplosporidium nelsoni*, the causative agent of **MSX disease** in oysters (see Figure 6.5). This has reached epidemic proportions along the East Coast of the United States; since 1986 it has killed about half the oysters in the Chesapeake Bay.

In open marine systems, toxins produced and released by **red tide** dinoflagellates such as *Gymnodinium* or some species of *Gonyaulax* can cause mass mortalities of all types of animals. Other species of *Gonyaulax* may not kill marine organisms, but their neurotoxins can build up in the tissues of filter-feeding organisms; if these filter feeders are then eaten

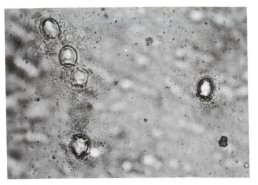

Figure 6.5   MSX spores. (Photo courtesy of Dr. Susan Ford, Rutgers University.)

by mammals, the result is **paralytic shellfish poisoning.** Blooms of some nontoxic dinoflagellates, such as *Heterosigma*, can be so dense that the fish's gills become covered, resulting in suffocation.

Metazoan Parasites   For many marine and freshwater organisms, **parasites** are a fact of life. Many of the parasites do not seriously affect the host and therefore go unnoticed. Some, however, can inhibit growth or reproduction, and others can be fatal. Parasites come in a variety of shapes and types; some have very specific needs in the host (e.g., they may infect only certain tissues of one particular species or genera), while others are more general and can attack an assortment of animals and spread through many of the host's organs. Some parasites have complex life cycles during which they require

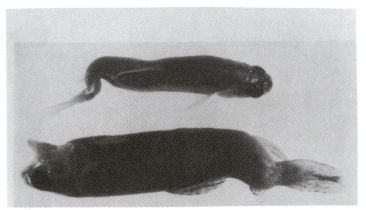

Figure 6.4   Young salmonids disfigured by whirling disease. (Photo from Wolf and Markiw, 1985.)

two or more hosts. The tapeworm, *Ligula*, for example, starts out as an egg that turns into a swimming ciliated larva; the larva is taken up by the copepod, *Diaptomus*, and the copepod is eaten by a fish (often a cyprinid) in which it develops into the tapeworm. When the fish is eaten by water fowl, the worm becomes reproductive and produces eggs that are excreted by the bird, thus completing the cycle. If the cycle is broken, for example, by preventing the birds from feeding on the pond fish, the parasite will not be able to reproduce.

Many of the most common parasites are "worms." Actually, there are several distinct groups of animals that are called worms. There are round worms, or **nematodes,** which, especially as larvae, are very common in fish. A few of the nematodes, such as *Anisakia* and *Contracaecum*, cause serious damage to the liver of many marine fish; in some instances larval anisakids, which are ingested by people eating infected fish, can invade the wall of the human digestive tract. **Trematodes (flukes)** are related to the tapeworms; they are also common parasites. The gill fluke, *Dactylogyrus*, attacks many freshwater fish; the fish may be successfully treated by salt baths and formalin. *Axine*, another trematode, causes severe damage to cultured yellowtail by sucking blood from the gill and thereby weakening the fish. Pompano are attacked in a similar manner by *Bicotylophora*, while other larval trematodes can cause mass mortalities in mussels. *Bucephalus* parasitizes oysters, making them sterile, although in this particular case sterility is not unwelcome because infected oysters will still have a rich taste owing to the high levels of fats and glycogen that they retain in the tissues instead of expending in reproduction. Segmented worms, the **annelids,** can also be parasites; the best known examples are the blood leeches. The leech *Pisciola* is particularly common in still waters, but parasitized fish can be treated by quickly dipping them into a lime bath. Another annelid, *Polydora*, causes "mud blisters" on the inner surface of oyster shells. The **acanthocephalid** worms ("spiny-headed worms") are also a pest for some fish such as flounder, mullet, and salmon.

Mollusks often serve as intermediate hosts, but in some cases they can be parasites themselves. There are parasitic gastropods, such as *Odostomia*, that attack bivalves (especially oysters).

There are numerous crustacean parasites. The most important group is clearly the **copepods.** There are many parasitic species that attack a variety of fish. Others live in invertebrates, such as *Mytilicola*, which is found in the digestive tract of the oyster. Most of these copepods are highly modified compared to the free-living species, consisting mostly of gonads and a head made for penetrating the host's tissue (e.g., the "anchor worm," *Lernaea*), while others like *Ergasilus* resemble the free-living copepods but have specially modified head parts for grasping and biting (see Figure 6.6). Some other crustaceans, such as the branchiuran *Argulus* (the "fish lice"), will parasitize fish. Parasitic rhizocephalan barnacles can live in crabs and shrimp, parasitic ostracods are found in crawfish, and some isopods parasitize fish, shrimp, and crabs.

"Diseases" Caused by Environmental Stress Poor diet or unsuitable living

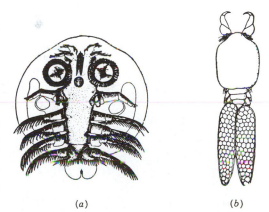

(a)   (b)

Figure 6.6 Freshwater parasites (a) *Argulus,* (b) *Ergasilus.* (From Wilson, 1903 and 1911.)

conditions will result in poor growth, weakened animals susceptible to infections, and sometimes direct tissue damage. When tissue damage occurs, it has the appearance of being a pathogen-related disease. There are several major causes.

If animals are subject to large differences in gas pressures in the pond waters, they may develop gas bubbles, causing tissue damage or death (see Figure 6.7). This begins when the animal's blood is highly saturated with a gas (a result of being in an equilibrium with water that has a high degree of gas saturation). When saturation is followed by a sudden drop in the water's gas pressure the gasses leave the blood quickly to maintain the equilibrium, resulting in the formation of bubbles in the body fluids. Changes in gas saturation are usually a result of using water that is supersaturated with N_2, failure of aeration equipment, or large temperature and pressure variations (since temperature and pressure affect the solubility of gasses).

Despite the fact the fish and invertebrates are cold-blooded, **thermal shock** can be fatal. Thermal shock is often observed when animals are transferred from one body of water to another; it is a good idea to try to slowly acclimate the organisms when transferring if the temperature difference is greater than 2°C.

In many areas of North America, **acid rain** has become a problem. As the pH of pond water decreases, the fish become stressed. At pH 5, many fish will die. Initial signs of acid water include increased mucus production and a darkening of the gill tissue. Problems arise when the pH is too high as well. In alkaline water conditions (above pH 9), the fins and gills may be injured.

Overcrowding animals will also lead to stress-related symptoms, probably because of hypoxia and ammonia exposure. In shrimp, some of the results of crowded conditions include **cramped tail,** which gives the animal a humped abdomen, sometimes inhibiting its ability to swim, and **muscle necrosis,** which first appears as white patches on the abdominal muscles.

Pollutant Contamination The poisoning of the waters by industrial and nonindustrial sources is a major concern for the

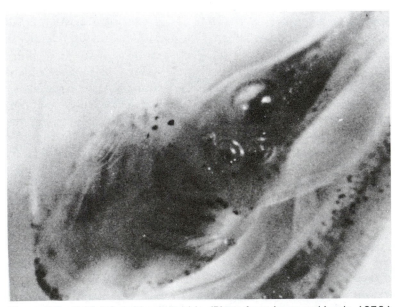

Figure 6.7 Catfish fry with gas bubble. (Photo from Jones and Lewis, 1976.)

culturist. Toxic substances in the water may mean

1. Aquatic organisms will incorporate these toxicants into their flesh, making them unfit for human consumption.
2. Inhibition of growth and development of the cultured species.

In general, toxicants of interest to the culturist can be divided into two groups of chemicals: **organic chemicals,** those that are carbon based, and include substances such as pesticides and petroleum products, and **inorganic chemicals,** including metals such as mercury or lead; sometimes metals are included in organic molecules, for example, methyl mercury. The danger of pollutants is partially in relation to their ability to be taken up and retained by the organisms in culture. Pollutants that are **water soluble** tend not to accumulate in tissues and will "wash in and out" of animals, but **fat-soluble** contaminants that are not soluble in water are retained, particularly in those tissues that have a high fat content.

There are two problems that the culturist must overcome somehow with regard to pollutants:

1. Preventing pollutants from entering the culture system can be done, to some extent, by treating the water before it enters the ponds; however, different pollutants have different chemical characteristics, which means some are best removed by one treatment method, and other pollutants must be removed by another method. (Of course, some organisms are more sensitive to one particular type of pollutant than are other organisms.) Also, discharge of contaminants into the water may not be steady, so there may be periods when the water coming into a system is much worse than at other times. Therefore, the culturist's treatment system should be designed to make the water acceptable, even if there are levels of pollutants present greater than those normally encountered. In practice, the prevention of entrance of pollutants is rarely done, except perhaps in small experimental closed systems; however, this may change in the future.

2. Another problem the culturist has, which is shared with fisheries and wildlife biologists, is that the toxicity of most pollutants is unknown; many common contaminants have been studied by state and federal agencies for short periods to establish *lethal* thresholds on particular *standard species*, but the effects of low levels of toxicants on the growth and reproduction of fish and invertebrates is unexplored in practically all cases.

Predators

Predators may find their way into semiclosed aquaculture systems, and are often present in open systems. Several of these predators have been discussed in Chapter 3, along with ways to reduce the damage they cause.

Some invertebrate predators include the oyster drills, such as *Urosalpinx*, which cause extensive damage to oyster beds. The boring sponge, *Cliona*, weakens the shell of the oyster. Many crabs, as well as the horseshoe crab (not really a crustacean), are important predators on beds of juvenile clams. The pea crab, *Pinnotheres maculatus*, lives in the body cavity of the blue mussel and can affect growth and shell development.

Other important predators for culturists are water fowl, such as ducks, herons, swans, and kingfishers, and mammals, such as the rat, muskrat, water shrew, and otter. Frogs can also be a problem, as can such fish as pike and some catfish, which are particularly voracious and therefore unwelcome in a fish pond. Insects, such as dragonfly nymphs, water boatmen, and water beetles, damage crops as well, by killing young fish (see Figure 3.9.)

ECOLOGY

Ecology refers to the interrelations between living things and their environment. Ponds can be considered restricted biotic communities, or ecosystems. As such, a few basic principles apply.

All the organisms in a pond, not only the species of interest to the culturist, establish their own ecological **niche.** The niche essentially describes where in the pond the organism is found, and how it gets whatever it requires. If two organisms in a pond have very similar niches, one will eventually outcompete the other. For example, if two species of green algae are found near the surface of the pond, and both require either ammonium or nitrate ions to grow, the species that is better adapted for acquiring these ions in that particular physical environment (temperature, salinity, pH, and so on) will become dominant and perhaps even eliminate the other from the pond ecosystem. But if one of those species could use nitrite ions or if it lived on the surface of the mud on the bottom of the pond, then both species of algae could probably survive because they occupy significantly different niches.

Ecologists often study **food webs** that describe the flow of energy and matter between organisms. Energy is transferred in the form of molecules in plant and animal tissues (lipids and carbohydrates are especially rich in energy). When these molecules are oxidized during digestion, there is a release of energy that then becomes available to the consumer. Each time energy in the form of tissue passes from one organism to another, some of that energy is lost because it is either partially or completely undigestible, or because it is used up during hunting, digestion, respiration, locomotion, or some

other function. Typically, only about 10% of the energy consumed is available for construction of new body tissue by the predator.

These simple ecological concepts have important implications in pond management. If a culturist can concentrate the pond resources into harvestable products, she will be managing an efficient pond. On the other hand, if the feeds that are being added are poorly utilized, the farmer may have little in the way of a marketable product to show for her effort, but instead may have invested most of her time and money into the production of by-products such as algae, snails, zooplankton, and worms.

The most well-known example of the application of ecological balancing in aquaculture is the Chinese system of carp polyculture. **Polyculture** refers to the technique of keeping several different harvestable animals (or plants) together so that all the space and nutrients are used to the fullest extent. That is, a food web is developed where most of the consumers are harvestable, and the transfer of energy is through short chains so less is lost. Consider this example of a Chinese polyculture system:

The farmer adds vegetable wastes and animal manure to a pond, and these are converted into useful plant nutrients by bacteria and fungi, or some fish will consume them directly. The released nutrients cause an increase in the biomass of phytoplankton and plants along the side and bottom of the pond. The phytoplankton can be consumed by the midwater feeding silver carp (*Hypophthalmichthys*) as well as by zooplankton such as *Daphnia* and copepods, and by filter feeding clams. These zooplankters are then eaten by the big-head carp (*Aristichthys*). The vegetation around the edge of the pond and along the sides is consumed by the grass carp (*Ctenopharyngodon*). The fecal material from these three species sinks to the bottom where (1) the nutrients can be rereleased by bacteria, (2) the feces can be used by benthic invertebrates such as worms or snails, or (3) it can be directly consumed by either the common carp (*Cyprinus*) or the mud carp (*Cirrhinus*) (both these species of carp can consume some of the soft invertebrates as well). Finally, the snails and small bivalves can be eaten by the black carp (*Mylopharyngodon*) (see Figure 6.8). All these carp species are marketable, and the system is highly efficient. Polyculture systems that support tilapia, catfish, prawns, and other organisms are used by culturists.

It should be obvious that bacteria and fungi play a major role in a pond ecosystem. The speed with which, and the form into which, organic matter is regenerated can be very important. It has been suggested by several biologists that growth of bacteria, fungi, and phytoplankton that remove wastes should be encouraged, and perhaps these organisms should even be added to ponds as a management technique. Several researchers have suggested that there are cases when it is less economically efficient to use prepared feeds for the fish than to "feed" manures and fertilizers to the microbial community that will in turn support the fish directly and indirectly.

GENETICS

Genetics can be considered the study of heredity, that is, the transmission of traits from one generation to the next. Genetic research is intertwined with aquaculture because we are interested in the improvement of strains, the development of special **hybrids** for different purposes, and the production of **polyploids.** To understand these ideas, some basic genetic concepts and terms must be discussed. These will be applicable to practically all the species of commonly cultured animals.

Animals and plants are made up of cells that contain a nucleus; inside the nucleus there are chains of **deoxyribonucleic acid (DNA)** called **chromosomes.** Most species of plants and animals have 10 to 50 of these chromosomes; humans have 46. These chromosomes are always found in pairs in the **somatic cells**; since the chromosomes are in pairs, they are considered **diploid.** (Somatic cells make up the brain, bones, muscles, skin, and most organs; all cells that are not involved directly in reproduction are somatic). In

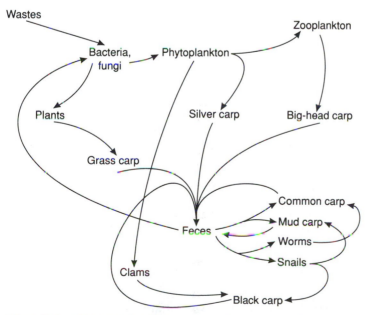

Figure 6.8 A food web in a Chinese carp polyculture pond.

every somatic cell in the human body there are the same 23 pairs of chromosomes.

Each chromosome contains hereditary factors called **genes.** These genes control the development and physiology of the organism because they govern protein synthesis. Body tissues composed of somatic cells must increase in number by the process of **mitosis.** During mitosis, these diploid cells replicate their chromosomes and then divide, leaving two diploid cells in place of the original one; these two cells are genetically identical to each other.

However, gonadal tissues in the reproductive organs go through a different process, known as **meiosis**; during meiosis diploid cells divide to produce cells that have unpaired chromosomes, so the cells are said to be **haploid** (see Figure 6.9). In humans, a haploid cell contains 23 chromosomes, none of which is paired. The chromosomes are derived from the diploid cells and distributed between the haploid cells randomly, so, for example, any human haploid cell can have 2^{23} potential combinations of chromosomes. The haploid cells are the eggs produced

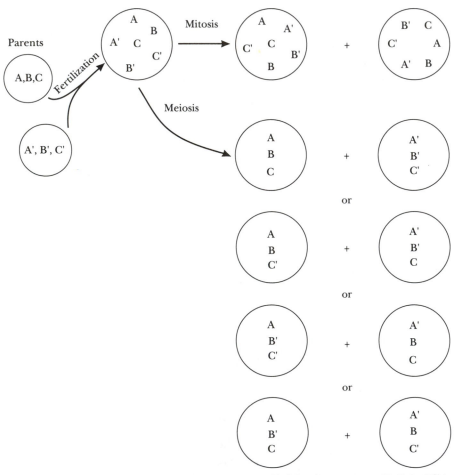

Figure 6.9 During fertilization one parent donates half of the chromosomes (A, B, and C) to the zygote, and the other parent contributes the corresponding chromosomes (A', B', and C'). The embryo grows by mitosis of diploid somatic cells containing three pairs of chromosomes (A + A', B + B', and C + C'). At maturity the gonadal tissues carry out meiosis; there are 2^3 (= 8) possible different haploid cells that can be produced by the gonads.

in the ovaries and the sperm cells that are made in the testes. At fertilization, the unpaired chromosomes in the egg combine with those in the sperm to yield a diploid cell, the fertilized egg or **zygote.** The paired chromosomes in diploid cells are not identical pairs since the contributing parents did not have identical genetic material. Corresponding genes on paired chromosomes are called **alleles.** Each allele pair influences the same trait, such as the color of the eyes, but the alleles may not be giving the same message; a pair of chromosomes *may* have some alleles that are identical, but certainly many others will be different.

Development begins as the fertilized egg cells divide mitotically and start to specialize for their future functions. Genes regulate development and the ultimate expression of traits (general body shape, hair and eye color, blood type, and so on). Since each separate fertilization results in a new combination of genes (one unique set from each parent), the offspring in a family tend to be similar, but not identical, to the parents and each other. In many instances, for each pair of genes one may be expressed (the **dominant** allele) and one may not (the **recessive** allele). For example, if one parent lobster contributes a gene for blue coloration and the other parent a gene for the normal green-brown color, the offspring will have a green-brown color because that gene is completely dominant and will therefore be expressed rather than the blue gene, which is recessive.

In some cases there will be partial dominance. Suppose that a certain aquarium fish may be yellow or red if it gets two yellow or two red alleles (that is, the fish is **homozygous** with respect to color genes); but, if the fish has one allele for yellow color and another allele for red color (it is therefore **heterzygous**), it may be an orange color if neither gene is completely dominant over the other.

Because each chromosome from the haploid cells must combine with its counterpart during fertilization, under most circumstances only haploid cells from two individuals of the same species can join to produce a viable zygote. In some cases, two closely related species can produce offspring that will develop completely, although these **hybrids** (the offspring of two different species) are usually unable to reproduce because their gonadal tissue cannot properly go through meiosis.

It is easy to see how the principles of genetics can be applied, by selective breeding, to the improvement of strains of particular plants, fish, or invertebrates. Selective breeding is based on the fact that all of the offspring from a single hatch are the result of many individual egg fertilizations; thus each of the offspring are genetically unique. Some will express "desired" traits more strongly than the others. A few of the traits that would be of interest for selection are fast growth, immunological resistance to certain diseases, high food conversion efficiency, or the ability to adapt to fluctuations in the temperature. For example, suppose the culturist is interested in producing a fish that grows quickly. Fish of that species would be spawned and the offspring observed; those that grow most quickly would be saved and spawned among themselves. Again, those that grow most quickly in the next generation would be saved for future breeding. However, **inbreeding** (matings of siblings) should be avoided since it tends to result in a large number of abnormal individuals. Just such a program of selected breeding has been carried out at the University of Washington since 1932. At the beginning of the breeding experiment, rainbow trout would spawn in their fourth year and at an average weight of 680 grams, producing 400 to 500 eggs. By the mid-1970s the females were maturing after only two years, weighing an average of 4500 grams, and producing about 9000 eggs. Selective breeding clearly will work, but not all animals will be equally promising targets. Ability of the culturist to control spawning, variability of the offspring, ability to rear the animals easily, and large numbers of larvae are all species traits that aid in the process; organisms that meet these criteria are good candidates for selective breeding.

Recently there has been considerable interest in the production of hybrids. The desired effect is **hybrid vigor**, a term used to describe an acquired resistance to disease or improved growth and survival in the face of environmental stress. For example, many trout farmers have lost animals to the virus that causes infectious hematopoietic necrosis (IHN), but coho salmon X rainbow trout hybrids seem to be resistant. ("X" indicates that the species are "cross"-bred.) The striped bass is a very popular food fish, but when crossed with the white bass, it grows more quickly and seems to have a higher survival rate. Red tilapia, which tend to bring higher prices than do darker tilapias, have been hybridized with cold-tolerant strains to give a red, cold-tolerant hybrid that can be raised in areas were the original red strain was unable to grow.

Tilapia are often hybridized for a different reason. One of the biggest problems with tilapia is that they reproduce too quickly, and ponds tend to fill up with large numbers of stunted fish that have little value. If reproduction could be halted, a farmer could stock the ponds with the ideal number of fish (for the maximum value after growout) and not have to worry about the fish increasing their numbers. It has been shown that several crosses of different species, such as male *Tilapia hornorum* X female *T. mossambica* produce all male offspring, so no unwanted reproduction can occur.

As was stated earlier, most animals have chromosomes in pairs so they are diploid, but **polyploids,** those animals that have more than two of each chromosome, can be produced by the culturist. Polyploids are generally either triploid or tetraploid. There are two characteristics of polyploidy that may in certain circumstances make this a desirable state. First,

polyploids in theory grow faster than diploids in culture conditions (although some culturists claim that they may actually grow a bit more slowly), and second, polyploids are sterile. Recently, the Pacific oyster has been subjected to experimental triploidy; this condition was induced because during the summer months, commercially grown oysters "waste" almost all their energy making eggs and sperm that make up the majority of their body weight, and the oysters are therefore an undesirable product at that time. Triploid oysters do not reproduce during the summers, giving them an important commercial potential.

Polyploidy is induced by **shocking** newly fertilized eggs. Shocking may take a variety of forms, but several factors are common in all cases: the shock must take place at a specific early developmental stage, the eggs must be in good condition to withstand the shock, and the shock must last for a specified length of time. For example, triploidy in catfish can be induced by cold shocking artificially fertilized eggs, 5 minutes after fertilization, for 1 hour. Later shocking will not cause triploidy, and shocking for a significantly longer period will result in a 100% mortality. However, triploid catfish not only grow more quickly than do sibling diploids, but have a better food conversion ratio. Triploidy can also be induced by heat shock and by increased hydrostatic pressure. Triploidy of the oyster can be chemically induced by **cytochalasin B,** a chemical that prevents the partitioning of the polar body and its genetic material during meiosis.

Tetraploids are produced with shocks in a manner similiar to triploids. Tetraploids are most important for use as brood stock animals, while the triploids are used for growout production. Tetraploids will produce eggs and sperm that are diploid, and when mated with normal diploid animals that have haploid eggs and sperm, the result is a triploid. Since many fisheries' biologists believe that the retarded growth of the young triploids is a result of the early shock to the eggs and is not an inherent fact of life for triploids, the use of tetraploid X diploid crosses may be an easy and practical solution to the problem.

The genetic makeup of an organism may also be altered by **genetic engineering,** the direct manipulation of the genetic material rather than through breeding. In time these molecular techniques may be used to produce better strains of culture organisms, as well as vaccines and hormones. This is discussed in greater detail in Appendix 3.

NUTRITION

Nutrition is one of the most important aspects of the culture process; feeds often constitute the major operating cost for aquaculture operations, and bad food choices result in poor growth, health, and reproduction. It is critical to know how much feed to use and what kind of feed is best.

Because animals cannot grow without feeds, it is important to be sure that they are not being underfed. Conversely, because feed is expensive, and because excess feed can result in disease problems and high BODs, overfeeding should also be avoided. Many fish feed manufacturers provide tables that describe the amount of feed needed, in terms of "pounds of feed/pound of fish," for many commonly grown fish, such as salmon, trout, and catfish. These tables are adjusted for fish of different sizes (since their metabolism changes as they grow) and for different temperatures (since temperature will strongly influence the metabolism of all poikilotherms). The culturist must therefore keep track of pond temperatures, as well as the size of the organisms and the density of animals in the ponds.

It is also advisable to understand some of the general principles of nutrition. There are several essential nutritional ingredients in an artificial diet, including proteins, lipids, carbohydrates, vitamins, and minerals (water may also be a limiting nutrient for terrestrial animals, but of course not for aquatic organisms). Beside each of these playing a specific role, the proteins, lipids, and carbohydrates must also supply the animals with energy.

Energy Energy is needed so the body can carry out its normal functions. Energy is measured in joules or calories. One calorie is the amount of energy needed to raise the temperature of 1 gram of water 1°C. Foods are usually discussed in terms of kilocalorie units (equal to 1000 calories) and sometimes in joules (1 calories = 4.184 joules). Energy in a feed is measured either directly with an instrument called a **bomb calorimeter** or can be estimated by knowing the components of the feed.

Lipids, in general, have about twice the energy of either proteins or carbohydrates. This is related to their structure; lipids have about half the oxygen per unit weight as proteins and one-fifth that of carbohydrates. Energy is produced by the "burning" of feeds, that is, the union of *external* oxygen molecules with carbon or hydrogen (the oxygen atoms that are already part of the molecules in the feed gave up their energy when that molecule was being formed), and therefore lipids have more potential energy for the animals to use. In carbohydrates there are enough oxygen molecules to combine with all the hydrogen molecules, so their energy potential is only a function of the combination of external oxygen with the carbon atoms of the carbohydrate. Lipids contain roughly twice the hydrogen atoms that proteins or carbohydrates do, and since the oxidation of hydrogen produces 34.5 kcal per gram, while carbon yields only about 8 kcal per gram, lipids contain more potential energy.

When the animal "burns" its feed or lipid reserves the energy is transferred, in most cases, to **ATP (adenosine triphosphate)** molecules. The combination of **ADP (adenosine diphosphate)** and a phosphate molecule to form ATP requires the energy released from the food or lipids. The energy is then temporarily stored in the form of ATP, and used by the organism when and where it is needed by reversing the processes (see Figure 6.10). If the energy from feeding is to be stored for an extended period of time, the ATP's energy is transferred by the organism to carbohydrates or lipids, and the resultant ADP molecules are free to continue trapping energy.

Proteins Proteins are long chains of molecules called **amino acids.** There are 20 kinds of amino acids that are used for protein building (see Table 6.4). Some organisms, such as green plants, can synthesize all the amino acids they need. All the commonly cultured animals are able to make some of the amino acids, but other amino acids must come from the ingestion of proteins because the animals either cannot synthesize those acids at all or cannot synthesize those particular acids quickly enough. The amino acids that are needed in the diet are called **essential.** In general, feeds made from animal tissues are richer in amino acid variety than plant protein–based feeds, and are therefore said to be of a higher quality.

Animals will use proteins by first breaking the chemical bonds that hold the amino acids together, called **peptide**

bonds, using specific and general enzymes. The amino acids will then enter the blood from the digestive system; at this point several things can happen to them, including

1. Their conversion to new amino acids
2. The release of energy as the acids are metabolized
3. The acids can be incorporated directly into new proteins that are being built by the animal

Lipids Lipids are a group of diverse compounds that are defined as being insoluble in water, but which can be dissolved in nonpolar solvents such as alcohols. There are a number of types of lipids including compounds known as

1. Terpenes
2. Prostaglandins, which regulate many metabolic reactions
3. Steroids, such as cholesterol, bile acids, and many hormones
4. Waxes, which are not easily digested by most animals

From the standpoint of nutrition, however, the most important are

5. Fatty acids
6. Fats

Fatty acids have a general structure consisting of a chain of carbon atoms with their associated hydrogen atoms; there is a carboxylic acid group (COOH) ending the chain (see Figure 6.11). If several of the bonds between carbon atoms are unsaturated (double bonds), the acid is said to be a polyunsaturated fatty acid (PUFA). Oils and fats, also called **neutral fats** or **triglycerols,** are made of a combination of fatty acids and glycerol molecules.

Some of the fatty acids are required in the diet of most animals because the

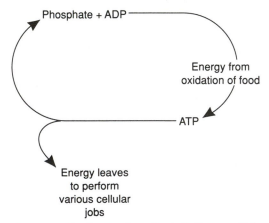

Figure 6.10 Energy from feeds is "trapped" in ATP. The energy is then brought to the cell sites that require energy and is released.

TABLE 6.4 The Seven Groups of the 20 Amino Acids Normally Found in Proteins

Acidic	Aliphatic	Aromatic	Basic	Hydroxyl	Imino	Sulfur
Aspartic acid (ASP)	Glycine (GLY)	Phenylalanine (PHE)	Lysine (LYS)	Serine (SER)	Proline (PRO)	Cysteine (CYS)
Asparagine (ASN)	Alanine (ALA)	Tyrocine (TYR)	Histidine (HIS)	Threonine (THR)		Methionine (MET)
Glutamic acid (GLU)	Valine (VAL)	Tryptophan (TRP)	Arginine (ARG)			
Glutamine (GLN)	Leucine (LEU)					
	Isoleucine (ILE)					

$$CH_3(CH_2)_{14}COOH \quad (a)$$

$$CH_3(CH_2)_5CH = CH(CH_2)_7COOH \quad (b)$$

$$CH_3(CH_2)_4CH = CHCH_2CH = CH(CH_2)_7COOH \quad (c)$$

$$CH_3CH_2CH = CHCH_2CH = CHCH_2CH = CH(CH_2)_7COOH \quad (d)$$

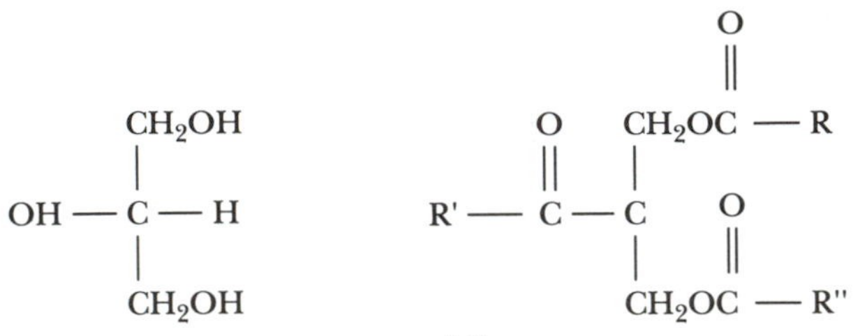

Figure 6.11 Lipids. Some fatty acids: (a) palmitic, (b) palmitoleic, (c) linoleic, and (d) linolenic. (e) Glycerol and (f) a triacylglycerol (fat), an ester of glycerol, where R, R', and R" are fatty acids.

animals are unable to synthesize the acids themselves. As in the case of the amino acids, these are called **essential fatty acids (EFA).** The two major groups of fatty acids that fall into this category are the linoleic acids and the linolenic acids (these groups are distinguished according to where the double bonds are found). In some cases, a fatty acid that cannot be produced from nonlipid components can be synthesized if the proper lipid building blocks are incorporated in the feed. For example, humans need arachidonic acid, although we cannot make it ourselves; if the EFA linoleic acid is in our diet we can use that to synthesize the arachidonic acid.

In addition to supplying essential fatty acids and energy to an animal, fats are also a good source of many vitamins.

Carbohydrates Sugars, starches, cellulose, and gums are examples of **carbohydrates.** Although there are a few exceptions, we can say in general that carbohydrates have a formula approximating $(CH_2O)_n$. There are small carbohydrate molecules, such as ribose or glucose, called **monosaccharides**; and small

chains of simple carbohydrates, sucrose for example, called **oligosaccharides** (see Figure 6.12); and **polysaccharides** are large complexes of monosaccharides, including starchs, glycogen, cellulose, and chitin. Glucose and glycogen are the most common animal carbohydrates, but in general, plants are richer in carbohydrates than animals.

How an animal uses the carbohydrates in its diet is related to the type of molecules the carbohydrates are, the species of animal, and the nutritional state of the animal at that time. If the carbohydrate can be digested by the animal, it will be used as

1. An immediate energy source
2. A precursor to glycogen or some other carbohydrate
3. A means of increasing the lipid content for long-term energy requirements

Since lipids contain more energy per gram than carbohydrates, it is more efficient to store energy that is not needed

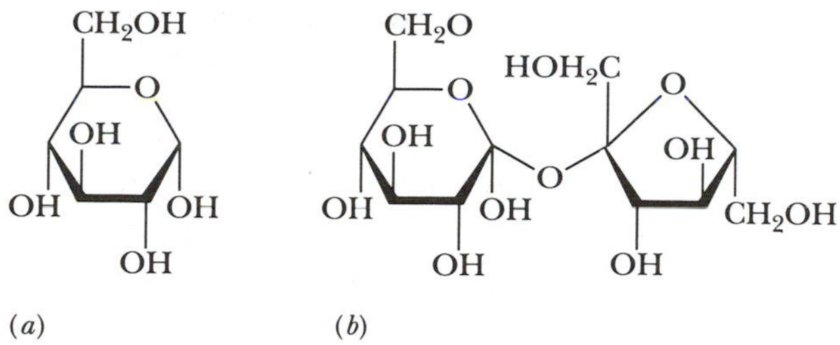

Figure 6.12 Carbohydrates (a) α-D-glucose and (b) sucrose.

immediately as fat. Glycogen is more quickly metabolized than fat, and simple sugars even more so; therefore they supply energy that is needed right away.

Vitamins Vitamins are loosely defined as substances that are required in very small amounts that are needed for proper development and growth, and for maintaining normal physiological activities. Vitamins often do this by acting as **coenzymes** (molecules that must be present for enzymes to work properly).

By definition, vitamins or their precursors are required in the diet. However, different animals have different requirements for vitamins, both from the standpoint of which vitamins are required and how much is needed. Vitamins can be grouped as either fat soluble or water soluble, which is a guide to the type of food in which they are most likely to be present. Vitamin deficiency symptoms are collectively called **hypovitaminosis,** while excess fat-soluble vitamins in the diet can also cause severe effects, a condition called **hypervitaminosis.**

The fat-soluble vitamins include vitamin A (which can be made by most animals if first given carotene from plants) and vitamins D, E (tocopherol), and K. The water-soluble vitamins are folic acid, thiamin (B_1), riboflavin (B_2), pyridoxine (B_6), cyanocobalamin (B_{12}), biotin, choline, pantothenic acid, inositol, niacin (nicotinic acid), para-aminobenzoic acid (PABA), and vitamin C (ascorbic acid). The functions of these vitamins are fairly well known in mammals, and moderately well understood in many fish. The roles that vitamins play in invertebrate physiology is a relatively unexplored field.

Minerals As in the case of vitamins, much is known about the requirements of **minerals** for mammals, but less so for fish and invertebrates, although it is not unreasonable to assume that many of their needs are at least similar. Scientists find it difficult to pinpoint the "minimum" amount of a mineral that is required, or even which minerals are required, because such minima vary with the condition and size of the animal, as well as the animal's ability to substitute one mineral for another.

In mammals, some of the most important minerals are calcium (Ca) and phosphorus (P), which are needed for building bones and soft tissues; magnesium (Mg), which is also needed for bones and soft tissues, but in addition it activates certain enzymes, including those regulating ATP formation; sodium (Na), potassium (K), and chlorine (Cl), which are used to regulate osmotic pressures and maintain acid–base equilibrium; iron (Fe), which is incorporated within the structure of hemoglobin and myoglobin; copper (Cu), which sometimes works with iron and is also part of some enzymes; cobalt (Co), which is needed for vitamin B_{12}; iodine (I), which must be present for thyroxine synthesis; sulfur (S), which is found in some of the amino acids; manganese (Mn), which is required for normal bone development; zinc (Zn), which is a cofactor for certain enzymes; selenium (Se), which is needed for normal development of muscle, liver, and pancreas; molybdenum (Mo), which is also a cofactor of some enzymes; fluorine (Fl), which is used in tooth and bone development; chromium (Cr), which must be present for glucose metabolism; silicon (Si), which is needed in the calcification of bone and in the synthesis of some sugars; and tin (Sn) and vanadium (V), which are needed for some enzymes.

Dietary Requirements for Fish

Most nutritional studies on fish have been confined to those species of the greatest commercial importance, such as trout, salmon, and catfish. In many respects the requirements of these animals are surprisingly like those of domesticated farm animals.

The value of any protein to a fish is a function of the energy it contains and the amino acids that compose it. Ideally, the proteins in the feed should be rich in the 10 amino acids the fish cannot synthesize: arginine, histidine, isoleucine, lysine, methionine, phenylalanine, threonine, tyrosine, valine, and leucine. This list of essential amino acids is very similar to that for mammals. The absolute amount of protein in a diet will vary with the species of the fish, their age, and the temperature of the water. Most fish feeds are 20% to 40% protein, but feeds for young fish and fry tend to be slightly higher.

Lipids in fish feeds must be digestible, which means that hard fats from animal sources should be avoided and replaced by oils. Some oils, such as cottonseed oil, contain toxins that can damage the fish, and these must be removed. Oxidation of the lipids in the feeds should also be prevented since not only are the oxidized lipids themselves toxic, but the oxidation process can also destroy some of the vitamins. Oxidation can be inhibited to some extent by the addition of vitamin E to the feed (although care must be taken since this vitamin is itself toxic in high amounts). Many fish, such as carp, eel, catfish, and tilapia, as well as mammals, seem to have similar requirements for fatty acids of the linoleic and linolenic groups, while other fish, including trout, do not seem to have any requirement for the linoleic acids. The level of lipid in most fish diets is between 5% and 10%. Shrimp by-products are sometimes included in diets of salmon and trout because they contain the pigment **astaxanthin,** which will give the cultured fish's muscle tissue the same pink color as wild fish's muscle; this is considered an attractive feature and increases the value of the salmon.

Little is known about the carbohydrate requirements of fish, although they do not appear to be as important as they are in mammals. Many of the carbohydrates included in fish feeds are actually indigestible by some fish. In fact, the livers of trout can be damaged by glycogen buildup resulting from excessive cornstarch or other carbohydrates in diets. Trout seem to be able to absorb simple sugars, such as maltose, more efficiently than bigger molecules, while just the opposite is true of catfish. Carbohydrates are generally kept at about 12% in trout diets, but catfish will grow well on diets containing as much as 40% starch.

Many vitamins are required by fish (see Table 6.5). Commercial dry diets usually contain a vitamin mixture. In contrast, little is known about the mineral requirements of fish, partially because dietary mineral needs will change as a function of the dissolved material in the water. Iodine is needed by trout and probably other fish as well, and most diets contain phosphorus and sulfur, as well as a trace element mix.

Crustacean Requirements

Crustaceans are the only group of aquatic invertebrates that have received much attention with respect to artificial diets. Mollusks have been only recently studied and information on them is sparse.

The essential amino acids for crustaceans are similar to those for fish, but they may not be able to produce tryptophan. Most diets for crustaceans are high in protein, although a recent study with

TABLE 6.5 Vitamin Requirements in Fish

Vitamin	Deficiency Symptoms	Excess Symptoms
A	Abnormal bone development; hemorrhages by fins or eye	Abnormal bone development; enlarged liver and spleen
D	Reduced growth; reduced levels of body ash, Ca, and K	Dark coloration in trout
E	Damaged blood cells; deposits on liver and spleen	Toxic liver reaction
K	Anemia; hemorrhages in gills, eyes, and tissues	
Folic acid	Fin damage; anemia; dark coloration; lethargy	
Biotin	Lesions; muscle atrophy; damaged red blood cells; erratic behavior	
Thiamin	Muscle atrophy; erratic behavior; edema	
Riboflavin	Eye damage; edema; erratic behavior	
Pyrodoxine	Erratic behavior; anemia; loss of appetite	
Vitamin B_{12}	Blood cell damage; poor appetite	
Choline	Hemorrhages of intestine and kidney	
Pantothenic acid	Gill and skin damage; loss of energy; poor appetite	
Inositol	Lesions and distended gut	
Niacin	Erratic behavior; edema	
Vitamin C	Abnormal development; hemorrhages; eye lesions	

lobsters has shown that much of the protein is used for energy and can therefore be replaced with cheaper carbohydrates.

Crustaceans are unable to make several of the lipids that they need. They require linoleic and linolenic fatty acids, as do the fish. They also require sterols in their diet to make the hormones that control reproduction and molting. In general, less than 1% of the diet needs to be composed of sterols (which are usually in the form of cholesterol).

The carbohydrate requirements of crustaceans are unknown, but they are clearly able to digest and use limited amounts; both simple sugars and starches may be utilized. Since the shell of crustaceans is made of a carbohydrate, **chitin,** and the levels of glycogen in the blood (used to make the chitin) fluctuate with the molt cycle, the animals must have a moderately well-developed mechanism for producing such molecules.

Vitamin studies on crustaceans have been limited. Vitamin C is believed to be needed to prevent **black death disease** and is probably, along with vitamin E, required for general cell functions. Pantothenic acid may be needed for reproduction. Vitamins D, E, and A may also be needed, but their role is not clear; vitamin K is currently thought to have no function in crustaceans. Other vitamins are believed by some biologists to be needed for growth and reproduction. Mineral requirements probably vary according to the environment. Calcium must be available, either in an edible form or dissolved in the water, so the animals can build their exoskeleton. Phosphorus is usually mixed with calcium and normally comprises 25% to 50% of the mineral supplement. Potassium and other trace metals are also thought to be important additions to a diet.

Artificial Diets

There are several considerations that must be recognized when a diet is being formulated. Not only should the diet be nutritionally sound, but it must also be in a form that is acceptable to the cultured organism. Some animals will readily take artificial feeds, others will slowly accept them, and some animals will not feed on pellets at all. Some factors that may affect an animal's acceptance of a feed are the size, color, smell, and hardness of the pellet. (See Appendix 4 for examples of feed formulas.)

Diets should be made from relatively fresh materials. If the feed ingredients become spoiled, or the oils are oxidized, the cultured animals will grow poorly, have an off-taste, or die. Unless feed materials can be stored properly, there is little sense in trying to prepare diets.

The materials that are combined to make the pelleted feed will sometimes hold together after processing and when added to the ponds. However, often carbohydrate **binders** such as alginate or carboxymethyl cellulose are required. The ingredients in the feed will depend on the nutrient formulation that is designed by the culturist based on the needs of the cultured animal. The price of the ingredients that are available is also considered. For example, squid, fish meal, or soy meal can be used to supply protein in a diet; soy meal is cheap, but squid or fish meal may be of a "higher-quality" protein and contain more of the essential fatty acids and

amino acids that are needed. The culturist may decide to use the soy meal and supplement it with certain amino acids and fatty acids, or to use one of the animal sources of protein, or perhaps use a mixture of all three protein sources, depending on the prices of the materials and his experience. Feed manufacturers use computer programs to generate **least cost** formulations, which are based on the required nutrients, a nutrient profile that is available for most of the common individual ingredients used in the feed, and the current costs of those raw materials.

There are several standard ways to prepare feeds, including dry meals, crumbles, blocks, and three primary types of pellets:

1. The **extruded, hard pellets** are the most widely used. These are easy to manufacture, store, and handle. Hard pellets can be made large or small depending on the animals being fed. These pellets sink quickly to the bottom of the pond and are easily used in automatic feeders (see Figure 6.13).

2. **Expanded pellets** are more difficult to make, requiring high tempera-

Figure 6.13 Pellets being fed to trout. (Photo courtesy of the New Jersey Division of Fish, Game, and Wildlife.)

(a)

(d)

Figure 6.14 Some of the pieces of equipment used in small-scale pelleted production. These are used for (a) grinding and (b) chopping and mixing the raw ingredients. After the feed is mixed, it is forced through (c) a pellet extruder and finally placed in (d) a drying oven.

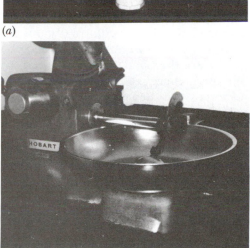

(b)

(c)

tures and pressures, which mean that some of the fatty acids, vitamins, and amino acids are destroyed during the process (although these nutrients can later be sprayed onto the feed before packaging). Expanded pellets float and are therefore preferred by many culturists who rear top-feeding fish, since the actual feeding activity can be observed. Direct observation of feeding reduces the chances of adding too much or too little feed. Expanded pellets are especially useful in cage culture situations.

3. **Moist** and **semimoist pellets** contain 25% to 35% water. These pellets must be chemically protected from spoilage, or else refrigerated. Moist pellets are also costly to manufacture, ship, and store, but their advantage is that their texture is similar to natural feeds, and many fish will accept them more readily than harder pellets.

Recently there has been another kind of feed produced, the **microcapsule.** The coatings of microcapsules are prepared from materials (such as a nylon-protein) which are digestible but retain all the nutrients inside the capsule when added to water (even those nutrients that are water soluble). Microencapsulated diets may have a significant impact in the future on the culture of larval organisms and perhaps filter feeding animals as well.

In some cases, only live feed organisms will be accepted by the cultured species. Rearing feed organisms is generally an expensive and difficult process, and therefore should be avoided if possible. However, in some instances feed organisms must be used, especially during early development. The culture of these organisms is discussed in Appendix 5.

SUMMARY

In many ways, aquaculture is a biological science, and those in the industry should have some appreciation of the general biological principles involved.

Crops can be reduced or lost because of predation and disease. The most important of the disease vectors are viruses, bacteria, fungi, and protozoans. Animals may also be attacked by a variety of parasites. In some cases there are methods for treating the distressed animals, but the problems of disease can be avoided in many instances by proper management techniques. Culture organisms may also succumb to predators, environmental stresses, and pollution.

Ponds are essentially finite ecosystems. A balanced pond system contains established niches for the culture organism as well as the other organisms and microbes that share the pond. The flow and distribution of energy (introduced as food) are of interest to the culturist. Energy waste can be minimized in some instances by growing two or more harvestable organisms together that occupy different ecological niches: the practice of polyculture.

Selective breeding programs can improve the genetic stock of cultured organisms. Special traits are selected which will make the animals more valuable or easier to grow. Hybridization and induced polyploidy have been used, in some instances, to produce more vigorous animals or animals that will not "waste" energy on reproduction.

A large portion of the operating budget of many aquaculture facilities goes toward feeding. When constructing or selecting a feed, the farmer should consider the stock's requirements in terms of energy, essential amino acids, essential fatty acids and other lipids, carbohydrates, vitamins, and minerals. The requirements of fish are less well known than those of most domestic animals, and those of the invertebrates are still less defined. Artificial diets are specially designed feed mixtures for cultured species that can be made in a variety of ways. Of the ways to introduce artificial diets, the pellets (either hard, expanded, or moist) are the most popular.

CHAPTER 7

SEAWEEDS AND SPIRULINA

INTRODUCTION

In this chapter we will discuss a cyanobacteria, *Spirulina*, and some of the important seaweeds. These are "lower" plants that either are eaten by people or whose products support some industry.

Seaweeds are multicellular, benthic forms of algae. Some are rather simple in shape, while others are more complex. The important seaweeds (in fact, almost all seaweeds) belong to one of three phyla: the Chlorophyta (**green algae**), the Phaeophyta (**brown algae**), or the Rhodophyta (**red algae**). The color of any particular seaweed may not agree with the common name given to the phyla.

Several seaweeds support major culture industries in the Orient. Other species are collected and eaten, but are not generally grown by aquaculturists. Only the principal cultured species are discussed in this chapter.

Seaweed Morphology

The seaweeds lack many of the specialized adaptations seen in higher terrestrial plants. Seaweeds have no leaves, flowers, fruits, seeds, or roots to take up water and nutrients. They are built on a simple, general plan that has, however, an almost limitless number of variations. Seaweeds may be small or very large, simple or branched, flat and leaflike, or thick and spongy. The parts of a typical seaweed include the holdfast, blade, and stipe (see Figure 7.1).

The **holdfast** looks something like the root system of the higher plants, but its function is only to anchor the plant to a surface and not to take up nutrients from the water. The individual fingers of the holdfast are the **haptera**. The shape and size of the holdfast are adaptations for the type of substrate (loose sand, hard rock, etc.) it must attach to, and for the wave action to which the plant is subjected. In some seaweeds, the holdfast is no more than a small disc or attachment point.

The **blades** can be leaflike or tubulous structures. They are similar to the true leaf of higher plants in that they are the principal sites of photosynthesis. Therefore, like terrestrial plants, they are often flattened so that they have a large surface area for absorbing as much light as possi-

143

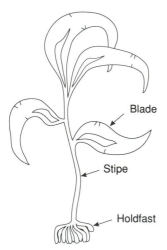

Figure 7.1 Parts of a typical seaweed.

ble. The blades are also responsible for taking nutrients out of the water. Associated with the blades of some of the brown seaweeds (for example, *Fucus*) are gas-filled sacs called **pneumatocysts**; the gasses in the pneumatocysts are those commonly found in seawater: oxygen, nitrogen, and carbon dioxide. These sacs act as floats to keep the blades from drooping and receiving less sunlight.

The **stipe** of the seaweed is a flexible, stemlike structure. A large seaweed would not be very efficient at gathering light if all the blades originated at the holdfast, because the blades would be crowded together and shade one another. One of the stipe's functions, therefore, is to serve as a site of attachment for the blades so that light will be maximally utilized. The stipe also acts as a sort of "shock absorber," preventing the waves from pulling the seaweed from its substrate.

Seaweed Ecology

Since seaweeds are attached to a substrate by a holdfast and need light for photosynthesis, they are found only in shallow coastal waters of less than 200 m (but are common only in about the first 35 m of water). The geographical distributions of the various species of seaweeds

are largely related to the temperature, light, and tides.

The vertical distribution of the seaweeds, that is, the depths at which they are found, has been explained only recently. For a long time scientists claimed that, generally speaking, green seaweeds were found at the shallowest depths, red seaweeds were dominant at the deepest depths, and brown seaweeds were found at intermediate depths. It was assumed that, because of the way that seawater filters out light of different wavelengths, this distribution of seaweeds was related to the *types* of chlorophylls and accessory pigments characteristically found in each phyla. New experiments and statistical analyses have shown that there is no difference in the depths of these seaweed groups. The quality of light that reaches the different depths may play some role in the vertical distribution, but not as large a role as other factors play. Other factors controlling seaweed distribution are the *amount* of chlorophylls and accessory pigments present, the type of substrate, the presence of certain herbivores, tolerance to high levels of irradiance or desiccation, and temperature or salinity fluctuations.

Seaweeds, from an ecological standpoint, have a variety of functions. They offer food to herbivores, and protection to many species of animals, including larval and juvenile forms of many commercially important species of fish and invertebrates. Seaweeds that break away from their substrate and are carried off by currents are important sources of detritus for other marine communities. Seaweeds also act as sediment traps and may affect wave action.

Reproduction

The life cycles of many species of seaweeds are rather complex and are not detailed in this book. However, some idea of reproduction in seaweeds is critical in understanding the culture of these plants. Reproduction may be simply vegetative (asexual), or it can be sexual, involving the fusion of sex gametes.

Many of the larger species of seaweeds go through an alternation of generations similar to mosses and ferns. They may be represented in the **sporophyte** phase, which is diploid, or the haploid **gametophyte** stage. The size, duration, and shape of the stages will vary with the species. The cycles of the important species are described in this chapter, but the reader should be aware that this is not a complete survey of seaweed life cycles.

LAMINARIA

Kelps are brown seaweeds that include some of the largest plants that have ever lived. Many have lengths of 50 m, with the longest reaching about 200 m. These giant kelps include the genera *Macrocystis*, *Nerocystis,* and ***Laminaria,*** the sugar wrack, which is cultured widely in the Orient and to a much lesser extent elsewhere (see Figure 7.2). Natural populations of *Laminaria* are limited to cool waters; the sporophyte begins to deteriorate at temperatures above 16°C.

Figure 7.2 *Laminaria* showing blade, stipe, and holdfast. (From Trainor, 1978.)

Life Cycle

The life cycle of *Laminaria* is similar to that of the other kelps. In cold waters these seaweeds may reproduce all year, but in warmer waters they spawn in the fall. *Laminaria* is an example of a plant with a prominent diploid stage, but a very small haploid stage. Specialized reproductive organs, **sori**, on the blades of the mature sporophyte kelp contain cells that undergo meiosis to produce microscopic, flagellated swimming cells called **zoospores.** Since these are the result of the meiosis of diploid cells, the zoospores themselves are haploid. These swim—usually for only a matter of hours—until they settle on substrates and germinate into male and female gametophytes.

The male gametophytes produce flagellated sperm cells that are released and begin swimming. The female gametophyte makes large, nonswimming eggs. The eggs are fertilized by the sperm, and the resultant diploid zygote then grows into the large sporophyte stage (see Figure 7.3).

Culture

Bottom culture is the oldest form of seaweed farming, but its practice is declining. The first step in bottom culture of *Laminaria* is to allow the zoospores to attach to stones, twigs, or bamboo baskets. This is usually done *in situ,* but zoospores released under controlled conditions are sometimes used. In the past, coral reefs have been dynamited to increase the amount of fresh substrate available for attachment, but this technique has been halted. The density of the seaweed bed is a function of the clarity of the water. The kelp is harvested by trained divers. Under good conditions, yields should exceed 10 metric tons per hectare.

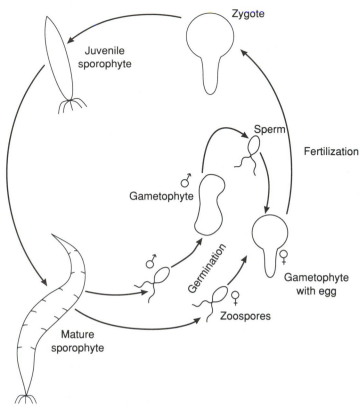

Figure 7.3 Life cycle of *Laminaria*.

The most important advance in the culture of *Laminaria* in recent years is the Chinese **forced cultivation** technique, the speeding-up of the life cycle of this cold water seaweed in temperate waters, where much of the culture takes place. In the spring, mature sporophytes either are shipped on ice from farms in the north were reproductive plants can be collected, or the plants are matured in cool seawater from wells. Thus, rather than waiting for the *Laminaria* to reproduce in the fall and harvesting in the summer before the plants start to deteriorate, the Chinese farmer grows the young sporophytes under controlled conditions for several months before planting them in the field in the fall; this extension of the growing season dramatically increases the annual yield.

Culture of kelp begins with the collection of the zoospores. This may be done *in situ* or in a controlled environment using the "dark dry" technique to produce the zoospores. To do this, the blades of the *Laminaria* with sori present are collected and hung in baskets out of water, in the dark. These baskets are covered with some sort of absorbent material to help remove the moisture from the sori. After several hours they are returned to seawater, and the zoospores are automatically released in a few minutes. Fertilization of the haploid zoospores takes place in the dark after they have settled. The young sporophytes grow best under specific light and temperature conditions; nitrogen levels must be maintained or the plants will be stunted. When the sporophytes have grown to a predetermined size, they may be transferred. The transferred sporophyte can be grown on the bottoms of shallow bays; however, most of the culture of this seaweed is now done on rafts or with long lines.

During growout, production is limited by nitrogen available to the seaweed. One solution to this problem has been to hang porous fertilizer cylinders (filled with sodium nitrate) along with the seaweeds. The nutrients that leach out of the container are replaced periodically. In recent years, as small power boats have become more available, farmers have begun to spray liquid ammonium nitrate once or twice a week. Alternatively, culturists may periodically immerse the kelp sporophytes in a nutrient solution containing high levels of available nitrogen. The length of immersion time is a function of the concentration of the fertilizer. The fertilizer solution may be reused several times.

Raft culture takes several forms:

1. **Basket culture.** Cylindrical bamboo baskets are tied together in rows and hung from a raft. The baskets are open at the top and contain a fertilizer cylinder. Along each side of the basket is strung a hemp rope

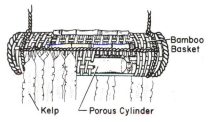

Figure 7.4 Basket culture of *Laminaria*. (After Cheng, 1969. Copyright © The New York Botanical Garden. Reprinted by permission.)

from which the *Laminaria* hang; the young seaweeds are attached by inserting the stipe between the strands of the rope (see Figure 7.4).

2. **Single-line tube raft culture.** Tubes of bamboo or rubber are strung together to form a raft; a rope runs alongside the tubes. The density of the sporophytes is related to the water condition. Until recently, this was the most popular method for kelp production because it gives a very good yield per kilogram of fertilizer (see Figure 7.5).

3. **Double-line tube raft culture.** The raft is a ladderlike structure with bamboo tubes as the rungs. A porous container of fertilizer is hung

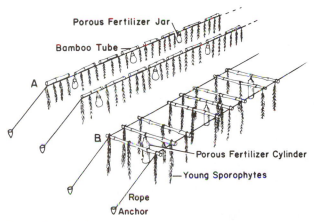

Figure 7.5 Raft culture of *Laminaria*. (a) single line, (b) double line. (After Cheng, 1969. Copyright © The New York Botanical Garden. Reprinted by permission.)

in the middle of each rung. Hemp ropes along the sides of the bamboo tubes hold the sporophytes (see Figure 7.5).

The raft techniques developed in the 1950s and 1960s have been replaced in many areas by the long-line technique. This change was largely the result of a shortage of bamboo. Hemp or synthetic ropes of polyethylene or polypropylene are supported by glass or plastic buoys (see Figure 7.6). The *Laminaria* are inserted into the ropes, which in most cases are strung perpendicular to the water flow.

After several months of culture, the sporophytes have grown to up to 6 m and are harvested in early to midsummer. The harvesting is done using small boats. In the spring, prior to harvesting, the distal ends of the blades are sometimes collected because this decreases the shading, and because the ends (which are the oldest part of the blades) erode as the water temperatures rise and would be lost by the summer harvest if not cut off in the spring.

UNDARIA

Like the kelps, **wakame (*Undaria pinnatifida*)** is a brown alga. It is found growing on stones, at depths of 7 m to 15 m, in cool waters off the coasts of Japan, Korea, and China. It does not tolerate low salinities and therefore is not found in estuaries. It has pinnate blades and a flat stipe; this species can grow to about 1 m. The wakame is eaten dry, semidry, roasted, and even sugared as a candy. Sometimes other species, such as *U. distans*, *U. peterseniana*, and *U. undarioides*, may be used in place of *U. pinnatifida*.

Life Cycle

The large sporophyte contains a specialized reproductive tissue, the **sporophyll.** The sporophyll develops zoospores during the winter, but these are not released until the late spring or early summer when the water has warmed. The zoospores settle on a hard surface after a short planktonic existence. As in the case of *Laminaria*, the zoospores give rise to the sexual gametophytes.

The microscopic gametophytes develop during the summer months, and as the water temperature begins to decline during the early fall, the male sheds the sperm that travels through the water and fertilizes the egg inside the female gametophyte's **oogonium** (the sex organ containing the egg or eggs). After fertilization, the seaweed begins to grow at the site of attachment of the female gametophyte.

Culture

Collection of the zoospores for culture is done during the spring and summer by individual culturists and by government

Figure 7.6 Long-line culture of *Laminaria* on the coast of China. (Photo courtesy of Liang Hongwu, Bureau of Aquatic Products, People's Republic of China.)

agencies. Mature sporophytes are placed in tanks, constructed of concrete, canvas, or plastic, that are kept out of direct light. In the tanks are placed frames (usually made of plastic) that are wrapped with a braided cotton string or a synthetic yarn. A square frame of 50 cm × 50 cm × 50 cm is wrapped at intervals of about 1 cm, so the whole frame will support about 100 m of substrate for attachment of the zoospores. After several hours, the frames are removed and placed in culture tanks.

The zoospores develop into the gametophytes when the temperature declines to about 20°C, and small sporophytes are produced with fertilization. When the ocean temperatures begin to decline in the fall, the strings of young wakame are removed from the frames. The strings are cut and attached to ropes at intervals that vary with culture conditions. The rope is suspended either from rafts or long lines. The sporophytes are normally grown in the first 5 m of water, but lines going down to 50 m may be used in clear oceanic waters. The plants grow quickly and may be harvested in midwinter. Growth of the plants is probably best at about 12°C, but will be depressed below 5°C.

The density of the culture is largely a function of the water clarity. Harvests per meter of culture rope are about 10 kg in the cold northern waters and about half that in warmer southern waters. Harvesting under high-density conditions may take place over a period of time, with the quickest-growing plants being removed first; this allows the remaining wakame, which are harvested later, to grow more rapidly because more light and nutrients are available after the crop is thinned. This partial harvesting can be repeated several times. When culture conditions are good, but the density of the plants is lower, harvesting is done by cutting off the ends of the seaweed and leaving the remaining plant to grow out again; this also makes more light and nutrients available to the plant tissue remaining in the water. When the culture season is short or conditions are not very good, the wakame are harvested all at once.

PORPHYRA

Probably the most popularly cultured seaweed is the red alga, **Porphyra,** called **nori** in Japan and **purple laver** in Europe (see Figure 7.7). There are about 25 species, most of which are found in the Pacific Ocean. The most commonly cultured species in Japan, the major producer, are *P. tenera* and *P. yezoensis*. Nori sheets are wrapped around rice to make a favorite Japanese dish, **sushi.** *Porphyra* are leafy seaweeds, mainly found in intertidal environments. Most grow on rocks and other hard substrates, although a few species are epiphytes on other algae. Many species are very resistant to dessication.

Life Cycles

There appear to be a number of different reproductive schemes within this ge-

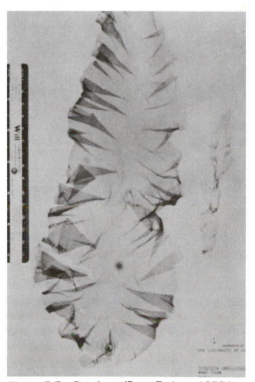

Figure 7.7 *Porphyra.* (From Trainor, 1978.)

nus. Most reproduce sexually, but a few are asexual. Some are dioecious and others monoecious. Only the cycle of *P. tenera*, which is very similar to *P. yezoensis*, is discussed in this text. In both these species the secondary growth occurs by asexual reproduction. They exist as leafy plants in the winter, but are filaments in the summer.

The most important discovery with reference to the life cycle of this genus was made by Drew (in England) and Kurogi (in Japan). They showed that the large leafy form of *Porphyra* was only one stage of its life history. A minutely filamentous, shell-boring stage, which had been incorrectly assigned to a separate genus, *Conchocelis*, is actually a different phase of the same seaweed. The **conchocelis stage** is the sporophyte, while the blades are haploid gametophytes.

The male sexual organ, the **antheridium,** produces small, nonmotile **spermatia** that are released in the water; the female sex organ, the **carpogonium,** contains the egg. When contact between the gametes is made, a zygote is formed. The zygotes begin to develop and are released in the form of **carpospores.** The carpospores will drift in the water, eventually sinking; if they land on a mollusk shell, they germinate, producing small fila-

ments that branch out and bore into the shell. The alga will pass the summer as a filamentous conchocelis; unlike the haploid seaweed, the conchocelis will vary in appearance with the environmental conditions and the species, though *P. tenera* and *P. yezoensis* are often purple.

From late September to early November, the **sporangia** organs release **monospores.** The monospores eventually land on a suitable substrate and tiny seaweeds, or **plumules,** begin to grow. In some species, the plumules simply continue to grow and in time produce the carpospores that develop into the conchocelis, but in some other species, including *P. tenera* and *P. yezoensis*, there is an extra "loop" in the life cycle. The plumules of *P. tenera* will release **neutral spores** from late September to early November that settle and form secondary plumules; the plumules continue to grow vegetatively. With further growth and a change in temperature, the production of the neutral spores stops and the production of carpospores will begin (see Figure 7.8).

By late November, the seaweeds may reach 20 cm or more. The population begins to decrease in the spring, and by the summer, the seaweeds have disappeared. The vegetative stage of *P. yezoensis* lasts

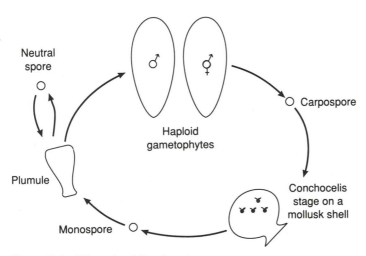

Figure 7.8 Life cycle of *Porphyra tenera*.

longer than that of *P. tenera*, in part, due to the long period of monospore release and the release of neutral spores by the former species until March.

Culture

The initial step in the culture process is the **seeding** or **spore fixing.** This refers to the inducement of the conchocelis stage to release the monospores or the plumules to release the neutral spores. Collection of the spores was originally done on branches and bamboo poles, but these have been replaced by nets made of a synthetic twine. There are two methods of seeding: natural and artificial.

During natural seeding, the nets are placed in special areas, called **taneba,** that are known to be good sites for collection of the spores. When and where the nets are set is a function of the species that will be cultured, the tidal cycle, and the temperatures of the water that particular year. The spores do not adhere well when the salinity has been lowered because of rain. The height of the net is also important; the newly settled *Porphyra* are not very resistant to desiccation, so the net cannot be so high that the plants are exposed during the low tide. However, if the net is too deep, it becomes covered with green algae and diatoms which also inhibits settlement.

When neutral spores are used for artificial seeding, there are two methods used for the collection of the secondary seaweed buds. The first method involves placing a clean net over another net that has the neutral spore-producing plants on it; the neutral spores are released into the water and adhere to the new net. The second method involves the crushing of the neutral spore-producing plumule; after the plant is crushed, a clean net can be immersed in the solution containing the spores.

Artificial seeding is also done by conchocelis culture. In a shallow tank of full salinity seawater, oyster or scallop shells are laid with the inner surfaces facing upward. To some extent, shells have been replaced by transparent vinyl films that are covered with 1–2 mm diameter calcite granules. Into these tanks are placed carpospores that were obtained from plants that matured with the approach of spring. Carpospore formation is induced by drying the parent seaweed overnight and immersing it in seawater for four to five hours the following day. This entire process is called **carpospore fixing.** The rate of boring into the shells by the spore is controlled by the culturist, who can adjust the temperature, pH, brightness, and salinity.

The shells with the conchocelis are used for seeding the nets. During field seedings, the shells may simply be spread under the nets, or they can be hung vertically in vinyl bags under the nets. To increase seeding efficiency, the nets may be laid out one above another, using up to 20 nets. Sometimes the nets themselves are put in vinyl bags with the conchocelis-bearing shells so the spores are not wasted.

The seeding of the nets can also be carried out indoors in tanks where the release of the monospores, and the adhering of the spores to the nets, can be observed. The spore-containing water is stirred up by moving the nets through the water, or the nets are simply immersed in the seeding tank and the water is mixed by air bubbles released from pipes on the bottom of the tank. The nets are then kept in seawater for a day (because the young *Porphyra* are sensitive to exposure just after settling) before being transferred to the growing beds. The time of the release of the monospores can be controlled by adjusting the light cycle and the temperature (the monospores will not be released when the temperatures are too high). After they are released by the conchocelis, the monospores can be stored if required, by holding them in a seawater and glucose solution at −5° to −15°C, and still maintain the ability to later adhere to a surface. The seaweed can also be maintained on the nets at $-20°C$ and will resume its normal growth when returned to a suitable environment.

Once the nets are seeded, the growout can begin. The growout beds are usually not placed in the same area as the beds used for natural seedings, so the nets with the partially grown seaweeds must be transported. Transportation is no problem once the vegetative phase is partially complete, since these plants withstand exposure well. Depending on the depth of the water, the tidal range, and the strength of the currents, the nets can be spread on fixed poles or floated using one of several methods. When the nets are on fixed poles, the height of the nets is adjusted so that they are exposed for four to five hours between dawn to dusk; this is generally the best way to get good growth of the nori while limiting competition from other algae and reducing the risk of infection by disease organisms. Under these conditions, the seaweeds grow to a harvestable size—15–20 cm—in less than two months during the fall and winter.

Harvesting takes place during the winter over a period of several months. Harvesting begins when the plants are large enough. Once done by hand, harvesting is now done by special machines that are built for each different culture system. After the first harvest, new growth will appear and the harvesting can be repeated every two to three weeks. Each net can be harvested several times, and when the production of a net begins to decline after repeated harvests, it can be replaced by a seeded net that has been kept cold; the new net will produce its first harvestable crop in 15 to 30 days.

After the nori is harvested, it must be processed. First, the seaweed is drained in bamboo baskets, and then machines are used to chop it up. The material is finally molded into sheets. There are several methods of producing sheets. Commonly, the chopped seaweed is mixed with water and then poured into a frame on a bamboo mat; the water runs through the mat and the remaining nori is either sundried or dried indoors with heaters. The dried sheets are stacked in bundles for packaging.

EUCHEUMA

Eucheuma is a red alga eaten in China, Malaysia, and other Southeast Asian and Pacific island countries. It is also cultured commercially for both the ι- and κ-carrageenan. Several species have been cultured experimentally, but *E. cottonii* has been brought into mass culture on farms in the Philippines. In 1970, the Philippines produced about 500 dry metric tons (export value about $\$1.68 \times 10^5$) of this seaweed, all harvested from wild stocks. By 1974, production had risen to almost 8000 metric tons (worth about $\$4.53 \times 10^6$) of which almost 99% was cultured on farms, and by 1985 production had risen another 400%.

The farms are located in coral reef areas, so although there is good water movement, they are protected from storms, and the water is shallow enough that poles can be driven into the bottom. The temperature is fairly constant throughout the year, ranging from 26° to 32°C.

Culture

This plant is farmed using both nets and rope. In both cases, the waters are tested for their ability to support growth of *Eucheuma* before major capital costs are incurred.

The net method requires four poles that are driven into the reef bottom. A net of polyethylene, about 2.5 m × 5 m, is stretched from the poles parallel to the bottom. The nets are "seeded" by attaching 100 g pieces of algae to the net with thin plastic strips. As the *Eucheuma* grows, the farmer must tend the crop to be sure that the plants are kept free of floating material and grazing herbivores. In addition, *Eucheuma* that is growing poorly is removed and replaced with other pieces. When a plant has increased to 1.2 kg to 1.5 kg by vegetative growth, the seaweed is partially harvested; about 500 g is left and allowed to continue to grow. The pruned seaweed is dried and sold.

Individual lines can be used in place of nets. This method yields less *Eucheuma* than the net method, but the plants are kept cleaner and free of significant amounts of silt. It is also easier to tend the crop.

Preliminary experiments have suggested that *Eucheuma* can be grown in a **spray culture** system, since this seaweed does not need much water as long as it is kept moist. Some tests indicate that these plants can be kept for at least nine months in such a culture system and still sustain good growth.

GRACILARIA

Gracilaria is a red alga. There are about 100 species in this genus, found in tropical and temperate waters, but only a few are important culture organisms. *Gracilaria* grow best in low-wave-action environments with salinities of 8–25 ppt.

These are fleshy seaweeds with a bushlike appearance. *Gracilaria* is eaten directly as food or is extracted for agar. In Taiwan, much of the cultivated seaweed is now being used as a feed in abalone farming.

Life Cycle

Like many of the red algae, *Gracilaria* display a complex life history. The haploid male gametophyte seaweeds produce and release the nonmotile spermatia from the **spermatangia.** The haploid female gametophyte's carpogonial filaments contain eggs that are fertilized and develop into diploid carposporophytes; the carposporophytes release carpospores that germinate into **tetrasporophyte** seaweed. The tetrasporophytes contain sporangia, the site of a reduction division, producing haploid tetraspores that eventually germinate into the male and female haploid gametophyte seaweeds (see Figure 7.9).

Culture

In Taiwan, *Gracilaria*, especially *G. confervoides*, is often cultured in old converted milkfish ponds. The ponds should have a sandy loam bottom, exchange water with the changing tides (100% of the

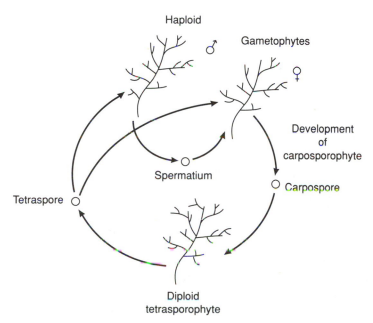

Figure 7.9   Life cycle of *Gracilaria*.

water should be replaced at least every three days), and be near a source of freshwater if the salinity gets too high because of evaporation. The depth of the pond during the spring planting should be 20–30 cm, but this may be increased to up to 80 cm during the summer when the temperature rises. A windbreak may be constructed to prevent the wind from piling all the seaweed at one end of the pond.

Gracilaria farmers purchase cuttings for planting in the ponds, often from other farmers. The farmer must inspect the seaweed and be sure that only healthy plants are being purchased. A 1-hectare area should have 3 to 5 metric tons of seaweed uniformly planted on the bottom in the spring (less intensive plantings result in phytoplankton blooms). The seaweeds may be covered with an old fishing net to prevent drifting, or they may be attached to bamboo poles driven into the pond bottom.

The *Gracilaria* is partially harvested every ten days from June to December by hand or by scoop nets. The seaweed is cleaned and sun dried. In production ponds about 1 to 1.2×10^4 kg are produced annually from 1 hectare.

The production ponds in Taiwan are fertilized using either organic or inorganic material. When the nutrient content (especially nitrogen) in the water is low, the plants lose their normal reddish-brown hue and begin to take on a straw color. The darker color is a reflection of the amount of the protein pigment, **phycoerythrin,** in the plant; it is believed that this pigment is the site of nitrogen storage by *Gracilaria*. In experiments conducted in Florida it was shown that *G. tikvahiae*, cultured in unenriched seawater, could be soaked in a solution of 100 micromoles of nitrogen (as NH_4^+) and 100 micromoles of phosphorus (as PO_4^-) for as little as 6 hours every 2 weeks and still maintain maximum growth by taking up the nutrients it would need. There are several advantages to this sort of culture:

1. The cost of the nutrients can be reduced since the plants take up only what they require, and none is washed away and lost in the production area.
2. Since the production area is not fertilized, there is no growth of other algae that would compete with the *Gracilaria*, nor development of epiphytes (a major problem in the ponds in Taiwan).

OTHER SEAWEEDS

Other species of seaweeds are being grown commercially or experimentally.

Many types of red algae are cultured. *Gloiopeltis* ("funori" in Japan), an edible plant also used for the production of glue, is grown in open culture. *Rhodymenia palmata,* or **dulse,** is eaten (raw, dried, or powdered) in Canada and Europe. It does not support an aquaculture industry at present, but processors of this seaweed hold harvested *Rhodymenia* in aerated tanks until they can be dried, and growth has been observed in these tanks. While healthy plants are being held in storage ponds 10 kg/m², production of over 60 g dry weight/m²/day has been recorded for this species. *Chondrus crispus,* sometimes called **Irish moss** or **carrageen,** has been experimentally cultured; this species is presently harvested from well-managed beds in eastern Canada, England, and Spain. It is eaten, and is also a source of carrageenan. Other experimentally or commercially cultured carrageenan-containing red algae are *Gigartina, Iridaea,* and **Hypnea.** When *H. musciformes* was grown experimentally in Florida, it was found that if the growth rate was stimulated by using a fertilizer, the percentage of carrageenan in the seaweed decreased compared to the slower-growing plants that did not receive the nutrients. *Gelidium* has also been cultivated experimentally because of its potential in agar production.

In addition to the red algae, green algae are also cultured. *Monostroma,* which is highly prized, may be cultured in a manner very similar to that of *Porphyra,* but is also sometimes collected as a by-product of nori culture, as is *Enteromorpha,* another edible alga. **Sea lettuce,** *Ulva,* a popular edible seaweed, has also been cultured experimentally. Individually or in commercial mixtures, *Ulva, Monostroma,* and *Enteromorpha* are called "green laver" or "aonori," although in some districts, these terms only refer to *Enteromorpha* and *Monostroma,* respectively. *Caulerpa* is grown in Japan and China in brackish ponds where this seaweed is cut into pieces that are used for seeding the new ponds, as in *Gracilaria* culture. *Caulerpa* is shipped to the Philippines as a food; in the United States, *Caulerpa* is one of the few seaweeds popularly kept in marine aquaria.

Among the brown algae, another kelp, *Macrocystis,* has been experimentally cultured, and wild stocks have been harvested for many years. (The "kelp forests" (see Figure 7.10) off the coast of California were first recognized for their potential during World War I when the supply of potash for farmers in the United States, which had been coming from Germany, was cut off. Potash was extracted from the burned kelp.) Methods have been developed to transplant *Macrocystis* to areas

Figure 7.10 A "forest" of *Macrocystis.* (From Nelson, Robinson, and Boolootian, 1970. *Fundamental Concepts in Biology,* John Wiley & Sons, Inc.)

where the native population has vanished because of warm waters, pollution, and/or grazing by sea urchins. Scientists have learned how to produce kelp embryos in the laboratory; the embryos can be used to seed new areas. *Hiziki* is also of interest to culturists because of its food value as well as the products manufactured from it.

SEAWEED PRODUCTS

Food

Seaweeds were probably first used by humans as food. This is still the fate of much of the macroalgae harvested today (see Table 7.1). Seaweeds are prepared in many different ways according to the traditions of the particular area. The composition of most dried seaweeds is predominately carbohydrate, with less protein and minerals (mostly sodium and potassium), and little lipid (see Table 7.2). Nori differs from most seaweeds because it is a relatively rich source of protein; generally, the distribution of amino acids is similar to that of vegetables, although it does contain rather high levels of arginine (like animal tissues) and little lysine. The lipids in seaweeds are sometimes rich in long polyunsaturated fatty acids. Many of the carbohydrates found in seaweeds are generally not digestible to any appreciable extent by humans, although people who consume seaweeds on a regular basis are better able to use these carbohydrates than are occasional consumers (perhaps because of the difference in the gut flora).

Seaweeds may also be a very good source of vitamins, although the absolute amounts contained may vary with the location of a species and with the season. Vitamins may be synthesized by the seaweeds themselves, or by bacteria living in the water or on the seaweed's surface. The vitamins that have been identified in edible seaweeds are A, B_1, B_2, B_{12}, C, D, E, folic acid, niacin, and pantothenic acid. Many of these are found in high levels;

TABLE 7.1   Partial List of Seaweeds Eaten by Humans*

Green Algae

Caulerpa (sea grapes, mouse plant)	Salads, spice
Chaetomorpha (kauat-kauat)	Dried, salads, sweetmeat
Codium (miru, sponge tang, fleece)	Garnish, salads, soups, tea
Enteromorpha (green laver, stone hair)	Dried, garnish, salads, soups
Monostroma (aonori)	Dried, salads, soups, soy sauce
Ulva (sea lettuce, aosa)	Dried, garnish, salads, soups

Brown Algae

Alaria (honey ware, wing kelp, tangle)	Salads
Analipus (sea fir needle)	Dried, salads
Dictyota	Garnish
Durvillea (bull kelp)	Dried
Ecklonia	Jelly
Eisenia (arame)	Salads, soup
Fucus (sea ware, bladderwrack)	Salads, tea
Hizikia (hiziki, hai tso)	Dried, salads, soup
Laminaria (kombu, hai dai, kelp, oarweed, sugar wrack, sweet tangle)	Curd, dried, salad, soup, sweetmeat
Macrocystis (giant kelp, sea ivy)	Garnish, salt substitute
Nereocystis (bull whip kelp)	Dried, garnish, pickled
Patina	Sweetmeat
Sargassum (mojaban, horsetail tangle)	Salads, soup
Scytosiphon (kayamonori, sugara)	Soup, substitute for wakame
Undaria (wakame, miyok)	Dried, garnish, salads, soups

Red Algae

Ahnfeltia (nibbles)	Garnish
Asparagopsis (limu kohu)	Garnish, salads
Bangia (cow hair)	Salads
Carpopeltis	Garnish
Chondrus (Irish moss, carrageen)	Bread, jelly, soup
Eucheuma (tosaka)	Garnish, salads
Gelidium (tengusa, genso)	Dried, garnish, salads, soup
Gloiopeltis (funori, chi tsai, gumweed)	Salads, soup
Gracilaria (ogo, hoi tsoi, gulaman)	Dried, garnish, pickled, salads

TABLE 7.1   (*Continued*)

Hypnea	Garnish
Iridaea	Fried
Laurencia (pepper dulse)	Garnish, salads
Nemalion (sea noodle, threadweed)	Dried, salads, soup
Palmaria (dulse, tellesk, darusu)	Bread, dried, garnish, salads
Porphyra (nori, purple laver)	Bread, dried, soup, various special products (cheese, wine, jam, etc.)
Rhodymenia (dulse)	Boiled, dried, salads
Suhria	Jelly

* Items listed as "salads" are sometimes also eaten as boiled vegetables.

for example, nori has about the same amount of vitamin A as spinach and more vitamin C than oranges. The trace metals in seaweeds can also be beneficial. Much of the impetus for *Laminaria* culture in China was due to its relatively high iodine content; iodine is particularly important in interior China, where goiter is a problem.

Besides their use as food for humans, many seaweeds have been used to feed animals, including cattle, sheep, pigs, goats, horses, and chickens. The seaweeds either are presented to the animals raw or dried, or are added as an ingredient to a feed mixture.

Drugs

Seaweeds are a part of the folk medicine of many cultures. Many of the seaweed treatments that have been used in the past were not actually curative, but rather acted to reduce the symptoms. However, new research has suggested that some biologically active substances from these plants may play a role in the cure of some diseases. For example, recent laboratory bioassay studies on mice and microbes have shown that various species of seaweeds in the genus *Dictyota* may have antifungal, antibacterial, and/or anti-inflammatory activities. Many other genera also include species whose

TABLE 7.2   Proximate Composition of Some Dried Seaweeds

Seaweed	% Protein	% Carbohydrate	% Lipid	% Ash
Enteromorpha	19.5	64.9	0.3	15.2
Hizikia	12.3	65.1	1.5	21.2
Monostroma	20.0	63.9	1.2	14.9
Porphyra	43.6	46.4	2.1	7.8
Ulva	26.1	51.2	0.7	22.6
Undaria	19.8	43.8	2.0	34.3

Source: Nisizawa et al., 1987. Copyright © Kluwer Academic Publishers. Reprinted by permission.

extracts display antimicrobial activities. These activities may vary with the location and time of collection.

Many seaweeds, especially the kelps and other brown algae, have been included in the diets of people who suffer from goiter (an enlargement of the thyroid gland due to a lack of iodine in their diet); in South America some seaweeds, probably **Phyllogigas,** are chewed for this reason and are locally known as the "goiter sticks." Other seaweeds are consumed because they are vermicides (i.e., they kill parasitic worms). The most widely used seaweed vermicide in Europe is the red algae **Alsidium,** the "Corsican worm moss." Other seaweeds used for this purpose are *Digenia, Hypnea, Sargassum,* and *Ulva.* The mucilage of **Cutleria** is used to treat stomach ulcers, and *Hypnea* has been used for other stomach troubles.

Extracts of seaweeds are also widely used. *Chondrus, Iridophycus,* and *Delesseria* extracts are blood anticoagulants, as is agar; agar has also been used to cure stomach disorders and as a laxative. Extracts of *Ascophyllum* are used to treat sprains and other muscle-related problems. Solutions of some *Laminaria* are used in China for relief from menstrual discomfort.

Fertilizer

Seaweeds are commonly used as fertilizer on coastal farms, where they are available. We have mentioned the use of *Macrocystis* as a source of potash. Other seaweeds, including *Ascophyllum, Fucus, Sargassum, Laminaria,* and *Ecklonia,* have been applied to agricultural fields; applications of the seaweeds may be wet, dry, composted, or extracted (see Figure 7.11).

There are several advantages to the use of seaweeds over other types of fertilizers:

1. Since some seaweeds contain antimicrobial compounds, they may inhibit the growth of certain bacterial

Figure 7.11 Eel grass and seaweeds that wash ashore can be used as fertilizer. (From Trainor, 1978.)

and fungal plant pathogens, and may even reduce insect damage.

2. They are as high in nitrogen, and three times as rich in potassium, as typical manures (although they may be low in phosphates, so there should be some supplemental fertilization). Excess salts in seaweeds, especially sodium, may be a problem in some cases since it can reduce plant germination and growth, as well as affect water infiltration rates and soil particle dispersion.

3. Seaweeds have a high organic content that can be converted to humus, thereby improving the physical properties of the soil.

4. Seaweeds are a good source of some trace metals, though excessive amounts of metals can be a problem.

5. Unlike animal manures, seaweeds contain no potential weed or pathogen contaminants that can damage the terrestrial crops.

6. Plant regulators, such as the growth hormones **gibberellin** and **cytokinin,** are found in seaweed, and may stimulate crop growth.

Colloids for Industrial Uses

The seaweeds, especially red algae, contain gelling and viscous substances, some of which have been extracted for centuries. There are three major macromolecular colloids that are of interest to industry today: agar, carrageenan, and algin. A fourth, **furcellaran,** which is similar to agar, is extracted from the red algae, *Furcellaria,* but is less important.

Agar is extracted, using hot water, from the cell walls of five principal red seaweeds: ***Acanthropeltis, Gelidiella, Gelidium, Gracilaria,*** and ***Pterocladia.*** It is a **polysaccharide** (a long chain of sugarlike molecules), but the extract's chemical structure will vary with the species and local strain of seaweed that is used. The extracted agar forms a gel that will vary in strength with the structure. The principal (and once exclusive) user of agar is the food industry. Because it forms rather temperature-stable gels, it has been used to thicken soups and jams; in Japan it is widely used in a jelly, **tokoroten,** that is sliced and eaten with soy sauce. Agar is also used in canned meat, clear pasta noodles, bakery toppings and fillings, and jellied candies.

Agar is also very important in the medical supply industry, where it is used in the laboratory culture of bacteria. Specific strains of *Gracilaria* and *Gelidium* are the principal source of the valuable bacteriological-grade agar. Other uses of agar are as clarifying agents in beers and wines and in gelatins on photographic films. In the past, agar was used in canned pet foods and in dental impression compounds.

Carrageenan is also extracted by hot water from red seaweeds, especially ***Chondrus, Eucheuma, Gigartina, Hypnea, Iridaea,*** and ***Phyllophora.*** The carrageenan molecules are polymers of the sugar D-galactose that vary in the degree and locations of sulfate esters (the average sulfate content is 20% to 30%). The differences in the chemistry of the molecules is again related to the species, strain, and location of the seaweed used in the extract. In addition, the chemistry may also change as the seaweed goes from one phase of its life cycle to the next. There are three main types of carrageenan, **λ-, κ-,** and **ι-.** The traditional method of producing the carrageenans is to dissolve them out of the seaweed and then to filter away the unwanted material, precipitate them from the solution, separate, then finally dry the carrageenan; this results in a fairly refined product, although the process is rather long and costly. A new method developed in the 1970s is much faster and simpler, but it yields a semirefined product. This semirefined carrageenan is made by dissolving away the noncarrageenan material, then drying and milling the remaining residue. This contains some cellulose, so it is not as clear as the refined carrageenan, but there are still many uses for this material.

Carrageenan was originally used in cooking and in the manufacturing of products like cough syrups. Active research in the use and extraction of carrageenan increased in the 1930s (with the result that it was found to be an excellent chocolate milk stabilizer) and continued steadily during World War II, when there was a shortage of agar. New uses were found, and harvests of carrageenan-containing seaweeds increased. By 1980, over 12,000 metric tons of dried carrageenan were being produced each year. Today, carrageenan has become an important part of the food industry. λ-carrageenan is nongelling, so it is used to thicken, suspend, or give "body" to products such as flavored milk and milk shakes, drink mixes, syrups, and sauces. κ- and ι-carrageenans will form gels that are thermoreversible; these are used to make jellies, jams, pet foods, frozen desserts, flan (an egglike custard), mousse and puddings, yogurt, and many other related foods.

Algin or **alginic acid,** unlike agar or carrageenan, is extracted from brown seaweeds. Although it is the major constituent of all the brown seaweeds, the principal seaweed genera used for commercial purposes are ***Ascophyllum, Durvillea, Ecklonia, Fucus, Laminaria, Macrocystis, Ne-***

reocystis, **Sargassum,** and **Turbinaria.** Algin is a polymer of either of two acids, D-mannuronate or L-guluronate, or it can be a mixture of both; salts of algin are commonly formed and are called **alginates** (unfortunately, algin is also sometimes called "alginate"). The parent algin, as well as the salts that are formed with divalent and trivalent metal ions (such as calcium), are not water soluble, but salts formed with alkali metals (such as sodium) can be dissolved in water. This is a fortunate situation, since mixtures of algins and their various salts can be designed to give substances different gelling properties and thus have a wide range of uses.

Because it is not water soluble, the extraction of algin is more difficult than isolating the other major seaweed colloids. Processing starts with the removal of water-soluble materials, followed by extraction with sodium carbonate, filtration, precipitation of the calcium salt with the addition of $CaCl_2$, conversion of the salt to the acid by treatment with hydrochloric acid, conversion of the acid to the sodium salt, and finally drying.

Perhaps the greatest consumer of algin and its salts is the cotton textile printing industry, where it is used as print paste thickener for dyestuffs. Algin also may be used as a surface coating agent for paper, in welding rods, and with plaster of Paris for dental impressions (it has largely replaced agar for this purpose).

Algin and alginates are used in the food industry (the second largest consumer after textile printing) in a variety of ways. It prevents the formation of ice crystals in ice cream (which would cause ice cream to lose its smooth texture), and it is a stabilizer for many dairy products (especially whipped cream, cream cheese, and processed cheese) and bakery products (such as glazing and some fillings). Propylene glycol alginate is a thickener and stabilizer (for sauces, gravies, salad dressings, and beer), and sodium alginate is found in quick-setting and room temperature-setting dessert puddings, as well as in fabricated foods, where it serves as a matrix. It is also used as a coating for frozen fish (see Figure 7.12).

SPIRULINA

Unlike the other microalgae, **Spirulina** has been included in this chapter because it is not primarily used as a feed for the culture of other organisms (see Appendix 5), but is directly consumed by humans. Exactly when this plant was first harvested is not clear, but Fray Toribo de Bonavente reported in 1524 that dry cakes, "tecuitlatl," were used as food by the Aztecs who gathered it with fine meshed nets from the surface of Lake Texcoco; the best evidence suggests that these cakes were composed of *S. maxima.* The Kanembu people along the shores of Lake Chad in West Africa eat a similar dried *Spirulina* cake called "dihe." Today this organism is commercially cultured in Japan, Israel, Taiwan, Mexico, Thailand, and the United States.

Figure 7.12 Products that contain seaweed extracts. The fruit pie contains agar, the chocolate milk has carrageenan, and the salad dressing contains an alginate.

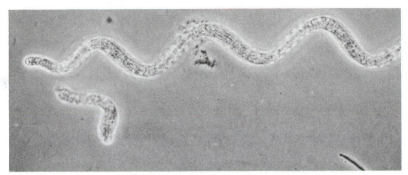

Figure 7.13 *Spirulina.* (Photo courtesy of The Oceanic Institute, Hawaii.)

Spirulina are cyanobacteria that grow as multicellular filaments that are helically coiled, hence its name (see Figure 7.13). The filaments, which are called **trichomes,** have a gliding motion. Many physiologically distinct strains have been collected from around the world and have been experimentally cultured as possible subjects for intensive production. Probably the most commonly cultured species are *S. platensis* and *S. maxima.* There are several reasons for the current excitement over *Spirulina*, including its productivity in culture (up to 25 metric tons of protein per hectare per year) and the value of high-quality, dried *Spirulina.*

The major market for *Spirulina* is the health food industry (see Figure 7.14). *Spirulina* can be consumed in the dried state or may be used as an ingredient in tablets and specially prepared foods. It has been marketed as a "miracle food"; although most scientists recognize that there is more poetry than truth to such claims, *Spirulina* should still be considered a good dietary supplement and may become more important as such in the future. There are several reasons why *Spirulina* is considered to have this potential when other species of microbes do not:

1. It contains high levels of protein (up to 70% of the dry weight).
2. Like some other cyanobacteria, it is a good source of β-carotene, a precursor of vitamin A (β-carotene *may* also be useful in preventing certain types of cancer).
3. In its raw form, it is rich in several vitamins, although this may not be true of all the processed products.
4. It has substantial levels of several polyunsaturated fatty acids, includ-

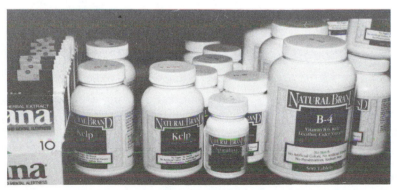

Figure 7.14 *Spirulina* tablets (along with kelp) have a market in the health food industry.

ing γ-linolenic acid, which is used in the synthesis of the prostaglandin PGE$_1$; dietary γ-linolenic acid *may* be of use in treating some cases of arthritis, heart disease, obesity, zinc deficiency, alcoholism, and mental disorders.

5. It may be high in minerals, including iodine, so thyroid hormone production is maintained.

6. The blue pigment, **phycocyanin,** may make up to 20% of the dry weight of the *Spirulina*; this pigment may be useful in stimulating the immune system in general, and there is some information that suggests that it *may* be useful in the treatment of cancer.

Figure 7.15 Culture of *Spirulina* in Israel. (Photo from Richmond, 1986. Courtesy of Dr. A. Richmond, Israel. Copyright © CRC Press, Inc., Florida. Reprinted by permission.)

Life Cycle and Ecology

When mature, trichomes develop specialized cells, **necridia,** along the length of the filament. The necridia lyse and the trichome is broken into short segments of two to four cells called the **hormogonia.** The hormogonia grow and develop into the coiled trichomes. Alternatively, a trichome may break mechanically, and the fragments will grow back into full trichomes.

Spirulina are found in soil, thermal springs, and waters that range in salinity from fresh to full seawater. *S. platensis* and *S. maxima* grow particularly well in alkaline lakes. In lakes that are dominated by *Spirulina*, the water generally has a pH of at least 10.0. Being a tropical organism, *Spirulina* is thermophilic; the optimal temperature for most strains is 36° ± 2°C.

Culture

Mass culture of *Spirulina* is done largely in raceways that are mixed with paddlewheels (see Figure 7.15). The raceways are shallow and turbulence is maintained so that light can be used more efficiently.

The most critical aspect of the culture during the summer (when there is maximum growth) is to maintain the proper density. This means that the *Spirulina* must be harvested at intervals so that the density is never allowed to become so great that it prevents the light from reaching the bottom of the raceway. Naturally, the *Spirulina* density cannot be reduced too much, since that would reduce the production to unacceptable levels. In the winter, temperature—not light—begins to limit the productivity of the raceways.

Spirulina is harvested by filtration. The relatively large size of the filaments, as well as the coiling that prevents the cells from becoming too tightly packed on the filter surface, make this method of collection the most practical (see Figure 7.16).

As in all mass algal culture, it is important to maintain a relatively monoalgal condition. Other species that become established in the pond not only rob the *Spirulina* of light and nutrients, but reduce the quality of the dried *Spirulina*. Maintaining a monoalgal condition in outdoor ponds is never easy, but in the case of this cyanobacteria it is less trouble than usual. Because *Spirulina* has such unique optimal growing conditions, most other species will not become established in the raceways. In addition to a high pH and temperature, it can survive in elevated concentrations of ammonia supplied by fertilization with urea; however, nitrates are taken up more easily than ammonia, so it may be advisable to use urea only when needed.

(a)

(b)

Figure 7.16 Harvesting of *Spirulina*. (*a*) Filtering apparatus showing culture water leaving the raceway, passing through the filter, and being returned to the raceway. (*b*) The filter from above. (Photos courtesy of The Oceanic Institute, Hawaii.)

SUMMARY

Seaweeds are macroalgae that fall into three major phyla: the chlorophytes (green algae), the phaeophytes (brown algae), and the rhodophytes (red algae). Many seaweeds look like higher plants but are physiologically quite different. They are benthic, found in shallow water, and often have life cycles that include an alternation of generations. During the culture operation, seaweeds are often induced, by environmental manipulation, to shed the spores (which are sometimes diploid and sometimes haploid) that will develop into the harvestable plant.

Some of the important seaweeds are (1) *Laminaria*, a kelp, grown on bay bottoms or (more commonly) using rafts or long lines; recently, forced cultivation, which adds several months to the culture period in temperate waters, has greatly increased production. (2) *Undaria*, "wakame," is grown from rafts or long lines. (3) *Porphyra*, "nori," used in sushi, is very popular. Nets are seeded with young seaweeds and then suspended on floats or strung between poles for growout. (4) *Eucheuma* is cultured on nets and ropes in coral reef areas. (5) *Gracilaria* is usually not cultured in open systems; rather it is grown in ponds. There are several other seaweeds that support smaller culture industries. The seaweeds are in some cases eaten directly, and sometimes extracts (agar, algin, and carrageenan) are used in food processing. Seaweeds are also used as medicines and fertilizers and for industrial purposes.

Spirulina is a cyanobacteria that is consumed directly as a vitamin–nutrient supplement. It is a spiral-shaped microorganism that is harvested from paddlewheel-driven raceways by filtration; very high production values have been recorded. The product is dried and eaten.

CHAPTER 8

COMMONLY CULTURED MOLLUSKS

Mollusks are among the most easily recognized of the aquatic invertebrates. They include such forms as the limpet, squid, clam, snail, and slug. Mollusks have a long history and are among the most successful of all the animals; there are over 80,000 living species. Their success is probably linked to the evolution of a hard shell that protects them from predators and harsh environmental conditions.

The mollusk's body is divided into three regions: **head, foot,** and **mantle.** The head is the anterior portion containing the sense organs and the mouth. The foot is muscular and contractile, and it is the primary organ of locomotion. The mantle is found on the dorsal side of the animal; it is a fold of skin that secretes the calcified shell.

There are six classes of mollusks:

1. The rare and primitive **monoplacophorans.**
2. The **cephalopods,** such as the cuttlefish, squid, and octopus, that support some modest-sized fisheries and are currently cultured for use in medical research.
3. The **amphineurans,** including the

chitons common in many intertidal ecosystems.

4. The **scaphopods,** known as the "tusk shells" or "tooth shells."
5. The **bivalves** or **pelecypods,** characterized by having two shells that fit together.
6. The **gastropods,** animals often with a twisting shells (like the snail or abalone), or a shell that is greatly reduced or absent (such as the sea hare, *Aplysia*).

Fertilization in the mollusks is usually external, with zygotes developing first into swimming **trochophore** larvae. The trochophore then becomes a **veliger,** an advanced swimming larval form that has a foot, shell, and other adult characteristics. The veligers of the bivalves are symmetrical, typically developing an oblong shape with one of the shell sides straightened (the **straight hinge** stage). Slowly, the shape of the shell becomes more like that of the adult. When the veliger's foot becomes functional, and settlement is imminent, the organism is referred to as a **pediveliger.** Most gastropod larvae pass through the trochophore stage before hatching and enter the water as free-swimming veligers with two cilia-bearing semicircular folds (**velum**) used not only for locomotion but also for feeding. Unlike the bivalves, the veligers of the gastropods are not symmetrical, having gone through a 180° twisting (**torsion**) in the later stages. Some gastropods and all cephalopods have eliminated the larval stages completely (this is termed **direct development**) and emerge into the environment as small juveniles.

Only gastropods and bivalves are extensively cultured. In this chapter we will discuss three groups of bivalves (mussels, oysters, and clams) and one group of gastropods (abalone). The culture of other gastropods and bivalves, as well as cephalopods, is discussed in Appendix 6.

MUSSELS

In the western hemisphere, the most commonly cultured species of **mussel** is *Mytilus edulis,* the blue mussel (see Figure 8.1). It grows wild on both sides of the Atlantic Ocean as well as the Pacific coast of the United States and Canada, usually in shallow water, which may be brackish or full salinity seawater. In the United

States, mussels were usually regarded as a pest until about World War II, but by 1986 the price of select-grade mussels had risen to $18 per bushel. Presently there is only a moderate amount of mussel culture in the United States, production being centered in Maine, but a much more extensive industry exists in Europe, where mussels are a more popular food.

Mussels are often found in large masses; the fact that this species grows well even under such crowded conditions has made it an ideal candidate for aquaculture. Like most other bivalves, *Mytilus* is a filter feeder. In experiments conducted with mussels taken from the Gulf of La Spezia (Italy), it was shown that the rate that water is filtered by *Mytilus* is related to the temperature of the water; at temperatures below 10°C or above 25°C there was minimal pumping, while between 15°C and 20°C the pumping was maximal. (Of course, this may vary between different genetic strains.) The rate that the water is pumped through the animal is also related to the amount of phytoplankton available. Generally, the filtration rate is inversely proportional to the concentration of algal cells in the water; at 5×10^5 algal cells/liter researchers found the mussels pumped at a rate of 3 liters/hour/mussel, while at a concentra-

Figure 8.1   A bed of the blue mussel, *Mytilus edulis.* (From Field, 1922.)

tion of 2×10^7 cells/liter the filtration rate was 0.5 liters/hour/mussel.

The life history of this species is well known. When environmental conditions are good, the mussel will reach sexual maturity at the end of one year. As in the case of most sessile invertebrates, fertilization is external; spawning takes place in the spring and summer as the water temperatures rise. After the egg is fertilized, it begins to develop into a trochophore larva, and then into a veliger as the shell starts to form. The swimming veliger feeds on phytoplankton and continues to grow for several weeks. When the veliger reaches a critical size, it begins to metamorphose to the benthic form when it comes into contact with a suitable substrate. Metamorphosis can be delayed if no adequate substrate is available. Settlement consists of two steps: (1) a crawling over the substrate and (2) attachment to the substrate with **byssal threads.**

Culture Methods

Several techniques have been developed for growing mussels. The choice of methods is related to the physical environment and the social traditions of the area.

Bottom Culture   The coasts of the Netherlands and Germany support **bottom culture** of mussels. Juvenile mussels are taken from natural beds and are transferred to leased beds in protected areas. In the Netherlands, the public seed beds are closed to fishing, except for the re-

moval of juvenile mussels during a short, well-defined period; the richest of these seed beds are in the Waddenzee. The seed-collecting industry is highly mechanized. Special boats are equipped with one or more dredges that can collect up to 15 metric tons of seed mussels in 1 hour; mussel farmers in the Netherlands are therefore able to compete with those farmers in other parts of Europe producing mussels in more traditional ways. In fact, the Netherlands is one of the two leading producers of mussels in Europe.

Bottom culture in Maine has replaced, to a large extent, hanging rope culture. Wild seed mussels are collected in dense masses from intertidal areas were growth is slow, and the mussels are separated and transferred to leased bottoms. Because they are no longer in thick aggregates, the animals are able to grow more quickly and are often harvested in one year. When the current is swift and there is enough food in the water, the annual production in these beds may be 12 metric tons per hectare.

Raft Culture   **Raft culture,** one of the most common forms of mussel aquaculture (see Figure 8.2), is used in the United States and in a number of other countries, and is practiced on a very large scale in the Glacician bays along the northern Atlantic Coast of Spain. Along with the Netherlands, Spain dominates mussels production in Europe. A method somewhat similar to the Glacician method is used in Venezuela to culture the rock mussel,

Figure 8.2   Mussel rafts in Venezuela. (From Jory and Iversen, 1985. Courtesy Dr. E. S. Iversen, University of Miami.)

Perna perna. A variation on the Spanish system can also be seen in the culture of *M. galloprovincialis*, which is grown very successfully above the muddy bottom of the Gulf of La Spezia, although the rafts are replaced by **vivai** (a network of poles that are linked together by ropes streched horizontally). This species is also grown on vivailike structures off the Istra coast of Yugoslavia.

The larvae required by the Spanish culturists are produced by the natural beds of mussels along the banks of the Glacician bays. There, larvae are found in the water throughout the warmest months of the year, but there are two peak periods of production in April and September.

Ropes for the collection of the larvae are suspended from rafts in the spring. The ropes are long and hang nearly to the bottom (if they touched the bottom, starfish could crawl up the rope and decimate the mussel population). At intervals along the rope are perpendicularly placed wooden pegs (**palos**), which prevent the mussel masses from sliding off under their own weight as they approach a harvestable size. Despite significant spawning in the fall, these larvae do not settle on the ropes as do the larvae produced earlier in the year, so spat must be collected from natural beds; these are placed along a rope and held there with cotton netting. The young mussels will soon send out new byssal threads and attach themselves to the rope. After a few months, the cotton netting rots and falls away from the rope. The young mussels that settle directly on the rope are eventually redistributed and transferred to other ropes to prevent overcrowding.

Originally, the Spanish rafts used to hold the ropes were old ships with attached wooden frames. Today, the rafts are specially designed for the purpose of growing mussels and are constructed of steel or concrete floats or some lightweight material like fiberglass. Above the floats are wooden frames that hold the ropes. Each rope requires only about $0.03 m^2$ of raft space, so each raft can hold hundreds of ropes. The rafts are held in place along the sides of the bays by large concrete moorings. A single typical raft will produce close to 23 metric tons of mussels!

As might be suspected, mussels will settle on certain surfaces in preference to others; this applies to different types of ropes as well as other substrates. In Spain, several different types of rope are used to collect spat. In experiments conducted in several parts of Puget Sound, Washington, mussels favored settlement on polypropylene rather than manila rope; after one year, a 3 m section of polypropylene rope yielded 36 kg, while manila rope produced only 6 kg of mussels. If a rope is used to collect the spat, the culturist should experiment with different types and sizes to see which rope will give the best results.

Bouchot Culture Mussel culture began in France. The **bouchot** system that is currently used, primarily along France's Atlantic coast, is similar to that country's early culture methods. Along the northern part of the coast the tidal flux is extreme, often greater than 10 m. The powerful tides help the growth of the mussels in culture (although their strength prevents the establishment of natural mussel beds, so collection of seed must be done along the southern coasts during the summers).

Traditionally, spat settled on collecting bouchots and were then transferred to growout bouchots. Since about 1960, this method has been largely replaced by the use of rope for collection. Loosely woven coco-fiber ropes are suspended in the water and the larvae alight in the depressions between the ropes' strands (see Figure 8.3). During collection of the spat, these ropes are stretched horizontally in two layers through the water on a series of poles connected to cross-beams. The ropes with the young mussels are transported to the growout area and wrapped

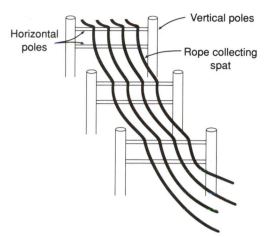

Figure 8.3 Collection of spat for bouchot culture. Coco-fiber ropes are suspended in the water. In this illustration, the ropes are shown only on the top horizontal cross-poles, but usually there are ropes hung across both the top and bottom cross-poles (two layers).

in a spiral fashion around oak poles (the bouchots) that are driven into the sediment in the intertidal zone (see Figure 8.4). The poles are arranged in rows; within a row, the poles are about 1 m apart and the rows are about 3 m apart. The absolute distribution of the poles is generally modified to some extent depending on the local tides and bottom topography. The ropes themselves gradually decay.

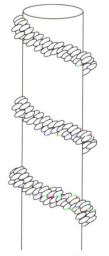

Figure 8.4 Bouchot with mussels on a rope wrapped around for growout.

The bottoms of the poles are covered with smooth plastic that prevents crabs and starfish from crawling up to the crop. Growth is so good that within a few months the outer layer of mussels must be removed to allow the inner layer to grow and to prevent those outer mussels from falling off the pole under their own weight. The mussels that are removed are placed in netting that can be strung between two poles or wrapped around another pole. The mussels are kept in culture for one to two years.

The continental shelf of northern France has a very gradual slope; in some areas the intertidal zone is over 1 km wide. Harvesting and tending to the bouchots is often done at low tide, when trucks, tractors, horses, and so on can be taken out over the exposed intertidal bottom. In areas in southern France where the tides are not as great, most of the work on the bouchots is conducted at low tides from small boats.

Philippine Culture In the Philippines, the green mussel or **tahong,** *Perna viridis* (sometimes called *Mytilus smaragdinus*), has been grown for several decades. The culture operation starts with the collection of small sessile mussels from natural beds. This is done by cutting the byssal threads and placing the mussels in a tray with either bamboo poles or oyster shells. The mussels will attach new threads to these substrates. If oyster shells are used, they are strung together and suspended from platforms into the water. If bamboo poles are used, they are driven into the bottom sediment in a manner similar to the bouchots of France. Some of the mussel farmers in the Philippines merely place bamboo stakes in Bacoor Bay, where the spat settle and are allowed to growout with no transplantation or thinning; the harvesting and setting of the stakes are done by divers hired by the culturists.

Long-Line Culture This method has become very popular in Sweden and has had some recent success in the United States. It is also the dominant method in China, which trails only Spain and the Netherlands in mussel production. While somewhat similar to the raft culture discussed earlier, there are no rafts to build and **long-line systems** sustain less damage than rafts in bad weather.

Rope is hung in the water both to collect the seed and for growout. The ropes hang vertically from another line that is held horizontally near the surface of the water by buoys. The buoyed lines are kept a few meters apart, while the hanging lines are about 0.5 m apart. The buoyed lines are held in place by large concrete anchors.

Genetic Variations

The study of mussel population genetics is particularly important because of the bottom culture practice of transferring seed from the intertidal zone of one area to deeper water in another area. As stated already, this is the most common culture method in Maine and northern Europe.

In the past, it was held that sessile invertebrates that broadcast their gametes into the water and had planktonic larvae were really large, genetically homogeneous populations distributed over a vast area; that is, the water currents worked to keep a uniform population spread out. Recent studies, however, have demonstrated that not only can there be different genetic populations along a large stretch of a coast, but there may be different populations within a single bay or even between mussels at different tidal levels!

Transplant experiments using several different *Mytilus* populations found along the Maritime Provinces of Canada have been conducted. In terms of both growth and survival, it was shown that some strains will do well at one location and poorly at another, while the reverse may be true for another genetic strain. It would be wise, therefore, for the culturist to "test grow" the seed mussels collected in one intertidal area in several deep areas to find the bottom location that gives the maximum yield.

Since mussels can be spawned in the laboratory, it may be possible to develop specific crosses that do well in particular environments. However, because the wild seed is so easy to collect, there has been little interest, so far, in developing sophisticated hatchery techniques.

OYSTERS

Oysters probably have been cultured longer than any other invertebrate, and they are still one of the most commonly grown saltwater organisms. A large portion of the oysters marketed today are cultured (for example, in Massachusetts about 90% of the oysters consumed are grown by culturists). The United States is the largest oyster producer (wild plus cultured); Korea and Japan each harvest nearly as much as the United States. These three countries account for about 70% of the worldwide crop.

In the United States, the primary cultured species are *Crassostrea virginica* on the East Coast and *C. gigas* on the West Coast. (*G. gigas* was originally imported by the United States from Japan and has recently been introduced into culture in Europe, Africa, South America, and the Indian Ocean.) *Ostrea* is also grown in the United States, but to a more limited extent. The exclusive use of these two genera in the United States is representative of affairs throughout the rest of the world. Essentially all oyster culture operations concern either *Ostrea* or *Crassostrea*; other genera, such as *Saccostrea*, which has been reared in Asia, have received only the most minimal attention.

Oysters are typically associated with brackish water rather than full salinity seawater. They are filter feeders and are found in temperate and tropical waters. Like mussels, oysters require a hard substrate for attachment under normal conditions. The requirements for water quality vary widely with the species being used, but in general it can be said that pollution of the coastal waters has reduced the production of oysters considerably over the past several decades.

The oyster, unlike most other animals, uses a carbohydrate, **glycogen,** rather than lipids for its primary energy storage. The stored glycogen is partially responsible for the oyster's unique taste, and is used by the oyster for gamete synthesis. Oysters should therefore be harvested before gametogenesis begins, when they are most flavorful. Stress resulting in the loss of stored glycogen means that reproduction will be hindered.

Spawning usually takes place while the waters are warming in the spring and summer, but even so, the temperature when release of the gametes occurs will vary with the species (for example, 13°–16°C for the Olympia oyster, *O. lurida,* and 30°–33°C for the Philippine slipper oyster, *C. eradelie*). In warm waters, oysters may spawn throughout the year; this is a disadvantage from the culturist's standpoint because it results in overcrowding in some cases, and because the oyster is putting its reserve energy into the production of gametes rather than into growth.

Oysters are able to change sex, starting as males and later changing to females as they grow larger. This adaptation is probably the result of the fact that sperm are small cells, and many can be produced by even small animals, but eggs are much larger and are therefore best made by large animals that can generate enough eggs in the ovaries to ensure a reasonable fertilization and survival rate. *Crassostrea* usually changes sex after the completion of the spawning season, but *Ostrea* may change during this period. The males of both genera release their sperm into the water; female *Crassostrea* also release their eggs into the water, but female *Ostrea* use their siphons to pump the sperm into the shell. The fertilized *Ostrea* eggs are retained near the gills in the **pallial cavity** where fertilization and early development take place.

As in the case of the mussels, oysters can be subjected to the culturist's genetic manipulation. Crosses between populations have in some cases yielded hybrids that appear to grow more quickly than pure strains. Oysters that grow quickly as larvae (that is, they settle to spat early) also have a rapid postlarval growth; thus, the early growth rate *may* be a simple trait to select for that will result in a greater yield at harvest. As in the case of most aquaculture species, much more work must be done in this area.

Hatchery Production

Traditionally, as with most broadcast spawning invertebrates, a crop is initiated with the collection of wild spat on various sorts of suitable substrates, such as shells or rope, that are placed in the water by the oyster farmer. However, there is a growing interest in oyster hatcheries that are able to produce a known amount of high-quality spat on demand.

Hatchery production starts with the collection of healthy brood animals; both large (females) and small (males) organisms are selected. If possible, the animals should be taken from known, fast-growing, genetically defined lines, although presently this is rarely done. Sexual maturation is encouraged by feeding the animals a concentrated algal diet and warming the water to the optimum temperature for gametogenesis (which, again, varies with the species). The release of the gametes takes place when the temperature of the water is rapidly raised and lowered or when oyster eggs or sperm are introduced into the water by the culturist. When the spawning begins, the oysters are placed in separate containers. At the conclusion of spawning, the eggs and sperm of the individuals of interest can then be mixed. If too few sperm are mixed with the eggs, the result is a poor fertilization rate, but if too many sperm are added, there may be several sperm entering each of the eggs (**polyspermy**), which will result in abnormal development. Recently it has been found that if the eggs are "aged" for about an hour in seawater before the sperm are introduced, the rate of polyspermy is greatly reduced.

When the larvae of the *Ostrea* are released by the female or when the *Crassostrea* larvae develop to the straight hinge stage, the larval culture begins (see Figure 8.5). The larvae are transferred to rearing tanks and are fed an algal suspension. The algae may be a mixture of local strains that are grown in nutrient-rich seawater or pure cultures of algae may be used. If local strains are used, the raw water is pumped into the facility to be filtered or otherwise treated to remove the larger species of phytoplankton (since the larvae will be able to ingest only relatively small cells). Some laboratories use local strains of algae, but supplement them with pure cultures. When pure cultures are employed, the genera of choice are often *Monochrysis* and *Isochrysis*, while *Tetraselmis* and *Dunaliella* are less commonly used; other algae have also been fed successfully. Cultured algae are more often used than are wild local strains. Researchers have recently shown that the algae may be partially replaced with artificial microencapsulated diets. While complete substitution of algae apparently results in suboptimal growth, these prepared diets may be used in the future in emergencies when algal cultures suddenly die ("crash").

Survival of the larvae to spat is probably at least partially related to the condi-

tion of the adults when they spawn. Conditioned oysters may be stimulated to release their gametes over a long period of time, but researchers have shown that the best survival of the larvae occurs when the spawning coincides with that period during conditioning when in the female the ratio (number of ova : number of ova plus oocytes) is maximal. This was demonstrated for *C. gigas* to be neither during the first month of the conditioning period nor after four months of conditioning.

Considerable attention has been given to the feeding of juvenile oysters. It is important to remember that the nutritional value of algae varies with the age of the algal culture and composition of the growth medium (see Appendix 5). Factors that may be related to the value of algae include

1. The absolute amounts of macronutrients (calories, total carbohydrates, total proteins, and total lipids)
2. The ratios of the macronutrients (for example, the carbohydrate : protein ratio)
3. The unsaturated fatty acids
4. Micronutrients (such as vitamins)
5. Minerals
6. The mollusk's ability to digest the algal cell
7. Cell toxins

There is no general agreement on the best species of algae or algal medium to use for the culture of young oysters. (The interested student can consult this chapter's Suggested Readings for several specific references on this problem.)

Given the proper temperatures and enough algae, the larvae will develop into eyed pediveligers, at which point they are ready to metamorphose into their benthic form, that is, to **set,** when they are transferred to a setting tank. Just before settlement, each animal's foot becomes prominent, allowing the oyster larvae to crawl

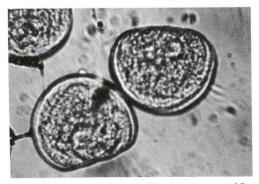

Figure 8.5 Larval oysters. (Photo courtesy of Dr. Susan Ford, Rutgers University.)

over the substrate while trying to find a suitable surface. There are two general methods for handling the set, depending largely on the growout environment that will be used: they will be allowed to settle on **cultch** material, or they will be **cultchless.**

The universally-used method until the early 1960s calls for all hatchery oysters to be raised on cultch, such as oyster shell or stones. The cultch materials are placed in bags in the setting tanks or are spread out over the bottom of the tanks. After settlement, the oysters are retained in the setting tank and are allowed to grow to a suitable size before being moved to the final growout location. Oysters cultured using cultch can be grown out on a variety of bottoms, including hard mud, or can be suspended in the water.

Often the oyster pediveligers that set on the cultch settle unevenly, and the resultant dense clusters lead to high mortalities. To prevent this sort of wastefulness, a so-called "cultchless method" has been developed. The oysters set on some material from which they may be easily dislodged (for example, mylar sheets) or on tiny shell fragments. It has also been demonstrated experimentally that the larvae, when exposed to the neurotransmitters epinephrine and norepinephrine, will sink and metamorphose to spat without going through the typical settlement behavior or attaching to a substrate. The cultchless oysters are kept in some sort of protective raceway or trough until they reach about 2.5 cm and are ready to be transferred to the growout environment. If these oysters are to be grown out on natural bottoms, then the surface must be covered with shell fragments or some other sort of hard surface.

The integration of a hatchery into a farming operation assures the oyster culturist that seed will always be available. However, seed production is somewhat costly and requires skilled personnel. It usually takes the farmer several years before the hatchery begins to "pay for itself"

since the oysters planted will not be immediately harvestable. A quicker return on the hatchery investment can be realized if other culturists are willing to purchase the excess seed.

Growout Procedures in the United States

Crassostrea virginica, the **American oyster,** is found along almost the entire East Coast, in both subtidal and intertidal estuarine communities. The most widespread of the culture methods starts with the distribution of *clean* shells or some other cultch over the bottom in an area that is known to produce oyster larvae (see Figure 8.6). Cultch shells may also be hung on lines or in wide-mesh bags (see Figure 8.7). Because *C. virginica* requires a clean substrate, the cultch is planted when the numbers of late-stage oyster larvae in plankton samples peak (rather than earlier, which would allow the cultch to be covered by biota before the maximum settlement period began). The oysters may be allowed to grow where they have settled, or the **seed oysters** may be mechanically transferred with the cultch to another area for growout. The growout period is generally two to five years. There are other alternatives:

1. Juvenile oysters may be purchased and planted for a single growing

Figure 8.6 Oyster bed. Veligers will settle on the clean shells that have recently been spread on the bottom. (Photo courtesy of Dr. Joseph Dobarro, New Jersey Bureau of Shellfisheries.)

Figure 8.7 Oyster spat may be collected by hanging wide-mesh bags of oyster shells in the water when the veligers begin to appear in plankton samples.

lems include the securing of water rights, the expense of building and maintaining the rafts, and even the high growth rate! (In this case, high growth rate means thin shells that are easily chipped; this, in turn, means leakage of the internal fluids and a high mortality during and after the harvest.) Only in Maine has off-bottom culture of this species been shown to be commercially viable.

Culturists of *C. virginica* have a great deal of experience with oyster predators (crabs, oyster drill snails, starfish, worms, and some fishes) and disease. However, nothing—except the pollution of the coastal ocean—has done as much damage to the industry as the protozoan *Haplosporidium* (formerly *Minchinia*) *nelsoni*, which causes MSX disease. MSX has all but crippled the oyster industry in the central eastern states, which have been the most productive areas for oyster harvesting. In some cases, 90% to 95% of a harvestable population may be lost to this parasite.

The **Pacific** or **Japanese oyster, *C. gigas,*** is the most commonly grown species on the West Coast. Culture starts with the collection of the spat during the summer on suspended oyster shells hung from rafts or racks, or on cultch material suspended in a setting tank and exposed to larvae that were either reared in a local hatchery or imported from Japan. Wild populations of this species have only been established recently in the United States, so the technique of collecting spat from suspended rafts and racks is relatively new.

Culturists of *C. gigas* may use either a bottom growout technique or off-bottom culture methods:

season. The risks are minimal and a return on the investment may be seen in a short period of time, but there is no control over the availability and the price of the young oysters.

2. Cultchless seed oysters can be reared on trays for a year to increase their survival rate and then placed in growout. This may be limited by the availability of a good location for the trays. A return on the investment is usually seen in about three years.

3. Seeded cultches may be purchased from a hatchery and spread over the bottom. Again, the return on the investment will take about three years.

4. Suspended cultures can be used, but this generally has not been profitable in the United States. Its prob-

1. During bottom culture, after an initial growth period in the water column, the cultch is moved to the upper intertidal areas in the spring, where it is planted in beds and left

for about two years. At the end of this period, the clusters of oysters may be broken up and placed in lower intertidal waters for fattening for one summer season.

2. Off-bottom culture employs stakes, or racks on posts, driven into shallow subtidal or lower intertidal bottoms, or moored rafts holding strings of spatted cultch. Suspended cultures allow for increased growth and reduced predation.

Ostrea edulis, called the **European oyster,** was successfully introduced into the United States in 1949. This species is now an important shellfish in Maine, although larvae are still strictly hatchery products. Because of this animal's success in Maine, there has been some interest in its production in Canada's maritime provinces. The seed oysters initially are kept in trays and then may be transferred to nets held by rafts or long lines. There is no bottom culture of this oyster in Maine.

The **Olympia oyster,** *O. lurida,* has been cultured in Washington and Oregon, where it supports a rather small industry. Both the growout and the collection of spat take place in shallow impounded intertidal areas; the impoundments are designed to reduce the stress on the oysters brought on by quick changes in depth (and therefore also in temperature) as the tides ebb and flood. Growout is done using bottom culture techniques, while spat are collected on cultches of suspended shells.

Growout Procedures in Japan and Korea

Oysters are highly prized in Japan and Korea, especially in the winter, when they are eaten boiled, fried, or raw. Although several species are being grown, the most commonly cultured oyster is *C. gigas.* Traditionally, oyster spat were collected on bamboo sticks driven into the bottom. The seed oysters would settle on the sticks

where they were either grownout and harvested or were collected at a suitable size and used in bottom culture. By the mid-1930s, the new technique of suspended culture began to significantly increase the production of oysters in Japan, not only because the growth per unit area was better, but also because places that could never have been used for the culture of oysters before, due to the characteristics of the bottom, became productive. Beside the introduction of suspended culture to Japan, procedures for producing seed oysters in hatcheries were developed at about the same time. Oyster production in Japan in 1960 was about 400% of production just 10 years before. The rate of growth of the oyster industry in Japan has slowed recently (the 1985 catch was about 125% of the 1975 catch), while the Korean industry's growth has increased (during the same period their catch increased by 166%), so that Korea now has a larger oyster harvest than does Japan.

Wild spat are still often collected. Spat collection takes place when the water temperature exceeds 23°C, using cultches of shells (either oyster or scallop) that are strung along a wire. These wires are hung from frames built of bamboo and pine. The seed oysters thus collected may then be cultured for the one-year market in the winter and spring; this requires that the seed be transferred to productive grounds for quick growth. Oysters may also be "hardened" for the larger two-year market. Hardening is a process in which the young seed are placed on wooden racks that are subjected to intertidal conditions (frequent air exposure and breaking waves). The hardened oysters grow slowly during this process, but the following summer they grow rapidly when transferred to growout grounds.

Rafts and long lines are the most commonly used devices for the growout of the oysters. (Of these two, the raft method accounts for more production in Japan, but long lines are used in areas where the

Pacific waters are deep and rough along the northern coast; long lines are used more often than rafts in Korea.) Both rafts and long lines, naturally, are anchored to the bottom to prevent drifting. Long lines are supported by floats (in the past, tar-painted barrels were used, but these have been almost completely replaced by hollow plastic buoys). Rafts are made of Japanese cedar or of bamboo supported by floats. Wires or ropes holding cultches of oysters are hung from the rafts; these wires and ropes may be up to 9 m long. The oyster cultches are separated by pieces of bamboo or plastic pipes placed along the line. Wire and nylon baskets are also used to hold oysters suspended from rafts.

Oyster Culture Elsewhere

In Europe, the oyster market is dominated by two species: *C. angulata*, which is grown in France, Spain, and Portugal, and *O. edulis*, which is cultured in France, Spain, the Netherlands, and the United Kingdom. The largest producer of oysters is France, where *O. edulis* is clearly the more popular species. The spat of *O. edulis* are collected on ceramic tiles coated with lime that are set in the rivers leading to the Gulf of Morbihan. The tiles are placed in the water in the summer and autumn; in the winter, the tiles are collected and the oysters are removed and planted on bottoms along the northern coast, usually in beds in the very lowest intertidal area. The beds that are used for this culture are called **parcs.** After a year in the parc, an oyster will be transferred to deeper water beds where they are allowed to grow for another two years and are then brought back into shallow ponds, called **claires,** for another year, during which time they gain a considerable amount of weight on a rich diet of diatoms.

In Taiwan, the major cultured species is *C. gigas.* As in Japan, the traditional culture method consists of simply driving bamboo poles into the sediment and collecting the spat; the poles are generally not moved for growout. Recently, raft culture has become popular in deeper waters, while a method best described as "line culture" is used in shallow estuaries. In preparation for line culture, bamboo stakes are set in the bottom, and plastic line is stretched between them. Oyster shells are placed along the line every few centimeters. The spat settle and growout on the line of shells. In raft culture, the oysters are always submerged, but the oysters cultured using the line method are exposed to air during low tide.

In China, *C. plicatula* and *C. rivularis* are the most commonly cultivated species. Spat of *C. plicatula* are collected during two peak periods, May and September, on bamboo stakes along the northern coast and on "stone bridges" along the southern coast. Estuaries are the site of *C. rivularis* bottom culture, where the spat are collected during the summer on gravel, shells, and cement plates.

Crassostrea madrasensis are collected on roofing tiles or shells and are grown out in the area of Tuticorin Bay in India. The oysters are reared using a rack-and-tray method; the racks are a series of teak poles forming a framework that supports trays for the last phase of the growout, and on which are hung oyster cages holding the young oysters just removed from the spat collectors.

In the Philippines, the production of oysters is dominated by large *C. iredalei* and moderate-sized *C. malabonensis*. The warm waters of the Philippines help the growth rate, and in many cases crops may reach a harvestable size in under a year. They are harvested throughout the year, but the ideal harvest times (when the glycogen levels are highest and the meat is the sweetest) are just before spawning or three to four months after spawning. There are four principal ways to farm oysters in the Philippines.

1. The **broadcast method** is simply the distribution of shells, logs, rocks,

and tin cans on a hard bottom. The oysters settle on these and are eventually harvested, or may be transplanted once before harvest.

2. **Stakes ("tulos")** are used in soft bottom environments. The stakes are usually made of bamboo and are driven into the sediment so that the upper part of the stake is above the water level at high tide. The surface area of the stake is sometimes increased by attaching pieces of wood horizontally between the vertical stakes or by attaching cultches of shells.

3. Oyster and coconut shells are used as spat collectors in the **hanging methods.** There are considerable variations on this theme, but it is basically similar to the Japanese technique.

4. The **lattice method** employs a frame made of bamboo splits that are positioned in the water; the frame is stuck in the bottom, supported by posts, or attached in any number of other ways depending on the local conditions.

The Sydney rock oyster, *C. commercialis*, is cultured in Australia. The spat are collected during the fall on tarred wooden frames that are piled one on top of another. The frames are left in the water until the end of the winter season, when they are transferred to growing areas. Then the frames are separated, so they are no longer in piles, and are placed on racks for the growout phase, which will last one to two years.

In South and Central America and the Caribbean, the oyster *C. rhizophorae* (so named because it is associated with the branches and roots of the mangrove trees, *Rhizophora*) (see Figure 8.8), has long been harvested. In Cuba, palm posts are driven into estuary bottoms and connected with pieces of wood that are horizontally nailed between the posts. From the transverse wooden beams are hung branches of the red mangrove that collect the spat; the oysters are harvested from the branches in five to six months (see Figure 8.9). In Jamaica, the oysters are collected and grown on pieces of old car tires. During growout, the tire pieces are strung together on a line; the tire fragments on

Figure 8.8 Oysters being harvested from mangrove trees in Peru. (From Coker, 1908.)

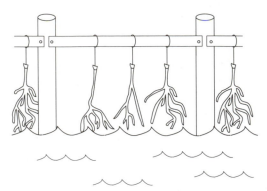

Figure 8.9 Cuban culture of *Crassostrea rhizo-phoroe.* Mangrove branches are hung from cross-beams between palm posts.

the line are separated from each other by 10 cm pieces of bamboo pole. These lines of bamboo and tire are hung from bamboo rafts held afloat by oil drums. The oysters are harvested after about six months.

Crassostrea corteziensis is the principal cultured oyster on Mexico's Pacific coast. The spat are collected using old oyster shells, roof tiles, and wire-mesh or plastic bags. These are initially held in a protected area for several months and then transferred for final growout. Both bottom culture and suspended culture are used.

Saccostrea echinata, the spiny oyster, has been grown experimentally in India and Indonesia. In Indonesia it was found that setting was most frequently seen on asbestos cement and coconut shell, less so on mollusk shells, and not at all on bamboo. Collectors were strung in the water, and a reasonable production was generated using this type of suspended culture. There is a high demand for this oyster in Indonesia.

CLAMS

Clams are bivalve mollusks with a relatively regular ovoid shape (compared to oysters). They are filter feeders, usually marine, and often burrow in the sediment with the aid of their foot. They are broadcast spawners and go through the typical larval trochophore and veliger stages like the mussel and oyster (see Figure 8.10). Recently metamorphosed juvenile clams are found on the surface of the sediment, but will soon begin to dig into the bottom.

The culture techniques used for clams of different species are actually similar in most cases; differences in rearing methods consist of variations in the length

Figure 8.10 Larval clams. (Photo courtesy of Mr. R. LeRoy Cresswell, Harbor Branch Oceanographic Institution.)

of time the animal spends in any particular phase of the culture operation or the temperatures used. There are several potential candidate species for large-scale commercial production; however, clam aquaculture has never gained the general popularity of many other forms of aquaculture, despite this organism's long history as a laboratory animal. Presently, the **northern hard clam** or **quahog,** *Mercenaria mercenaria,* is the only species in the United States whose culture has made a significant economic impact. *Mercenaria* is marketed according to the size of the animal: hard clams of about 5–6 cm across are called **littlenecks** (as is another clam species, *Protothaca staminea,* which makes the use of this term a bit confusing); those of 7–8 cm are called **cherrystones**; larger clams are called **chowder clams.** Commercial fishermen usually will select only one particular size to harvest. Culture of other species of clams is discussed in Appendix 6.

Spawning and Larval and Juvenile Culture of the Quahog

Mercenaria mercenaria is conditioned to spawn by feeding brood stock animals a high-quality algal diet and controlling the water temperature. These animals normally spawn in the summer, but can be conditioned to spawn at other times by slowly raising the temperature of the water to 18°–20°C; likewise, they may be prevented from spawning in the summer by placement in cool water just before the gonads become completely ripe. Spawning in the hatchery takes place on a **spawning table,** which is a shallow trough that receives cold and warm seawater (see Figure 8.11). (If a spawning table is not available, the clams simply may be placed in dishes of saltwater, and the dishes placed in cold or warm tap water.) The actual release of the gametes is initiated by suddenly increasing the temperature to 28°–30°C, then lowering it again to about 22°C; this heating–cooling cycle is repeated every 30 minutes until spawning

Figure 8.11 A spawning table. (Photo courtesy of Mr. R. LeRoy Cresswell, Harbor Branch Oceanographic Institution.)

begins. If raising the temperature does not result in a spawn, the addition of phytoplankton and sperm to the water will probably produce the desired results. Alternatively, the neurotransmitter serotonin may be injected into the gonads or adductor muscles of ripe clams to induce spawning.

The eggs and sperm are mixed, and the zygotes are transferred to warmed larval rearing tanks where they develop into the veligers. The larvae are generally reared on a diet of cultured phytoplankton (commonly used for this purpose are *Isochrysis, Monochrysis, Pavlova, Tetraselmis,* and *Dunaliella*). The length of the larval period, usually lasting one to two weeks, is a function of temperature and the amount and quality of the phytoplankton. The water in the larval rearing tanks is changed regularly, during which time the sides of the culture vessel are cleaned and the larvae sorted by size.

The water used in larval culture should be of a high quality, having been filtered and otherwise treated to remove bacteria and other potential pathogens.

The newly metamorphosed benthic clams are often held in the larval containers or some other vessel without flowing seawater; the water, however, is changed often, and the clams, fed a rich phytoplankton diet, grow quickly.

After a few weeks, the clams are moved to a nursery facility. At this stage, the clams no longer need the constant care they have been receiving, though they are not yet large enough to be transferred to the field for growout. There are two basic designs for nurseries. The nurseries may be **land based,** consisting of raceways, troughs, or upflow tanks (which are very popular in Europe). These systems generally use freshly pumped seawater containing sufficient phytoplankton and oxygen for good growth. Recently, extremely good production data have been generated using experimental upflow systems; water flow rate seems to be positively correlated with seed growth and total production, which in these experimental situations has reached 62 kg of 7 mm seed clams/m^2 over a three-month period.

Nurseries may also be **constructed in bays and other calm water environments,** which are designed so that the clams are protected from predators, excessive wave action, and sediment, but are exposed to ambient conditions.

The best size for transfer to the field from the nurseries so that a maximum profit can be realized is somewhat ill defined, with suggestions generally being 7 ± 3 mm (see Figure 8.12).

Growout of the Quahog

The transfer of seed to the field is the last step before harvest. There is considerable variation in the growout method. The site selected for growout should have a modestly strong current to bring in O_2 and take out wastes, and to provide a supply of food that can support the crop's growth. (The culturist may wish to see how the local *Mercenaria* populations are faring before choosing a site. Thick shells are an indication of poor growth. Shell length, not thickness, is an indication of the amount of edible tissue. Therefore, thick shells suggest that the animals have been in the area a long time but the food supply has not been very good since the

Figure 8.12 Seed clams that can be put into growout. (Photo courtesy of Mr. R. LeRoy Cresswell, Harbor Branch Oceanographic Institution.)

growth of the animal has not kept pace with the thickening of the shell.) Once a site has been selected, the seed is usually simply spread on the bottom of the leased area.

The planting area should have a firm mud–sand bottom. Densities of about 500 seed clams/m^2 are desirable for the production of 5–6 cm clams; higher densities result in reduced growth as the clams near harvestable size. The less stressful the environment, the better the growth and health of the mollusk; areas that are intertidal, subject to large fluctuations in temperature or salinity, or that see substantial wave action (which buries clams and damages protective devices put out by the culturist) should be avoided. The water should flush regularly and support enough phytoplankton to sustain growth, and should also be shallow enough to allow the farmer to work the bed easily.

An increased yield often results if the clams are protected from predators (crabs, gastropods, starfish, and rays) and vandals. Baffles, cages, stones spread around the beds, and fences are used to keep out predators. It has been pointed out that as clams increase in size and depth of burial, their predators change, and perhaps the devices used to protect the clam bed should also change as the crop ages. Another consideration may be the location of the predators; if the clams can be grown where there are few predators, then the net production will be greater. Some research in Georgia has shown that if the clams are situated in sandy creek beds that flow into a bay or estuary, they are subjected to reduced predation but still grow rapidly.

Off-bottom culture has also been used for growout. Rafts are constructed that support trays containing sand, or the trays are mounted on posts driven into the bottom. The trays are filled with the seed clams. The culturist should avoid overcrowding the clams, as this results in stunted growth. Better growth and economic return is normally seen with this sort of off-bottom technique, but the capital costs are greater.

Perhaps the biggest problem with the production of clams is the extended growout period. Much consideration has been given to ways of decreasing the amount of time that a clam is in culture, including

1. Induction of polyploidy
2. Hybridizing *M. mercenaria* with the southern hard clam, *M. campechiensis*
3. Selection of quickly growing parent stock

The southern hard clam grows more quickly than does *M. mercenaria* in moderate or warm temperatures, but it keeps poorly out of water even when chilled. It has been suggested that in cooler waters, rapidly growing *M. mercenaria* should be selected to develop a stock, but in semitropical and tropical waters, a hybrid, with at least two-thirds of its pedigree being *M. campechiensis*, would be better.

ABALONE

The abalone is a gastropod, although to the nonbiologist the shell of the animal looks like it might be half of a bivalve shell. Careful examination of the shell, however, reveals that there is a good deal of torsion even though the shell is relatively flat. The shell also has a row of openings that are used by the animal for respiration, excretion, and the release of gametes. Abalone are herbivores, grazing on hard surfaces for algae. The characteristic colors of the outer shell are derived from the algal pigments; the inner shell has a pearly shine.

Many species of abalone are cultured, all belonging to the genus *Haliotis* (see Table 8.1). The muscular foot of the abalone is one of the most valuable of seafood products; a large foot muscle may be cut

TABLE 8.1   Commercially, Experimentally, or Potentially
Culturable *Haliotis*

Species	Range or Area Where Cultured
H. assimilis (threaded abalone)	United States, Mexico
H. australis (silver paua)	New Zealand
H. corrugata (pink abalone)	United States, Mexico
H. cracherodii (black abalone)	United States, Mexico
H. discus (kuro awabi, or black abalone)	Japan, Korea
H. discus hannai (ezo awabi)	Japan, Korea
H. diversicolor supertexta (small abalone)	Taiwan, Japan
H. fulgens (green abalone)	United States, Mexico
H. gigantea (madaka)	Japan, Korea
H. iris (black paua)	New Zealand
H. kamtschatkana (pinto abalone)	United States
H. laevigata (green-lip abalone)	Australia
H. midae (perlemoen)	South Africa
H. roei (Roe's abalone)	Australia
H. ruber (black-lip abalone)	Australia
H. rufescens (red abalone)	United States, Mexico, Chile
H. sieboldii (megai)	Japan, Korea
H. sorenseni (white abalone)	United States, Mexico
H. tuberculata (ormer)	France, Ireland
H. virginia (white-footed paua)	New Zealand
H. walallensis (flat abalone)	Canada, United States, Mexico

up into several "steaks." Its popularity has resulted in overharvesting and a decline in the size of natural stocks. Removal of an abalone from the rock it is grazing on is normally a simple procedure if the animal is taken quickly, but if the abalone is alarmed, it will contract the foot muscle and hold tightly to the surface, making removal by hand impossible.

One feature making abalone culture attractive to prospective farmers is that these animals usually have no significant disease or parasite problems, although infections of the bacteria, *Vibrio*, and pro-
tozoan and trematode parasites have been reported. Wild abalone populations are limited by available grazing space and predators. Newly settled abalone are preyed upon by a host of crustaceans and worms. Juvenile and adult abalone are eaten by octopuses, sea otters, crabs, various other gastropods, and starfish.

The sexes are separate in the abalone. For most species, breeding takes place during the warmest months of the year, the exact time being a function of the species and the location of the population within the range. Animals in a single population all spawn at the same time, probably as a result of the release of a pheromone. The fertilized eggs sink from the water and begin development. The eggs hatch in about a day and release planktonic trochophore larvae. In another day the trochophores develop into veligers, a stage that normally lasts for one to three weeks. The abalone larvae are unusual in that they do not feed during their planktonic life; rather, they carry with them reserves from the eggs that will support them until they settle and begin to feed.

The settlement of abalone larvae has been studied in recent years by Daniel Morse and coworkers at the University of California at Santa Barbara. For many species of abalone, settlement and metamorphosis is triggered by contact with certain red algae. The evolutionary advantage here is obvious—the algae is a good source of food for the young abalone, so this triggering reaction ensures a suitable source of nutrients and energy at a time when the animal is less able to search effectively for food (it moves slower than the adult does, and does not have reserve energy to travel far looking for food). The red algae that cause the settlement are also a good source of camouflage because the plant's pigments are taken up by the abalone. Settlement can also be induced by **GABA (γ-aminobutyric acid)**, an amino acid and neutrotransmitter. Settlement has also been reported in the presence of benthic diatoms (including *Navicula*, *Cocconeis*, *Amphora*, and *Nitzschia*), some cyanobacteria, and the slime produced by juvenile and adult abalone. The factor(s) from the cyanobacteria and the red algae is not GABA; rather it is an unknown molecule(s) that weighs 640 to 1250 daltons (a **dalton** is a unit of weight very close to the weight of a proton).

Control of Reproduction

Once they have reached sexual maturity, brood animals are selected from wild or cultured populations. The sex of the animals can be determined by the color of the gonad, which is examined by pulling the foot away from the right side of the shell. Ovaries are brown, gray, dark green, or violet, while the testes are white, yellow, pink, or very lightly tinged brown or green.

The gravid animals are held in temperature-controlled tanks with flowing seawater and fresh macroalgae for grazing. The abalone are judged to be ready for induced spawning by the amount of swelling of the gonads. Two methods can be used to induce the release of the gametes. Other methods have been suggested, but are generally held to be inferior—including dissection, stripping the gametes by applying finger pressure, and thermal shock. The two methods of choice are

1. The abalone are exposed to flowing heated seawater that has been irradiated with UV energy, and spawning takes place in 3–16 hours.
2. The abalone are placed in a container of seawater with the pH adjusted to 9.1 and containing a 5 mM concentration of hydrogen peroxide (H_2O_2); after about $2\frac{1}{2}$ hours, they are transferred back into clean seawater where the gametes are released within an hour.

These two methods are probably more similar than it would appear at first. It is thought that UV energy causes the conversion of some water molecules to the hydroperoxy free radical (HO_2^-) and the peroxy diradical (O_2^{-2}), which are also products of H_2O_2 in water. The equilibrium between the radicals is pH dependent, which is why an increase in the pH to 9.1 increases the efficiency of this method. These radical oxidizing agents are thought to stimulate enzymes (by acting as donors of charged oxygens) responsible for the synthesis of key prostaglandins that control spawning.

Larval and Juvenile Development

As stated, the larvae of abalone do not eat, so they are free of many of the problems associated with larvae of other animals. The larval rearing vessel may be any shape but should contain free flowing water kept at the proper temperature for the particular species; the water must be filtered to remove bacteria that may grow on the veligers. If flowing water is not available, the water in the container must be changed periodically; this water should be disinfected, using UV treatment. Antibiotics are also sometimes used in these static cultures.

When the veligers approach metamorphosis, as determined by a microscopic examination of the foot and **cephalic tentacles,** they are transferred to settling tanks. The larvae are presented with substrate plates on which a film of diatoms and sometimes red algae have developed. The plates may also have had juvenile abalone on them that were removed just before the substrates are presented to the larvae; the juveniles leave a slime that induces settlement.

The young abalone (see Figure 8.13) are switched from a diet of diatoms and other microalgae to seaweeds when they reach about 0.5 cm; at this time they are transferred to tanks for the first step in the growout process. Abalone are usually presented with brown algae, such as *Mac-*

rocystis, Egeria, and *Laminaria*, but may take green algae like *Ulva* or red algae such as *Gracilaria* and *Gelidium*. Artificial feeds are sometimes used in Japan. As the animals grow, they become more sensitive to light, so growout tanks may include habitats that not only increase the growing area in the tank, but also shelter the abalone (see Figure 8.14).

Final Growout

Abalone, once they reach about 3 cm, can be used to reseed an area where the natural population has declined. Temporary transplantation cages or habitats aid in the transition to the natural environment. Predators such as starfish and octopuses may also be removed from the area before reseeding. If sufficient algae is available, the abalone may be harvested as adults after three to five years. Recapture rates of adults have been reported to be 0.5% to 10% in Japan. In California, to offset some of the costs of this sort of **ocean ranching,** culturists have experimented with using much smaller seed abalone; although the percentage recapture is much smaller, so are the costs, and this may turn out to be a more effective way to repopulate parts of the California coast.

Abalone have also been reared in raceways, round tanks, and suspended cages and baskets. In Taiwan, 1.5 cm juveniles are stocked in intertidal ponds and are harvested (with an annual production of 4.0 kg/m^2) with a 70% survival rate.

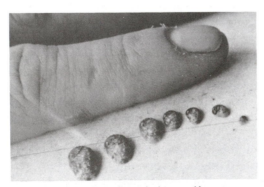

Figure 8.13 Juvenile abalone. (Anonymous, 1986. Copyright © Archill River Corporation. Reprinted by permission.)

DEPURATION AND PUBLIC HEALTH

Bivalve mollusks are filter feeders, so their tissue is a reflection of the condition of the water. For this reason, there has long been a concern about the link between public health and bivalve shellfish. As pollution has become an increasing problem, the quality of filter feeders har-

Figure 8.14 Young abalone on a habitat designed to give shelter and increase the grazing surface area in culture. (From Hahn, 1988. Courtesy of CRC Press and Dr. E. E. Ebert, California Department of Fish and Game.)

vested from the coastal environment has declined.

Depuration is the cleaning process that occurs when the bivalve is placed in clean water. As the animal pushes the clean water through its system, some bacteria, viruses, and toxins may be removed. In many areas, shellfish can be grown but not legally marketed unless there is a subsequent transfer of mollusks from the areas of poor water quality to depuration areas just prior to sale. This ensures the safety of the product for the consumer.

The dangerous types of shellfish contamination are

1. Infectious diseases, which include cholera, typhoid fever, hepatitis, and gastroenteritis.
2. Pollutants, both inorganic (forms of lead, mercury, cadmium, and other heavy metals) and organic (such as pesticides and PCBs).

3. The neurotoxins of **red tide**–producing dinoflagellates, particularly *Gonyaulax* and *Protogonyaulax*, that cause **paralytic shellfish poisoning (PSP);** there is a 5% to 10% mortality rate for PSP.
4. The toxins from the dinoflagellates of the genus *Dinophysis* that cause **diarrhetic shellfish poisoning (DSP),** which, although less serious than PSP, makes large harvests of mollusks unfit for consumption.

Depuration can be conducted by simply moving the shellfish to cleaner natural waters, or they may be placed in water that has been disinfected. Disinfection can be carried out using UV energy, chlorine, or ozone. Ozone has an advantage because it will inactivate PSP toxins in the shellfish. The length of time that organisms are depurated varies with the nature and extent of the contamination. Normally, depuration takes one to four days. For example, oysters taken from Galveston Bay, with fecal coliform levels of $4.6 \times 10^4/100$ grams of tissue, were transported to a petroleum drilling platform 6 km offshore; in two days the bacterial levels had dropped to 430/100 g, and by day 7 there was no detectable bacterial level.

In some countries, depuration is required. For example, in Spain, where mussel production is so important, mussels must have a seal showing that they have been "purified" before they are sold in quantity.

SUMMARY

The mollusks are a large and diversified group; most have a planktonic larval stage (trochophore and veliger). They have a long history of culture and are currently reared throughout the world. This chapter focuses on four of the most commonly reared mollusks.

Mussel culture begins with the collection of small seed animals from natural beds or with placing some sort of substrate (shells, rope, bamboo, and so on) in the water for the larvae to settle on. Growout is accomplished using either bottom culture or an assortment of suspended culture techniques, including rafts, large posts with mussels in a net wrapped around them (bouchots), trays, and long lines.

Oyster larvae of *Ostrea* or *Crassostrea* are obtained either by settlement on collectors placed in shallow water (in particular, clean shells that are spread on the bottom or hung from a raft), or by spawning adults in a hatchery. Hatchery-produced larvae are raised on microalgae and can be induced to settle on a substrate called a "cultch" or may be "cultchless" (that is, unattached and not in aggregates). In the United States and Europe, bottom culture is the growout method of choice, although stakes, racks, or rafts are used too. In Japan and elsewhere, rafts and other suspended techniques are more popular.

The northern hard clam, *Mercenaria mercenaria*, can be collected as seed from sediment, or adults can be conditioned and spawned in the laboratory. A few weeks after the larvae metamorphose, they are transferred to nurseries where growth continues; nurseries may be open or semiclosed systems. Growout is generally done on bottoms, with site selection being critical. Rafts supporting trays can also be used.

Unlike mussels, oysters, and clams, which are all bivalves, abalone are gastropods. Conditioned adults may be spawned when exposed to UV irradiated water or low levels of hydrogen peroxide. Larvae do not feed and may be stimulated to settle after sufficient development using several natural triggering sustances. Juveniles are raised on benthic diatoms and then seaweeds. Young abalone can be reared in raceways or released in the marine environment and collected after several years.

CHAPTER 9

COMMONLY CULTURED CRUSTACEANS

The **crustaceans** are a very successful group of animals that share the phylum Arthropoda with such other organisms as spiders, scorpions, horseshoe crabs, centipedes, millipedes, and insects. The **arthropods** are characterized by having an exoskeleton and jointed appendages. The crustaceans (considered, in this text, a subphylum rather than a class) are primarily aquatic and segmented, and use gills to breathe. The heart pumps blood (**hemolymph**) through several large arteries. Oxygen is carried by the pigment hemocyanin.

There are many kinds of crustaceans, including such dissimilar organisms as the brine shrimp, copepod, lobster, and barnacle. In this chapter, we will focus on three of the important cultured crustaceans, all of which belong to the class Malacostraca. All adult members of the Malacostraca have 19 segments (segments 1 to 5 are the head, 6 to 13 are the thorax, and 14 to 19 are the abdomen), although some of the segments may be partially fused. In addition to being malacostracans, all three of the major cultured organisms are members of the order Decapoda (having 5 pairs of walking legs on segments 9 to 13). These legs may be **che-** late (clawed). Appendages on the abdomen, **pleopods,** may be modified for swimming and/or holding fertilized eggs.

The exoskeleton of the crustaceans, the carapace, is composed of a carbohydrate polymer called **chitin,** minerals (mostly $CaCO_3$), and proteins. In order to grow, the animal must shed the old carapace and secrete a new, larger one that it will grow into and eventually shed, too. This process of shedding is called **ecdysis.** There are a number of important physiological steps leading to, and immediately following, ecdysis; these include breaking down the old carapace, storing the $CaCO_3$, regenerating missing limbs, building up reserve carbohydrates before ecdysis that can be used for chitin synthesis, and others. These events, along with the actual act of ecdysis, are called the **molt.** The molt cycle is normally triggered by external stimuli (light and temperature), and there are probably some internal signals as well. The molt is initiated by the secretion of the hormone **ecdysone** from an endocrine gland called the **Y-organ;** ecdysone is rapidly converted to a more active compound, 20-hydroxyecdysone, which is responsible for the molting events. The Y-organ is prevented from constantly secreting the ecdysone by another hormone, a peptide called **molt-inhibiting hormone (MIH),** that is released into the hemolymph by the eyestalk's **X-organ–sinus gland** complex (a center of endocrine activity for the Crustacea). Thus, removing the eyestalk will often initiate molting since it removes the source of MIH (see Figure 9.1).

Under the carapace, in the epidermis, are color pigment cells, the **chromatophores.** These may expand or contract, allowing the crustacean to change color. The chromatophores are under control of peptide hormones, both stimulatory and inhibitory, that are produced in the eyestalk.

The gonads are paired, elongate organs in the thorax and/or abdomen. The oviducts are tubules, generally terminating near the last walking legs of the males and the third from the last pair in the females. Like molting and color change, reproduction also appears to be under control of hormones. There is a **gonad-inhibiting hormone (GIH)** in the eyestalk and there seems to be **gonad-stimulating hormone (GSH)** produced in the thoracic

Figure 9.1 Removing the eyestalks, the site of MIH production, leads to accelerated molting and growth (if enough food is available). The lobsters shown here were grown from the same clutch of eggs; the upper lobster had both eyestalks removed two months after hatching, while its sibling below was not operated on. These animals are about one year old. (Photo from McVey, 1983. Copyright © CRC Press, Inc., Boca Raton, Florida. Reprinted by permission. Courtesy of Drs. Ernest Chang and Douglas Conklin, Bodega Bay Marine Laboratory, California.)

ganglion. During the reproductive season, the removal of the eyestalk (and therefore GIH) stimulates egg protein synthesis. Another hormone, **methyl farnesoate,** is produced by a gland called the mandibular organ; it has been strongly suggested that this too plays a role in reproduction. Once the ovaries are fully developed and the male is ready to copulate, some crustaceans release chemical signals called **pheromones** into the water to alert potential mates.

During copulation, the male releases its gametes, often in the form of a sperm mass called the **spermatophore.** Males have specially modified appendages to aid in the transfer of the spermatophore. Females often have a receptacle that is used to hold the male gametes; fertilization generally occurs when the eggs are laid. Freshwater prawns and crayfishes carry the fertilized eggs until they hatch, but the more primitive penaeid shrimps shed the eggs quickly after they are laid.

The developmental stage of the crustacean at hatching is variable. In the more advanced species, the animals are generally more fully developed when they emerge. In many cases, the larvae are planktonic, metamorphosing from one stage to the next until giving rise to a benthic postlarval or juvenile stage. The development of each of the three species will be discussed separately.

The adult decapod may be an active predator, a scavenger, or an omnivore. Some, like the mole crab *Emerita*, are filter feeders. In all crustaceans, the mouth parts break up the food, which passes through a short esophagus, then into the cardiac stomach, pyloric stomach, and finally intestines. Between the cardiac and pyloric stomachs is the **gastric mill** that helps to grind up the food. Digestive enzymes are released by an organ called the **hepatopancreas.** Some biologists think that, in at least a few species of decapods, the hepatopancreas is also the site of formation of egg yolk proteins (**vitellogenins**) that are transported to the oocytes during maturation.

In this chapter, the penaeid shrimps, the crawfish *Procambarus*, and the prawn *Macrobrachium*, will be discussed. Other crustaceans are reviewed in Appendix 7.

PENAEID SHRIMPS

Shrimps of the genus *Penaeus* are considered in this section. Other **penaeid shrimps** (members of the family Penaeidae), such as *Metapenaeus*, *Parapenaeus*, and *Sicyonia*, support small fisheries and have been used in some laboratory studies, but are not important in terms of shrimp culture (see Table 9.1). The term "shrimp" will be used in this text when referring to the marine penaeids, and "prawn" will be used to discuss carideans such as *Crangon*, *Palaemon*, and *Macrobrachium*.

Penaeids, based on anatomical studies and the fact that they are the only group to retain the free-swimming nauplius stage, are among the most primitive of the decapods. The genus *Penaeus* can be subdivided into two groups:

1. The **grooved shrimps** have a pair of parallel grooves, starting near the rostrum and running posteriorly along the back. These shrimps also have **closed thelyca.** The thelycum is part of the female copulatory structure that is a modification of the sternal surface. In species with a closed thelycum, such as *P. Merguiensis*, *P. japonicus*, and *P. monodon*, the part of the spermatophore that contains the semen is fitted into a seminal receptor while the semen-free portion of the spermatophore swells on contact with seawater during copulation and acts as a plug to protect the opening. The spermatophore is held tightly to the female. Copulation in these shrimp must take place while the female is still in a soft-shell condition immediately after ecdysis and well before ovarian maturation and spawning.

2. The **nongrooved shrimps,** such as *P. setiferus* and *P. vannamei*, have **open thelyca.** In these species, the semen-free part of the spermatophore acts as an anchor, holding it to the female. Copulation takes place just before spawning, probably a behavioral adaptation resulting from the fact there is no true seminal receptacle and the spermatophore is therefore easily dislodged. The spermatophore also may fail to attach during mating of nongrooved shrimp, resulting in a spawn of unfertilized eggs.

Regardless of the thelycum type, spawning must take place after insemination and before the next molt or the spermatophore will be lost with the shed carapace. The spermatophore is transferred to the female by a membraneous structure

TABLE 9.1 Some Experimentally and Commercially Grown Penaeid Shrimps

Penaeus
P. aztecus (brown shrimp), *P. brasiliensis, P. californiensis, P. canaliculatus, P. duorarum* (pink shrimp), *P. esculentus, P. indicus, P. kerathurus, P. latisulcatus* (western king prawn), *P. longirostris, P. marginatus, P. merguiensis* (banana prawn), *P. monodon* (sugpo or tiger prawn), *P. occidentalis, P. orientalis, P. paulensis, P. penicillatus* (red-tailed shrimp), *P. schmitti, P. semisulcatus, P. setiferus* (white shrimp), *P. stylirostris* (blue shrimp), *P. teraoi, P. vannamei* (white shrimp)

Metapenaeus
M. affinis, M. bennettae, M. brevicornis, M. burkenroadi, M. dobsoni, M. ensis, M. joyneri, M. monoceros, M. stebbingi

Others
Hymenopenaeus robustus (royal red shrimp), *Parapenaeopsis acclivirostris, P. stylifera, Sicyonia brevirostris* (rock shrimp), *Trachypeneus similis, Xiphopeneus kroyeri* (sea bob)

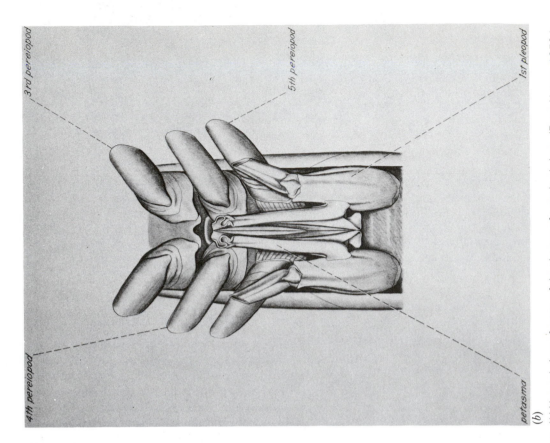

3rd pereiopod

5th pereiopod

1st pleopod

4th pereiopod

petasma

(b)

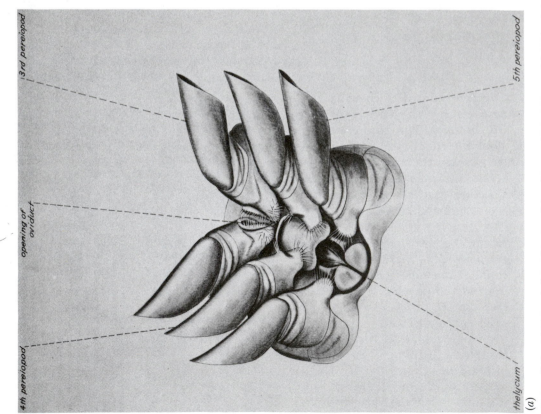

3rd pereiopod

5th pereiopod

opening of oviduct

4th pereiopod

thelycum

(a)

Figure 9.2 (a) Ventral view of parts of the thorax of a nongrooved female shrimp. (b) Ventral view of parts of the thorax of a male shrimp. (From Young, 1959.)

called the **petasma,** a modified pair of the male's pleopods (see Figure 9.2).

Techniques have been developed for removing the spermatophores from the male shrimp and artificially fertilizing the female. For example, in *P. japonicus*, an AC current of 5 volts applied briefly at the base of the fifth pair of pereiopods will cause the spermatophore to be partially expelled; the spermatophore can then be removed with tweezers and placed into the thelycum. This technique has allowed researchers to produce interspecies hybrids.

The penaeid life cycle is characterized by a migration between brackish and oceanic waters. Maturation and spawning take place offshore; the newly hatched shrimp nauplii begin development at sea,

but when they reach the postlarval (PL) stage, they begin to migrate toward the coastal estuaries that will serve as nurseries for them. The shrimp will eventually move from the estuaries, heading back offshore to mature and spawn. The environmental signals controlling these migrations are probably temperature, salinity, and tidal current.

Several authors have used different terms to describe the developmental stages of penaeid shrimp. The stages (see Figure 9.3) described in this book follow the most widely used nomenclature in the United States:

1. A **nauplius** hatches from the egg. There are generally five planktonic naupliar stages, each slightly more

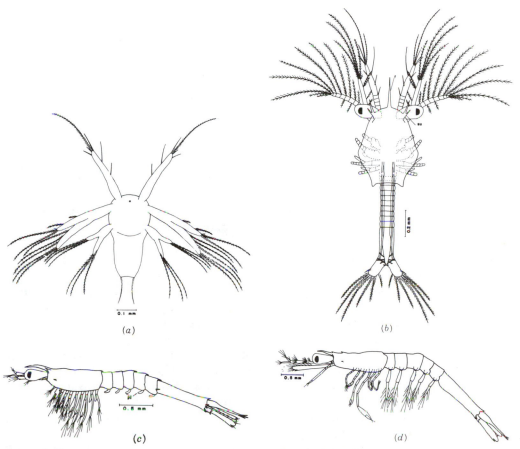

Figure 9.3   Some of the early developmental stages of a penaeid shrimp. (*a*) Nauplius, (*b*) protozoea, (*c*) mysis, and (*d*) PL. (From Dobkin, 1961.)

complex than the last. During these stages the shrimp does not feed, existing on yolk.

2. The **protozoeal stage** is critical because feeding on phytoplankton begins. The shrimp is more elongate now than it was during the nauplius stage, with compound eyes and a rostrum developing during the three protozoeal molts.

3. The **mysis stage** is characterized by greater elongation. During the three mysis molts, the telson and pleopods appear. Feeding on zooplankton begins.

4. Shrimp in the **postlarval stage** are similar to adult shrimp. They spend much of their time on the bottom.

Maturation and Spawning

Until recently, maturation of penaeid shrimps generally was not attempted. Rather, the culturist would take one of two options:

1. Collect mature females with spermatophores intact and hold them for spawning.

In most penaeid shrimp, a mature ovary is pigmented and rather large, and is therefore visible through the carapace along the dorsal portion of the abdomen. These shrimp are placed in tanks and spawning is monitored; some biologists feel that the addition of the chelating agent, **EDTA,** will increase the success of the spawning. Care must be taken when these shrimp are initially collected from the sea, since gravid shrimp may reabsorb their eggs if handled roughly.

2. Collect PLs and juvenile shrimp from the wild for rearing.

Collection of either mated adults or larval shrimp has proved to be costly and unpredictable; penaeid culture was limited by the supply of larval shrimp available to the shrimp farmer. Although both these alternatives are still used, great strides have been made recently in the development of maturation techniques which, coupled with improved hatchery procedures, have reduced some of the risks associated with shrimp culture.

Unilateral eyestalk **ablation** of the female shrimp, the technique of removing one of the eyestalks, is commonly used in culture operations. This is done to partially remove the source of GIH; removal of both eyestalks has been used, but the mortality rates are high and the differences in success of maturation are marginal. Eyestalk ablation of males may also improve reproductive performance; *P. vannemai* males that were unilaterally ablated were shown by researchers to have a higher sperm count without diminishment of sperm quality.

Many (but not all) biologists hold that the quality of embryos produced by ablation of the female shrimp is, to some extent, inferior to the quality of those embryos taken from naturally maturing shrimp. However, at present, ablation is the most reliable method available to ensure maturation and spawning. Maturation, but sometimes failed spawning, has been accomplished by manipulating the temperature, light (photoperiod and/or light intensity), and diet of the shrimp. *P. merguiensis* and *P. japonicus* spawn easily by manipulating environmental conditions; *P. monodon* requires ablation; and *P. stylirostris* and *P. vannamei* will mature and copulate if environmental conditions are right, but generally are ablated as well to ensure a greater percentage spawn. Each year there are more reports of maturations and spawnings without the aid of ablation. Even greater success is probably close at hand, and will be achieved as we learn more about the natural spawning events.

Much more attention must be paid to the diet of the brood stock shrimp than to those shrimp in growout. Brood shrimp may be fed as much as four times each day. Dried prepared feeds used in growout are partially or completely replaced with fresh and frozen food. The most important of the live–frozen supplements are polychaete worms, squid, and bivalves; trash fish, shrimp heads, and other crustaceans have also been used. It is suspected that some sort of nutrient or hormonal factor that is destroyed by the heat or pressure of pellet manufacture is critical for gonadal development; among the suggested candidates are sterols and polyunsaturated fatty acids (particularly those that may be used in the synthesis of prostaglandins).

Brood shrimp are kept in large, round spawning tanks; a 1 : 1 sex ratio is commonly used. There is no sediment on the bottom of the tank (unless *P. japonicus* is the species of interest), and airstones and other objects in the water are placed in the center of the tank so they do not interfere with the chasing courtship behavior around the tank's perimeter. Tanks should contain a high-quality seawater that is replaced several times per day; brackish water is not acceptable for many species. The tanks may be kept indoors and be equipped with overhead fluorescent lamps, or kept outside but shaded. The photoperiod and intensity requirements vary from species to species.

Once the shrimp are placed in the spawning tanks, they should be observed at least daily for chasing behavior and maturation of the gonads. The spawning itself is not always obvious, and therefore the culturist may wish to have an egg collector by the tank outflow to catch eggs that have been released by the female shrimp. The eggs will develop rapidly, usually hatching within a day, so the eggs collected in the collector or off the bottom of the tank should be moved to the hatchery without delay.

Collection of Larvae

The alternative to controlled spawning or collecting gravid females is to collect young shrimp for culture.

Ecuador, the major shrimp producer in the western hemisphere, still depends on the collection of young *P. vannamei* for much of its seed supply (although this is changing). The PLs, called **semilla,** enter the estuaries from the sea to mature. Fishermen, known as **semilleros,** concentrate on the collection of the PLs. The most productive collecting time is during the spring tides **(aguaje),** when the tide is ebbing and the PLs must struggle against the current to move into the safety of the mangroves. Each semillero can collect 20,000 to 50,000 PLs on a single tide. Often the semilleros sell their catch to a seed broker who deals directly with the culturist. The PLs are usually placed in nursery ponds **(precriaderos)** for a month until the juveniles weigh 2–3 g; then they are transferred to a growout facility.

In the western Pacific area, one technique used for the collection of seed shrimp involves the construction of clay ponds in a swamp that is known to contain penaeids. These ponds have sluice gates that allow water to flow in and out. When the water outside the gate is roughly 60 cm above the water level in the pond, the gates are opened and the water, full of shrimp—it is hoped—enters. Shrimp already in the pond are unable to escape because of the pressure of the water flowing in. As soon as the tide begins to ebb, a wire screen is placed in front of the gate to retain the animals.

In the Philippines, the PLs of *P. monodon* have been collected either using dip nets or lures called **bon-bon.** The lures are bundles of twigs and grasses that are held in place in the rivers by strings running to the banks; the shrimp settle on the lure and can be removed simply by shaking it.

Hatchery Production of Larvae

Penaeid hatcheries for the production of PLs are divided into two basic types, the **Japanese variety** and **Galveston variety,** although some hatcheries are hybrids of the two. The Japanese hatcheries are less labor and capital intensive, so although the survival is lower, the cost of the shrimp larvae is a bit less than those produced in Galveston hatcheries. However, the Galveston hatcheries are far more predictable in terms of production, so the decrease in risk often compensates the culturist for the increased cost.

The Galveston method makes use of relatively small round tanks and a strictly controlled set of culture parameters. The temperature is controlled to within 2°C of the ideal for the particular species, salinity is kept (for most species) oceanic or slightly less, and oxygen is kept at or near saturation using aeration. The water entering the tanks is filtered and/or otherwise treated to remove the potential pathogens and contaminants. The eggs are stocked at a particular density and the number of larvae per unit volume is carefully monitored. All the food in the Galveston-type hatchery is added by the hatchery workers; this usually begins with specific algal cultures (often diatoms such as *Chaetocerus* and *Skeletonema*, or flagellates such as *Isochrysis* and *Tetraselmis*) while the eggs are hatching so the larvae will be able to feed immediately when the nonfeeding nauplii metamorphose to the protozoea. The algal diet may be followed by brine shrimp, rotifers, and other feeds as the larvae grow.

The Japanese hatcheries use larger, aerated tanks, often rectangular in shape. If these are outdoors, they may be covered with large plastic or fiberglass sheets to prevent heat loss and introduction of airborne materials. Spawning shrimp are placed in the tanks and release their eggs; the number of larvae, therefore, is not strictly determined. The water entering the tanks is filtered to remove only the zooplankton, the smaller phytoplankton being allowed to enter. The tanks are fertilized to encourage the phytoplankton (preferably diatom) growth. Water exchange is usually minimized to help maintain the phytoplankton bloom, although *some* water exchange may be good for both larvae and phytoplank-

ton. Additional feeds may be introduced later, including brine shrimp, soybean meal, copepods, larvae of other invertebrates (bivalves, barnacles, worms, etc.), crumbled feed, and so on.

Harvesting of the PLs should be done by the time they switch from being free swimming to benthic, especially if concrete tanks are used, since the young shrimp may injure their legs and telson on the rough surfaces, which will result in infection. The shrimp are generally taken from the tanks using a siphon and mesh apparatus of some sort. The age of the PLs, in the parlance of the shrimp culturist, is given in terms of the number of days since the animal has been a mysis; a PL that has metamorphosed from a mysis 2 days previously is called a P_2, 10 days before makes it a P_{10}, and so on.

A considerable amount of attention has been given to larval diet. Different hatcheries not only rear different species and strains of shrimp, but probably use different strains of algae, have a different water chemistry, purchase their brine shrimp from different wholesale sources, and do myriad other things differently as well, so it is wise for each culturist to experiment a bit while deciding on a diet for the young shrimp. During this experimentation period, the published scientific literature should be used as a guide rather than the final authority. The literature on this subject is quite extensive; a few examples are given in Table 9.2.

Growout

Growout is usually discussed in terms of the stock density. Once the density is established, then the choices a culturist has regarding such practices as water exchange, feeding, aeration, and harvesting methods will more or less fall into place. This text will focus on the high (**intensive** and **semi-intensive**) and the low (**extensive**) stocking density protocols. These include a range of less than 1 shrimp/m^2 to over 200 shrimp/m^2 in some of the facilities in China.

TABLE 9.2 Examples from Published Reports of Shrimp Diets

Species and Stage	Dietary Comments	Reference
Larval Diets		
P. stylirostris and *P. vannamei*	*Chaetoceros* (diatom) gave survival rate of 85% from protozoea to mysis stages.	Simon, 1978
P. monodon	*Chaetoceros* gave better growth than *Tetraselmis* (flagellate) in early protozoea stage, but no difference in mysis stage.	Tobias-Quinitio and Villegas, 1982
P. aztecus, P. setiferus, and *P. vannamei*	*Panagrellus* (nematode) may be used in some cases as a substitute for brine shrimp.	Wilkenfeld, Lawrence, and Kuban, 1984
P. vannamei	*Artemia* of the San Francisco strain were superior to the Colombian strain for postlarval growth.	Landau and Eifert, 1985
P. vannamei	*Isochrysis* (flagellate) was superior to *Prorocentrum* (dinoflagellate) and *Bacteriastrum* (diatom) through the mysis stage.	Sanchez M., 1986
P. monodon, P. vannamei, and *P. stylirostris*	Microencapsulated diets were as good as live feeds (diatoms, *Artemia,* and clam).	Jones, Kurmaly, and Arshard, 1987
Growout Diets		
P. japonicus	An artificial diet was developed.	Kanazawa, Shimaya, Kawasaki, and Kasiwada, 1970
P. japonicus	Diet high in protein (60%) and similar in amino acid composition to shrimp gave best growth results.	Deshimaru and Shigeno, 1972
P. japonicus	P, K, and trace minerals, but not Ca, Mg, and Fe, were needed in diet.	Deshimaru and Yone, 1978
P. vannamei and *P. setiferus*	Growth improved when more animal protein and less plant protein was used, especially for juveniles.	Chen, Zein-Eldin, and Aldrich, 1985
P. japonicus	Squid protein was superior to fish protein.	Cruz-Suarez, Guillaume, and Wormhoudt, 1987

Extensive Systems Most shrimp growout facilities are normally rather "low-tech" semiclosed systems; the term "extensive" is often used to describe them because they begin with low stocking densities of 1 to 5 juvenile shrimp/m^2 in earthen ponds. There is no supplemental aeration in these facilities since the BOD is rather low. Growth of the shrimp may be supported simply by the natural biota in the pond, though the growth rate can be increased with fertilization of the water using either defined chemical nutrients or (more commonly) organic materials such as manures, plant cuttings, or even treated sewage. Some commercial shrimp feeds may be added to the ponds to increase the rate of growth, but unlike the nutritionally complete pelleted feeds used in more intensive systems, this is a protein supplement rather than the exclusive source of nutrition for the shrimp, and is therefore much less expensive than the feed used in intensive culture.

Because most of the nutrients that the shrimp get in these extensive systems are derived from ingestion of pond organisms, some biologist have suggested that culturists should "feed the pond, not the shrimp." That is, the pond should be seeded with organisms that are most beneficial for the shrimp to eat, and then fertilized in such a manner as to encourage the growth of the organisms for grazing by the shrimp in culture.

Besides the amount and quality of food available to the shrimp, the other factors that could theoretically determine production are water quality, temperature, and stocking density (although it is unclear how much the stocking density actually affects these extensive systems). The assumption is that the concentration of shrimp must reach some critical density before growth is inhibited. For example, in experimental extensive culture of *P. vannamei*, the growth rates of the shrimp were similar in ponds of 1 shrimp/m^2 and 10 shrimp/m^2. Likewise, survival was equivocal for *P. monodon* PLs stocked at 1, 2, 3, and 4 shrimp/m^2. On the other hand, biologists have carried out experiments clearly showing that salinity and temperature can significantly affect the growth and survival of *P. vannamei* and *P. aztecus* in extensive culture.

Production in extensive systems is not great, usually ranging from several hundred to more than 1000 kg/ha/year. Production of over 100 kg/ha/year usually requires that the ponds be fertilized and/or the shrimp be given some supplemental feed.

Shrimp are harvested with large seine or cast nets or are simply drained with the pond water into a basket structure or bag net. Harvesting is generally done early or late in the day to avoid the heat that may damage the crop. The shrimp may be sampled before harvest to determine molting activity; if a large number of shrimp have soft exoskeletons, the harvest may be postponed for a few days. The ponds are partially drained before the nets are used, and in enclosures exposed to tidal fluctuations the harvesting period often coincides with low tide to take advantage of the natural drop in water level. As soon as the shrimp are harvested, they are frozen or placed in ice water until they can be processed. If they are to be processed with the head on, the culturist may wish to suspend feeding for a few days before the harvest.

Semi-Intensive and Intensive Systems

When shrimp are stocked at middle and high densities (20 to over 200 PLs/m^2), the culturist cannot rely on natural production to feed the shrimp or simple gas diffusion to supply all the oxygen that will be needed. Growing shrimp in high density requires skill and a constant monitoring of the system. Outdoor earthen ponds are sometimes used for high-density culture, but tanks and raceways (both indoors and outdoors) are more commonly employed.

In semi-intensive and intensive systems there is often a clear inverse relationship between stocking density and growth when all other parameters are held constant. To sustain the increase in the stocking density, culturists most often change diets and increase water flow.

In Japan, the culture of *P. japonicus* is carried out in large circular concrete tanks. Running water is used to keep the O_2 levels high and the concentration of metabolic wastes low. Stocking will generally be about 160 juveniles/m^2 and production averages are 4.5 to 24 metric tons/hectare, although rates of up to 35 metric tons/hectare have been reported. In experimental 3–12 m³ tanks, *P. vannamei* has been grown to a harvestable size quickly on pelleted feeds, giving production values of up to 40 metric tons/hectare/year (although pond studies by the same research group yielded somewhat smaller harvests). Researchers in China have recently reported that *P. penicillatus,* cultured in small earthen ponds with concrete sides, with initial densities as high as 286 PLs/m², yielded two crops taken over a 272-day experimental period that totaled over 23 metric tons/hectare.

In semi-intensive culture yields are somewhat lower, but overall the economics are often better. In Taiwan, the culture of *P. monodon* in small earthen ponds stocked at 20 to 30 juveniles/m² may produce two crops a year, averaging 1.4 to 9.6 metric tons/year. These ponds have blowers or paddlewheels to increase the DO, and the shrimp are fed pelleted diets supplemented with trash fish and mussels. When a similar system was used experimentally in Israel with *P. semisulcatus,* ponds having paddlewheel aerators and receiving a high-grade feed once or twice a day, produced on average nearly 1 metric ton/hectare/harvest (with a growing period of 71 to 79 days). In the United States, *P. vannamei* has been grown experimentally using stocking densities of 40–45 shrimp/m² in rectangular ponds in South Carolina equipped with paddlewheels and fed a high-protein (40%) diet; over periods averaging 144 days, crops of 3 to 3.75 metric tons/hectare were harvested by the researchers. Recent reports indicate that these ponds, and some large round concrete ponds in Hawaii, may be able to support much higher densities, resulting in yields twice as great.

Currently, there are a number of feeds that are being marketed as adequate for high-density culture. Some are clearly better than others under certain conditions, but none is ideal. Research on larval diets has largely centered on finding the best feed organisms, while growout research is concerned with pellet produc-

tion (since live feeds for hundreds of hectares of ponds would be very impractical). The greatest number of studies have been carried out on *P. japonicus,* and results show that growout diets must contain cholesterol and polyunsaturated fatty acids, minerals, and certain vitamins. Simple sugars may inhibit growth, but disaccharides and more complex carbohydrates are good sources of energy. Not only is total protein important, but so is protein quality. A few examples of studies on growout feeds are listed in Table 9.2.

CRAWFISH

The **crawfish** (or **crayfish** or **crawdad**) is the most important of the cultured crustaceans in the United States. The center of the industry is Louisiana (over 5.2×10^4 hectares in 1986 were dedicated to crawfish farming, and the area has been expanding), with some culture being carried out in other southern states, as well as California, Oregon, and the Great Lakes states. Wild crawfish are found in almost all of the continental states. In 1985, the United States produced almost 3×10^4 metric tons of crawfish, over 270% of the production seen in 1980; by 1987 production rose to over 150% of the 1985 harvest. Most of the production goes for food, but small crawfish are also sold as bait for sportfishing.

The most important of the cultured crawfishes is *Procambarus clarkii,* the red swamp crawfish. It is an opportunistic omnivore that eats plants and their epibiota, worms, insect larvae, and detritus. Others that are used in the United States and elsewhere include the white river crawfish, *P. acutus,* and members of the genera *Astacus, Orconectes,* and *Pacifastacus.* The descriptions that follow refer to *P. clarkii* in its southern habitat (although it can be found as far north as Illinois).

Unlike the European *Astacus* industry, which has been severely damaged by the fungus *Aphanomyces,* disease organisms have made no significant impact on crawfish culture in the United States. Predators of *Procambarus*, such as predaceous insects, raccoons, and birds, cause some mortality problems but are not major factors in most instances. Fish may cause losses of crawfish, but this is normally controlled by filtering the water as it enters the pond; if a fish population does become established, the pond should be drained after harvesting and treated with rotenone or another drug that will kill the remaining fish and their eggs.

Reproduction

P. clakii is typically found in swampy environments that often dry up for part of the year. The crawfish dig burrows and enter the moist mud for the dry period, while other aquatic organisms are unable to survive.

Sexually mature crawfish are said to be in **form I,** meaning that their morphology is slightly altered to make mating more efficient. The male crawfish, easily distinguishable from the female, has swollen claws and well-developed organs **(gonopods)** for transferring the spermatophore to the female (gonopods are present but less developed while the male is in the **form II** phase). The sperm may not be used by the female for up to six months. Females usually, but not always, lay their eggs while in a burrow. Fertilization takes place at laying, and the newly extruded fertilized eggs are attached to the pleopods by a cement called **glair.** The eggs are moved back and forth by the pleopods to assure that they get enough oxygen, and will hatch after two to three weeks at 20°–25°C. The number of eggs extruded is a function of the size of the female, but in general the eggs are relatively large, so the number of offspring does not exceed 700 (and is often less than half that) (see Figure 9.4).

Upon hatching, the little crawfish remain attached to the mother. During the next few weeks, they go through two molts and then are able to leave the mother's protection and begin their own benthic existence, though they may re-

Figure 9.4 A female crawfish holding recently laid eggs on her pleopods. (Huner, 1988. Copyright © Archill River Corporation. Reprinted by permission.)

turn to the mother periodically. *P. clarkii* will mature in three to nine months and may live up to four years. In the warm waters of Louisiana, the reproductive cycle lasts all year long, but this is not the case in colder regions. Following mating, the male crawfish will molt back to the nonreproductive form II. After several molts at form II, the crawfish can return to form I and the cycle is repeated.

Under typical Louisiana culture conditions, mating takes place most often when the weather begins to warm, and the eggs are often laid and fertilized in the fall, although as stated, this can occur at any time.

Culture

Although crawfish can withstand rather poor water quality for short periods of time, best growth results are seen when the water quality is good. There are several factors to consider:

1. Water should be 20°–25°C if possible, since most crawfish grow quickest at these temperatures (however,

since most culture takes place in open and semiclosed systems, the farmer has no control of temperature). Growth slows down as the temperature deviates from this range; when it gets very warm, the crawfish retreats into the coolness of its burrow.

2. There should be sufficient calcium (about 100 ± 40 ppm) in the water for the animals to easily take up all they need to build exoskeletons and avoid ionic stress; however, salinity should not exceed 5 ppt. Soils low in calcium can be treated with limestone.

3. High DO levels mean good growth; the water should have at least 3 ppm O_2. At low DO levels, the crawfish will come to the surface to get atmospheric O_2 that they can use as long as their gills remain wet. The water in many Louisiana facilities is mechanically aerated by pumping it through perforated plates.

4. In many cases, the water for crawfish culture is taken from a surface source, such as a stream. In this case, the water must be filtered through a mesh to remove potential predators.

5. The pH should be 7.0 ± 1.5, so very acid soils for ponds should be avoided. Again, limestone can be added to raise the pH as well as add calcium.

6. The soils should have a high enough clay content to hold water and high enough levels of nutrients to support pond production. Clays and nutrients can be added to the soils if needed.

A number of different types of facilities can be used to grow *P. clarkii*:

1. Typical **rectangular ponds** may be constructed. The dam height should be adjusted so that the water level is normally about 0.5 m, al-though deeper ponds can be used. These ponds are large, averaging about 16 hectares, but if the DO drops to dangerous levels, they cannot be aerated quickly enough to prevent some mortality that might otherwise be avoided.

2. **Wooded ponds** are similar to other large aquaculture ponds except that trees and shrubs are not removed. That is, these are basically areas of low woods surrounded by dams. These wooded ponds are usually less productive than cleared ponds and are more difficult to harvest. The high organic content of these ponds will sometimes result in low DO levels.

3. **Marshes** are used in some coastal areas, but must be selected carefully to be sure that the influx of salt water is not too great. Although growth is not hampered by slightly brackish water, reproduction will halt.

4. There has also been some **intensive culture** of crawfish. While high-density rearing of crawfish is less difficult than intensive culture of many other aquatic organisms, this technique is still not popular since this animal can be produced cheaply using the less complex and less capital-intensive methods that are currently employed. Tanks are used and low water levels are employed (depths of 15–20 cm); crawfish are very adept at crawling out of tanks on air hoses, drain pipes, etc., so precautions must be taken to reduce escape losses.

5. **Rice fields** can be used to grow alternating crops of rice and crawfish, a procedure called **double cropping.** In Louisiana, rice is usually planted in March or April, grows for about 100 to 120 days to maturity, and is harvested in August; the rice field is flooded in the fall so the crawfish can graze on the rice stub-

ble. When growing rice, the culturist should be careful about using insecticides, many of which are highly toxic to the crawfish. Harvesting of the crawfish in this system begins during late November and continues until it is time to plant another crop of rice in the spring; at this time, the pond water is pumped out (either partially or completely) for planting the rice. The remaining crawfish that were not harvested serve as the brood stock for next year's crop (see Figure 9.5).

When starting a new pond/farm, or restocking an old pond, brood crawfish are spread around the edge of the pond in spring or the early summer months. A 1 : 1 sex ratio is desired; the pond is stocked at the rate of 20–65 kg of brooders per hectare, depending on the amount of available cover and the number of crawfish already present in the ponds. The newly added crawfish will burrow into the dams over a period of several weeks. The ponds are typically drained after the crawfish are in their summer burrows, and filled with water in the fall, mid-September to mid-October.

Crawfish are normally not fed immediately after the fields are flooded; rather, crops for crawfish to forage on are planted in the spring or summer and then covered with water in the fall. Natural vegetation will account for some of the nutrient base, but rice plants and the rice straw (decomposing with many microbes present) are probably more important in sustaining the crawfish population. (Note: Rice is often planted simply as forage and not as an alternate crop.) Planting takes place after the crawfish have burrowed and the water is being drawn down. Under some conditions rice may not grow, so some research has gone into finding alternative forage crops. A recent study has indicated that sorghum sudangrass can be used because of the high biomass produced and the slow rate of decomposition (rapid decomposition can lead to a food shortage).

When the crawfish emerge from their burrows, they will live on the rice and other plants for awhile, but eventually the food supply will dwindle and growth of the crawfish will slow or stop. When no more forage is available, the farmer may switch to commercial feeds, or additional hay can be added to the ponds. Agricul-

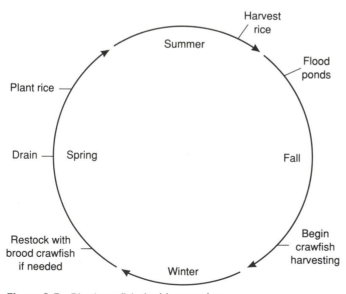

Figure 9.5 Rice/crawfish double cropping.

tural by-products have been used experimentally to support crawfish populations; researchers found that sweet potato trimmings supported good growth and could be fed directly to the crawfish, while rice stubble and rye hay supported fair growth but should be partially composted so microbes could improve their nutritive value. Dried sugar cane and soybean stubble did not support crawfish growth.

Harvesting is typically done using wire-mesh traps baited with gizzard shad, some other fish, or an artificial bait. There are several trap designs; the most popular traps are hexagonal wire-mesh devices that stand vertically and have several funnel entrances around the base of the trap. The harvesting is done from a boat, operated by one or two people; again, there are a number of boat designs that can be used depending on the physical properties of the culture system. Harvesting begins in the winter, often less than two months after the ponds are flooded. The old brood stock is thus harvested, although the culturist should examine the crawfish at this time to be sure that females holding eggs or about to extrude their eggs are not taken. Traps are normally set daily, with a harvest of 0.5 to 1.0 kg/trap; production may drop off after the initial harvests, when most of the brooders have been taken but the young are not yet large enough to be caught in the traps. Trapping may be temporarily halted at this time. Once harvesting has resumed, it will continue until the late spring or summer when burrowing behavior begins, or the catch drops dramatically, or the crawfish take on a dark color that denotes that growth has stopped. In Louisiana, crawfish are usually sold live to consumers who boil them in seasoned water, or are sold to processing plants that remove the exoskeleton and sell the abdominal muscles ("tails").

One of the most popular, new, and expensive crawfish products is the **soft-shelled crawfish,** which, like the soft-shelled crab, is simply an animal that has recently gone through ecdysis. The larger immature crawfish taken from ponds are the starting material for soft-shell production; mature animals molt too infrequently. These immature crawfish will go through ecdysis, usually between dawn and dusk, within 30 days of being captured. The soft-shell production facility should be indoors in a shed or a greenhouse. It must be well lit, and light-colored tanks or trays should be used to make it easier to select the crawfish. Warm water of a high quality is required. During the early premolt stages the animals must not be too crowded (or they will never complete the molting cycle), are fed floating trout pellets that may be supplemented with duckweed, and are observed daily. Later, they may be crowded during late preecdysis when they have pasted their "physiological point of no return" and must complete the molt. The late preecdysial crawfish can be identified using several microscopical techniques, or by characteristic color changes. (For example, 7 to 10 days before ecdysis the top of the carapace begins to darken to a brown color; by 2 days prior to shedding the carapace is uniformly brown and parts begin to weaken, becoming somewhat flexible.) Finally, when the crawfish are near ecdysis, they should be separated to prevent cannibalism and placed in shedding trays or tanks; these are highly compartmentalized to make observation easier (at this point the crawfish should be observed twice daily). In the molting tanks, no food is given and the shed exoskeleton is removed from the tray when shed. It has been suggested that ecdysis is best done in deionized water, since this will retard the hardening of the exoskeleton. The soft crawfish should be either processed immediately or chilled or frozen until it is processed.

MACROBRACHIUM

The giant Malaysian freshwater prawn, *Macrobrachium rosenbergii* (see Figure 9.6), and to a lesser extent, other

Figure 9.6 *Macrobrachium rosenbergii.* Note the longer, more robust pair of second walking legs on the mature males in the foreground. (Photo courtesy of the University of Hawaii Sea Grant Extension Service.)

members of this genus, are important fisheries and aquaculture organisms in many areas, including parts of the United States. *Macrobrachium* is a large genus, having about 125 species, living in fresh- and brackish waters throughout the world (although most species are found in the Indo-Pacific region). In this section, only *M. rosenbergii* will be discussed, though other species such as *M. acanthurus* and *M. nipponese* have been grown experimentally. The culture of this organism is not an old industry, but it has spread quickly over the past two decades, and *Macrobrachium* is now the major aquaculture product in many countries. In the United States there has been a great deal of experimental interest in *M. rosenbergii*, but only in Hawaii has its culture made a significant impact on the local economy. Efforts are now underway to bring aquaculture of the prawn to temperate localities (for example, South Carolina).

Macrobrachium are **caridean** "shrimp" and can be distinguished from the more primitive penaeid shrimp most quickly by the second abdominal segment of the exoskeleton that overlaps both the first and the third segments. Adult *M. rosenbergii* are often a bluish color and have a greatly enlarged pair of claw-bearing walking legs (the second pair), which gave rise to the name of the genus (macro = large,

brachium = arm). In mature males, these legs are much longer and thicker than are those of the immature males and females. Mature males also weigh more than the females, reaching over 200 g. (This had led to some researchers suggesting that all-male cultures may be more profitable than traditional mixed-sex culture; in an experimental growout of 140 days, starting with juvenile prawns of about 2.6 g, all-male ponds yielded bigger animals and a greater biomass with estimated revenues 24.5% greater than did mixed-sex ponds and 85.7% greater than all-female ponds.) Mature ovigerous females are easily identified because the ovaries, large orange-colored masses when ripe, can be seen through the carapace along the dorsal and lateral portions of the cephalothorax.

Reproduction and Life Cycle

Mature *M. rosenbergii* are usually found in freshwater rivers and lakes. During the breeding season, adults copulate soon after the females go through a pre-spawning molt (the actual ecdysis often takes place at night, and is preceded by a few days of reduced activity and feeding). The eggs are extruded by the female within a day of copulation and become attached to the pleopods. As the embryos mature, the female migrates into estuarine waters where the eggs hatch into swimming larvae. The larvae pass through 11 stages before becoming PLs that are benthic and begin to migrate upstream into fresh water. The different stages of larval development can be recognized by the presence of stage-specific characteristics (for example, only stage 1 larvae do not have stalked eyes, pleopod buds appear at stage 6, teeth on the rostrum appear at stage 10, and so on) (see Figure 9.7). Larval development lasts from three to seven weeks, depending on the temperature, water quality, and food.

The PLs molt into adults and continue to grow. There appear to be several forms into which the male prawns may fall, and

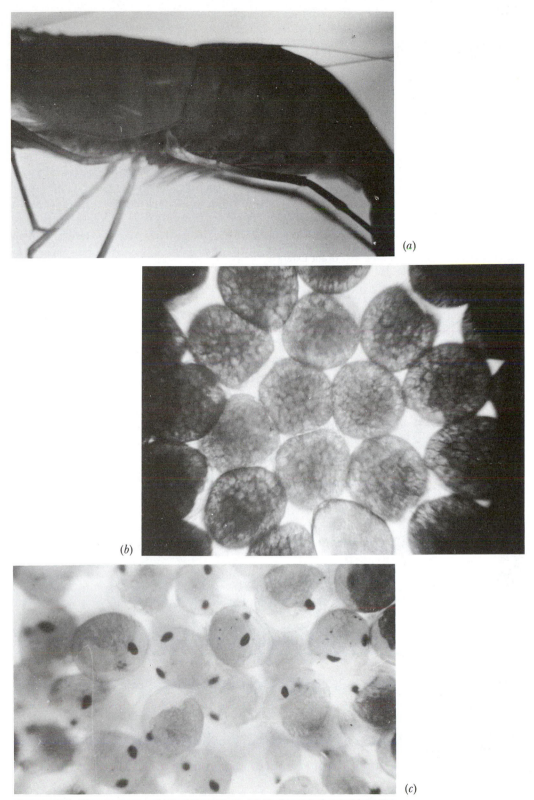

(a)

(b)

(c)

Figure 9.7 Development of *Macrobrachium*. (*a*) Embryos being held on the pleopods, (*b*) embryos in early development, (*c*) embryos in later stages of development, (*d*) early larva without stalked eyes, and (*e*) later larvae with stalked eyes. (Photos courtesy of Dr. M. Hartman, Florida Institute of Technology.)

(d)

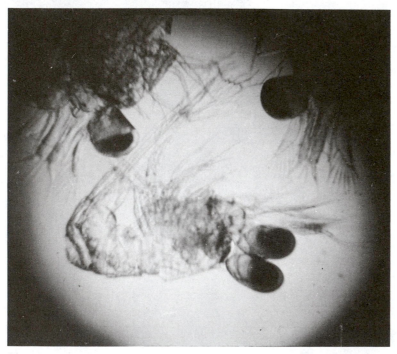

(e)

Figure 9.7   *(Continued)*

research seems to indicate that the form, to a large extent, is a function of the population's social structure and the rapidity of larval growth. These male forms are

1. The large, blue, big-clawed (BC) males, **bulls,** which are sexually active and most aggressive; they will form harems of females.
2. Small males (SM), which are capable of increased growth when isolated from other types of males and will copulate with females in harems when the BC is distracted.
3. The orange-clawed (OC) males, which are less aggressive and not sexually active; these prawns channel their energy into growth rather than reproduction, so they grow more quickly than the SM or BC prawns.

Because of the aggressive behavior of the BC males, production in ponds is somewhat hindered. Some biologists have cited this as the reason why prawn culture has not been more economically successful.

Mating, as stated earlier, occurs soon after the female's ecdysis. Mating can take place throughout the year if conditions are right, though in their natural environment there are peak reproductive seasons for *M. rosenbergii*. In ponds, the male may build a depression in the substrate for mating. Unlike many aquatic organisms, this prawn will breed readily in small laboratory tanks if conditions and feed are acceptable. Copulation is often, but not always, preceded by a mating behavior that includes the male "reaching" for the female and then turning her over so that he may clean her thoracic sterna. The spermatophore is then transferred to the female's sperm receptacle. Spawning takes place soon after, with the eggs being pushed from the female's gonopores and over the spermatophore where fertilization takes place. Depending on the age and general health of the female, 5000 to 100,000 eggs are produced per spawn. The fertilized eggs are finally cemented to the pleopods, where they are incubated until hatching. Under good conditions, ovaries will ripen again while the female is still carrying the fertilized eggs on her pleopods.

Culture

As was mentioned in the introductory chapter of this book, the discovery that the larvae of *M. rosenbergii* require brackish water was a major advance for aquaculture. Juveniles and adults are kept in freshwater, but salinities of 12–16 ppt are used for the larvae. DO levels should be kept high. Mass mortalities have been reported in very hard water, but when too little calcium is present the exoskeleton may be thin and soft. Since *M. rosenbergii* is a tropical organism, it is not surprising that the ideal temperatures for culture are rather warm, 28°–30°C, for all stages in the life cycle. *Macrobrachium* is very sensitive to water quality, especially the larval stages; filtration, disinfection, and water exchange should be given special consideration when a facility is being designed.

The first step in the culture of the prawn, the hatchery phase in which the larvae are reared to PLs, is conducted in brackish water. It is therefore convenient to have the hatchery near a source of saltwater, although artificial seawater can be used. Under hatchery conditions, this phase of development will take 25 to 40 days. The larvae are obtained from brood stock animals. Experimentally, eggs have been removed from females and hatched *in vitro,* but this is not a general practice among culturists. Two hatchery systems are used to produce PLs: the AQUACOP type and the Anuenue type. In either type of hatchery, the initial density may be reduced as the animals grow to increase the efficiency of space utilization. The operation of these is as follows:

1. The **AQUACOP hatchery** makes use of small, round, conical bottom

tanks that house high densities of larvae (100 to 200/liter). Such high densities require very careful management, not only because water quality can easily become a problem, but also because cannibalism may occur. The systems may be closed, requiring extensive filtration, or semiclosed, in which case water is exchanged once a day; if a good source of seawater is available, the semiclosed design is probably the easiest to manage, but when artificial seawaters are used, a closed system may be required for economic reasons. Tanks should be sterilized between larval batches.

2. The **Anuenue hatchery** makes use of large rectangular tanks stocked with larvae at lower densities (30 to 50/liter). These may also be closed or semiclosed. Because this type of hatchery uses lower larval densities, the amount of time spent in management is reduced, which in turn has made this type of hatchery more popular. Frequently, an Anuenue hatchery will utilize **green water** (that is, algae is batch cultured and added to the prawn tanks). Green water may consist of a variety of algal species and even some protozoans. At the Anuenue Fisheries Research Center in Hawaii, where this technique was developed, the dominant organisms are *Chlorella*, *Palmellococcus*, and *Nannochloris*. Among other phytoplankters that have been reported to successfully enhance PL production are *Isochrysis*, *Tetraselmis*, and *Phaeodactylum*. While phytoplankton is not directly consumed by the prawns, it has been suggested that the algae improve the water quality and are ingested by the zooplanktonic feed organisms that in turn are eaten by the prawn larvae. The phytoplankton may also prevent blooms of unwanted and dangerous organisms.

Newly hatched *M. rosenbergii* do not feed until the second day; therefore, many culturists do not begin adding food on the first day, while others like to include brine shrimp (*Artemia*) nauplii as soon as the young prawns are placed in the hatchery tanks. It appears that, until the larvae matamorphose into PLs, the capture of food is a function of chance encounter, and ingestion takes place only if the captured particles "taste right" (that is, if the correct chemosensory cues are present). The most commonly used feed for larval prawns for the duration of their stay in the hatchery are brine shrimp, though artificial feeds have been used in conjunction with *Artemia* in some facilities. Fish flesh (pollack or skipjack tuna, forced through steel screens at high pressure) may also be used. Fish flesh may foul the water, so the amount of the fish the larvae are eating must be monitored carefully to prevent buildup of uneaten food. It is better to feed small amounts frequently than to present larger amounts of food infrequently, especially during the early phases of hatchery production. Other feeds have been used, including copepods, fish eggs, worms, rotifers, insect larvae, albumin, soybean, and chopped bivalves and squid. Uneaten food should be removed from the tank bottoms daily.

Once the larvae have metamorphosed into PLs, they can be transferred and the nursery phase begins. The nursery is for the production of juveniles that can be used later for growout. In temperate climates the nursery is enclosed in a heated facility, but in the tropics these may be outdoor facilities. In some tropical farms there is no nursery phase and the PLs are stocked directly into growout ponds; however, production is both more efficient and predictable when a nursery is used. Tropical nurseries are small ponds stocked with 70 to 800 PLs/m^2. The animals are fed a diet similar to that received in the growout ponds; unlike the larvae, the PLs feed very efficiently using chemical, visual, and rheotactic senses to locate and capture food. The prawns will be large enough to be transferred to growout ponds in one to four months, depending on the desired size for stocking. Tanks in temperate climate nurseries, because they must be constructed indoors in restricted space, are stocked at densities greater than those seen in tropical nursery ponds; typically, cultures of the PLs in indoor nurseries reach densities of 1000 to 1500 PLs/m^2. The tanks may include netting or some other material that increases the surface area on which the benthic PLs may crawl.

As in the case of the penaeid shrimp, the different growout systems are classified according to the stocking density, being either intensive or extensive. Most *Macrobrachium* culture operations are extensive and require less management than their intensive counterparts. Intensive culture is carried out on some small farms and by experimentalists. Productions of up to 10,000 kg/ha/year have been reported, but the cost and technical skill needed to operate such a culture system have, so far, restricted their popularity. However, it is likely that as more is learned about the dynamics of prawn culture under high-density conditions, these facilities will become more widespread.

In extensive culture, the ponds are normally earthen and 1 hectare or less in size, although some farms do use larger ponds. Large ponds have the advantage that the shallow area around the edge of the pond is a small proportion of the total pond area, and it is in this shallow water that the prawns are most susceptible to predation by birds. (Prawns that are near ecdysis, or that have just completed ecdysis, tend to be found in shallow water, while animals in **intermolt** prefer the deeper water; it is thought that vegetation for protective cover around the edge, and the absence of the intermolt animals that attack and eat newly molted prawns, are the reasons that molting takes place along the shallow edges of the pond.) Ponds are kept at about 1 m depth; shallower water may not afford adequate protection for the prawns, and the reduced water vol-ume may be subject to rapid changes in temperature, but ponds much deeper than 1 m become difficult to harvest. A moderate phytoplankton bloom is often encouraged by fertilization of the ponds. Phytoplankton reduces visibility into the water, resulting in decreased predation by birds, as well as preventing the growth of macroalgae that hinder the harvesting process; however, as in any pond system, a heavy bloom can cause DO problems. When ponds are being constructed, the pH of the soil should be checked and adjusted to 6.5–7.0. When refilling ponds, the bottom should be treated with chlorine or rotenone to kill predators and parasites.

In theory, since *Macrobrachium* are benthic organisms, the amount of available culture space can be increased by increasing the surface area in a pond that the prawns can occupy. Because of this, some farmers place nets and branches in ponds, claiming that these have increased the yields. Biologists in Israel constructed earthen ponds, some of which included three layers of plastic nets and attached corrugated pipes, and stocked the ponds with juvenile prawns at a density of 15/m^2; at harvest in 184 days the ponds with the added substrates had a 10% better survival rate and had 24% more prawns reaching a marketable size.

The stocking rate of earthen growout ponds is extremely variable, depending on the supply of seed, the type of culture facility, and whether or not PLs or juveniles are used. In temperate environments, with a short growout season, it is best to use relatively large juveniles and a low stocking density to be sure that when the prawns have to be harvested as cold temperatures approach, they are large enough to be marketed. Two general management techniques can be used in extensive prawn culture: harvesting may be in **batches** or **continuous.**

Batch culture means PLs or juveniles are stocked at one time, and harvesting takes place after six to nine months, depending on the growth rate and the economics of the local market. Harvested prawns are usually 20 to 60 g each. At harvest time, the ponds are partially drained and the prawns are removed all at once with nets. The ponds may then be completely drained and dried and are then refilled and new seed prawns are added. Because of the very variable growth rates of the male prawns (see Figure 9.8), this method does not yield a very uniform crop. In a temperate environment, with its restricted growing season, this management technique must be used for extensive culture. Production in these ponds typically varies with the length of the growout period and the stocking density, ranging from 1000 to 3000 kg/ha/year.

In continuous culture, the ponds are drained every few years. Clearly, this method can be used only in the tropics where growth of the animal extends throughout the year. An initial seeding of the pond is made, and after four to nine months the harvesting begins. Harvesting consists of pulling of large-mesh seine net through the pond and collecting only the largest animals; the ponds are thus *selectively* harvested, giving a uniform crop. Seining takes place once or twice a month. Care should be taken to be sure that the seine net reaches the bottom of the pond because large males may avoid the net by burrowing into the circular breeding depressions they build. After the initial har-

vest, PLs or juveniles may be added periodically. Continuous culture produces a larger and more uniform crop than batch culture, ranging from 2000 to 4000 kg/ha/year.

Feeding animals in growout is necessary to get a good yield. Several artificial diets, of about 25% protein, are available that support good growth; poultry broiler feeds have been used with some success but are clearly inferior to feeds developed for crustaceans. Catfish rations have also been used. Many farmers also use rice, vegetable and animal farm wastes, mollusks, trash fish, and prawn wastes, although these may reduce water quality. Feeding, as in most culture operations, is based on the estimated biomass, with the understanding that as the animals grow they use less feed per unit biomass. *Macrobrachium* is able to derive some of its dietary requirements from the pond benthos.

There has been much interest in polyculture of the prawn. They are grown with common carp, grass carp, and tilapia, which not only increase the marketable biomass from the ponds but also seem to help maintain water quality. These fish, however, may make removing the prawns from the pond with a seine net more difficult if continuous harvesting is desired. Channel catfish have also been used experimentally; it is interesting in that, although the prawn and channel catfish are both benthic omnivores, the presence of one has little effect on the production of the other at the stocking densities tested so far.

SUMMARY

Crustaceans are members of the phylum Arthropoda; all the important cultured species belong to the order Decapoda. They are covered with a hard shell that is periodically shed (ecdysis) so the animals can grow. During mating, the male transfers a spermatophore to the fe-

Figure 9.8 A large male *M. rosenbergii*. (Photo courtesy of the University of Hawaii Sea Grant Extension Service.)

male who later fertilizes her eggs while they are being extruded. Hormones, many that are produced in the eyestalk, control ecdysis as well as reproduction.

Production of penaeid shrimp is limited by the production of seed stock. In the past, culturists would either use captured wild seed, or collect mature females and hold them until spawning. Hatcheries now often induce maturation and mating by using specially built tanks, controlling light and temperature parameters, using specific diets, and performing unilateral eyestalk ablation. Fertilized eggs are dropped by the female and soon hatch to nauplius larvae, which go through additional protozoeal, mysis, and postlarval stages. Throughout their larval existence, the shrimp are fed plankton (in most cases) and are harvested for growout as PLs. Extensive (low-density) growout in ponds requires little or no supplementary feeding, while semi-intensive and intensive (high-density) growout in ponds, tanks, or raceways may require a high-grade feed and aeration.

Crawfish are the most commonly grown crustacean in the United States, with Louisiana being the center of the industry. Culture is generally extensive, taking place in large ponds. The animals are harvested daily in traps. Harvesting starts in the winter and continues through the late spring or early summer, at which time the crawfish dig burrows in the mud. After the crawfish have entered the summer burrows, the ponds may be drained and planted so that there will be forage plants for the animals next season. The females often extrude their eggs while in the summer burrows; fertilization takes place during extrusion. In the fall, the ponds are flooded and the crawfish leave the burrows. The embryos hatch soon after the flooding, but stay with the mother for at least two molts. One of the new and popular ways to market crawfish is in a "soft-shell" state; methods have been developed to mass produce these newly molted animals.

The prawn, *Macrobrachium*, is reared throughout the world in tropical and some temperate environments. These animals are found in rivers and lakes, but females with developing embryos must migrate to brackish water because the newly hatched larvae cannot live in freshwater. The young animals develop into PLs and migrate back into rivers and streams. The prawn reproduces easily in culture; the female molts, and this is immediately followed by mating and then extrusion of the eggs for fertilization. The larvae are reared, often on brine shrimp, to PLs using either the AQUACOP method (high densities, extensive water treatment, small conical tanks) or the Anuenue method (lower densities, larger tanks, "green water"). When the prawns metamorphose into PLs, they can be transferred into a nursery pond, although not all farms use this step. Juveniles or late PLs are finally moved to growout ponds that are either extensive or intensive systems. Feeding of pelleted diets and/or plant and animal matter is required. Harvesting may be either in batches at the end of the growing season, or in tropical areas with a continuous growing season, the prawns may be harvested selectively (large animals being removed and replaced by PLs or juveniles).

CHAPTER 10

COMMONLY CULTURED FRESHWATER FISH

INTRODUCTION TO THE BIOLOGY OF FISH

Fish are the most common and diversified of all the vertebrate groups (see Table 10.1). The term "fish" generally refers to aquatic animals, often with paired fins, that obtain at least some of their oxygen via gills throughout their entire life cycle (see Figures 10.1 and 10.2). Aquaculturists are interested in only a small group of these organisms, all belonging to the order Teleostei.

Movement

Most fish have a density only slightly greater than that of water. Nevertheless, some mechanism must exist so the fish can maintain its neutral buoyancy; for most fish, this mechanism is the **swim bladder.** There is a connection between the esophagus and the bladder in some fish, although not in many adult marine species. The bladder wall is a densely woven structure with a layer of overlapping guanine crystals making it essentially impermeable to gas. A **gas gland** removes gas from the blood and shunts it into the bladder. As the fish swims down, the pressure on the

gas bladder is increased so it is compressed and the fish becomes more dense; to compensate for this, more gas is shunted into the bladder so it will expand. When the fish swims toward the surface, some of the gas must be released or the bladder would enlarge under the reduced pressure and the fish would ascend uncontrollably to the surface like a balloon, with the bladder continuing to expand as it rises.

Fish move through the water with minimal effort because they have a body shape that reduces the resistance of the water. Swiftly moving fish are especially well adapted to reduce drag and turbulence. The relatively cylindrical shape of the body, with a rounded front end and a tapering rear end, are streamlining modifications. Other adaptations to reduce drag include hydrodynamically shaped scales and a coat of slime (which also protects the fish from disease).

Fish move themselves through the water by the **caudal** (tail) fin. As the caudal fin is moved back and forth, it pushes against the water and the body of the fish makes alternating S-shapes. This motion is a result of the contraction and relax-

TABLE 10.1   A Classification of Living Fish

I. Agnatha—no jaw or paired fins, cartilaginous skeleton (e.g., lampreys)
II. Gnathostomata—jawed fish
 A. Elasmobranchiomorphi—cartilaginous skeletons
  1. Holocephali (e.g., chimaeras)
  2. Elasmobranchii (e.g., sharks, skates, and rays)
 B. Teleostomi—bony fish
  1. Crossopterygii—lobe fins (e.g., the coelacanth)
  2. Dipnoi—lungfish
  3. Actinopterygi—rays in the fins
   a. Polypternini (e.g., bichirs)
   b. Chondrostei (e.g., sturgeon)
   c. Holostei (e.g., garpikes)
   d. Teleostei (contains about 90% of all living fish, including herring, salmon, eels, cod, tuna, seahorses, sea bass, barracuda, puffers, perch, flatfish, and others)

ation of muscles, the **myomeres,** on the sides of the fish. If the myomeres on the left side of the fish contract and those on the right side relax, then the front and back of the fish will bow to the left; this is followed by a relaxing of the left side and contraction of the right side.

The shape of the caudal fin also affects the way the fish moves. Fish with a rounded or squared-off tail, such as the killifish, *Fundulus,* are capable of rapid acceleration and quick maneuvers; fish with a caudal fin that is tall and deeply forked,

such as pompano or dolphin fish, can swim for extended period of time at high speeds, but are not able to move slowly or make quick turns efficiently. **Pectoral** and **pelvic** fins are used for maneuvering and slowing down when extended, but are normally held along the fish's side when swimming so that resistance will be minimized.

Respiration

Like all animals, fish need oxygen. A few fish can get a significant amount of their oxygen directly from the air; the fishes that are best adapted to do this are native to shallow water that is either slow moving or stagnant. When the fish needs oxygen it rises to the surface, its mouth extending out of the water, and gulps in air that is passed to the **air-breathing organ.** There are a variety of these specialized organs; in some fish the inner walls of the swim bladder have alveoli, like our lungs, while in the cyprinoid loaches, most of the gut wall contains a rich blood supply so it can be used for breathing rather than digestion. There are other such adaptations.

For most fish, the primary mode of respiration is uptake of oxygen from the water by way of the **gills.** The fish takes in water through the mouth when it expands the **buccal cavity,** thus reducing the pressure in its mouth so the water is sucked in. The mouth is then closed, the buccal cavity contracted, and at the same

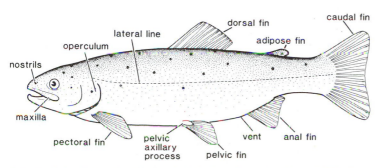

Figure 10.1   External features of a fish. (Laird and Needham, 1988. Copyright © Ellis Harwood, Ltd. Reprinted by permission.)

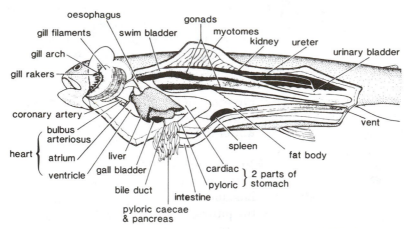

Figure 10.2 Internal anatomy of a fish. (Laird and Needham, 1988. Copyright © Ellis Harwood, Ltd. Reprinted by permission.)

time, the **operculi** are expanded to reduce the pressure in the branchial cavity pulling the water over the gills. Then the branchial cavity contracts and the water is pushed out the opercular openings. The gills are supporting arches to which are attached filaments. On these filaments are numerous small, fingerlike extensions called **lamellae,** that are sites of gas exchange. These lamellae are rich in tiny blood vessels. As the oxygen-poor blood passes through the gills, the oxygen in the water diffuses into the blood; the efficiency of this gas transfer may be as great as 85% (our lungs take up only about 25% of the oxygen we breathe in).

Some of the oxygen that is taken up is simply dissolved in the blood plasma; the colder the water, the greater the amount that is dissolved. But most of the oxygen that is carried in the blood is bound to the respiratory pigment, **hemoglobin,** in the red blood cells. The hemoglobin molecule is constructed so that it will combine with O_2 in an oxygen-rich environment, such as the gills, but it will release the O_2 when the blood moves through the oxygen-poor inner tissues. In addition, high levels of CO_2 in the oxygen-poor tissues reduce the pH, which also causes the oxygen to disassociate from the hemoglobin.

Senses

Vision is important for some fish, and less so for others. In general, those fish found in slow-moving water, murky with suspended sediments and algal blooms, rely less on sight than fish found in clear water. In many ways, the eyes of fish are different from those of land vertebrates. The lens of the fish eye bulges through the pupil, but in the other vertebrate eyes, the lens lies behind the pupil. The cornea of the fish eye does not bend light, as does the human eye's, since the cornea has the same density as water; rather, the light is not bent until it reaches the lens. The lens of the fish eye bends light more than the lens of other animals, since the cornea does none of the focusing. In most fish, the eyes protrude from the side of the head to allow the fish to have expanded field or **all-round** vision. All-round vision is important because fish have no neck to swing their head around; with all-round vision, fish can see what is on either side of them without having to rotate their body.

Fish can also hear. The inner ears are in the skull, just behind the eyes of most fish. In many ways they are similar to the ears of other vertebrates, containing three **semicircular canals** lined with sensory hairs that send messages to the brain. These canals also control the fish's balance, which is why when the canal is attacked by whirling disease (see Chapter 6), fish lose their sense of direction.

Fish can also "hear" using their **lateral line system.** Along the sides of the fish are canals with pores that receive signals from the environment. Inside the canal is a series of small cells containing cilia; if there is a change in the water pressure on one side of the fish—as might be caused by something approaching—nerve cells from the canals signal the brain as a result of cilia having been bent by that change in pressure. The lateral line system, along with vision, are used in fish schooling behavior.

Taste and smell of some fish are very acute. **Nasal sacks** are housed in bony capsules that are found on either side of the snout. These are lined with sensory cells that are in communication with the brain by way of the olfactory nerves. The water enters the anterior nare of the sack, passes over pleated tissues (which increase the sensing contact area of tissue), and the water leaves through the posterior nare. **Taste buds** are found in the mouth, on the lips and gill arches, and on the barbels if present.

Reproduction

In some fish the sexes can be easily distinguished from each other, but in other cases an internal examination during the spawning season is needed. At spawning, the male will have smooth, white, long testes that will make up no more than 12% of the animal's weight, while the ovaries are large, granular structures, often with a yellowish appearance, that are 30% to 70% of the fish's weight when reproductive.

The fish ovary is a mass of eggs and follicle cells, bound together by connective tissue. The eggs are usually very small, but as the breeding season approaches they enlarge with yolk; when they are ready to be laid, the eggs are released from the follicles. The testes are a mass of tubes and loops of sperm-producing cells.

Much is still to be learned about the reproductive endocrinology of fish, but progress is being made rapidly. Sense organs send signals to the fish's brain that the breeding season is approaching. The production of gametes and sex steroids is under the control of a **gonadotropin** produced by the pituitary gland in the brain. In "higher" vertebrates, the **hypothalamus** produces a **LH/FSH-releasing hormone** (LH/FSH–RH) that acts on the pituitary gland. There are two gonadotropins: the **follicle stimulating hormone (FSH),** which stimulates the growth of the ovarian follicles, and the **luteinizing hormone (LH),** which, along with the FSH, stimulates ovulation and production of sex steroids, including **androgens** in the male. In the fish, one hormone seems to carry out the jobs of both LH and FSH. In the laboratory, mammalian FSH has been shown to have no effect on fish, but LH elicits gamete production and steroidogenesis, which are associated with both hormones in the higher vertebrates. The fish gonadotropin regulates not only gamete production, but also metabolism, behavior, and the development of gonadal endocrine tissue.

The sex steroids include the male androgens, such as **testosterone,** and the **estrogens** and **progesterone,** which are important for the female. The gonadal endocrine tissue hormones eventually take over some of the functions of the gonadotropin, such as continued gamete development, reproductive behavior, spawning, and sometimes parental care.

In fish culture, reproduction is sometimes stimulated by injection of hormones. These may be fish gonadotropin, usually extracted from the carp pituitary gland, or mammalian hormones. Intramuscular injection of pituitary suspensions or extracts into the fish is called **hypophysation;** the dosage is a function of the size, sex, and species. Pituitaries can be removed from the carp, dried and pulverized, then dissolved in a saline and glycerine solution. Three mammalian preparations also can be used:

1. Serum from **pregnant mares (PMS)**
2. **Human chorionic gonadotropin (hCG),** a placental hormone prepared from urine samples, which acts like LH
3. LH or synthetic analogs of the LH–RH

These hormones are used to promote the final stage of egg maturation, called **ovulation,** the discharge of the ova from the ovary.

Fish display an array of behaviors and strategies that are related to the mode of reproduction, the fish's morphology, and the ecology of the species. These behaviors can be classified according to the system of E. K. Balon as follows:

1. **Nonguarders**—fish that leave the eggs or young to fend for themselves with no parental protection. These may simply spawn in an open area (e.g., carp) or may first hide their eggs (e.g., salmon).
2. **Guarders**—fish that protect the eggs until hatching or through the larval stages. The eggs may be laid on a natural substratum, or the parents may build a nest.
3. **Bearers**—fish that carry the embryos with them; this sometimes extends to the young fish after hatching as well. The young may be carried externally, as in the case of the seahorse, where the young develop in a "pouch" on the male, or the mouth-brooding tilapias. Embryos may also be carried internally following internal fertilization. The eggs can be laid soon after fertilization to hatch in the environment, or they can feed on yolk sac nutrients and be born "live" after the yolk sac is depleted, or they can dis-

play a mode of reproduction, **viviparity,** in which the fish are born live after an internal development period during which they obtain nourishment directly from the mother rather than from an individual yolk sac.

Growth and Development

The fish starts its life as a fertilized egg that begins to cleave and take shape, becoming an **embryo.** The embryo has clearly developed vertebrate characteristics but is still living on the yolk sac nutrients (except in the case of viviparous fish) that are supplied by the mother. The fish may hatch and begin to swim after the yolk sac is fully absorbed, or they may retain the sac for a short period of time after hatching. When hatching is completed, the fish begins its **larval period,** marked by the beginning of feeding. The larval stage gives way to the **juvenile stage,** gradually in some fish and with a distinct metamorphosis in others. Early juveniles are sometimes called **fry** and late juveniles **fingerlings;** these terms often have different meanings to different fish farmers. Juvenile fish are miniature adults, with fully formed organs, but often have distinctive color patterns. The juveniles become **adult** fish when the gonads begin to mature during the approach of the fish's first spawning season. Adult fish may enter a senescent period if they live long enough, when growth stops and the gonads degenerate.

The growth of fish, and most other animals, can be graphically expressed as an S-shaped curve (see Figure 10.3). Small fish grow slowly, in terms of absolute weight gains (although they often have a rapid relative weight gain) because a small fish has a small mouth and gut, and is therefore unable to take in enough food to gain weight quickly. As the fish gets larger, it is able to eat more, and it grows more rapidly. The most rapid abso-

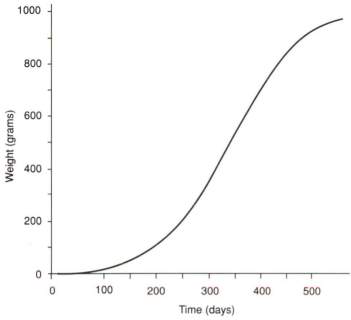

Figure 10.3   The typical sigmoid growth curve. (Laird and Needham, 1988. Copyright © Ellis Harwood, Ltd. Reprinted by permission.)

lute growth usually takes place during the later juvenile stages when a large portion of the nutrients and energy consumed go to building new tissue. In adult fish, more energy goes to maintaining the animal and less to growth; in addition, a considerable amount of energy is used to produce gametes. As the fish approaches senescence, all the energy that is consumed goes toward maintaining the animal, with little or none going toward growth or reproduction.

Because of this growth pattern, the best time to harvest a fish is just before the animal matures. Up to this point, the fish has been growing quickly, and a large portion of the food it has consumed has gone to increasing its size. Often, because of the demands of the market, it is not possible to harvest juvenile fish; rather, fish must be raised to adults before being harvested. If adults are to be the end product, they should be harvested before the reproductive season; the farmer does not want to feed fish that do not increase in fillet size and are only using the feed to make gonadal tissues.

Culture of Freshwater Fish

Aquaculture probably started with the culture of freshwater fish, and the continued commercial rearing of these animals makes this form of aquatic culture the most economically important to this day. In this chapter we will consider the most commonly reared species, although many other freshwater fish are being grown commercially and experimentally, for food and as aquarium pets (see Appendix 8).

CATFISH

Introduction

Catfish are the most commonly grown fish in the United States (see Figure 10.4). In 1960 there were 400 acres devoted to their culture, but this has grown to tens of thousands of acres, located primarily in the southeastern states. Since most species of cultured catfish will cease to eat (and therefore grow) when the water is cold, there are only a limited number of catfish farms in the northern states. Conse-

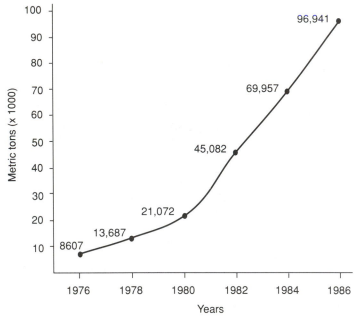

Figure 10.4 Farm-raised catfish production in the United States.

quently, catfish have a much greater public acceptance in the South, though recently the industry has been opening new markets around the country with an advertising–information program.

The catfish industry in the United States utilizes several types of farms:

1. Farms specializing in the production of fingerlings and/or broodfish for sale to other farms.
2. **Fee lakes** that are stocked with catfish and are open to the public for fishing, charging either a flat rate or a rate based on the number or weight of the fish caught.
3. Farms concentrating on production of fish for human consumption; catfish are usually stocked as fingerlings and grown out to 0.4–1.4 kg, harvested on a large scale, and sold in bulk

Many farmers have combined all these types of farming to increase the profitability of their operation.

Species

There are over 1200 different marine and freshwater catfish, of which only a few freshwater species are important to the commercial farmer. In Hungary and Yugoslavia, the culture of the European catfish—sheatfish, or **wels** (*Silurus glanis*) —is fairly well developed. These tasty and fast-growing fish may exceed 2 m in length. Wels are spawned in ponds, the eggs are removed, and hatching takes place after several days. The larvae are fed plankton, then a diet of trash fish, liver, and spleen. They are stocked in ponds that often contain spawning carp and tench, the fry of which serve as food for the growing wels. These catfish are often harvested at 0.5–1.4 kg after two years of culture, but are sometimes kept for three years (1.5–2.0 kg).

Two catfish of the family Clariidae, *Clarias batrachus* and *C. macrocephalus*, are raised in Thailand, Vietnam, India, Pakistan, and Malaya. These fish are very hardy, rapidly growing animals; *Clarias* is a good air-breather (therefore easily transported when kept moist and cool), which sometimes comes ashore after dark to hunt for food. Fry are stocked in ponds or concrete tanks and are fed trash fish, pellets, and rice bran. *Clarias* are generally harvested when they are small (about 150 g), but there are two to three harvests per year, and they are stocked at 50 fingerlings per square meter, ensuring a high annual production.

Catfish belonging to the family Pangasiidae, in particular *Pangasius*, are cultured in Thailand, Vietnam, Laos, and Cambodia. These fish are usually grown in floating cages and fed trash fish, aquatic vegetation, rice bran, and other plant products. *Pangasius* cages may be densely stocked (although there is a high mortality rate). These fish grow from 80 g to 1 kg in 8 to 10 months.

All catfish cultured in the United States belong to a family of the Ictaluridae. Some farmers grow the **blue catfish,** *Ictalurus furcatus*, or the **white catfish,** *I. catus*, but by far the most commonly cultured species is the **channel catfish,** *I. punctatus* (see Figure 10.5). **Bullhead catfish** are considered a pest by most catfish farmers; bullheads, or mudcats, includes the speckled bullhead, *I. nebulosus marmoratus*, brown bullhead, *I. nebulosus*, black bullhead, *I. melas*, and yellow bullhead, *I. natalis*.

It has been suggested that bullheads themselves may be suitable for culture. They require less management than the more commonly cultured species and accept a wider variety of feeds. Bullheads usually can survive very cold temperatures, although they will stop feeding. However, there are problems with bullhead culture; the flesh is red rather than white like the more popular species, and they tend to reproduce quickly, so ponds become overpopulated with stunted fish. When kept in warm water, the flesh tends to become soft and is therefore not often eaten. Many problems associated with bullhead culture probably can be overcome with improved manage-

Figure 10.5 A channel catfish. (Photo courtesy of the New Jersey Division of Fish, Game, and Wildlife.)

ment techniques, and these species may have a future in aquaculture.

The following discussion pertains to the channel catfish.

Spawning

If fingerlings are to be produced by the farmer, broodfish must be selected carefully. Female catfish are characterized by a flat, rounded genital opening, a wide abdomen, and a small head. A large rounded belly and soft ovaries are characteristic of a good spawner. Farmers stock one or two males for each pair of females. Males have a nipplelike **genital papilla,** a wide head, and a relatively thin abdomen. The characteristics of a male brooder are a darkening under the body and a protruded genital papilla. Although catfish are capable of reproduction at 0.3 k and can grow very large before reproduction ceases, broodfish generally weigh 0.9–4.5 kg because they produce more eggs than smaller fish, but are easier to handle than the very big catfish. Broodfish must receive special care year round, and their diets are supplemented with meat, fish, tadpoles, or crawfish. Farmers may also have minnows in the ponds as forage for brooders; this is especially important during the winter months. (Recently, research has suggested that tilapia may also be used as catfish forage, although they must be restocked each year because winter temperatures will kill them). Females usually produce 7500 to 10,000 eggs per kilogram of body weight, but the rule of thumb is to assume there will be about 5000 eggs per kilogram of body weight surviving to fry.

Catfish spawning takes place naturally in the early summer when the water warms. Spawning can be encouraged to occur slightly earlier by using small brooding ponds that can be warmed by the farmer. Hormone injections can also be used to induce spawning behavior, if needed. Some culturists place artificial nests in ponds; these are usually large cans, kegs, or sections of pipe (with one end closed or placed against the pond bank). Other farmers may use pens specially constructed for spawning, equipped with receptacles for the eggs (see Figure 10.6). Pens allow the farmer to select the pairs of fish that will mate; males should be slightly larger than the females, or the female's aggressive behavior will pre-

vent successful spawning. The female should be removed from the pen as soon as spawning has taken place.

In natural situations, the eggs are guarded by the male catfish after fertilization. In culture, the eggs can be left to hatch in the pond, or they can be removed and hatched mechanically. Removal frees the nest to be used by another pair of catfish for spawning. If the eggs are removed, however, great care must be taken not to damage them physically or subject them to temperature shock or direct sunlight, for damaged eggs mean high mortality rates and deformed fry. When the eggs are removed from the nest, they are either placed in jars receiving running water or hatched in troughs. In either case, it is very important that the water be oxygenated and maintained at 26°–28°C, at which temperature the yellowish eggs turn pink, then red, and hatch in about six to eight days.

Recently, there has been much interest in the genetics of catfish, led by biologists at Auburn University. The production of different strains of channel catfish and species hybrids could be important to the industry. The most desirable trait to select for may be rate of growth. Using four strains of channel catfish, growth was experimentally selected for over two gener-

Figure 10.6   Catfish spawning pens. (From Stickney, 1979.)

ations by breeding the 10% heaviest males with the 10% heaviest females. The result were fish that were, on average, 17% to 20% larger than the original catfish. In another experiment using a hybrid, produced by male blue catfish × female channel catfish, grown for a period of 220 days in earthen ponds, it was found that the harvest of the hybrids was 13.5% greater than that of channel catfish. Additionally, the hybrids were more easily captured by seining, more uniform in size, and had a greater **dress-out percentage** (weight of the fish after the head and skin are removed and it is eviscerated, divided by the live weight of the fish).

Culture Conditions and Growout

Fry are removed from the hatching facility and transferred to a rearing trough or pond where the fish are given a special prepared diet (see Figures 10.7 and 10.8). The water must be warm (24°–29°C) and the DO should be at least 6 ppm. The stocking rate during rearing is guided by the desired size of the fingerlings to be harvested.

Catfish are grown out from fingerlings in a number of different ways, depending on the nature of the catfish farm, its size, and the local environment. Growout of-ten takes place in large ponds, 4 to 8 hectares or larger (smaller ponds are usually constructed for holding and spawning brooding fish or for growing fry into fingerlings). Growth is best in warm water, ideally at about 29°C. The pH of the ponds should be kept at 6.3 to 7.5, and moderately hard water is often encouraged. The DO must be kept at 4 to 5 ppm 6 inches below the surface; DO below 1.0 ppm is normally fatal in a short time. Nighttime DO is one of the most critical problems for catfish farmers. When the levels of O_2 are too low, the fish come to the surface and gulp air; the water should be aerated immediately if this is observed. Paddlewheels are commonly used in catfish ponds; some other aerators, including the turbine types, are used to increase circulation and inhibit stratification.

The stocking density and temperature will naturally affect both the growth rate and the feed requirement. During the past three decades, major changes in stocking and harvesting practices have occurred. Three pond techniques are the following:

1. **Batch harvesting** is the oldest method and is sometimes still used. All the fish are harvested at once.

Figure 10.7 Harvesting catfish fingerlings. (Photo courtesy of the New Jersey Division of Fish, Game, and Wildlife.)

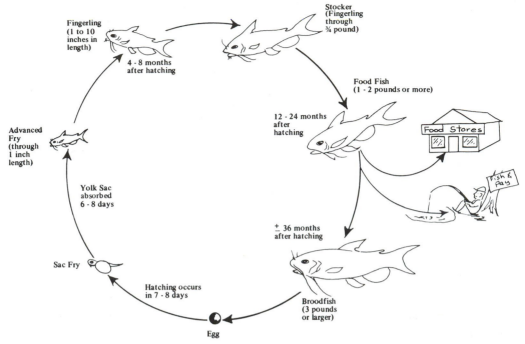

Figure 10.8 Stages of catfish growth. (Lee, 1981. Copyright © The Interstate Printers & Publishers, Inc., Illinois. Reprinted by permission).

Ponds are best stocked with fingerlings in the spring at 6500 to 9000 fingerlings per hectare.

2. Ponds are stocked with 12,000 to 24,000 fingerlings per hectare. When 10% to 20% of the fish reach marketable size, the ponds are partially harvested with size-selective seine nets (**grader** nets); the large fish are removed, allowing the remaining catfish to grow more quickly. This procedure is then repeated regularly.

3. **Topping off** is the current method of choice. Ponds are stocked with 12,000 small and 12,000 medium-sized fingerlings per hectare. Again, when a certain percentage of the fish reach harvestable size, grader nets are used to remove them; these large fish are replaced with fingerlings.

There are alternatives to the traditional methods of pond culture which are used experimentally and commercially. When the size of the farm is restricted, methods to consider are the following:

1. Cage culture is popular with many modern farmers. Production in cages is approximately 1700 kg/ha. A high stocking density in the cages inhibits growth, but a low density allows the fish to display aggressive behavior, which results in a poor feed conversion and high mortalities.

2. Sometimes catfish are grown in raceways and in large round or rectangular tanks.

Catfish are omnivores, and in the wild normally eat plants, algae, insects, crawfish, and other fish. Catfish feeds that are produced commercially for pond culture usually contain 25% to 35% protein, but those used in raceways and tanks tend to have as much as 40%; feeds for fingerlings have 30% to 36% protein. Catfish receive 3% to 5% of their weight in feed per day when first transferred from a rearing pond or trough to a growout pond; this drops to 0.5% to 1.25% near harvesting, depending on the temperature. A good food conversion ratio for fingerlings is 0.9 (that is, 0.9 kg of dry feed is converted to 1.0 kg of live fish tissue), and for second-year fish, this is about 1.5 to 2.5. Most farmers feed their fish at the same time every day, usually midday when the DO is greatest. Fish in tanks, raceways, and small ponds are usually fed by hand, but for large ponds the feed must be dispensed using floating automatic feeders, large mechanical blowers, boats, or even airplanes.

Catfish are generally grown in monoculture. However, sometimes other fish are added to increase the pond's production. The big-head carp, which consumes catfish wastes and uneaten pellets, may be added to the ponds as fingerlings at 200 to 350 per hectare and be harvested after one season weighing 1.0 to 3.5 kg. Phytoplankton-grazing tilapias and the silver carp have been used to reduce the chances of unwanted algal blooms developing.

A sex ratio of 1 : 1 is normally found in catfish ponds. But since males grow slightly quicker than females, an all-male pond should be more productive. All-male populations have been generated experimentally by feeding fingerlings a diet containing male sex hormones, and by using sex-reversed females (genetically males) to produce all-male offspring.

Harvesting

After the fingerlings or stockers have been transferred (see Figure 10.9) to the rearing facility and then grown to the desired size, they must be harvested. For fee fishing operations, this simply means allowing the customers to catch the fish one at a time. Traps, baited with cakes of soybean or cottonseed, can also be used when partial harvesting is to be carried out. When a large portion of a pond is to be harvested, or large fish are to be selectively removed from a pond, seining nets are used. Seines are about 70% to 85% efficient and are therefore used for limited harvesting. Although the cost of a large seine net can be high, it is inexpensive to use, requiring a few **pullers** and/or a motorized winch. Some farmers will drain the ponds; this is extremely efficient and allows the farmer to examine the condition of the pond bottom. Draining is often done instead of seining

Figure 10.9 Truck for hauling catfish for stocking. (Photo courtesy of Peterson Fiberglass Laminates, Inc., Wisconsin.)

in ponds that have obstacles (such as tree stumps) that prevent harvesting by seine nets. However, pumping the water and refilling the ponds after draining can be expensive, so it is best suited to smaller ponds.

Catfish will sometimes develop an off-flavor, often due to blooms of cyanobacteria. Before a harvest, several fish should be taken for cooking. If the off-flavor is present, harvesting should be postponed, and the fish placed in vats of fresh running water and fed only a diet that is known not to cause the off-flavor, or they can be starved while purging the offending chemical from their system. The condition usually clears itself up in a few days.

CARP

Although not often grown in the United States, **carps** are probably the most commonly cultured fishes in the world. Brought to the United States in 1877 to be cultured for food, they never became popular with the consuming public, but the original carps became established in the natural water systems, and today wild carp are found throughout most of the country.

There are many different types of carps, which belong to a large group of fish called the cyprinids. The most popularly grown species is the **common carp, *Cyprinus carpio,*** which is raised all over the world. Two important varieties of this fish are cultured: a **fully scaled strain** (see Figure 10.10) and the **partially scaled mirror carp** (see Figure 10.11) that is reared in Europe and Israel.

Other species are also commercially grown, namely,

1. The European **Crucian carp,** *Carassius carassius.*
2. The **Indian carps,** of which the most important are the catla (*Catla catla*), rohu (*Labeo rohita*), calbusa (*L.*

Figure 10.10 A strain of the common carp used in culture. (Photo courtesy of Dr. Ferenc Muller, Fish Culture Research Institute, Szarvas, Hungary.)

Figure 10.11 Mirror carp. (Photo courtesy of Dr. Ferenc Muller, Fish Culture Research Institute, Szarvas, Hungary.)

calbasu), and mrigal (*Cirrhinus mrigala*).

3. The **Chinese carps,** raised not only in China but in Japan and throughout Southeast Asia as well, are the sandkhol carp (*Thynnichthys sandkhol*), ca choi (*L. collaris*), ca ven (*Megalobrama bramula*), mud carp (*Cirrhinus molitorella*), the snail or black carp (*Mylopharyngodon piceus*), grass carp or white amur (*Ctenopharyngodon idellus*) (see Figure 10.12), big-head carp (*Aristichthys nobilis*) (see Figure 10.13), silver carp (*Hypophthalmichtys molitrix*), and others. Of these, the grass, silver, and big-head carps have begun to interest European and Israeli culturists; African fish farmers may also find these warm water species profitable.

Figure 10.13 Big-head carp. (Photo courtesy of Dr. Ferenc Muller, Fish Culture Research Institute, Szarvas, Hungary.)

Breeding

Carp are sometimes collected from wild stocks for culture, but normally they are bred in captivity (many strains of carps are generations removed from the wild). The common carp breeds throughout the year if kept warm; otherwise, breeding is seasonal and begins with rising spring temperatures. Breeding usually takes place after dark or in the morning. The eggs are laid on plants or branches near the water surface.

Just prior to spawning, the sexes of the common carp can be distinguished. The belly of the female swells as the ovaries develop, and the male releases milt easily when subjected to gentle pressure on the abdomen. At this time, the farmer can separate the sexes to prevent unwanted spawnings. Segregated brooder ponds should be rich in natural food, or a high-protein diet can be used. In the female brooder ponds, all large plants must be

Figure 10.12 Grass carp. (From Stickney, 1979.)

removed or spawning may take place even without the males. Brood stock should be healthy and show no signs of deformity. Fish that are 2–3 kg are often used as brooders and can be bred over several years until they reach about 8 kg, when they are too difficult to handle.

There are several related breeding techniques, all based on the common carp's natural reproductive behavior. The culturist can stimulate spawning by transferring the carp to shallow ponds with slightly elevated temperatures that contain vegetation (such as grasses, hyacinths, or other water plants) onto which the eggs can be laid (see Figure 10.14). In India, a cloth tank, called a **hapa,** is placed in the spawning pond instead of the plants, and in Indonesia a fiber mat, called a **kakaban,** is used. Israeli farmers use a plastic green fibrous material that serves as the site for egg laying.

In tropical environments, well-cared-for common carp will spawn two to four times a year, but in temperate water, the spawning is less reliable, so pituitary gland extracts are used. (Unlike most other fish, the common carp must receive the pituitary extract of its own species only.) After injection, the fish are either allowed to spawn naturally in ponds, or are hand stripped and the eggs and milt are mixed. When spawning takes place in ponds, the eggs should be moved to a **hatching pond** if possible, but if the eggs are allowed to hatch in the **spawning pond,** the adult fish should be seined out to prevent them from feeding on the newly hatched carp and to decrease the chance of diseases being transferred from the brooders to the sensitive larvae. If the brood fish are stripped rather than allowed to spawn in the pond, a small amount of water is added after the mixing of the gametes to stimulate fertilization, and the eggs are transferred to a surface for hatching. The eggs may also be incubated in hatching containers, but first they must be treated with salt and urea, or salt and milk, to eliminate their adhesive properties.

Selection of quickly growing fish for spawning can lead to dramatic increases in productivity for the carp farmer. However, inbreeding (even between cousins) can cause significant stunting of common carp. Several hybrids of the common carp show considerable promise with respect to growth, cold tolerance, and resistance to disease; these include crosses with the mud carp, Prussian carp, Amur carp, and several others. Generally speaking, simply taking common carp of different genetic lines and using these as brood fish will result in higher yields, better growth rates, and fish with less fat in the tissues.

Spawning of Chinese carps takes place naturally in swiftly moving rivers. In China, for a long time, the fry collected in the spring and summer were the only source of these fish. In other locations, there is either irregular spawning or none at all. A method for conditioning and hormone injection has been developed so the production of fry, despite the fish's delicate nature, is no longer a great problem. Artificial spawning of Chinese carp should be done in water that is 23°–29°C, clear, well-aerated, and slightly alkaline. The size and age of reproductive animals varies considerably with the climate and environment; for example, female grass carp reared in the central USSR become mature after 8 to 10 years, while in Malaysia it takes 1 to 2 years. As in the case of

Figure 10.14 A pond used for spawning carp. (From Hepher and Pruginin, 1981.)

the common carp and Indian carp, the eggs and brood fish are separated after spawning.

In some cases it may be considered an advantage that Chinese carp will not spawn unless treated, since unwanted fry can stunt the growth of the stocked fish. Even in environments where the fish may reproduce naturally, this can be prevented by using monosex fish for stocking. Monosex silver carp can be generated by treating young females with androgens that will turn them into **sex-inverted males;** the sex-inverted males, which are genetically female, are then crossed with normal females to produce all-female progeny. In the United States, triploid grass carp have been used to control growth of aquatic weeds, but are sterile because of the third set of chromosomes that causes meiosis to fail, so again there will be no unwanted spawnings that might damage the environment; triploidy of grass carp can be induced by temperature shock soon after fertilization.

Indian carp will normally only breed in the moving water of a stream or river. Culturists in India have therefore built large ponds or reservoirs called **bunds.** The bunds are constructed in such a way that during the rainy season, an inlet is opened so that water can circulate and spawning is thus stimulated. Areas of the bund are planted with grasses during the dry season; the grasses keep the water from becoming too murky and serve as a site for the attachment of eggs. After spawning is complete, the eggs are removed by dragging a seine net through the area. The eggs are transferred to nearby **hatching pits.** When a bund cannot be built, for topographical reasons, Indian carp can be induced to spawn by using pituitary extracts.

Hatching and Rearing

The eggs of all carps are highly prone to infection and predation. The fungus *Saprolegnia*, for example, is a particular problem for the common carp, but can be treated by flushing the eggs with malachite green or some other fungicide. If hatching ponds are to be used, potential parasites and larvae of predators should be eliminated by treating the pond in some way, such as adding lime or drying the ponds prior to filling.

Better survival can be expected if the eggs are exposed to only clean flowing water in tanks, jars, baskets, or trays, until they are just ready to hatch (see Figure 10.15). For the common carp, the ideal hatching temperature is 20°–25°C slightly warmer temperatures are used for Indian and Chinese carp. The newly hatched fry begin to feed on zooplankton such as protozoans, rotifers, and small crustacean nauplii. This is followed by copepods and other benthic or planktonic crustacea. If the fry are to be reared initially indoors, brine shrimp nauplii can sometimes be used for the common carp.

If the carp are hatched in ponds, the ponds must be built in sunny areas to encourage the growth of phytoplankton and zooplankton. Hatching ponds are often fertilized to increase production. These ponds should not be too crowded; if the density is kept low, the newly hatched carp pass through the most vulnerable stage (from hatching to 3 g) relatively quickly. If the density is too high, the growth is slowed and there are higher mortalities. Different fertilization schemes are used for hatching ponds of

Figure 10.15 Containers of carp eggs in a hatchery. (Photo courtesy of Dr. Ferenc Muller, Fish Culture Research Institute, Szarvas, Hungary.)

each of the Chinese carp; culturists also supplement the ponds with a variety of feeds, including soybean and peanut products, small snails, duckweed, flour, and eggs. Indian carp culturists often use powdered oil cakes and rice bran, although it is not clear whether the fish feed directly on them or if they serve to increase the production of zooplankton.

Immediately after hatching, the larvae do not begin to swim, but rather they attach themselves, using a cement gland, to the walls of the rearing containers, rocks, or plants. In less than a week, the yolk sac is absorbed and the fish begin to swim and feed. The fry are then transferred to nursery ponds soon after swimming has begun.

Growout and Harvest

After common carp begin to swim in the hatching pond, they can be transferred into slightly larger and deeper ponds called the **rearing ponds.** The stocking density of the rearing pond is a function of the desired size of the young fish when they are harvested. In Israel, if the fish to be harvested are 10–15 g, then the rearing pond is stocked at 10^5 carp/hectare, while fish to be grown to 100 g are stocked at 10^4 carp/hectare; generally, the final biomass at harvest will be about 1 metric ton/hectare. At this stage, the fish are fed ground cereal grains and later the whole grains.

Finally, the carp fingerlings are moved to the large **production ponds** for growout to marketable size. Again, the size of the fish that is produced is a function of the amount of space that is available, and whether the water is still or moving. Circulating water will greatly increase the production per unit volume of the ponds, perhaps by as much as an order of magnitude. Sometimes floating or submerged cages in rivers or streams can be used to increase production (see Figure 10.16). Recirculating systems can also be used, but these are costly to operate.

Figure 10.16 Floating cages for carp culture. (Photo courtesy of Dr. Ferenc Muller, Fish Culture Research Institute, Szarvas, Hungary.)

Production can also be increased by polyculture. The Chinese system of carp polyculture has been discussed in Chapter 6; the common carp is often grown in Europe with one or more species of Chinese carp or crucian carp, goldfish, tilapia, mullet, bream (*Abramis brama*), tench (*Tinca tinca*), or roach (*Rutilus rutilus*). Sometimes pike and trout are cultured with common carp; these are used to reduce the pond's population of young carp, a result of unplanned spawnings in the production pond, or to eliminate trash fish that have established themselves in the carp ponds. Like the catfish in the United States, small carp fingerlings are sometimes mixed with larger carp for continuous harvesting.

Common carp are not only fed a prepared diet, but ponds are often fertilized as well, so that the fish's natural foods (zooplankton, phytoplankton, and small benthic animals) are abundant. These fertilizers can take a number of forms, including manures from farm animals, sewage effluent, and wastes from food-processing mills. The artificial feeds should supply a balanced diet; if the ponds are fertilized, the feeds often contain high levels of cheap and energy-rich carbohydrates, but relatively little (10% to 15%) expensive protein. Recent experiments in Israel have indicated that the addition of **growth promoters,** commonly added to diets of poultry and cattle, may also stimulate the growth of the common carp; these growth promoters seem to act by selectively affecting the gut microflora.

Chinese carps are largely supported by the natural production in the fertilized ponds. If supplemental feeding is carried out for the grass carp, it is normally in the form of vegetable matter (especially duckweed) added to the pond; these fish eat pellets and grains, but use them inefficiently. The silver carp normally feeds on phytoplankton and very finely ground feeds that may be added to the ponds.

Carp are harvested at 0.5–2 kg, depending on the local market. The length of time before harvesting will vary greatly depending on how the fish are cultured, temperature, water quality, and feeding. Untreated ponds may yield 25–400 kg/ha/year, while under very intensive conditions in Japan rates of 220 kg/m² have been claimed! For the Chinese carps, production averages about 4000 kg/ha, but much higher production has been reported. In India, production of carps is about 900 kg/ha. Carp ponds are harvested by draining or seining; silver carp are a problem during seining since they tend to jump over the tops of the nets.

In Europe, where temperatures can fall below freezing, and the demand is for larger fish, the carp must overwinter. Deep wintering ponds are used. These ponds are kept dry during the summer and should have little organic material on the bottom that could rot (thus using up oxygen during the winter when the ponds are filled). In addition, water should be exchanged every 10 to 12 days to ensure that there is enough oxygen for the fish; cutting holes in the ice will not suffice. During this period, fish will feed on what is naturally produced in the pond, but at a greatly reduced rate. For this reason, snow should be removed from the ice when possible to let light enter the pond and support primary production.

TROUT

Trout have been grown on a commercial scale in the United States longer than any other group of fish, and their culture in Europe dates back over 400 years. Trout are either grown as a food fish or are produced in hatcheries to be released into natural waters for sport fishermen. They are popular because they are attractive, active fighting fish, and have a very-high-quality meat. Trout have been released and cultured in waters all over the world.

A number of species of trout are grown. The three most common are the **rainbow trout,** *Salmo gairdneri* (=*Oncorhynchus mykiss*), the European **brown trout,** *S. trutta* (see Figures 10.17 and

Figure 10.17 The brown trout. (Photo courtesy of the New Jersey Division of Fish, Game, and Wildlife.)

Figure 10.18 Rainbow trout brood fish. (Photo courtesy of the New Jersey Division of Fish, Game, and Wildlife.)

10.18), and the **brook trout**, *Salvelinus fontinalis*. There are a large number of variants, or subspecies, within each of these species; for example, *S. gairdneri*, has a continental form that is nonmigratory, but there is also a coastal, migratory subspecies, called the steelhead trout. There has also been some recent interest in the culture of the lake trout, *Salvelinus namaycush*.

Spawning

In their natural environment, trout build nests and spawn in streambeds. In this process, the female deposits the eggs in the nest; this is followed by external fertilization of the eggs by the male. The eggs are then covered and begin the long incubation period. Although there is considerable variation, as a function of climate and genetic strain, in general wild brown trout and brook trout spawn from October to January and wild rainbow trout from January to May.

Culturists do not let trout spawn naturally; fertilization is always artificial. Streambed fertilization results in rates of hatching much lower than artificial fertilization. It is also important to be able to spawn animals throughout the year so a constant supply of young fish are generated. There are genetic strains of cultured

trout, especially the rainbow, that breed at different times of the year, and manipulation of the photoperiod and/or water temperature can be used to induce gonadal development.

Most trout begin to spawn at the end of their second year, but environmental factors play a significant role in determining exactly when that spawning will take place. Male sperm, or **milt,** according to the species, is best taken between years 2 and 4, while eggs from females should be taken between years 3 and 6. The larger the female, the greater the number and size of the eggs. The size of the eggs at spawning is also genetically controlled; fish with large eggs should be selected since this increases the chances of survival for the fry. Females of 2 years do not make many eggs, and females older than 6 years frequently have sterile offspring. When older females are bred with younger males, the descendants are predominantly male. Ripe females are very delicate, and should be subjected to as little handling as possible before spawning.

Brook and brown trout brooders sometimes are wild fish that are caught at or near maturity as they are swimming upstream; they may also be raised to maturity in ponds. Rainbow trout brooders are nearly always pond cultured. At maturity, the broodfish are placed in small tanks or ponds that should have flowing water and are often covered with netting to prevent jumping (see Figure 10.19). A low stocking density is used for brooders. The sexes must be separated to prevent the males from fighting; sometimes, to stimulate milt production, males are placed in a holding facility downstream of the females. More brood females are kept than males because the milt from a single male can be used to fertilize approximately two females' eggs.

In some cases, sterile or monosex cultures of trout may be desired. Fish in which gonads ripen not only "waste" the culturist's feed, but reproductive males may be unmarketable because of their appearance and the quality of the meat. All-female spawns may be produced using sex-reversed fish (as described earlier for carp) and sterile triploids can be produced by temperature or pressure shocking the eggs.

During the spawning period, the sexes of trout are easily recognized. Males become more brightly colored and the lower jaws of older individuals develop a hooked beak appearance. Females develop extended bellies and the anus (genital papilla) becomes larger and reddish.

The reproductive stage of the animals must be checked frequently. This is done by lightly pressing the abdominal vent; this pressure results in the emergence of eggs or milt. The animals are ripe when the milt is a creamy white and the eggs are removed with relative ease. Trying to strip under- or overripe trout will lead either to damage to the fish, a poor hatch, or weak offspring.

When the trout is judged to be ripe, the animals are stripped (see Figure 10.20). This operation is best done by two people. One person firmly holds the trout over a pan, head up and tail down, while the other begins "pushing" the eggs or milt out by applying pressure to the anterior portion of the abdominal area and continuing with a downward motion to the rear. Care should be taken to avoid breaking the eggs, as the contents of a broken egg will coat the other eggs and inhibit fertilization. The milt is added to the eggs, and mixed, and then water is added. Water should not be added before the mixing has taken place since the motility of the sperm is greatly reduced in its presence, and the water causes the eggs to swell, resulting in a closing of the **micropyle** (the site of entry for the sperm). Therefore the trout are often dried be-

Figure 10.19 Broodstock building. (Photo courtesy of the New Jersey Division of Fish, Game, and Wildlife.)

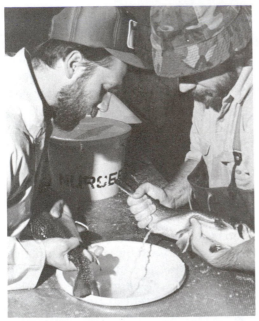

Figure 10.20 Stripping brooders. (Photo courtesy of the New Jersey Division of Fish, Game, and Wildlife.)

fore they are stripped so that water does not drip into the pans.

Hatchery Technique

In the past, fertilized eggs were held in baskets, but these have been replaced by **incubation trays.** In the United States, **vertical flow incubators** are the most popular because they allow efficient use of hatchery space; as the name suggests, water is pumped to the top of the incubator and flows down from one tray to the next.

Water used in the vertical flow incubators (see Figure 10.21) may be recirculated if it is heated or chilled (however, some culturists do not recommend using recycled water), and the speed of the flow is adjusted according to the number of eggs that are held. The trays are composed of a shallow basket in which the eggs are held and a cover; the basket itself rests inside the tray. These trays are made of stainless steel, fiberglass, aluminum, or plastic. The water is introduced to the tray in such a way that it flows *up* through the

basket containing the eggs, then down to the tray below and up through that basket, and so on through the incubator. This upward flow of water through the eggs allows increased aeration and facilitates removal of metabolites.

During the hatching period, the eggs are highly sensitive. Newly fertilized eggs can be killed in just a few minutes if directly exposed to sunlight. In addition, water used during incubation must be moving and have a high oxygen content. The length of the hatching period for the eggs is a function of the species and the water temperature. Incubation normally takes place in water of 8°–12°C, and the incubation period is expressed in **degree days,** the average temperature multiplied by the number of days before hatching takes place. For example, if hatching starts to take place in 11°C water after 30 days, then it took 330 degree days. Hatching is often variable, normally extending over 50 degree days for eggs in the same fertilization batch. The average number of degree days is relatively constant for

Figure 10.21 Trout eggs in a vertical flow incubator. (Photo courtesy of the New Jersey Division of Fish, Game, and Wildlife.)

each particular species, being 400 to 460 for the brown trout and 290 to 330 for the rainbow.

Incubation of the eggs takes place in three phases:

1. Fertilization to the appearance of the eyes
2. Appearance of the eyes to hatching
3. From hatching until the yolk sac is absorbed

After the eyespots appear, the eggs can be gently handled without fear of injury. When the eyes are clearly visible, the eggs are subjected to some sort of mild mechanical shock, such as stirring or siphoning them into a pail; the infertile eggs, which rupture and turn white, should be removed from the culture since they are subject to fungus infection that could spread to the healthy embryos. (Some facilities do not remove the dead eggs to prevent fungus; rather, the eggs are treated to prevent infection.) The infertile eggs once were removed one at a time with long metal tweezers, but this has been replaced by several other methods:

1. Removal with a long pipette that may be fitted with a siphon.
2. Transferring the eggs to a salt solution that is adjusted to a particular concentration such that the dead eggs will float and the eyed embryos will sink.
3. Light-sensitive sorting machines, based on separating the clear fertile eggs from the opaque dead eggs, which can sort up to 10^5 eggs per hour.

During the last phase, the fry, or **alevins,** hatch. The fry can be held in the trays until they become active and are able to begin to feed (when the sack is three quarters absorbed). The covers on the trays prevent the fry from escaping. At this point, they can be released for restocking natural waters; however, they are often released later as fingerlings or yearlings.

Growout

If the trout are to be further cultured, they are transferred to a small rearing trough before they have completely absorbed their yolk sac, and introduced to either live or artificial feeds (see Figure 10.22). If they are started on live feeds, they are switched to a prepared diet as soon as possible. Getting the young fish to eat is sometimes a problem, but this is less often the case when domesticated strains are used. The density of the fish in the troughs may be adjusted over time.

Once the fry take prepared feeds and seem to be healthy, the trout can be moved to ponds or raceways. Starting when the fingerlings are put into the growout facility, the fish are **graded** (separated according to size) (see Figure 10.23). Grading is done throughout the culture of the fish because a large trout will eat another trout that is half its length. Grading reduces cannibalism as well as competition, and fish of a uniform size make it easier to calculate the amount of feed to be used. Additionally, fish harvested all at once that are approximately the same size are more marketable than nonuniform fish.

Figure 10.22 Moving trout fry into rearing troughs. (Photo courtesy of the New Jersey Division of Fish, Game, and Wildlife.)

(a)

Figure 10.23 Sorting of trout according to size. (a) Using a grading box in the hatchery. (b) A grader on top of a fish pump for sorting larger fish going into raceways. (Photos courtesy of the New Jersey Division of Fish, Game, and Wildlife.)

(b)

Ponds used for culturing trout tend to be elongated to allow for swifter water movement. Circular and oval tanks are also used. The most popular design is probably the concrete raceway. These are usually about 30 m × 3 m, with a depth of about 1 m at one end and about 70 cm at the other end. The flow of water varies according to how many raceways are in series: the more raceways linked in series, the higher the flow must be. There may be an aerator used midway in the raceway series.

Silos, or vertical raceways, seem to be better for trout production, but may be harder to manage and have a greater capital cost (although the cost per kilogram of trout produced is lower). In experiments conducted by the New Mexico Department of Game and Fish comparing traditional raceways to silos, it was shown that although total productions in raceways and silos were about the same, the raceways required 3.3 times more water.

Cage culture of trout is growing in popularity. In an experiment carried out in simple dugout ponds, trout were caged and held for one season. Fingerlings stocked at 1.4 kg/m^3 in cages, fed 3% of their body weight daily, had a food conversion ratio of 1.6, survival of 100%, and a daily weight gain of 3.78% over two months, growing from 27 g to 88 g. This seems to be one way to use an untapped water resource for aquaculture.

Feeds

The use of dry prepared feeds has become popular; increased growth and production have been reported with their use. In addition, feed storage is simplified, growth is uniform, and feeding can be automatic. Many commercial brands of trout feed are on the market, many of which are being improved constantly. Fish cultured on prepared feeds have a different taste than wild fish, and the flesh of the cultured fish is white rather than pink, although food additives may be used to "correct" this phenomenon. Those raised on pellets also differ in taste from trout fed fish offal and slaughter-

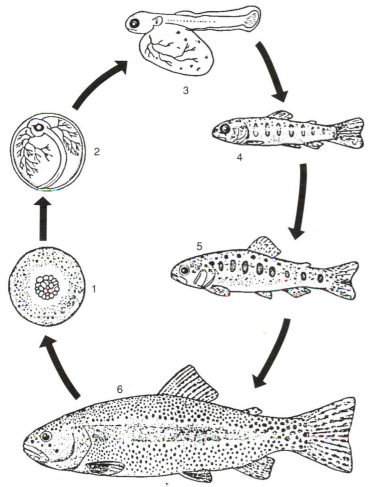

Figure 10.24   The life cycle of the rainbow trout. (1) Embryo, (2) eyed egg, (3) sac fry, (4) advanced fry, (5) fingerling, and (6) adult. (From Kafuku and Ikenoue, 1983. Copyright © Elsevier Scientific Publishing Company, New York. Reprinted by permission.)

house by-products. Trout receive a high-protein diet since, as carnivores, they can utilize proteins well. Fats are fairly well digested by trout, but these fish cannot utilize most types of carbohydrates, and their inclusion in prepared diets is very limited.

Commercial feeds usually come with charts describing the amount of feed to be used for any particular species, given an average size and water temperature. These recommendations are based on many feeding experiments conducted by those who produce the feed and by other scientists; however, these amounts may be modified slightly by the farmer according to the strain of fish being used. Feeds for trout containing growth-promoting hormones may become popular in the future, as they are presently for beef; trout that experimentally received these steroids in their diet were better able to use proteins than trout that did not receive the hormone treatment.

Trout are fed three or four times daily, but fry may be fed much more often. Feeding fish more than recommended may increase growth slightly, but the crop will have a reduced uniformity, the water quality may decline, and there will be an increased cost since the fish will not use the nutrients as efficiently.

There are a number of ways of getting the feed to the trout. **Hand feeding** trout refers to scooping feed out of a bag or tub and flinging it into the pond. A good farmer can tell just how much feed to add to the water by the reactions of the fish. Hand feeding is slow and not generally used for large trout culturing facilities. Most large commercial trout farms use semiautomatic feeding devices; these are placed on a truck, or pulled by a tractor, and shoot the pellets into the ponds or raceways. Other farms use **demand feeders** that are controlled by the fish; when the trout bumps into the trigger, feed is released into the water (see Figure 10.25). Automatic feeders can also be used (see Figure 10.26); these will dispense a fixed weight of dry feed at regular intervals.

BAIT FISH

The culture of **bait fish** is unique. While most of the aquaculture industry has grown because commercial fishing efforts have declined, bait fish sales are dependent on there being wild fish for the

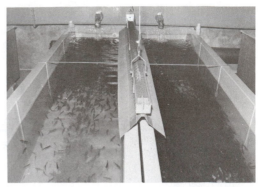

Figure 10.26 Automatic feeder between two tanks in a trout nursery. (Photo courtesy of the New Jersey Division of Fish, Game, and Wildlife.)

commercial and recreational fisherman. If the streams and lakes near a bait farm become polluted so that the sport fish die or are unfit for consumption, then the sales of the bait fish will naturally decline. The culture of bait fish seems to have started in the United States during the 1920s and has been most important in the southern states, especially in Arkansas, which produces about half the nation's annual 10^4 metric ton crop.

Species Cultured

About 100 species of freshwater fish are raised for bait, and there is also some interest in the production of saltwater bait fish. The most important of the freshwater species are the golden shiner, *Netemigonus crysoleucas*, the fathead minnow, *Pimephales promelas*, and the goldfish, *Carassius auratus*.

The **golden shiner** is popular with many fishermen because of its easily seen bright gold- and silver-colored scales (see Figure 10.27). Local strains may have slightly varied colors. Small shiners are used for crappie bait, while larger ones are used to catch largemouth bass and catfish. These fish are commonly found in lakes, but also occur in the shallow water of large streams. They are principally grown in the mid-South. Golden shiners can live up to eight years, grow to 25 cm, are omnivores, and readily accept prepared diets small enough for them to con-

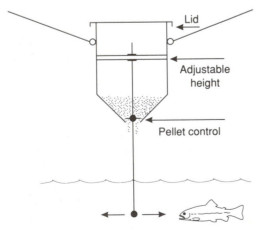

Figure 10.25 Demand feeder. (From Stickney, 1979.)

Figure 10.27 Golden shiner. (From Dobie, Meehean, Snieszko, and Washburn, 1956.)

Figure 10.29 Goldfish. (From Dobie, Meehean, Snieszko, and Washburn, 1956.)

sume. Domesticated strains are much easier to use in culture than wild varieties. In general, golden shiners are rather delicate and do not handle well in hot weather.

Fathead minnows do not usually grow much over 8 cm (see Figure 10.28). They are commonly cultured in Minnesota and the Dakotas, but are also farmed in other locations. Besides their use as bait, the fathead is used as forage in catfish ponds. It is an omnivore, like the golden shiner, and also accepts prepared feeds. Its major advantage over the golden shiner is that it is fairly tolerant of high temperature. These minnows live only two to three years, so they grow quickly and are very easy to spawn. They do have one major disadvantage—many adults die after spawning.

Goldfish may be the characteristic metallic orange color (see Figure 10.29), but can also vary considerably from this.

Figure 10.28 Fathead minnows. (Martin, 1986. Copyright © Archill River Corporation. Reprinted by permission.)

Their diet is much like that of the other bait fish. They can grow to 30 cm if not crowded and are handled and transported with relative ease. They are less active than the golden shiner and are less widely accepted by fishermen for that reason.

Propagation

Goldfish and golden shiner are cultured in much the same manner. Water should not be fertilized during the spawning season, since high pond productivity seems to inhibit spawning behavior. If there does not appear to be enough natural food for the fish, they should be given prepared feeds. Goldfish will begin to spawn at about 15° or 16°C, while the golden shiner will start spawning at about 21°C. Shiners will mature at about one year, with eggs being laid on plant blades or roots; a female will produce up to 10^4 eggs, deposited throughout the day, which will hatch in four to eight days in water of 24°–26°C. Female goldfish will spawn several times, each yielding 2000 to 4000 eggs, which will hatch in two to eight days, depending on the temperature; spawning takes place mainly in the morning. The brood stock is selected according to the preference of the fishing community. Brood fish are chosen because of shape and color; a number of genetic strains may be purchased. There are three propagation methods: free spawning, egg transfer, and fry transfer.

Free spawning begins by dropping the water level in the pond. This encourages plant growth along the sides; rye grass may be planted on the dry pond bottoms.

When the grasses have grown, water is added back, along with the brood stock, which will lay their eggs on the plants. If there are no plants, hay or straw can be anchored in shallow water, or mats of Spanish moss (**spawning mats**) held together by wire can be used. Roughly 25–50 kg of ungraded shiner brood stock is used per hectare; for goldfish the number of spawners will vary with the size of the fish from 250 to 750 per hectare. If spawning activity declines, it can be restimulated by adding cool water that raises the level of the pond. Both of these fish can be chemically shocked to spawn by the addition of potassium permanganate. After the eggs are laid, the brood stock should be removed. The eggs are hatched and reared in the same pond.

The **egg transfer** spawning method requires a pond that is completely free of any natural vegetation. Shiners are stocked at 450–550 kg per hectare, and goldfish at 900–1100 kg per hectare. When spawning is imminent, spawning mats are placed in shallow water, and the fish may be stimulated to spawn by rapidly increasing the water level. The mats should be removed when they are evenly covered with eggs, but must not be left in the water so long that they become crowded, since this results in fungal infections. As mats are covered, they must be replaced by new, clean mats. When the egg-covered mats are removed, they are transferred to a rearing pond.

Fry transfer starts with the hatching of eggs resulting from free spawning or egg transfer. When the fish reach 1.5–2 cm, they are collected with fine seines or lift traps and transferred to growout ponds. The advantage of this method is that a relatively accurate count of the fish can be made so stocking densities can be controlled. Stocking will vary according to the amount of time the fish are to be in culture and their desired size when sold as bait. For golden shiners, the stocking density is 1.2 to 5×10^5 per hectare, while goldfish can vary from 6×10^4 to $2.5 \times$ 10^6 fish per hectare. Fry transfer makes more efficient use of the pond's space.

Fathead minnows are treated differently since they have a dissimilar spawning behavior. These fish become mature after one year; the best brooding size is 6–8 cm. The spawning season starts in the spring when the water temperatures reach about 18°C, and continues until the early fall. If the water temperature rises above 29° or 30°C reproduction will cease, but may resume when the water cools. During the spawning season the two sexes look very different; the male is black with vertical bands of golden brown, while the female is olive to silvery and has a more pointed head. The eggs are deposited on the undersides of boards that are floated or suspended below the surface. For each 100 male fish, the farmer should use a board about 0.5 m². The eggs will hatch in 5 to 7 days, depending on the water temperature.

If the young minnows are allowed to growout in the pond with the adults, the density of the adult fish should be kept low, 1200 to 4800 per hectare, or stunting will result. For intensive culture, a brooding pond is stocked with about 5×10^4 minnows per hectare, with a female:male ratio of 5:1. The young fish are removed soon after they hatch and are reared in separate ponds stocked at 2.5 to 7.5×10^5 minnows per hectare.

Feeding

Fry are supported by the natural productivity of the water. Before the young fish are added, the pond should be fertilized; nutrients are typically added as inorganic fertilizers or as a mixture of inorganics and organics. Organic fertilizers for bait fish culture are often manures, dried hay, and meals that are high in nitrogen (such as soybean and cottonseed meals).

Production can be increased by using artificial feeds. The **starter feeds** for bait fish tend to be higher in protein than the growout feeds and have a more complete vitamin mixture. The prepared diets are given to fry as a slurry of very fine particles that is spread evenly around the entire edge of the pond. The fish can be fed several times a day; they must be fed more often if the plankton bloom dies off. It is not a problem if the fish are moderately overfed for the first few weeks, the excess food simply serves as a fertilizer for the pond.

The **grower feeds** may be given to the fish after several weeks. It is more coarsely ground than starter feed and has a lower protein content. The dry feed can be broadcast over the pond from the downwind side, rather than poured around the entire outline of the pond like the starter diet. Commercial feeds for bait fish are available, and chicken broiler diets are also adequate. The fish are fed about 3% of their body weight per day; if the water is checked on the downwind side of the pond about two hours after the fish are fed, the farmer can observe any uneaten pellets. The amount of uneaten pellets will be an indication as to whether the fish are being given enough food or are being overfed. If the farmer is going to change the feed presented to the fish, the old and new feeds should be blended together and presented as a mixture to the fish for a few days (shiners are particularly sensitive to changes in the feed and may not accept a new diet easily).

The feeds for all the freshwater baitfish are similar, but there are some differences in the ways the different fishes are fed. Golden shiners can be fed more than once a day, resulting in faster growth but a reduced conversion ratio. If the shiners stop feeding, it is an indication of low DO, high natural productivity, disease, or pesticide contamination. Goldfish fry usually require heavy fertilization and/or feeding; natural blooms are discouraged after the fish reach 2–3 cm, when the starter diet is replaced with a grower diet. Small goldfish can be given a golden shiner ration, but part of the original low-grade protein ingredient (e.g., feather meal or cottonseed meal) in the shiner diet is replaced with commercial-grade egg yolk for the goldfish. Fathead minnows are fed a starter feed until they are 1.5–2 cm and then are given pelletized grower feed. The minnows cannot swallow the pellets, but will continually nibble at them; these fish seem to grow better on the pellets than on crumbles. The pellets are often placed in little piles in shallow water by the farmer rather than broadcast over the pond. The piles should be checked after two or three hours; if pellets remain, the amount should be reduced at the next feeding.

Harvesting

There are several general rules for harvesting bait fish, regardless of the species. Harvesting should be done in the morning or on cloudy days, because as the water gets warm, the fish become more difficult to handle and there are greater mortalities. Temperatures can be dropped quickly by adding cooler well or spring water, but golden shiners and minnows do not tolerate quick drops of 8°C or more as well as goldfish; this is a particular problem since in warm waters the shiners are a "nervous" fish. Excessive handling should also be avoided. Handling can be reduced by harvesting one day, grading the fish the following day, and transporting them on the third day. Golden shiners are more difficult to harvest than the other species because they tend to shed scales, which results in injury and death shortly after transportation.

Bait fish are normally harvested with a seine net. If the pond is small, a net that stretches the width of the pond can be pulled from one end to the other. In large ponds, a bait (food) can be placed in one corner of the pond to attract the fish, which can then be removed with the seine net from the restricted area (this technique works especially well for the fathead minnow). The mesh of the net should be 0.3–0.6 cm, depending on the size of the fish. A small mesh will be harder to pull and will pick up more debris in the water, but it will cause less injury to the smaller fish. The fish are removed from the seines with dip nets. After several seine hauls at full or even half the pond water volume, the water level should be reduced so the last remaining fish are in the shallow water in the deepest part of the pond; these fish can then be removed easily. The ponds should be dried before being refilled so there are no remaining fish.

Another effective way to harvest, if the pond is suitably constructed, is to quickly drain the pond and allow the fish to be flushed out with the water through the drainpipe (see Figure 10.30). The fish can be collected on a screen at the outfall. This method requires less labor and is quicker than seining. There is also less stress on smaller fish, particularly shiners.

Although less effective, baited lift nets and traps can also be used to harvest. Nets and traps can be checked every 15 to 30 minutes and the fish removed. These can be set several times in the same location until fish no longer enter, and are then moved to a new location in the pond. This method is usually not used for golden shiners.

After the fish are removed from the ponds, they are carried to trucks in gal-

Figure 10.31 Truck with cylinders for hauling fish. (Photo from Lee, 1981. Copyright © The Interstate Printers & Publishers, Inc., Illinois. Reprinted by permission.)

vanized buckets. (Note: Some fish are sensitive to the zinc used in galvanization.) The trucks often have hauling tanks of about 1000 liters (see Figure 10.31). Because the fish are very crowded, these tanks should not be allowed to get too warm and should be aerated if possible. The fish should be moved to holding tanks as soon as possible. The number of fish that any hauling tank can carry will change with the temperature.

Holding, Grading, and Transportation

Tanks used for holding the harvested bait fish are typically concrete with a smooth interior surface, 90–150 cm wide, and 3–12 m long (see Figure 10.32); these are sometimes divided at intervals of about 2.5 m. The tanks have a standpipe fitted so that the water depth is less than

Figure 10.30 Baitfish ponds with a catch basin for rapid draining. (From Dobie, Meehean, Snieszko, and Washburn, 1956.)

Figure 10.32 Holding tanks for baitfish. (From Giudice, Gray, and Martin, undated.)

60 cm. Holding tanks are constructed in open-sided buildings to protect the fish from the sunlight that can quickly heat the water and stimulate algal growth. The tank water is kept cool by slowly exchanging it with well water, a practice that also keeps the levels of ammonia low in the crowded tanks. Water is exchanged about every two hours depending on the temperature and the number of fish being held. The tanks should be aerated because the fish use a lot of oxygen and the well water has a low DO.

Since bait fish are sold to fishermen for specific purposes, the fish must be graded to meet the demand. Small fish are used to catch one type of fish, while large bait fish are used for another type of fish. Several types of graders are available; the choice of graders will be a function of the number of fish, the species, and the facilities. Graders are of either the panel or the box variety (see Figure 10.33). **Panels** are used in several ways; the simplest way to use a panel grader is to pull it through the trough, from the head to the drain end, and all the fish that are too large to move through the bars on the grader will be

kept in the confined section near the drain. **Box graders** may be free standing or floating. The fish are placed inside the box, and the small fish swim out while the big ones are retained. The fish in a grader box should be at a high enough density so the small fish will try to swim out, but not so great that the fish are injured or so the BOD gets too high and the fish that remain are overly stressed.

Small numbers of fish are best transported in plastic bags filled with cool water and oxygen. The bags must be made of a very sturdy material and the water quality must be good. Fish should not be fed before they are shipped; rather they should empty their gut before being put in the bag so the water quality remains acceptable. The bags are placed in cardboard boxes that are lined with styrofoam (to prevent the temperature from rising), and are then shipped.

When large numbers of fish are to be transported, a specially designed truck is used. These may be simple vehicles with temporary tanks or large semitrailer trucks equipped with liquid oxygen. The trucks have tanks made of a marine-grade

(a)

(b)

Figure 10.33 (a) Panel grader. (b) Cylindrical box grader. (From Giudice, Gray, and Martin, undated.)

plywood or aluminum insulated with styrofoam. Some of these trucks have compartments to hold ice that is used for cooling the water on hot days or during long
hauls. There are either several small tanks
or one large tank divided into sections,
each with its own drain of about 10 cm for
quick removal of the fish.

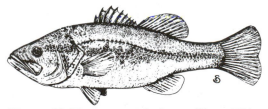

Figure 10.34 Largemouth bass. (From Whitworth, Berrien, and Keller, 1968. Courtesy of Connecticut Department of Environmental Protection.)

Bait Fish Egg Culture

Although less popular than the pond-
spawning species, several fish that normally require a small hatchery are grown
for bait. These are the white sucker, *Ca-
tostomus commersonii*; the creek chub, *Semo-
tilus atromaculatus*; the blacknose dace,
Rhinichthys atratulus; and the stoneroller,
Campostona anomalum.

The white sucker is a relatively large
omnivore, reaching 0.5 m, found in clean
waters in eastern North America. It will
spawn in gravel beds in the spring, but in
culture the adults are caught and the eggs
fertilized artificially. Hatching is done in
jars held at 12°–15°C for about 10 days.
The fry are transferred to rearing ponds
that are sometimes manured if there is a
low natural productivity.

Creek chubs are somewhat smaller
than the white suckers. The chubs are
found in relatively calm waters in the eastern United States, but reproduction takes
place on gravel beds in running water
during the spring. Adults are captured
and may be given hormones to induce
ovulation. The fry are grown in ponds at
densities up to 1.4×10^6 per hectare.

LARGEMOUTH BASS AND ASSOCIATED SPECIES

Freshwater bass belong to a family
of fish called the centrarchids. The
largemouth bass, *Micropterus salmoides*,
is one of the most popular sport fish in
North America and Europe (where it has
been introduced) (see Figure 10.34).
These fish are found in soft-bottom lakes,
ponds, and streams, particularly in weedy

areas. Two subspecies are used in culture,
the northern and Florida strains, that differ in temperature tolerance and temperature of spawning. Along with the
largemouth, there is also the **smallmouth
bass, *M. dolomieui*,** and the **spotted bass,
M. punctulatus,** which can also be adapted
to culture. The smallmouth is found in
cooler waters than the largemouth, although their natural distributions overlap
considerably. Smallmouth bass are found
in waters that are more quickly moving
and less vegetated than the largemouth
bass. Largemouth, smallmouth, and spotted bass are known collectively as **black
bass.**

Bass are exclusively carnivorous fish.
The young feed on rotifers, cladocerans,
copepods, and insects. At 6–8 cm they
switch to a diet of fish, frogs, and crayfish.
Largemouth bass seem to require larger
prey than other species. The growth of
bass is related to the amount of food available and temperature. They may cease to
feed when the water temperature drops
in the winter. Some work has been done
on the intensive culture of the bass, but it
is most commonly part of a polyculture
pond system geared for low-level production and sport fishing.

Farm Pond Culture

Farm ponds are dugouts used by agricultural farmers to store water for irrigat-
ing crops, watering livestock, and erosion
control. For a long time, bass fingerlings
were available from state or federal agencies for stocking in farm ponds to provide
an extra source of food (as well as recreation) to the farm community. Free fin-

gerling are still available from some government facilities, but they also can be purchased commercially.

Since the bass are carnivorous as adults, they must be stocked with other fish that can serve as forage for them; the most common forage fish is the **bluegill**, **_Lepomis macrochirus_,** that can also be harvested itself as a food fish (see Figure 10.35). The bluegill eats aquatic invertebrates, insects, and (to a limited extent) algae. The bass will also feed on some invertebrates as well as young bluegills. The concept is that the very fecund bluegills will produce offspring that graze on the pond products that cannot be utilized by the larger bluegills or the bass, and these offspring will be eaten by the bass so the bluegill population will be held in check.

There has been considerable debate as to how to stock the farm ponds, and whether the bluegill is the best fish to stock with the bass. The stocking scheme will change from one location to the next. An example of suggested stockings for bass ponds in Texas is given in Table 10.2. Besides "what" and "how much" to stock, the season in which the stocking takes place is important; time of stocking varies with the location, the species, and the size of the fish being used.

Several general formulas have been generated to guide the farmer in maintaining the pond. For example, it is rec-

TABLE 10.2 Suggested Forage Stocking in Texas Ponds of 0.4 Hectares*

Zone 1
Generally at least 102 cm of rain per year

15 adult bluegills plus 15 adult redear sunfish *or* 250 bluegill fingerlings plus 250 redear sunfish fingerlings

Optional: 200 threadfin shad, 500 golden shiners, 500 fathead minnows (if fingerling bass are stocked)

Zone 2
Generally 76–102 cm of rain per year
30 adult bluegills or 500 bluegill fingerlings

Optional: 15 adult or 250 fingerling redear sunfish, 200 threadfin shad, 500 golden shiners, 500 fathead minnows (if fingerling bass are stocked)

Zone 3
Generally less than 76 cm of rain per year
30 adult bluegills or 500 bluegill fingerlings or 500 golden shiners

Optional: 15 adult or 250 fingerling redear sunfish, 200 threadfin shad, 500 fathead minnows (if fingerling bass are stocked)

* Largemouth bass are stocked either as 20 juveniles (15–20 cm) or 50 fingerlings (2.5–7.5 cm)
Source: Anonymous, 1986.

ommended that the pond should contain, by weight, three to six times as much biomass in the form of forage fish as bass. In a healthy farm pond, most of the bluegills are 15 cm or larger and the bass are 0.45–0.9 kg; this indicates that there are enough healthy bass to keep the bluegill population from getting too large. As the pond gets farther from the ideal, the bluegills become more numerous and smaller while the bass become fewer and larger.

Most farm ponds are essentially unmanaged and tend to be overpopulated with bluegills. There are several reasons for this, the first being improper stocking and the second being what William Mc-Larney has called the "sport fishing mentality," referring to the fact that these ponds are harvested by individuals using fishing poles, not nets, primarily for recreation. When fishing, people want to

Figure 10.35 Bluegill. (From Whitworth, Berrien, and Keller, 1968. Courtesy of Connecticut Department of Environmental Protection.)

take the largest and most challenging fish in the pond (the bass), while the bluegills are often ignored or thrown back. In addition, since the sport fisherman fishes when and as much as he wants to, some years the farm pond may be overfished, while other years it will be underfished. Depending on the productivity of the pond, 55–220 kg/ha should be harvested each year; of this, only about 20% of the total fish biomass removed from the pond should be bass so the proper ratio of bass and forage animals is maintained. If small bass are stocked in the ponds, they should not be caught until they are in the second or third year. The total productivity of a pond can be increased with light fertilization. Large bluegills should be removed when caught since they cannot be eaten by the bass and have a very high reproductive potential.

Alternative Fish in the Bluegill–Bass Farm Pond

The bluegill–bass combination has been very popular and requires only minimum management. But for an assortment of reasons, some other fish have been used in farm ponds in addition to, or in place of, these species. These may or may not be appropriate, depending on the location and function of the pond.

The **white crappie,** *Pomoxis annularis,* and the **black crappie,** *P. nigromaculatus,* have both been used in farm pond culture. They are popular food fish that may reach 30 cm and do well in ponds that are too turbid to support good growth of some other fish; they require prey somewhere in size between what is needed by the bluegill and bass. Crappies seem to do best in larger ponds; *P. annularis* especially tends to reproduce quickly and is subject to overpopulation in small ponds. Reproduction may be limited by lowering the water level in the summer and raising it again in the winter. Hybrids of the two species may also be used to help prevent uncontrolled reproduction. The crappies are not a substitute for bluegills because

they will compete with the bass for the available food and will even eat small bass.

The most popular substitute for the bluegill is its relative, the **redear sunfish (***L. microlophus***).** This fish has a lower reproductive potential than the bluegill, so overpopulation is less likely. Recently there has been an increased interest in hybrids of this genus (which also includes the **green sunfish,** *L. cyanellus,* the **pumpkinseed,** *L. gibbosus,* and the **longear sunfish,** *L. megalotus***).** Since uncontrolled reproduction is a problem, hybrids that are sterile, have low reproductive rates, or have offspring that are far from the 1 : 1 sex ratio are possible solutions. But again there is a problem: since their reproduction potential is low, hybrids may be unable to support the bass population as well as the bluegills and redears. Also, the offspring of the hybrids, if they are produced, may not have the same desirable characteristics as the parents. Only a few of the hybrids have received attention from farm pond owners, in particular the female bluegill × male green sunfish cross, which grows more quickly than the parent strains. Research continues in this field, and improved and less expensive crosses may be available in the future.

Bait fish, such as the golden shiner and fathead minnow, have been used for forage. Golden shiners may be stocked when the bluegill's reproductive potential is reduced, as it will be in clear alkaline waters; they may also be a good addition when water levels fall and predation on the bluegills becomes too great for sustained production. The fathead minnow is subject to high predation because it is a slow swimmer. Therefore, fatheads are not stocked with established, large bass, but can be used in ponds stocked with small bass.

The **threadfin shad,** *Dorosoma petenense,* is a small fish, rarely exceeding 17 cm, found in lakes. Like the fathead minnow, it cannot withstand bass predation for long periods of time unless other forage fish are present in the pond. They are rather sensitive to cold temperatures and are best used in Florida and southern Texas. The production of the threadfin shad can be increased with fertilization of the pond that increases its food supply and decreases its visibility to bass (which hunt by sight).

Largemouth bass have been introduced in many places outside North America. In many tropical and semitropical regions some species of tilapia work well as forage, and the tench, **Tinca tinca,** is used for this purpose in South Africa. In Europe, the common roach, *Leuciscus rutilus*, and the rudd, *Scardinius erythrophthalmus*, are used for forage.

In some areas, channel and blue catfish are stocked in farm ponds. These serve as sport fish but do not interfere with the bass–bluegill balance. Catfish-spawning devices can be placed in the ponds, but most of their offspring serve as food for the bass. Other species of catfish should not be used since they tend to overpopulate the ponds, may compete with the bass for forage, eat the young bass, and stir up the mud on the bottom (this is called **roiling**), which reduces feeding and reproduction of both the bluegill and bass. Carp should not be stocked in farm ponds because they too will roil the water.

There are substitutes for largemouth bass in farm ponds. Smallmouth bass will do better in cooler waters, and the chain pickerel, *Esox niger*, unlike the bass, feed all year almost regardless of the temperature.

Cultivating Bass

Black bass may be cultured under conditions other than the farm pond. While not intensively cultured as a food fish, they are grown for stocking programs. Fingerlings of 20–25 cm can be stocked directly into small impoundments for fishing.

There are several ways to sex largemouth bass brooders. Some workers sex the fish by the shape of the urogenital opening, but this is not a very reliable method. In the spring, the presence on females (and absence on males) of the genital papilla, a reddish protuberance that surrounds the urogenital opening, can be used to sex the fish, but this method is less reliable outside the spawning and prespawning seasons. Perhaps the best method is the use of a 1 mm diameter probe, which in the female goes deep into the ovary at an oblique angle, while in the male passes only a short distance perpendicular to the body axis, into the urinary bladder.

Breeding is carried out in ponds during the spring and/or summer, depending on the location. The fish mature slowly as the water warms, so spawning can be inhibited, if desired by the culturist, by keeping the fish in cool water or by separating the sexes. The male builds a circular nest in shallow water and will later guard the eggs and the fry. The eggs, 1.5–2.5 mm in diameter depending on the species, are laid by the female and fertilized immediately by the male. The females each lay 2000 to 10,000 eggs. When one female has finished laying her eggs, she may be pushed away by the male and another female may use the same male and nest.

The fertilized eggs stick to stones and rocks in the nest. Gravel is normally added to these breeding ponds so the nests can be properly constructed. The piles of gravel must be far enough apart that there is no competition between the males. In Europe it is common to place walls between the piles of gravel; the farmer therefore may crowd the male nests closer together with no interaction between the fish. These brooding stalls may also be fitted with screens on the open sides so the fry are trapped and can be easily harvested. Without the stalls, bass should be stocked at about 100 males per hectare, with a male : female ratio of 2 : 3 to 1 : 2. Brooding stalls allow more bass to use the brooding pond, but the stocking rate of brooders must be reduced if the fry are to remain in the same pond as the adults until the end of the summer.

The bass can be raised to fingerlings using either a one-phase or a two-phase method:

1. In the **one-phase method,** the fry are left in the spawning pond until they are harvested in the fall; this method requires little work but does not yield many fish. The spawning ponds are fertilized and the number of brood fish used is low.

2. The **two-phase method** requires that the fry be moved from the spawning pond to a rearing pond within two weeks of hatching. The rearing pond may simply be fertilized, or small forage fish may be added to the pond, or the bass may be intensively cultured on cladocerans and then switched to a prepared diet when they reach 2–10 cm, depending on the species and the facility. The change from living to nonliving feed is critical and difficult, probably because even at 1 cm the bass are visual feeders with little or no reliance on other feeding cues. Some culturists feel that the best results are obtained by progressing from live feed to ground trash fish, sliced meat and liver, or moist pellets, before presenting a dry floating diet to the bass.

ANGUILLID EELS

Perhaps the most distinctive feature of the anguillid eels is that they are **catadromous.** The life histories of the **American eel,** *Anguilla rostrata* (see Figure 10.36),

Figure 10.36 American eel. (From Whitworth, Berrien, and Keller, 1968. Courtesy of Connecticut Department of Environmental Protection.)

and the **European eel,** *A. anguilla,* have been studied, but there are still many questions to be answered (including whether these are really two separate species or the same fish).

The adults, in the freshwater rivers and lakes of Europe and North America, mature after 5 to 10 years, changing from **yellow eels** to **silver eels** with large eyes and the characteristic dark and silver colors of midwater marine fish. The eels migrate toward the ocean, heading in the direction of the Sargasso Sea, where they spawn in deep water and die. A similar pattern is seen in the Japanese species, *A. japonica,* with spawning probably taking place in some location south of Taiwan. The eggs hatch to **leptocephalus** larvae, which are clear, leaf shaped, and about 5 mm in length. These drift near the surface for one to three years, depending on the distance that must be traveled, and metamorphose into juvenile **elvers** (**glass eels** are the early elvers that have no pigmentation), which enter the rivers during the spring as the waters warm to 8°–10°C. Once in freshwater, the female eels grow more quickly and larger (up to 1.8 m) than the males and will generally move further upstream.

Culture of eels is most important in Taiwan and Japan. There is a smaller culture industry in Europe, and one even smaller in the United States. Young eels are susceptible to a great variety of diseases, but under good culture conditions, the older an animal gets, the less likely it is that there will be any sort of infection.

Capture and Transportation of the Elvers

The 5–20 cm elvers are used for eel culture. Naturally, the larger the elvers that are caught, the shorter and more profitable the culture. These animals are captured in Japan and Taiwan using dip nets, plankton nets, and traps, primarily in winter as they move into the estuaries. The elvers are delicate, so handling should be kept to a minimum. The eels come in with the tide, usually at night, and are caught and may be temporarily stored in floating boxes or plastic containers.

In the Atlantic, the elvers are taken as they migrate upstream during the late winter through early summer; the best time to catch the elvers varies with the latitude (for example, they are captured in Florida as early as November and in Canada as late as August). Interestingly, the elvers move at night and hide in the bottom sediment during the day while migrating, but can still be attracted after dark by artificial lights. After the elvers are captured, they may be held in large boxes that are attached to the fishing ship; these boxes have holes that allow some water to move in and out, ensuring that the elvers have enough oxygen.

The elvers can be transported to the eel farm in large plastic bags that have received extra oxygen. Elvers may also remain out of water for one to two days if kept moist because they can get supplementary oxygen through their skin. They may also be transported in moss or wet straw. Elvers can be held at rather high densities (up to 1 per 3 cm^3) during this short shipping period.

Rearing of the Elvers

In Japan and Taiwan, the elvers are stocked at high densities in small ponds. The population is then thinned and redistributed to larger ponds kept at lower densities. They are fed minced fish, fish meal, tubifex worms, and minced oysters and clams; silkworm pupae have also been added, but this is less common now that the silk industry is declining and eel culture is expanding. Baskets are used, especially in Taiwan, to present the food to the elvers. Feeding takes place at night; elvers are at first unwilling to feed, but they are slowly trained to accept this ration. After about three weeks, they come to the basket as soon as it is lowered into the water. Minced fish and commercial eel diet gradually replace the elver diet. The fish may be briefly boiled in water to soften their skin before being presented to the young eels.

The first elver pond may be replaced by large circular tanks. Recent research has shown that circular tanks are superior to rectangular tanks and ponds; growth is quicker in circular tanks, and although there are higher levels of cannibalism, this is not economically important since the elvers that are eaten are the slow growers that often have not learned to accept the formulated feed. These tanks are sometimes kept indoors and warmed (25°C) so the elvers will grow quickly. The temperatures should be increased slowly to avoid thermal shock, and the DO should be above 5 ppm. Under the best of conditions, elvers may be stocked in circular tanks at 5000/m^3. Feeding again starts with small live feed and finely minced fish, and the animals are slowly trained to accept trash fish and a prepared diet. Experiments have shown that the addition of blood and spleen to the elvers' prepared diet increases their growth, because they make the prepared feeds more attractive, or they contain some nutrients that are important, or they are more digestible by the elvers than some of the other sources of protein.

Depending on the size of the elvers when captured, water conditions, and the feeding, the elvers will metamorphose into small black eels in 1 to 10 months. At this time they may be harvested for bait, but are normally transported to rearing ponds.

Growout

Eels are cannibalistic and therefore must be sorted according to size throughout their culture; this requires that there be several ponds in production at any time. Ponds are often aerated with paddlewheels that are operated in the evening and early morning. The water quality must be good, pH should be 6.5 to 8.0, and the temperature should be above 15°C (20°–23°C seems to be the ideal, depending on the species). Eel ponds are shallow (1–1.5 m deep), with a muddy or sandy bottom, and the walls are as steep as possible, with concrete or brick walls that can be built perpendicular to the water surface being preferred. (The perpendicular wall is used because young eels will try to leave the pond on rainy nights. Because the elvers are so light, they are able to climb up even perpendicular walls for some distance, so a lip may be added to the top of the wall.) Escape of larger eels is also well-known.

Feeding is an extremely important part of growout since natural production in the ponds will not support these animals. Traditional feeds for growout are high-protein products such as trash fish, silkworm pupae, worms, crabs, mollusks, and offal from slaughterhouses and fish-processing plants. This material is ground up and made into a paste that is typically presented in a basket or a trough, with holes, that is lowered into the water in the morning. The eels will learn when and where they are to be fed. Unlike many fish, eels are "sloppy" eaters that bury themselves in the paste; much of the food is wasted as it falls away from the basket because of the violent writhing of the feeding eels. Partially because some of the food is lost, and partially because of a poor feed conversion ratio, the feed presented is 5% to 15% of the total weight of the eels in culture.

Recently, much work has been done with artificial feeds for eels. These seem to allow better conversion ratios than the traditional feeds, and are easy to use, although they may be too costly for some farmers. Therefore, these prepared diets may be turned to only when trash fish are not available or the cost of these fish rises because of a shortage. The eels will get 1% to 3.5% of their weight per day. The artificial feeds are higher in carbohydrates than the old diets and can be produced as a dry powder that is mixed with fish oil and water to make a paste. Some eels will never accept the paste diet, but in most cases, once a few eels begin to feed on the paste the rest will quickly join them.

In most countries eels are monocultured, but in Taiwan they are grown out in ponds that are sometimes also stocked with mullet, several species of carp, and tilapias. It has been suggested that eels could be profitably cultured with catfish in the southern United States. Depending on the size of the pond, the source of the water, the size of the eels or elvers that are used, and the number of polycultured fish, the stocking rate for eels can be 10^4 to 3×10^5/hectare. By the time the eels are harvested, this can translate to up to 2 kg/m^2.

Eels at 100 to 120 g are harvested in Japan after the first year, but elsewhere eels of about 200 g are produced for consumption in 12 to 24 months. There is a great amount of variability in the growth rates of eels, and females grow more quickly than males (therefore, if a monosex technique could be developed it would be a great advance).

Harvesting eels is somewhat unique. Because they dig into the mud, draining the ponds is not effective. Capture is done with traps or by a net that is set below the feeding platform. In some cases the harvested eels may have a muddy taste, but this can be eliminated by holding the animals in clean water for several days. Ponds are drained annually to remove and segregate by size the remaining eels (if possible), and to clean and disinfect the bottom of the pond.

TILAPIA

The **tilapias** are members of the family Cichlidae. They have long dorsal fins and may be brightly colored. These fish are native to Africa and the Middle East where they were found exclusively until about 50 years ago, but have since spread and breeding populations are now found in many tropical and semitropical bodies of water in the western hemisphere and the Orient. Tilapias are, after the carps, the most popular fish for culture in the world, being reared in Africa, Europe, throughout the Pacific, Japan, China, Israel (where they are called "Saint Peter's fish"), and the Caribbean. There is some culture of tilapias in the United States, but it is limited. The reasons that tilapia culture has not been more successful in the United States are unclear, but are probably related to introducing this "new" fish to the consumer. There are also legal restrictions on tilapia culture: 13 states prohibit the commercial culture of these fish, and 3 states even prohibit transport of tilapia through the state.

Species and Hybrids

There are many species and varieties of tilapias, a large number of which have been used, at least experimentally, in culture. The tilapias, for a long time, were all lumped into a single genus, *Tilapia*, but many have recently been removed from this genus by scientists who study the systematics of fish. Biologists have distinguished two groups:

Figure 10.37 Male *Tilapia aurea*. (Photo courtesy Dr. Robert Winfree, U.S. Fish and Wildlife Service.)

1. *Tilapia* construct nests on the pond bottom and guard the eggs and young, have coarse teeth and few gill rackers, and are herbivorous and macrophytophagous.
2. *Sarotherodon* and *Oreochromis* brood the eggs and the larvae in the mouth (and therefore have a lower reproductive potential), have fine teeth and more gill rackers, and are microphagous and omnivorous.

The most commonly reared species of tilapias are members of the second group, but many culturists still refer to the single genus *Tilapia*, so this will be the format used in this text. The three most important species for the aquaculturist are *T. aurea*, *T. mossambica*, and *T. nilotica*.

T. aurea, normally a drab-colored fish, turns blue with a red margin on the caudal and dorsal fins as it approaches breeding age, and is therefore sometimes called the **blue tilapia** (see Figure 10.37). This fish naturally feeds on both plankton and benthic organisms, but normally in culture it also consumes detritus. It is more tolerant of salt and low temperatures than many other species. *T. aurea* sometimes buries itself in mud or lies on its side in nests during seining, so it is difficult to harvest; other management problems are a result of the relatively large number of eggs (for a mouth brooder) that can be produced and the fact that these fish will mature in warm waters at 5 to 6 months of age.

The head of *T. mossambica*, and **Java tilapia,** is slightly concave in profile. They are olive gray, brown, or blackish fish except during the breeding season, when the females turn gray with black spots. The males change to dark black with some white by the mouth and lower parts of the head, and red along the margins of the dorsal and caudal fins. *T. mossambica* are omnivorous fish eating mostly algae, but also invertebrates and detritus. Like the blue tilapia, it breeds under optimal conditions when still relatively young (2 to 3 months), and if not interrupted by a cold spell will continue to breed throughout the year every thirty to forty days. These fish withstand relatively high salinities. There appear to be many separate genetic strains of this fish, each with distinct growth rates, maximum sizes, and temperature tolerances.

T. nilotica, and **Nile tilapia,** is also a grayish species, looking much like *T. aurea*, but during the breeding season, it turns black with a flushing of red. It will feed on invertebrates and detritus, but primarily consumes diatoms and other phytoplankters. Sexual maturity can be reached in 5 or 6 months. Like *T. mossambica*, there seem to be many genetic strains that differ in growth rates.

Among the other cultured species are *T. galilea, T. hornorum* (the Zanzibar tilapia), *T. macrochir, T. melanopleura* (the Congo tilapia), *T. nigra, T. rendalli, T. sparmanni, T. tholloni,* and *T. zillii* (Zill's or the redbelly tilapia).

Each of the species of tilapia has advantages and disadvantages, depending on the culture conditions (temperature, salinity, stocking density, food, and the presence of other species in a polyculture situation). Hybrids may be used in place of "pure" species. The hybrids have been produced for a variety of reasons:

1. Often a species with good culture characteristics can be improved by changing one trait. Hybrids of *T. aurea* × *T. nilotica* are easier to harvest than the burrowing adult *T. aurea* and are more temperature tolerant than the parent *T. nilotica.*

2. Some hybrids have different feeding habits from those of the parents, which results in a more efficient use of the pond ecosystem because new trophic niches are filled; for example, *T. aurea* feeds mostly on bottom material, but the hybrid produced by a cross with *T. nilotica* feeds mainly on phytoplankton.

3. Hybrids may have different environmental tolerances than the parents. *T. spilurus* grows well even in the high saline waters of the Red Sea and crosses with *T. aurea* yield a hardy hybrid that grows moderately well in full seawater. Many crosses that include *T. aurea* may also be more tolerant of lower temperatures than the other parent species.

4. Hybrids sometimes have superior growth rates. In Taiwan, the male *T. nilotica* × female *T. mossambica* is called the Fu-shou Yu (blessed fish) because it grows more quickly than the reciprocal hybrid or either of the parent species.

5. Other traits may be desired from the hybrids. In some places, consumers prefer fish of a particular color, often reddish, to other fish; red tilapias have been produced by crossing *T. mossambica* × *T. hornorum.*

6. Probably the most important reason for producing hybrids of tilapias is to control reproduction. Tilapia ponds that are not carefully managed tend to be overpopulated; this results in many stunted fish that are of no commercial value. Several species' crosses result in mostly or completely male offspring, although the hybrid males themselves may not be sterile. All-male populations are produced rather than all-female populations because the females grow more slowly. (This slow growth is probably genetic, but it may also be related to the fact that the females do not feed while they are brooding the young in their mouth.) Some tilapia crosses (male × female) that have often, but not always, resulted in all-male offspring are *T. hornorum* × *T. mossambica, T. aurea* × *T. nilotica* (some

strains), *T. macrochir* × *T. nilotica*, *T. mossambica* (Zanzibar strain) × *T. nilotica* (Lake Albert strain), and *T. hornorum* × *T. nilotica* (the Ivory Coast strain results in low fry production). A cross that produces *mostly* males is *T. aurea* × *T. vulcani*.

Spawning and Its Control

Several modes of spawning are utilized by the tilapias. *T. zillii* and *T. sparmanni* are not mouth breeders, but lay their eggs on clean surfaces where they are fertilized. Both these species will guard the eggs and the newly hatched fish. Most other cultured species of tilapias are mouth breeders, but even so, there is room for some variation. For example, the females are responsible for the brooding in many species, but it is the male *T. heudeloti* that is the brooder, while these duties are shared by both sexes in *T. galilea*.

For *T. mossambica*, *T. aurea*, and *T. nilotica*, the spawning processes are fairly similar. A male fish, after developing its breeding colors, will "claim" a particular territory on the bottom of the pond that he will guard. He makes a **breeding nest** by digging a round or oval hole. A ripe female will enter the nest and spawn. In *T. aurea* and *T. nilotica* the spawning may be followed immediately by fertilization, after which the fertilized eggs are taken up in the mouth of the female. (Another sequence may sometimes be seen in *T. mossambica*: the spawning of the eggs is followed by the female taking them up in her mouth, where the male deposits the sperm; the mouth is the site of fertilization.) The female leaves the nest after she has the fertilized eggs in her mouth and a new female may enter the nest to join the male. The greater the number of females in a pond, the greater the number of fry produced. The number of eggs that a female produces is a function of the species and the size of the fish; a Nile tilapia of 0.1 kg will lay about 100 eggs per spawn, while one that is 1.0 kg may lay up to 1500 eggs per spawn. If there is no cold

period during which spawning is repressed, the female may spawn continuously.

The female broods the fry, at least until the yolk sac has been absorbed. The young tilapias then leave her mouth to begin to feed, but return temporarily when they sense danger. Slowly, fewer and fewer fry return, and brooding is finally terminated. While the female is brooding, she eats little or nothing. In the laboratory, young tilapias survive without the female, so it is assumed that the brooding of the fry is solely for protection.

Brooding ponds are dried out before being filled to be sure that no unwanted fish are present. Most species spawn in shallow water of about 0.5–0.6 m, so the pond should not be any deeper than this, or only the shallow edge of the pond will be used for nest building. These ponds are stocked with a male : female ratio of 1 : 2 to 1 : 4 (although 1 : 1 may be needed for producing hybrids), and the absolute number of males that is used is 1000 to 3000 per hectare, depending on the size of the females (big females mean more fry so fewer males are needed). The females do not spawn simultaneously, so the fry must be continually harvested from the brooding ponds. The fry are removed at about 0.5 g by seining the pond with a 4–5 mm mesh net; seining should be repeated every two weeks.

Some work has been done on breeding tilapias in hatcheries (see Figure 10.38). The advantages to this are that specific individual fish can be used to make hybrids and that spawning can be carried on throughout the year in areas where it is inhibited by cold seasons. Hatchery spawning is done in two ways:

1. One male and 7 to 10 females are selected and placed in a tank that is large enough for the male to establish a territory; it is best to use fish just on the brink of maturity. A nest is built in the gravel on the bot-

Figure 10.38 Tilapia hatchery on Maui, Hawaii. Fish to be used as tuna bait. (From Hida, Harada, and King, 1962.)

tom of the tank, but if no gravel is present—only the glass or slate tank bottom—the male will go through the digging motions anyway. Spawning takes place naturally. The eggs are left in the female's mouth for the initial developmental period, and are then transferred to a shaking incubator. The embryos are removed from the female's mouth because the tanks would quickly become overcrowded if the fry were left in the tank, and this would lead to cannibalism of the newly hatched fish; in addition, the removal of the embryos means that the females will be ready to spawn again sooner. The fry are left in the shaker incubators after hatching until the yolk sac is absorbed and are then placed in larger tanks, ponds, or cages for rearing.

2. Stripping is important when working with species that do not naturally hybridize in the hatchery. No hormone injections are needed; ripe fish are recognized when the fish display characteristic breeding colors and have swollen genital papilla, and the scales are erected by the swelling of the body. The gametes are stripped and mixed, followed by the addition of some saline, and then are mixed again. The fertilized eggs are then transferred to the shaker incubator.

The biggest management problem, as stated earlier, is that tilapias will spawn in growout ponds, which become overcrowded with stunted fish. There are several ways to prevent this. The use of all-male hybrids, already discussed, is commonly employed. Another technique is to use predator fish that limit the number of fry; catfish, largemouth bass, eels, and even other cichlids have been used to this end. Tilapias that are held in cages cannot reproduce since the spawned material will fall through the bottom mesh.

Monosex cultures are used sometimes even when hybrid tilapia are not. Some farmers have grown fish for a short period of time and then visually sexed them, putting the faster-growing males in rearing ponds for growout. Tilapia sexes may be identified because, although an anus is present in both sexes, the male has a urogenital opening, and the female has urinary *and* genital openings. The problem with sorting the fish is that it is very time-consuming and the sex of young fish is often misidentified.

A management technique that shows some promise, and is already in use in some farms, is **sex reversal.** When androgenic hormones, such as methyltestosterone, are given to the fry, most (if not all) become males. Injections and submerging the fry do not appear to be the best method of delivering the hormones, but when the hormones are given orally (in the feed), the labor and handling are reduced and the results are better. This method should be used in tanks, not ponds, since it is important that the fry not be exposed to any natural food. The fry are fed the prepared diet with the hormone until they reach at least 18 mm; the length of time that it takes for the fish to reach this target size varies with the stocking density of the tank. It is important to use fry of a uniform size, which means that they are obtained in one of two ways:

1. Fry are produced in a hatchery.
2. A breeding pond is seined for adults; the seined females will spit out the newly hatched fry under this stress, and the fry will school together at the surface and can easily be removed with a fine net.

Growout and Pond Management

There are a number of ways to growout tilapias. These vary with species, temperature, and market. Probably the most important growout procedure is the polyculture of one or more species of tilapias with other fish. While tilapia polyculture systems are less sophisticated than those applied to carp culture, tilapias are generally good in mixed ponds. In the United States, tilapias have been grown successfully with the channel catfish, in Africa with the common carp, in India and the Philippines with milkfish, and in Taiwan with mullet and carp.

In Taiwan and some other areas, tilapias are cultured with ducks and hogs. In the early 1970s, a significant number of rice farmers converted their paddies to fish ponds situated adjacent to hog or duck houses; the wastes from the animals are washed into the ponds, or added after fermentation. The wastes fertilize the ponds by widening the base of the food chain, and some wastes can be used directly by a few species of tilapias. Tilapia often share these ponds with carps. Rice paddies were first used for culture of the Java tilapia in the early 1950s, but this practice declined with the increased spraying of pesticides (which killed the fish) to protect the rice, a shortage of labor, and poaching (fish in rice paddies were traditionally held to be public property by a large portion of the population).

Tilapia may be stocked as forage for larger carnivores. In experimental ponds in Alabama, better production of harvestable-sized *T. mossambica* was achieved when it was cultured with largemouth bass than when grown in monoculture. The best results were seen when *T. mossambica* were stocked in the late fall or early summer, and the stocking of the bass was delayed until the end of the summer. This gave the tilapia a chance to grow to be too large to be eaten by the bass fingerlings, but by the time *T. mossambica* began to reproduce, the fry could serve as forage for the bass; therefore, the pond did not become overcrowded and the original tilapia added to the pond were able to grow without stunting.

Tilapias can also be cultured in cages. As we have stated, one significant advantage to this is that there will be no spawning since the eggs and sperm that are shed will drop down through the bottom of the cage. Cage production of tilapias, as in the case of other fish, is related not only to the number of fish that are placed in each cage, but also on the volume of water that the cages are in.

The age of the tilapia may play a role in management; several pond growout systems have been developed based on age (size):

1. Fish of different ages may be kept together in a pond, while some fish are removed periodically to prevent crowding. After 8 to 12 months, the ponds are drained and some fish are harvested for sale, while others (ungraded) are used to restock a new pond.

2. A pond is stocked with small tilapia, approximately the same size. The fish grow and are allowed to reproduce; when the second generation reaches 60–100 g, the large fish are removed for harvest and the small fish are used for restocking.

3. Fish are allowed to spawn in a separate brooding pond; the fry are harvested at a few centimeters in length and transferred to growout ponds.

Depending on the species and the size of the fish that the farmer desires, tilapias can be stocked in low densities of 1000 per hectare or more than 20,000 per hectare. Higher densities have been used, but smaller fish are produced. Naturally, if the farmer is using other fish in the pond, the number of tilapias should be reduced.

Ponds are often fertilized to increase the production. Most ponds that are fertilized will yield 1000–3000 kg/ha each year. Fertilization, of course, is most important for species that feed on plankton. Several fertilizers have been used, ranging from inorganic mixtures, to lime and superphosphate or phosphates, to manures and oil seed cakes. Sewage also is used in some Asian countries. In Taiwan, sewage wastewater is pumped into the ponds before the tilapia are added to the

ponds. The wastewater is allowed to dry, and this procedure is repeated three or four times. In March, freshwater and tilapia are added, but new wastewater is added every few days to replace water that has evaporated. Such sewage enriched ponds, which are selectively harvested about every two weeks after the first 40 days, have an annual yield of up to 7800 kg/ha.

Feeds have also been used for tilapia culture. In many countries, this supplemental feeding is simple. The feeds include agricultural and kitchen wastes, oil cakes, rice bran, plants, coffee pulp, and corn meal. These seem to increase production, but whether the fish are feeding directly on this material, or it is being used to increase the ponds' productivity, or both, is unclear and probably varies from one situation to the next.

As tilapia culture has moved to some of the more technologically advanced countries, intensive culture in tanks, cages, and raceways has replaced the pond, and formulated prepared diets have been used rather than (or in addition to) fertilizers or simple feeds. During growout, tilapias should be fed at the same time and location seven days a week because the fish learn the feeding schedule and will therefore not waste time looking for the feed; two daily feedings give better yields than one feeding.

In intensive culture, artificial feeds allow tilapia to be grown in high densities, giving productions that are significantly greater than those seen in less intensive systems. In Israel, fish in monoculture were fed a diet of 25% protein; when stocked at a density of 80,000 per hectare, a production of 16,750 kg was realized in 100 days. But the development of a "complete" artificial feed is still very much in progress. Laboratory studies have shown that tilapias need about the same amount of energy in their feed as carp and channel catfish (8–9 kcal/kg of diet), but they have a slightly lower requirement for protein (reports vary, but 20% to 36% high-quality animal protein gives the best results; the variability reported is probably a result of the different protein sources used and the species of tilapia). The amount of feed varies from 3% to 10% of the body weight per day; fish that are warmer need more feed, while those stocked at greater densities need less.

Experimentally, it has been shown that certain amino acids and Kreb's cycle intermediate acids will stimulate feeding, at least of *T. zillii*; these may be used as **feeding enhancers** that would ensure uptake of prepared diets and thereby increase production.

Harvesting and Marketing

When and how to harvest is determined by the market. Harvesting is generally done one to three times each year, depending on the species, the length of the growing season, and the desired size of the final product. Fish are generally seined from the ponds; seining may be done after the water volume in the pond has been reduced. After the first seinings, the water is pumped from a pond so the remaining fish are left in a very small volume in the pond's deep end. Once these fish are concentrated, they may be removed easily with small seines or dip nets. Although tilapias are less sensitive than many fish to low oxygen concentrations, attention still must be paid to this problem when the water level is reduced during harvesting.

Like the catfish, tilapias can develop an off-flavor, especially when they are reared at high densities in ponds or grown using high feeding rates. Several fish should be captured a few days before the planned harvest, cooked (unseasoned), and tasted to be sure this is not a problem. If the off-flavor is present, harvesting should be delayed until the flavor is acceptable. If the fish can be transferred to raceways with quickly running water, the off-flavor will be lost in a few days to two weeks.

Weed Control

Tilapias have been used to **control the growth of aquatic plants.** The species that seem to be the best at this are

1. *T. melanopleura*, a strict herbivore with a preference for higher plants, and generally considered one of the best weed controllers (along with the grass carp).
2. *T. nilotica*, which reduces the levels of filamentous algae in a pond and will also feed on some of the higher plants.
3. *T. zilli*, used to eradicate aquatic plants in the United States and elsewhere; they often show a preference for soft-textured plants, and will ignore those of a coarser texture if soft plants are available.
4. *T. mossambica*, which seems to prefer the filamentous algae.

The Nile and Java tilapias are sometimes used in **malaria control,** because the mosquito larvae live in algae they consume; *T. nilotica* may be especially useful for controlling these pests because it also appears to eat the insect larvae.

SUMMARY

Fish are well suited to the aquatic medium. They have adaptations for movement (swim bladder, hydrodynamic design, myomeres, caudal fin), for respiration (gills and/or accessory breathing organs), and for sensing the environment (eyes, lateral lines, and taste and smell cells). Reproduction can take a variety of forms and seems to be controlled by a single brain gonadotropin and gonadal hormones; in culture, ovulation is very often stimulated by hormone injection. Fish develop from embryos, to fry, fingerlings, and adults; the greatest growth takes place immediately before sexual maturation.

Many types of catfish are cultured throughout the world; in the United States the channel catfish is reared more often than any other fish. Brood catfish are spawned and eggs either are allowed to hatch in the spawning pond or are moved to jars or troughs for hatching. Fry are cultured to fingerlings and are then transferred to growout ponds (in most cases), although troughs, cages, raceways, and tanks are sometimes used. Harvesting is done in batch by netting or draining, or by selectively harvesting only the largest animals over time.

The most frequently cultured fish in the world are the carps (common, Crucian, Indian, Chinese, and others); the different species fill a variety of ecological niches. They will breed in shallow water attaching the eggs to plants, or they may be stripped, fertilized, and the eggs incubated in containers. The carp can be transferred to ponds for growout soon after hatching; the growout techniques will vary with the environmental requirements of the species, as well as the fertilization and feeding schemes.

Trout are grown for direct consumption as food and for restocking of natural waters. The most commonly used species are the rainbow, brown, and brook trouts. Natural spawning takes place in stream beds, but cultured animals are always stripped and artificially fertilized. Incubation of the eggs takes place in trays. After hatching, the trouts are moved to growout facilities (often raceways) and fed high-protein diets.

The major bait fish in the United States are the golden shiner, fathead minnow, and goldfish. Brood fish are introduced into a pond and allowed to spawn. The eggs may hatch in the spawning pond after the brood fish are removed or moved to a separate pond for hatching; the fry can be transferred to another pond after hatching. Growout ponds are fertilized, and feeds can be used to increase production. Bait fish are generally harvested by seine nets, although other techniques can be used. The fish are held for a short period after harvesting before being shipped to market.

Largemouth bass are usually cultured with the bluegill in "farm ponds," which are largely recreational. Both types of adult fish are harvested by hook and line. In theory, the bluegill reproduce and their offspring are the forage for the bass; this prevents the bluegill population from getting so large that the fish are stunted. In practice, these ponds are often unbalanced because of a lack of management. Other fish may be substituted for bass and bluegill. Bass themselves are spawned in ponds in warm waters and grown out to fingerlings for stocking.

Eels are catadromous fish that are captured as elvers as they enter freshwater from the ocean; they are not spawned. The elvers are reared on a diet of minced fish and invertebrates, followed by a prepared ration. The elvers metamorphose to juvenile eels that are transferred to growout ponds. They are fed a paste diet that is high in animal protein or a formulated diet with less protein and more carbohydrate. The eels are harvested with special traps or nets, not by seining or draining the pond.

Tilapia are tropical fish grown all over the world. Different species have different properties, and hybrids can be produced with modified traits. Perhaps the most important use of hybrids is to produce all-male populations (to prevent uncontrolled reproduction). In ponds, the males build nests where fertilization takes place; females often brood the young in their mouth. Growout is usually in fertilized ponds, but formulated feeds and tanks, raceways, and cages have become increasingly popular.

CHAPTER 11

COMMONLY CULTURED SALTWATER FISH

Marine fish are farmed far less frequently than those that inhabit freshwater. This is because saltwater environments are harder to manage, the artificial propagation of many marine fish is difficult, and the larvae are often delicate. However, several species have established themselves as candidates for aquaculture. Those that are most commonly grown are usually able to survive in estuarine waters and do not require strictly oceanic conditions. Some of the less commonly cultured saltwater fish are discussed in Appendix 8.

MULLET

The most commonly cultured member of the family Mugilidae is the **striped mullet, *Mugil cephalus*** (see Figure 11.1). Other species of interest include *M. capito*, *M. dussumieri*, *M. corsula*, *M. engeli*, *M. tade*, *M. macrolepis*, *M. troschelli*, *M. saliens*, *M. falcipinnis*, *M. grandisquamis,* and the gray mullet, *M. auratus*. These are sleek fishes with two dorsal fins, a terminal mouth, and medium-sized scales. *Mugil* are among the most common estuarine fishes in the world, being found in full seawater as well as freshwater streams, in most tropical and semitropical regions.

Mullets are good candidates for culture because the nutritional requirements of adults are easily met, they are hardy animals withstanding a wide range of environmental conditions, they are not hard to rear once they reach fingerling size, and they produce a good-quality meat. They are therefore expected by many scientists to make a significant contribution to the total harvest of cultured fish in the future. The major problem in mullet culture is the production and rearing of larvae. Until recently, when controlled spawning by hormone injection became possible, fry for ponds were collected from the wild. Natural fry are still a very important part of the international mullet production scenario because the spawning fish are difficult to handle and the larvae require a specialized diet. The fry can be collected in shallow estuaries when they are about 2 cm in length as they are migrating from the sea to freshwater. Culturists have reported declining numbers of available wild fry for many years.

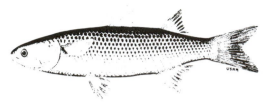

Figure 11.1 The striped mullet, *Mugil cephalus.* (From Whitworth, Berrien, and Keller, 1968. Courtesy of Connecticut Department Environmental Protection.)

Spawning

Mullets live in fresh and brackish water, but enter the sea to spawn when reaching sexual maturity at 2 to 5 years. Spawning takes place at different times of the year depending on the species and the location. Females produce up to 3×10^6 straw-colored eggs, each about 1 mm in diameter, with a large oil globule that keeps it afloat and supplies energy. To obtain gametes, biologists in the 1930s stripped fish that were captured just before spawning. But the capture of ripe, prespawned fish is difficult. The need to catch these fish was eliminated when a technique to induce spawning by hormonal injections was established.

Hormonal injection techniques that result in successful ovulation in the striped mullet were developed in Taiwan in the 1960s. Using either synthetic or mammalian hCG hormones, extracts from the pituitary of mullet, or extracts from any of several other fishes including salmon and carp, a high rate of spawning can be expected shortly after the injection. The usual sequence of events begins with the gravid female fish, usually 4 to 6 years old, collected by fishermen and brought to the laboratory (or the initial stages of ovarian development can be stimulated to occur by manipulating the temperature and photoperiod). The females are then injected several times with the hormone solution to induce spawning; adult male fish captured during the reproductive season do not require injections. Mullet do not handle well, so normal injection techniques may traumatize them. Any method

that can be used to reduce the amount of handling, such as the use of a catheter placed in the body cavity, or implanting pellets that slowly release hormones, should be explored by the culturist. Administration of the hormone must be done at the right time during ovarian development; to do this, the stage of the eggs' development must be determined directly by collecting eggs from the oviduct using a cannula or other device. After the second or third injection, the fish are placed in holding ponds while ovulation takes place. If the fish are captured in seawater, they are placed in seawater ponds, and if captured in freshwater before the injections, they are placed in freshwater where the salinity is increased slowly to that of seawater. Fertilization may then take place in the ponds or be carried out artificially using the dry method.

Rearing the Fry

After the fertilized eggs are collected, they are transferred to hatching tanks kept at nearly full seawater salinity with temperatures of 21°–24°C. Antibiotics may be added to prevent disease problems. Hatching takes place in about one and a half days, at which point the mullet are about 2.5 mm long.

Fry begin to feed during the third day. They are planktonic, and little is known of their natural dietary habits. At this point the largest losses in culture take place, but this is not a great problem since each female produces so many eggs, and the culturist has not yet invested much time and effort in the offspring.

Several methods for rearing mullet fry have been tried; diet seems to be the biggest problem for these planktivores. The best results have been recorded when feeding the fry a mixture of different types of natural zooplankters such as copepods, mollusk larvae, rotifers, sea urchin larvae, and newly hatched *Artemia*. The zooplankton are mixed with phytoplankton, especially diatoms. The survival of the fry may be up to 5%, though it is often considerably below this and highly variable. Once the animals reach 0.2 to 0.5 g, an extruded pellet may be used, although the formulated feeds may be supplemented with live feed organisms for better survival.

Among the other factors playing an important part in survival of the young fry are stocking density, species of mullet used, methods of aeration, and temperature.

Young striped mullet, at least to 35 mm, are carnivores living largely on zooplankton. As they grow, algae and detritus become increasingly important in the diet; a bacterial population is built up in the gut of the older mullet that allows it to rely on these nonanimal nitrogen sources.

Growout

Whether raised in the hatchery or captured in estuaries, mullet fry of 20 mm or more may be moved into ponds and allowed to grow for one to two years before harvesting. These ponds may be slightly brackish or near seawater's salinity. Alternatively, keeping the animals in a smaller nursery pond for one to two months and then transferring them to the growout pond makes better use of available space. In Israel, mullet are raised in monoculture until they are 30–70g; then they are placed in freshwater polyculture ponds with tilapia and carp. Harvesting of all fish can take place in four to five months.

Mullet are commonly reared with other species, such as milkfish and silver carp (the Philippines), shrimp (India), and eels (France and Italy). In properly fertilized ponds, there will be a good deal of feed available for mullet. However, to increase production, the diet may be supplemented with rice, peanut and other offals, vegetable wastes, and some artificial feeds.

Mullet are harvested by nets, often after the volume of water in a pond has been reduced, when they reach a marketable size (about 400 g). The rates of production vary greatly, depending on water conditions, feed, and stocking density. Reasonably good production has been observed in ponds with very limited management.

YELLOWTAIL AND POMPANO

Marine fish belonging to the family Carangidae include the jacks, pompano, and yellowtail. These are medium to large fish, oceanic, and piscivorous. Carangids are most common in semitropical waters. Many are taken by sport fishermen as well as being commercially sought. They have been actively cultured for about three decades.

Yellowtail

Yellowtail, *Seriola guingueradiata,* are among the most popular of cultured fish in Japan, with an annual production in that country of over 90,000 metric tons. This fish was first raised in Japan in 1928, but it was over 30 years later that its culture became economically important. Yellowtail have been taken by fishermen for centuries; however, since they are active, open-ocean fish, fish farmers for a long time felt they would be unsuitable for rearing in confined environments.

Capture and Rearing of the Fry Scientists have been unable to induce ovulation in this fish, although artificial fertilization of eggs is possible. For the purposes of large-scale culture, the wild fry still must be captured and reared. In the spring, the Kuroshio current carries the delicate larvae to the coast of Japan. Here they are captured in nets by specially licensed fishermen who are limited in the number they may take. The fish, at this time, are about 15 mm long and weigh about 1 g. The fry must be separated from each other according to their size; if small, medium, and large fish are not kept in individual enclosures, there may be mortalities of up to 50% in a few days due to cannibalism.

After the fish have been separated according to size, the process of rearing them begins. Culture may take place in diked enclosures with sea gates that permit exchange of water with the ocean when the tides rise. This old method of culture is no longer popular since stocking densities are only a fraction of those that can be realized when net cages are used.

The net cages used for culturing the yellowtail juveniles are rectangular, made of nylon, and may range from 2 m³ to 150 m³. These are placed in rows, often with walkways on the side for workers. Optimum temperatures for growth are 24°–29°C. It is thought that the growth of larvae in the sea is not very rapid, but after the fish metamorphose to juveniles (at 10–14 mm) during the spring and early summer, and become acclimatized to their net cages, they may quickly double or triple in size. In an ideal laboratory environment, the juvenile growth rate may be expressed as

$$SL = 3.27e^{0.048x}$$

where SL is the standard length, x is the number of days after hatching, and e is the base of the natural log system. During their initial period in the cages, they are fed a diet of finely chopped fish (white flesh fish with little oil) and shrimp. This may be supplemented with zooplankton that are attracted to the cages by lights. During this stage of culture the fish may reach up to 10 cm in length and 50 g in weight, although they are usually smaller. Having gone through their initial culture, the fingerlings may be graded again according to size and shipped for growout.

Growout Growout is usually carried out in large nylon or metal cages that may be hundreds of cubic meters, and stocked at densities up to 1 fish per 100 cm² of surface area. Water conditions must be good, with dissolved oxygen never falling below

3 ppm, temperature never going below 19°C, and salinities staying above 16 ppt. Cages may be at the surface or down to 30 m depth if there are strong surface currents that may damage the nets.

The fish in the pens are voracious, and much of the cost of the entire culture operation goes to feeding the yellowtail until they are harvested. Trash fish that are low in oil are shoveled into the pens in the morning and afternoon. Some pellets are used, but these pellets are mostly ground white trash fish too, with a vitamin and mineral mix added. Iron and cobalt are considered to be important for preventing anemia. The feeding rate decreases as the size of the fish increases and harvesting is approached. In six months, fingerlings that were stocked in the growout cages reach 1 kg or more.

The harvested yellowtail (see Figure 11.2) are often sold live, being shipped to markets in iced water. Up to half of all the cultured yellowtail may be marketed in

Figure 11.2 Yellowtail of about 1 kg taken from floating pen. (From Bardach, Ryther, and McLarney, 1972.)

this manner. About 95% of all the fish cultured finally end up in restaurants, with only the small fish going to super-markets.

Pompano

Pompano are fishes of the genus ***Trachinotus*** (see Figure 11.3). The Florida pompano, *T. carolinus*, is the major species of interest and has been explored as a potential culture organism since the mid-1960s because of the great demand for this fish.

Figure 11.3 Pompano. (From Thompson, Weed, and Taruski, 1971. Courtesy of Connecticut Department Environmental Protection.)

A method has been developed to produce pompano throughout the year. Temperature and light are manipulated so that gonadal development is stimulated. The size and stage of the eggs is established, and if the eggs are in the proper developmental stage, the fish are then treated with hormones and stripped of their gametes within 32 hours of the first injection. Fish so treated can be returned to the broodstock tanks and used again in six months. The spawning success rate over a 1-year period, of hormone injected pompano, has been reported to be over 60%. The newly hatched fish can be raised on a diet of brine shrimp and rotifers. The larvae will reach a length of 15 mm in about 22 days and can then be trained to accept artificial foods.

If facilities are not available for spawning and rearing of newly hatched larvae, young pompano can be caught in the spring and summer for culture. On beaches with a gradual slope, small seine nets are used to collect the young pompano. Fry can be common during the breeding season, although the number that can be taken may be restricted by law.

The natural spawning season of the pompano is believed to be rather long because fry can be caught for an extended period of time (in Florida, from April to November). Females mature in two years, and ripe males can usually be obtained after they are 1 year old. A second-year female produces about one-half million eggs.

Most of the experimental culture of these fish has been in ponds and tanks, although a net cage system like that used for yellowtail may hold more promise. In ponds, the fry have been stocked at densities of almost 2.5×10^5/ha and growth rates are such that fish of the ideal marketing size, 300–500 g, can probably be produced within one growing season. They require oxygen concentrations of 3 ppm or greater.

Feeding of the pompano has been the subject of several experiments; originally, cultured pompano were given trash fish, but because of the variable supply, several prepared diets have been tested. High-protein trout diets have been used with some success for young pompano, but as the fish grow, these must be supplemented with natural foods. Recent experiments with artificial feeds that include menhaden oil as an important ingredient show that a complete and acceptable diet can be produced. In a commercial facility in the Dominican Republic, it was shown that a prepared diet supported fish to 150–200 g with a feed conversion of 3.5 : 1, but once they reached this size, the conversion ratio changed to 6 : 1, an unacceptable level. This probably can be corrected by further research on the change in requirements of the growing fish.

Besides rapid growth and a strong market, there are several reasons why pompano are attractive fish for the potential culturist:

1. They are hardy, with little mortality when being shipped as fry, and adults heal quickly from wounds and other injuries.
2. They grow best in seawater, but can adapt to much lower salinities.
3. They can tolerate temperatures of 12°–38°C, although the best growth takes place at 25°–34°C.
4. They are able to withstand almost any natural changes that are likely to take place in the pH or turbidity.

Despite early promise, pompano culture has not been carried out successfully on a commercial scale. Some of the problems that faced the first pompano culturists remain unsolved, so the fate of the industry remains uncertain for the time being. Among these problems are the following:

1. A low-cost food must be developed that allows quick growth and low mortality.
2. There are still problems with diseases and parasites.
3. The fish are susceptible to cold shock in water below 10°C.

FLATFISH

Flatfish are members of the evolutionarily modern order Pleuronectiformes. These fish are greatly flattened laterally, with both eyes on one side of the head as adults. The larval flatfish look much like other larval fish, with one of the eyes migrating around the skull as the fish metamorphoses to the juvenile stage. These important commercial fish, found in shallow temperate and subpolar waters, include flounder, sole, halibut, turbot, and plaice.

Flatfish were among the first marine fish to be cultured experimentally. A great deal of this work was done in the late nineteenth and early twentieth centuries in the United States, but much of the activity has shifted to Europe, especially the laboratories of Lowesoft, England, and Port Erin on the Isle of Man, and recently spreading to other facilities in France, Scotland, Spain, Norway, and Denmark.

Among the flatfish that have been reared in the laboratory, the principal species are the **sole,** *Solea solea,* and **plaice,** *Pleuronectes platessa.* Recently, there has been increased interest in the **turbot,** *Scophthalmus maximus,* and **halibut,** *Hippoglossus hippoglossus.*

Spawning and Hatching

Broodstock sole are brought indoors to tanks of clean water with a layer of sand on the bottom that is used by the fish to bury themselves (see Figure 11.4). During the breeding season, in the late winter and early spring, spawning takes place in these indoor tanks naturally with no special manipulations being carried out. The eggs float to the surface, are collected with nets, and are reared in a separate facility. The eggs are hatched in troughs held at about 6°C; hatching takes about three weeks.

Technology for the production of turbot ova is such that they can be produced

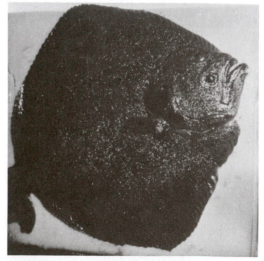

Figure 11.4 Brood turbot. (Ingram, 1987. Copyright © Archill River Corporation. Reprinted by permission.)

almost at will. Females will produce about 2.5×10^5 ova during a three- to four-week period during the spawning season. These eggs are removed during several separate strippings with new ovulations taking place about every 70 hours. The success of fertilization depends upon it taking place within hours of ovulation. If fertilization has been successful, the embryos can be transferred to incubators kept at about $14°–15°C$; posthatch larvae can be shipped after about eight days.

Rearing the Larvae

When the larvae hatch from the eggs, they are planktonic and are fed a diet of newly hatched *Artemia,* or in the case of the turbot are started on a diet of algae-enriched rotifers and then switched to algae-enriched brine shrimp nauplii. During this premetamorphosing stage, most mortalities take place. Water quality is critical. Bacterial infections were a major hindrance to any sort of flatfish culture work until James Shelbourne, in 1962, began to use penicillin and streptomycin in the hatching tanks.

The fish are weaned from brine shrimp as soon as possible. For turbot, this takes about 40 days (see Figure 11.5). Weaning may be started with small worms. Later, after metamorphosis, the fish will accept a diet of chopped mussels and fish. There has been some recent success using prepared diets for weaning. Two-month-old sole, experimentally fed a diet of extruded pellets to which had been added a mixture of amino acids, grew well while they were being weaned from live food. The amino acid mixture stimulated feeding behavior and resulted in high survival and growth, even when compared to fish given a similar diet that included brine shrimp, polychaetes, or mollusks in place of the amino acid mixture.

Growout

One of the biggest problems with the culture of flatfish is that they use only the bottom of the rearing facility, not the water above, so production is more limited than in the case of fishes occupying the entire volume of water. Most of the past interest in flatfish was simply in their culture to restock natural waters, but ponds, tanks, and dammed loches can be used to rear flatfish to a marketable size for commercial purposes.

Turbot have been reared experimentally in plastic-netted floating cages with flat bottoms. The fish, on average, reached the marketable size of 0.75 kg in two summer growing seasons and 1.5 kg after three years in culture (some were as large as 3 kg). The techniques of growout are expected to improve in the near future, and commercial success probably will not be far behind.

MILKFISH

The **milkfish, *Chanos chanos,*** is found in the tropical and semitropical coastal waters of the Indian and Pacific oceans (see Figure 11.6). It belongs to the order Clupeiformes along with herring, sardines, and anchovies. Milkfish reach 1.5 m in length and weigh up to 18 kg. This fish is not easily taken by most commercial fishing methods, but can be cultured from larvae. Milkfish have been grown for hundreds of years in the Pacific; its popularity with culturists results from its tolerance to temperature flux ($15°–40°C$), ability to survive in low-oxygen environments if necessary, tolerance to a very wide range of salinities

Figure 11.5   Small juvenile flatfish. (Reprinted by permission of *Nature* Vol. 330 pp. 337. Copyright © 1987, Macmillan Magazines Limited.)

Figure 11.6 Milkfish. (Photo courtesy of The Oceanic Institute, Hawaii.)

(from freshwater to nearly 70 ppt), and its diet. Milkfish are also among the most disease-free of all cultured aquatic animals.

Obtaining Fry

Milkfish spawn in clear oceanic waters, usually less than 30 m in depth over sand or coral; the females produce millions of eggs that hatch within 24 hours into planktonic larvae. The larvae migrate toward coastal waters and metamorphose into fry. The young fish stay inshore until they reach 200 to 4500 g (1 to 4 years of age); then they return to the sea, where they survive on a diet of plankton and eventually spawn.

Until recently, milkfish were not spawned in culture facilities, so in most locations, the collection of wild fry is still very important. In Indonesia the fry are collected in coastal ponds and by the use of a **blabar,** a long rope with plant leaves attached to it; the blabar is set on the water and the fry are attracted to the floating cover. In the Philippines and Taiwan, most of the fry are simply collected with dip nets when the milkfish move into the narrow tidal streams, while seine nets and traps are used in deeper water.

Some milkfish have been reported to spawn when kept in earthen ponds. In the Philippines, during a series of experiments, large net cages (270 m^3) were used for spawning. The cages held both hatchery-raised milkfish and those caught in the wild. For all fish, spawning took place spontaneously; this occurred near midnight, during the natural spawning period. There was a high fertilization rate, and up to 80% of the eggs that were collected hatched. The problem with using this system is that in waters with salinities above 32 ppt, the eggs sink and are lost to the currents after going through the bottom of the cage. Alternatively, when kept in tanks, the fish can be induced to mature early using lights to increase the number of hours of "daylight."

Hormones have also been used to induce spawning. This is followed by stripping both sexes of their gametes. The hormones used have included injections of fish pituitary extracts and hCG. However, stripping generally results in a low fertilization rate and the loss of the mature broodfish. The loss of the broodfish is a particularly severe problem since it takes at least five years for the female to reach maturity. However, stripping can be eliminated. Spawning can be induced in tank-held milkfish by using **LHRH-a** (a superactive analog of luteinizing hormone-releasing hormone); in addition, **17-α-methyltestosterone** may be used to increase the number that reach maturity (both males and females). These hormones may be injected into the fish; implanted pellets that slowly release the hormones into the blood also can be used.

Rearing Fry

Larvae survive on a diet of unicellular algae, rotifers, copepods, and newly hatched brine shrimp. The fry are then moved to ponds.

Before the fry are introduced, the ponds are dried, plowed, limed (if needed), and filled with a dilute seawater. These ponds are fairly shallow to encourage the growth of plants. The fry should be introduced into the ponds at night when the water is cool. Milkfish are introduced into the ponds only when there is sufficient flora to support them. The term for the flora, **lab-lab,** refers to the rich growth of cyanobacteria and associated organisms that line the pond and serve as the basal ecosystem that supports the fry. In laboratory experiments, the cyanobacteria, *Oscillatoria* and *Chroococcus,* supported growth of the young milkfish, while cultures of unicellular algal species were inferior.

Sometime algal growth in the ponds is encouraged by manuring, but care must be taken since oxygen depletion may be a problem in such shallow ponds. In Indonesia, ponds are sometimes supplemented with rice bran, wheat starch, and hard-boiled egg yolks; in Taiwan fry may also be given peanut and soybean meals.

The water in fry culture should be kept warm, and may exceed 38°C. Quick drops in temperature caused by rains are a source of extensive mortalities.

Growout

After several weeks, when the fingerlings are 4–8 cm, they are transferred to growout ponds. There are several methods used to growout the milkfish:

Indonesia Culture is carried out in ponds called **tambaks.** These may be a number of kilometers from the coast but still receive seawater through canals, ditches, and tidal streams. Salinities in the tambaks may range from 10 to 35 ppt.

The ponds are fertilized with plant wastes and stocked with 2000 to 10,000 fingerlings/ha. Since 1952, the use of freshwater ponds that receive dilute sewage for fertilization has become popular.

There may be some supplemental feeding, but the milkfish live largely on the algal mats that grow in the pond. The mats consist primarily of cyanobacteria, along with some benthic diatoms and associated flora. Other organisms may find their way into the ponds, including mullet and penaeid shrimp.

Milkfish in culture may grow to be several kilograms, but they are normally harvested at 300–800 g. The number of crops per year may be 1 to 3, depending on conditions and the skill of the aquaculturist. Harvesting is done at high tide because the fish tend to be concentrated near the sluice gate at that time and can be netted with little difficulty. If there is no strong tidal influence, the ponds are partially drained and cast nets are used.

The Philippines Fingerlings are placed in transition ponds for about one month before being moved to growout ponds. The transition ponds should have a heavy growth of lab-lab. The fingerlings are transferred to the growout ponds when they are 10–15 cm long and are stocked at a density of 1000 to 2500 kg/ha. The ponds are fertilized with chicken manure, inorganic nutrients, tobacco wastes, and/or plant wastes for good algal growth. The ponds may be supplemented later with added algae if the primary production is not great enough to support the fish. Decayed and dried water plants, and fresh red algae (*Gracilaria*), are used. Polyculture, especially with prawns, is not uncommon.

A successful scheme, developed in the mid-1960s, is to stock the ponds with a combination of half-grown milkfish and several sizes of smaller milkfish and then partially harvest every two weeks; the purpose of this is to maintain a fixed biomass in the ponds and make better use of the space. In the case of a single harvest, the fish are collected after six to nine months at about 0.5 kg. Gill nets and seining can be used to collect the fish. If the lab-lab is too thick, it may be damaged by the nets during the harvesting, in which case the ponds can be partially drained and the fish collected in a specially constructed catch pond (this is also done in Indonesia).

Taiwan The ponds are fertilized with manures to support algal growth; but when the rainy season starts in May, the salinity becomes too low to support good algal growth, and there is supplemental feeding. The feeds used are rice bran, soybean, or peanut meals.

The ponds are first stocked with large overwintering fingerlings that were too small to harvest during the previous year. In about a month, wild fry become available, and these are added to the ponds. These new fry are added every two to four weeks. Milkfish harvesting starts in May, and there are about 18 harvests until November. The biomass of the ponds is greatest during June and July.

The fish are harvested by gill nets. Before harvesting, a rope with bamboo pieces attached to it is towed across the pond to scare the fish into emptying their stomachs, because fish with empty stomachs ship and keep better after harvesting.

SALMON

Marine salmon, along with their relatives the freshwater trout, are among the most popular fish in the United States and Europe. Their culture has spread over most of the world and continues to grow. There are two important culture methods:

1. *Salmon ranching* The release of the salmon into the environment and their subsequent return.
2. *Net cage culture*

The salmonids are considered relatively primitive fish that can be distinguished by a fleshly dorsal **adipose fin.**

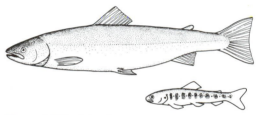

Figure 11.7 Atlantic salmon. (Reproduced with permission from *Salmon and Trout Farming* by Laird and Needham, published in 1988 by Ellis Horwood Limited, Chichester.)

Salmon can withstand considerable salinity fluctuation and most are anadromous breeders. Two types of salmon that are of interest to the culturist:

1. The **Atlantic salmon,** *Salmo salar* (see Figure 11.7)
2. The various species of **Pacific salmon, *Oncorhynchus,*** that include
   a. **Chum salmon, *O. keta,*** which is not one of the preferred species in the United States, but is ranched in Japan and the Soviet Union (see Figure 11.8).
   b. **Pink** or **humpback salmon, *O. gorbuscha,*** which has pink rather than red meat (see Figure 11.9).
   c. **Coho** or **silver salmon, *O. kisutch,*** perhaps the easiest to culture (see Figure 11.10).
   d. **Sockeye** or **red salmon, *O. nerka,*** which has a deep red color in the ocean and is highly

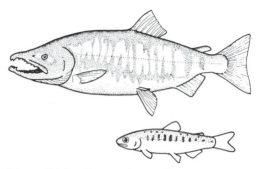

Figure 11.8 Chum salmon. (Reproduced with permission from *Salmon and Trout Farming* by Laird and Needham, published in 1988 by Ellis Horwood Limited, Chichester.)

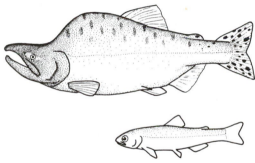

Figure 11.9 Pink salmon. (Reproduced with permission from *Salmon and Trout Farming* by Laird and Needham, published in 1988 by Ellis Horwood Limited, Chichester.)

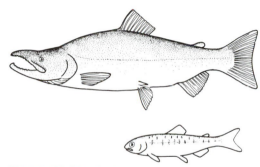

Figure 11.11 Sockeye salmon. (Reproduced with permission from *Salmon and Trout Farming* by Laird and Needham, published in 1988 by Ellis Horwood Limited, Chichester.)

regarded for canning (see Figure 11.11).

e. **Chinook** or **king salmon,** *O. tshawytscha,* the largest of the Pacific salmon (see Figure 11.12).

In addition, some strains of rainbow (or **steelhead**) trout are anadromous with extensive oceanic migrations; while most rainbow trout culture is done in freshwater, some European farms do move adults to sea for the final growout in pens. The brown trout (or **sea trout**) can also be reared in marine pens.

Salmon go through a remarkable spawning migration. After several years at sea, they return to the rivers and streams they lived in as fry; these migrations may be hundreds of kilometers up-stream. During this time, the fish cease to feed and develop secondary sexual characteristics such as changes in color, hooking of the jaw, and a humping of the spine. The fish select a suitable site and clear a space; the female releases her eggs, which are immediately fertilized by the male(s). The female then covers the eggs with gravel. This nest is called a **redd** (see Figure 11.13). The eggs hatch and the salmon emerge as **sac fry** or **alevins,** and when the yolk sac is absorbed they are simply called **fry.** As they grow, vertical lines develop on the sides; at this stage they are known as **parr.** The parr eventually go through a process called **smoltification,** which includes a change to a silvery color, and an assortment of physiological adaptations that allow them to live in seawater. These **smolt** then migrate downstream and enter the ocean. Small males that return for spawning are

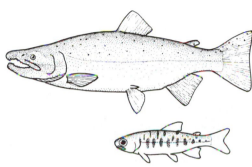

Figure 11.10 Coho salmon. (Reproduced with permission from *Salmon and Trout Farming* by Laird and Needham, published in 1988 by Ellis Horwood Limited, Chichester.)

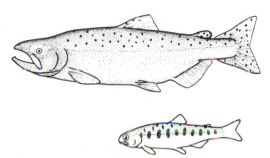

Figure 11.12 Chinook salmon. (Reproduced with permission from *Salmon and Trout Farming* by Laird and Needham, published in 1988 by Ellis Horwood Limited, Chichester.)

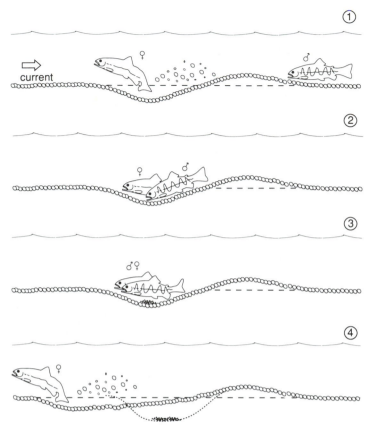

Figure 11.13 Spawning of the chum salmon. (1) Female digs a nest in the stream bed while the male stands guard. (2) Male quivers his body to stimulate spawning by the female. (3) The release of gametes by both the male and female in the nest. (4) Female fills the nest with gravel. (From Kafuku and Ikenoue, 1983. Copyright © Elsevier Scientific Publishing Company. Reprinted by permission.)

called **jacks** or **grilse,** and salmon that are **spawned out** are referred to as **kelts.** For the Pacific species, spawning is followed by death, but the Atlantic salmon may return for several spawns.

The homing mechanism that these anadromous fish use to return to the streams that they left as smolt is not well understood. When at sea they may use electrical currents, characteristics of water masses, or some other signals to return to the general area of the stream or river. Once near the mouth of the river, it is likely that they are at least partially affected by the "smell" of smolt and fry that are upstream, as well as the general chemistry of the water.

Fertilization

Fertilization may take place artificially after stripping, or the salmon may spawn naturally in streambeds or in spawning channels. Spawning channels are specially constructed with a particular gravel and a controlled water flow. These work best for pink, chum, and sockeye salmon.

Adults are normally collected for artificial fertilization as they return to freshwater to spawn. Alternatively, the reproductive cycle may be advanced or retarded by up to 6 months, at least in the case of pink and coho salmon, by the culturist who alters the photoperiod of the held salmon. Advances have also been made using hormone injections.

Because Pacific salmon die after spawning, they are not stripped. Females are cut and allowed to bleed for several minutes; this reduces the chance that the eggs will be in contact with the blood that reduces the fertilization percentage. After the bleeding, the fish is lifted over a bucket and an incision is made with a special **spawning knife;** the eggs should flow freely into the bucket if they are well developed. Males are also killed and the milt is hand stripped into the bucket; the milt and eggs are gently mixed by hand.

The female Atlantic salmon is placed in a double-walled rubber sack with the end of the fish protruding for the collection of eggs. The sack may be adjusted to fit each individual fish. The use of this mechanism decreases the chances of damage to the fish. Water fills the space between the walls of the sack, causing ripe eggs to be released. The milt from the male is gently removed by drawing it into a glass tube with suction as the sides of the fish are gently pressed. The milt may be stored in test tubes. Eggs and milt have a limited storage period, the length of which varies with the species and the storage temperature.

Incubation, Hatching, and Rearing to Smolt

Incubation of salmon eggs is very similar to incubation of trout eggs. The number of eggs produced is estimated by any of several methods based on weight, volume, or using a mechanical counting device with a photoelectric cell. When eggs reach the **eyed stage,** the underdeveloped and infertile eggs are removed; these are identified by giving the eggs a moderate mechanical shock, such as striking the egg tray sharply. When the undeveloped eggs are thus shocked, they rupture, water enters, and the yolk coagulates, turning it a whitish color which can be seen easily. These may be picked out by hand, or an electronic egg sorter can be used. Another method for separating the live and dead eggs is flotation. The eggs are placed in a container of water and NaCl or sugar is slowly added; when the water reaches a certain density, the lighter dead eggs will float to the surface where they can be skimmed off, but those with a viable embryo will remain on the bottom. The exact amount of salt or sugar to add depends on the stage of development.

As in the case of many other fish eggs, direct light—especially in the blue-violet and UV ranges, typical of most fluorescent lights—is the most damaging; other lights, especially the pink fluorescent types that emit mostly yellow to red light waves, can be used, but it is safest to keep the eggs covered.

The eggs can be hatched in jars, trays, or troughs. Good results have also been achieved when the eggs are allowed to hatch in a gravel environment. Gravel is believed to be important because

1. After hatching, the alevins will use the gravel to maintain their position (thus expending less energy swimming).
2. When they are disturbed, the alevins react by moving into the substrate (if there is no gravel, they waste energy in a vain attempt to seek shelter).

The number of alevins per unit area also seems to affect the condition of the newly hatched salmon; in experiments where chum salmon eggs were incubated at two different densities, with and without a gravel substrate, it was found that alevins that were kept at low densities with a substrate were larger and healthier than those under other conditions.

As in the case of trout, hatching time is a function of the temperature at which the eggs are incubated, but it is probable that the ideal temperature will vary slightly between different genetic stocks. Chinook salmon have been incubated at temperatures of up to 16°C without significant mortalities, although most incubations are conducted at several degrees below this. Coho salmon show best egg and alevin survival at 1.3°–10.9°C, with mortalities increasing very quickly at lower temperatures and above 12.4°C.

For ranching, chum and pink salmon are generally released upon emergence, although they may be kept for a short period of time; chinook, coho, and sockeye salmon are kept for longer periods of time before they are released to the environment. These fish are stocked as fry in ponds, raceways, troughs, and silos. Salmon to be grown in pens are initially treated in a similar fashion. Under normal conditions it takes one to two years for the smolt stage to be reached, but when good water conditions are maintained and the fish are fed a high-protein diet, some species can be acclimatized to seawater in four to five months.

After the salmon fry hatch, they may be weaned onto artificial feeds before the yolk is depleted. Some fish may never learn how to accept feed, and starvation results in high mortalities. After the fish have learned to feed and deaths from starvation have declined, the fish are graded according to their size; grading continues throughout their culture. Smolts are finally transported to pens in boats with wells. For the Atlantic salmon, the boat wells contain oxygen-rich chilled freshwater or oxygenated water of the salinity to which they have been acclimatized in the hatchery.

Growout

Salmon growout takes two generally successful forms: fish are either pen cultured or ranched. Structures such as raceways, concrete ponds, and circular ponds have been adapted from trout culture for salmon growout, but their use is minimal. **Ranching** of salmon refers to the release of the salmon to natural waters, followed by their development at sea, and subsequent return to freshwater at maturity to be harvested. The great advantages of this technique are that the salmon do

not require the culturist's care during the long growout, so money spent in other culture operations on labor, feed, and facilities is negligible for the salmon rancher. Salmon are self-herding, so expenditures for boats and fuel to recapture the fish are minimized. The salmon rancher's biggest problem is that there is no way of knowing how good a recapture harvest will be until the fish actually return, several years after their release.

How to get the best return of released salmon has been the subject of much research. Returns of up to 20% have been reported, but normally many fewer fish make the final spawning run. The size of the salmon that are released affects the success of the program; it has been reported that coho salmon fry of about 13 g were released and showed a return of 1% to 2%, but when their release size was doubled, the rate of return was 7% to 8%. It is also important that the quality of the water to which the salmon are returning remains good.

In the late 1960s, researchers at the National Marine Fisheries Service at Puget Sound made an important discovery for future salmon ranchers. It was shown that salmon will return to a saltwater site just as though it was a freshwater site if they are **imprinted.** There are two substantial advantages to this:

1. The cost of waterfront property connected to a stream or river is very high.
2. After the fish enter the freshwater, the quality of the meat declines.

A good example of imprinting for salmon ranching is carried out by Sea Run, Inc., of Kennebunkport, Maine. This company chose to rear Pacific pink and chum salmon from eggs; at about 4 cm, the fingerlings are taken from the hatchery environment and placed in pens for several weeks. The pens are near the discharge of an oil-fired power plant; the plant discharge is untreated seawater used as a coolant to condense steam. The water has an identifiable temperature and probably also a distinctive smell after passing through several miles of piping. In addition, while the fish are in the pens, a mixture of synthetic organic substances, morpholine and phenylethyl alcohol, can be added to the water to aid in the imprinting (see Figure 11.14). The value of this extra imprinting in this environment is questionable and has recently been discontinued, although this method has been used successfully in the Great Lakes. After the imprinting period, one side of

(a)

(b)

Figure 11.14 (a) Recently smolted salmon being held in pens before they are released into the environment. (b) Additional imprinting of the water can be made to help salmon return to the release area by adding a mixture of morpholine and phenylethyl alcohol while the fingerlings are in the pens. (Photos courtesy Sea Run, Inc., Maine.)

the netting is dropped to release the fish to the sea; the salmon return over the following two to three years.

The use of **sea (net) pens** for the culture of salmon has proved, during the last 20 years, to be one of the most successful aquaculture techniques. Sea pen culture is done to some extent along the Pacific coast of the United States and Canada, but is most important in France, Scotland, and the Scandinavian countries, especially Norway. In Norway, Atlantic salmon are grown in hatcheries until they are parr and are then moved to outdoor tanks. After one to two years, they become smolt and are then transferred to floating pens for one to two more years until they have reached a market size of 2–6 kg. The rainbow and sea trouts are handled in a similar way, although they are usually larger when moved into the pens.

Pen sites are generally well sheltered (to prevent wave damage) and near shore (because of frequent feedings); fjords in Norway often house salmon pens. There should be good water circulation, and the pens must have a minimum depth of 10 m.

A variety of pens are used for salmon culture. Most consist of a net bag open at the top and hung from a floating collar. Weights are attached to the netting so the shape is maintained (see Figure 11.15). While floating cages are very popular, sea pen **enclosures** have also been used, al-though to a much lesser extent. Enclosures are portions of the seabed, with poles driven in and nets extending above high water.

Feeding penned fish is one of the greatest expenses for the salmon farmer. In Europe, salmon may be fed fish processing wastes and trash fish such as herring, capelin, menhaden, and anchovies up to six times a day. In the United States, dry and semimoist pellets are used; pellets are also becoming more popular in Europe.

Before the salmon are harvested from the pens, they are starved for a week or more; starving generally firms up the flesh, decreases the amount of fat, and aids in sanitary shipping. Harvesting is done by lifting the net and removing the fish with dip nets. The salmon are killed

1. By a blow to the head with a blunt instrument called a **priest.**
2. By first being tranquilized in seawater through which CO_2 is bubbled; then they are bled in seawater by cutting one of the gill arches.

Choosing a Species

Each of the different species of salmon has advantages and disadvantages. The species selected for culture depends upon the environment where the culture takes place and the current market. Pink and chum salmon migrate to the sea while very small, so little time and effort is expended in ranch culture; holding them for some extended period while they grow to be several grams and are imprinted helps ensure greater returns. The coho salmon grows well under culture conditions, and is the major pen-cultured species along the Pacific coast of North America. Coho can be grown in either freshwater or saltwater. In pens, they can be grown to pan-sized fish (about 35 g) in six to eight months, but extended growouts of this fish may be impractical since it will start to mature while still rela-

Figure 11.15 Round net cages moored in a line in Scotland. (Reproduced with permission from *Salmon and Trout Farming* by Laird and Needham, published in 1988 by Ellis Horwood Limited, Chichester.)

tively small, converting feed to gonadal material rather than muscle. This problem may be overcome in the future by selecting and developing strains that mature late or by using polyploid fish. The sockeye salmon has received less attention from culturists than other species, but some of its traits suggest a potential for culture. It can be raised in freshwater and is the least piscivorous of the Pacific salmon. It also matures late, so no energy is wasted making gametes if the fish are harvested before they get too large. Chinook salmon are being cultured in pens, but to a limited extent; they mature late and can be grown to a very large size.

STRIPED BASS

Morone saxatilis, the **striped bass,** is an anadromous fish native to the East Coast of the United States and the Gulf of Mexico (see Figure 11.16). There also exists an established fishery on the West Coast that developed from fingerlings released in San Francisco Bay in the 1890s. Freshwater populations were first described in 1954, probably as a result of impounding the Santee and Cooper Rivers in South Carolina in 1941. Much of the present aquaculture effort is geared toward providing sport fish for freshwater lakes.

Spawning and Hatching

A technique for inducing laboratory maturation of striped bass has been developed that is based on photoperiod and temperature conditioning; however, this technique is far from perfected at this time. Culturists still rely on collecting naturally matured broodfish.

Broodfish are caught during the spawning season in areas where the fish are known to congregate. There are a number of ways to capture broodfish, including several types of nets, hook and line, and electrofishing gear. Because the number of viable eggs collected depends

Figure 11.16 Striped bass are popular game fish. (Photo courtesy New Jersey Department of Environmental Protection, Division of Fish and Game.)

on the physical stress to which the fish is subjected, electrofishing is the most popular method (there is no struggling during capture and the fish is immobile for several minutes during the pickup and initial transportation). The number of wild broodfish seems to be declining, so developing a simple method for culturing the striped bass through its entire life cycle has grown in importance. Sexual maturity in males grown from seed comes in less than 3 years, although females may take three to five years and be more difficult to spawn than wild captured brooders.

The fish are placed in transport boxes and taken to the hatchery as soon as possible; both sexes are injected with hCG (this may be done before shipping). Clean water at 16°–19°C should be used, and the sexes must be kept separate to prevent spawning in the holding area. A sample of the eggs is taken from the female about one day after the hormone injection, and the time of ovulation is estimated. Verification of ovulation can be obtained by manual palpitation of the abdomen; if ovulation has started, some of the eggs will flow from the vent when pressure is applied. The female is anesthetized prior to stripping of the gametes. Some culturists encourage tank spawning since it means less handling, which may increase the chances of survival of both the broodfish and the fry. Natural spawning requires larger facilities than stripping, although less labor is involved.

Spawning of the female should take place as soon as possible after ovulation is complete. The female is held down by one or two workers while another applies pressure to the abdomen and forces the eggs into a pan. The milt of the male is then stripped into the pan, water is added, and a thorough mixing ensures complete fertilization. The eggs are placed in hatching jars after fertilization. As the eggs hatch and the fry begin to swim, they are discharged from the jar into a small tank. Again, it is important that the water quality be good in the hatching jar and tank.

The sac fry can be held in tanks for the first 4 to 5 days until the yolk is used up and the gut develops. At this point, they may be transferred to troughs with dark-colored interiors and held there for 9 to 15 days before being moved to ponds. The fry are fed a diet of newly hatched brine shrimp and, if available, other zooplankters. Since the fry at this stage have very limited swimming abilities and rely on visual cues to capture zooplankton or brine shrimp nauplii, these feed organisms must be present in high concentrations.

Fingerlings

Older fry may be released directly into the nursery ponds, or younger sac fry may be placed in ponds in cages of a fine, nontoxic, smooth material. For direct stocking, ponds usually start with 2×10^5 to 8×10^5 fry/ha. The use of cages for sack fry reduces the amount of labor and handling and eliminates the need for a holding facility, although there is less control of environmental conditions than if tanks are used. Fry are released from the cages at about 6 days of age when the yolk sac is absorbed.

There should be an adequate supply of zooplankton for feeding when the striped bass are initially released in the ponds. The production of zooplankton *in situ* is encouraged by fertilizing the pond. Fertilization techniques vary, but hay, grass cuttings, manures, and soybean and cottonseed meals have all been used, along with inorganic fertilizers. If the zooplankton bloom is lost, the fingerlings may be offered a prepared salmon diet. Overfertilizing should be avoided since the DO should remain above 4 ppm. The pH should be kept neutral or basic, but should not exceed 9.5, and the temperature should remain between 14.5° and 21°C. Freshwater or brackish water (to 10 ppt) can be used when fry are stocked in ponds, and after the fingerlings are fully scaled (30 to 40 days) this can be changed to full seawater.

Fingerlings may be harvested from the ponds in one or two months and shipped for restocking elsewhere, or they may be intensively cultured. The most common method of harvesting is simply to use a catch basin, although a device called a **glass V-trap** and seine nets are also used. The fish weigh 0.45–0.65 g at harvesting. The greatest mortalities while transporting fingerlings have presumably resulted from osmoregulatory problems induced by stress; this is partially overcome by adding some salt to the shipping water if the fish have been grown in, and are destined for, freshwater (if the fish are sent in seawater, no extra salt is added) along with a fish anesthetic, tricaine, (MS-222).

Fish that are to be intensively cultured may be kept in ponds, placed directly into tanks in which they will be grown out, or kept in a cage. Once the fish reach about 3 cm in length, they easily accept prepared feeds and are less sensitive to physical and chemical shock. The feed should be about the same size as the zooplankton or the brine shrimp, and should be added initially with the live food. The transition from live to prepared foods is critical, and a large mortality is often experienced at this time. This transition can be made less traumatic if fish oils or fresh fish are added to the feed. When the prepared diets are being taken readily, they should be offered hourly at least. This can be reduced to three times a day over a period of a few weeks. In an experiment, fertilized ponds with 15-day-old fish received a supplemental prepared feed either twice a day or hourly; ponds that received hourly supplements had a survival that was 151% and a weight production that was 152% that of the other ponds.

The size of the pellets should be increased as the fish grows. Salmon and trout diets can be used, and semimoist pellets made from processed fish and trout feeds, extruded in strings, are also popular. All feeds should be at least 38% protein.

Hybrids

Hybrids of striped bass and white bass, *M. crysops,* are very popular with culturists. The most often used variety, the **original hybrid,** is the female striped bass X male white bass, while the **reciprocal hybrid** of the male striped bass X female white bass is less frequently seen (see Figure 11.17). However, since white bass are much smaller fish, it is easier to handle the female white bass than the large female striped bass; there also seems to be less of a problem collecting brooding female white bass than striped bass (at least there is less public resentment associated with culturists taking white bass from public waters and moving them to hatcheries than there is for removal of striped bass). The males of either species are smaller than the female striped bass, so they present little trouble handling. These

Figure 11.17 A white bass × striped bass hybrid. (Photo courtesy of the New Jersey Division of Fish, Game, and Wildlife.)

crosses produce fertile offspring that may be used to produce F_2 generations. The striped and white basses do not hybridize naturally, so the fish must be stripped, and fertilization is artificial. The eggs of the white bass are highly adhesive, so if the reciprocal cross is made, the fertilized eggs should be treated with tannic acid to prevent clumping before they can be placed into hatching jars.

Both the original and reciprocal hybrids have proved to outgrow the striped bass, at least during the first two years, and are more hardy and easier to produce than the striped bass.

Apart from restocking programs, there is significant research and interest in growing out the hybrids for market. When 48 g hybrids were stocked (200/m^3) in net pens in estuarine waters in South Carolina, and a floating trout diet was given to the fish two or three times each day, there was a conversion ratio of 1.58 : 1. The average survival of the hybrid was 96% after 8 months and the mean size was 311 g, resulting in a production of about 50 kg/m^3. When 194 g hybrids were stocked in earthen freshwater ponds at 10,000 fish/ha and fed a commercial salmon diet two or three times daily, there was an 84% survival after 10 months with an average size of 656 g; that is a production of 3881 kg/ha.

SUMMARY

Marine fishes are cultured less often than freshwater species. Those living in estuarine environments are more often reared than are those requiring true oceanic conditions.

Mullet are common in semitropical and tropical environments. They may be spawned using hormones, and the fry can be reared on zooplankton. However, controlled spawning is a difficult technique to use, so fry are normally collected rather than produced in a hatchery. The fingerlings are often raised in brackish ponds with other harvestable species.

The yellowtail is an important crop in Japan. Juveniles are collected in the spring and initially reared in net cages being fed fish and shrimp. For growout, they are moved to larger cages and fed trash fish and pellets. Pompano, a relative of the yellowtail, has also been considered for culture. They can be induced to mature in the laboratory and spawn with hormone treatments; young pompano can also be collected with little trouble. Small pompano are easily reared, but large-scale production of harvestable size fish has been hampered, largely because a cheap and acceptable feed has not been developed.

The flatfish that have been of the greatest importance to the aquaculture community are the sole, plaice, turbot, and halibut. Spawning flatfish has been relatively simple. Larvae are fed a diet of zooplankton; after they metamorphose to bottom fish, they may be fed fish and mollusks, and pellets are being developed. Flatfish may be grown out in ponds, tanks, or floating cages with hard flat bottoms.

Milkfish are commonly cultured in the Pacific. Fry may be collected, the fish may spawn naturally in cages or tanks, or spawning can be induced with hormone treatments. The larvae feed largely on zoo- and phytoplankton, but later growth is supported by larger flora (especially cyanobacteria) growing in fertilized ponds.

Salmon eggs hatch in freshwater. The young fish develop and go through a smoltification that allows them to enter the oceans and grow to adults. At maturity they will return to freshwater streams to spawn. Most culturists artificially fertilize salmon eggs and hatch them like trout eggs. For all salmon, except chum and pink, the fish are held for one to two years until the smolt are ready to enter the marine environment. There are two popular methods of growout: the salmon can be released to the sea and collected in several years when they return to spawn, or the salmon can be grown out in floating pens.

Striped bass are collected during the spawning season, injected with hormones, and shipped to hatcheries. The bass may be stripped, or they may be allowed to spawn naturally. The eggs are placed in hatching containers; the fry may either be kept initially in tanks and troughs, or they may be placed in very fine net cages in ponds. After further development, the striped bass are released into fertilized ponds to grow to fingerlings. The fingerlings may be used for restocking, or they may be grown out in tanks, ponds, or cages. There is much interest in striped bass × white bass hybrids.

PART THREE

BUSINESS

CHAPTER 12

LAWS PERTAINING TO AQUACULTURE

As aquaculture has assumed greater and greater economic importance, it has become clear that laws and regulations are needed to protect both culturist and consumer. International, federal, state, and local laws, as well as common law, all come to bear. Yet, until very recently, little legislation had been written specifically for aquaculture. New legislation is now being considered and passed into law at an accelerated rate, and it is therefore unlikely that any culturist will be familiar with all the laws pertaining to their project before they actually begin the process of getting "all the paperwork done." As William McLarney has stated, for the person in such a situation, "the key piece of advice is this: ASK."

The sometimes overwhelming number of aquaculture-related laws has occasionally resulted in confusion, and many culture operations face long delays because of the excessive number of permits that must be obtained from different agencies. A California study stated that "up to 42 federal, state, and local permits and licenses would be involved in a single aquaculture venture," and many other states require similar numbers. Some of the types of laws that even small fish farmers may find themselves faced with are: those that regulate the granting of leases for the use of public lands; health regulations regarding the processing of the product; laws concerning the regulation of drugs and feeds that can be used; zoning ordinances and permits; building permits; laws restricting the diversion of water as well as the discharge of water; restrictions on the species that may be held or imported; laws relating to the health of the fish; even old fishing laws that sometimes apply to the aquaculturist.

The laws that are applicable and the regulatory authorities that are designated to exercise jurisdiction can vary as a function of the type of activity that is carried out by the culturist and whether the culture is to take place on the high seas (as in the case of salmon ranching), on the territorial sea, in the intertidal zone, or in inland waters that are navigable or nonnavigable. Some activities, such as water discharge, are often subject to both federal and state laws.

Several states have begun to address this "permit crisis." California now provides a clearly written publication containing the regulations governing marine aquaculture, and the California Department of Fish and Game, Marine Resources Division, will assist in completing and filing applications. The Mississippi Aquaculture Act, which went into effect in July 1988, was in part an effort to amend existing regulations and improve coordination between state agencies involved in aquaculture.

RIPARIAN LAW AND BASIC COMMON LAW

Property rights that apply to ownership and use of water are known as **riparian rights.** Riparian law is derived from English common law; it became a recognized principle in the United States in the nineteenth century, and has been incorporated, in various forms, into many state laws. Riparian rights are acquired by obtaining title to land adjacent to a natural body of water, and concern the right to use water rather than the actual ownership of the water. The riparian owner has the right to use a reasonable amount of available water as long as that use does not interfere with the rights of other riparian owners. While the idea is clear, in practice such concepts as "reasonable" may be difficult to interpret. "Reasonable" depends on the circumstances of the owners and the conflicting interests. In its 1938 decision (*Dunlap* v. *Carolina Power and Light Co.*), the Supreme Court of North Carolina stated that reasonable use, with reference to riparian law, "is a question of fact having regard to the subject-matter and the use; the occasion and the manner of its application; its object and extent and necessity; the nature and size of the stream; the kind of business to which it is subservient; the importance and necessity of the use claimed by one party, and the extent of the injury caused by it to the other."

Clearly then, the more water being used, and the extent that the water quality is changed, are extremely important. Cooking, washing, or drinking would constitute "reasonable" use in most cases, but the use of large parcels of water for manufacturing purposes or irrigation could easily be considered "unreasonable" under certain circumstances. Generally, the diversion of water out of the riparian land is not considered reasonable, and the term "reasonable" takes on a new meaning during times when water is scarce. Riparian rights may be lost by condemnation, improper use of an upstream user for a specified period of time, changing stream channel location, or transfer of water use rights without conveyance of riparian land. Riparian rights may be reduced by the state or federal government as a result of regulations pertaining to health, safety, or environmental protection.

Although the riparian laws in the eastern United States seemed workable at first, when the arid western lands where water is more scarce became populated, problems began to arise. In particular, questions about diversion of water, use of groundwater, and inefficient use of water by some riparian landowners sparked controversy. Some western states only partially recognize riparian rights, and several others no longer recognize riparian principles at all and are guided by the **appropriation doctrine** of water rights. That is, one who has been using a source of water has the superior right to keep using the water over a new user, as long as the original user continues to use it for a beneficial purpose. A few eastern states have also modified their laws concerning riparian rights.

The natural uncertainty associated with riparian rights, because of the concept of reasonable use, has probably hindered the development of aquaculture. The growth of the aquaculture industry depends to large extent on expanded use of water, and this often includes the diversion of water to an aquaculture facility. Since the availability of the water for the culturist may be in doubt, many investors are hesitant to involve themselves in aquaculture projects. New industries such as aquaculture may have a difficult time securing rights to water since courts generally have not considered such factors as priority, nonuse, or the type of nondomestic use the water is going to be used for, when trying to establish what is "reasonable." Many people in the aquaculture industry feel, as have some of the western states, that riparian rights are not the best solution to modern problems associated with greater water needs, pollution, and the rights of the public.

Another question of paramount importance to the aquaculture community concerns the property rights of the culturist with regard to fish and shellfish. Wild animals, legally *ferae naturae*, were considered by the Romans to be the property of anyone who could take possession of them; this is the concept of *res nullius*. In England, common law held that all wild animals belonged to the crown. Most of the U.S. laws lie somewhere between these two concepts—wild animals are considered to be held as part of the public trust by the government. When something such as a parcel of land is claimed, it is normally relatively simple to fix the associated property rights; however, when the property in question is alive and can move about, there are clearly considerations that must be taken into account that are not otherwise called into play. "Possession" then becomes more difficult to define, especially when we are considering animals that are not traditionally considered to be a domesticated variety (legally referred to as *domitae naturae*, such as horses, chickens, or sheep). A third category of animals, *animus revertendi*, refers to animals that are released but will return to the original owner, such as racing pigeons or hunting hawks; these are considered to be property even when not in the possession of the owner.

With particular reference to ocean farming, then, it may be possible to show that fish released from a hatchery into the open sea are considered *animus revertendi* and therefore absolute property if

1. The culturist can show that the fish have an intrinsic value.
2. They can be recognized as not belonging to a natural local population.
3. Fishermen can be informed that such fish, found in a specified area, are the property of the culturist.
4. The fish are part of an industry and are therefore considered to be taxable.

The principle of *ratione impotentiate,* which concerns qualified property rights, is also applicable to mariculture, especially in the case of sessile species. Wild animals, such as mussels on rafts, may be considered the property of the culturist when planted in public waters by a private citizen, if the rafts are clearly marked and are in an area where that species of mussel does not normally grow. Placing the animals in public waters does not constitute abandonment as long as there is an intention of reclamation.

INTERNATIONAL LAW

History

Migratory species subject to ranching, such as salmon or sea turtles, may receive state or federal protection up to a point, but in many instances will stray into international waters. While this is not a significant problem at the moment, it may become one in the near future if the aquaculture industry continues to grow and fishing technologies associated with fish finding and harvesting continue to improve.

International law, as it pertains to aquaculture, has its roots in the **maritime laws** governing activities on the high seas. The earliest such law was the Phoenicians' Rhodian Sea Law (300 to 200 B.C.), which was later accepted by both the Greeks and the Romans. This was followed by the Byzantine "Basilika," the Italian "code of the Amalfi," and the Crusaders' "Assizes of Jerusalem." In 1160, "The Rolls of Oleron" was promulgated by Queen Eleanor of Aquitaine, and its principles were soon adapted by most of the countries in western Europe in some form, such as the Wisby Sea Laws of Amsterdam (1505), the German Hanseatic Code, and the laws of England's Black Book of the Admiralty. After the collapse of the Roman Empire and its aftermath, the Mediterranean countries established the Consulato del Mare.

During the second half of the Middle Ages, as shipbuilding technology advanced, some countries began to claim portions of the oceans as their own. Portugal claimed all of the Indian Ocean and the Atlantic south of Morocco, while Spain declared her sovereignty over the Gulf of Mexico and the Pacific Ocean. Freedom on the open sea was finally established after Sir Walter Drake sailed to the Pacific, and Queen Elizabeth declared to the Spanish Ambassador Mendoza, who had protested Drake's voyage, that the seas were rightfully free to all who were equipped for commerce, transportation, or fishing, by reason of nature and regard for the public use.

With specific reference to mariculture and the high seas, the principle of the freedom of the seas has been modified so as to include the concept of **territorial waters.** In 1945 the United States issued the **Truman Proclamation,** which stated that the continental shelf adjacent to the coast of the United States was part of the country's territory, although at that time this did not include the water above the shelf. This in some ways, was the start of a new era in oceanic law. In 1958 the conventions of modern maritime law were negotiated in Geneva, Switzerland, and ratified by many nations. The basic points of the **Geneva Convention** were

1. A country's sovereignty extends to the adjacent territorial sea, its sea bottom, and the air above the territorial seas, but that nation must not interrupt the passage of innocent foreign vessels through those waters.

2. The continental shelf was defined as starting at the seaward limit of the territorial sea, and a coastal country was given the exclusive rights to explore and exploit the resources of the seabed. (However, this did not include the water above the continental shelf.)

3. The high seas begin at the seaward limit of the territorial seas, and all nations are free to navigate, fish, lay cables, conduct research in, and fly over the open seas, but reasonable regard for the interests of other countries must be exercised.

4. All nations have the right to fish the high seas, but are subject to treaty obligations, the rights of coastal countries, and conservation rules specified by the Convention on Fishing and Conservation of the Living Resources of the High Seas.

Unfortunately, many deficiencies existed in these 1958 laws. The limits of territorial seas and continental shelves, and the rights of the coastal countries were not defined. The Geneva Conventions were very vague and tended to ignore the rights of landlocked countries.

Current International Law

The United Nations has made efforts to improve the laws that govern the seas. The Third United Nations Conference on the Laws of the Sea (**UNCLOS III**) convened in 1969, and although aquaculture was not dealt with per se, the implications are clear. A convention was adopted in 1982; a large majority of the countries ratified it, although four, including the

United States, did not. The following are among the items covered in the new **Laws of the Sea Treaty:**

1. The rights and duties of coastal countries in the **exclusive economic zone (EEZ)** were defined; this includes the rights of "exploring and exploiting, conserving and managing the natural resources, whether living or nonliving . . ."
2. The maximum size of an EEZ was established as being 200 nautical miles.
3. Conservation of living resources is called for by using the best scientific data available to establish allowable catches in the EEZ.
4. Coastal countries are required to promote optimum utilization of the EEZ living resources, and if the coastal countries cannot harvest an optimum catch, other countries have the right to harvest, as long as they come to an agreement with the coastal country.
5. Countries whose fishermen harvest highly migratory species shall cooperate to maximize the catch, both within and out of the EEZs.
6. Countries from which anadromous fish originate have the primary responsibility for the stock.
7. All countries will reduce or prevent pollution of the seas.

The Food and Agriculture Organization of the United Nations (**FAO**) is also concerned with aquaculture and fisheries. The FAO organized the first world conference on aquaculture in 1966 and has been active in the establishment of other conferences and smaller symposia ever since. It has also supported aquaculture projects in many countries.

Treaties between countries are recognized as international laws. Several of these pertain to aquaculture even though they may not be mentioned directly. These include agreements on fishing, conservation, and ocean dumping of wastes and sewage. The United States has such agreements with Mexico, Japan, and Canada, among others. For example, when the United States first established a 200-nautical-mile zone in 1976 (by passing the Fisheries Conservation and Management Act), American shrimp fishermen in the Gulf of Mexico, who had been able to fish Mexican coastal waters without significant restrictions, were compelled to recognize Mexico's jurisdiction over the shrimping grounds along its coast. Shrimping by U.S. vessels was phased out over a negotiated time period. (It is worth noting that many fisheries scientists believe that much of the Mexican shrimp stocks originate in the coastal waters of the United States, but do not reach a catchable size until after they reach Mexican waters.)

Although the United States did not ratify the Law of the Sea Treaty (because of provisions regarding mining rights), in 1983 it did establish, by proclamation, an EEZ around the states and controlled islands in the Pacific and the Caribbean. In many ways it is similar to the Laws of the Sea Treaty and the Magnuson Fishery Conservation and Management Act of 1976. Agreements are made with other countries concerning harvesting of fish within the EEZ; for example, in 1984 Poland was allocated 10^4 metric tons of whiting off the northwestern U.S. coast, and Japan signed a pact with the United States that would allow it to continue to fish in the U.S. EEZ if it purchased about $\$10^8$ of fish products from American companies.

FEDERAL LAWS

Many countries have laws that directly or indirectly affect the culturist. As aquaculture has grown, so has the number of laws. Reviewed here are a few of the laws of two federal states, Canada and the United States, that pertain to aquaculture.

Canada

The legislative authority of the federal and provincial governments in Canada was established by the British Parliament in 1867 by the **British North America Act (BNA).** Section 95 of the BNA authorizes both parliamentary and provincial agriculture legislation; questions remain regarding the interpretation of the word "agriculture," and most of the decisions concerning this concept have emphasized terms such as soil, livestock, and land. Also considered in Section 95 are property laws and civil rights.

Section 91 of the BNA deals with a number of factors that more closely concern the Canadian aquaculture community. For example, it embraces the concept of the protection and conservation of public fisheries. This *may* be relevant to the introduction of new aquaculture species that could interact with the native organisms, inspection of cultured organisms for diseases or parasites that might affect established fisheries, use or space for culture facilities in areas traditionally used for fisheries, and the public's rights to use tidal waters. The concept of "public property" is also included in Section 91, and it is felt that lands, such as public harbors, that existed at the time of confederation are public property (although much of this land would be ideal for aquaculture). The federal authority over navigation and shipping may also impinge on the culturist's use of nets, rafts, or cages.

The BNA also confers on the federal government the power to regulate certain types of trade. If a culturist intends to sell his product outside of the province it was reared in, it is subject to federal regulations such as the Fish Inspection Act.

The 1936 Mollusc Agreement was established between the Canadian federal government and Nova Scotia. Federal regulatory power was established. The federal minister of fisheries was to control and administer culture of bivalves, including the granting of leases for the culturist. Later, the Federal Fisheries Act delegated power to the provinces. Other regulations encompassed under this act by the federal cabinet include the Atlantic Coast Marine Plant Regulations, the Fish Health Protection Regulations, the Pacific Coast Marine Plant Regulations, the Pacific Shellfish Regulations, the Salmonidae Import Regulations, and the Sanitary Control of Shellfish Fisheries Regulations.

During the past few years, the different provincial governments and the federal government have signed **Memoranda of Understanding** on the division of responsibility and jurisdiction relating to aquaculture. Those provinces that have active aquaculture industries have recently passed **Provincial Aquaculture Acts** that establish regulations and objectives.

United States

In the United States, federal laws that apply to aquaculture are in many ways similar to those of Canada. Federal regulations pertaining to the use of waterways, or trade and tariff, are all of interest to the U.S. aquaculture industry, just as similar Canadian laws are to the Canadian industry. There are over 20 federal agencies or organizations in the United States that are somehow related to aquaculture, in particular, the Environmental Protection Agency, the Department of Commerce, the Department of the Interior, and the Army Corps of Engineers play important roles. The Food and Agriculture Act of 1977 assigned the U.S. Department of Agriculture (see Table 12.1) the status of lead agency for aquaculture. (A **lead agency** directs research, provides the culturist with assistance, informs them of regulations, and/or issues permits and licenses.)

The use of coastal land for aquaculture must be viewed in the light of the **Federal Coastal Zone Management Act (CZMA).**

TABLE 12.1   Branches of the United States Department of Agriculture That Affect Aquaculture

1. *Agricultural Marketing Service*
     Federal–State Marketing Improvement Program. Improves marketing of products.
     Market Research and Development Division. Conducts research on marketing of products.
2. *Agricultural Research Service.* Conducts research on important problems, much of which has centered on the catfish industry.
3. *Animal and Plant Health Inspection Service.* Diagnosis of diseases and prelicense testing of fish vaccines.
4. *Cooperative State Research Program.* Awards grants, administers regional aquaculture centers, works with state agencies.
5. *Extension Service.* Educational network interested in helping farmers by introducing recently developed technology and information.
6. *Farmer's Home Administration.* Makes and guarantees farm ownership and operating loans and provides management assistance.
7. *National Agricultural Library.* Has established a National Aquaculture Information Center.
8. *National Agricultural Statistics Service.* Publishes reports on production and sales, especially for catfish.

The CZMA was enacted to encourage the states to use their coastal land wisely; incentives are given to states that can show they are able to plan and partition the use of this resource and are able to enforce the plan. The use of submerged lands—that is, both the leasing and the regulation of the lands—is dealt with in the Outer Continental Shelf Lands Act and the Submerged Lands Act.

The **Environmental Protection Agency (EPA)** and the **Army Corps of Engineers** have programs that coordinate the granting of permits for point sources of pollution. Some culture operations can significantly affect the quality of the nearby water and are therefore subject to federal regulations. Waste products may include dead organisms, metabolites and fecal material, chemicals, and uneaten feed. The Federal Water Pollution Control Act Amendments of 1972 are administered by the EPA and some cooperating state agencies. Point discharge of any pollutant is prohibited unless a permit is granted by the EPA or the state agency given the authority by the EPA. This pertains to raceways and ponds, but cages and rafts in open water are not considered sources of point discharge. In addition, some aquaculture projects under state or federal supervision have permits to discharge into navigable waters effluents that exceed standards because the facilities are considered to be a "benefit to society." The amount and type of the discharge are subject to guidelines and standards, if available. There are no federal standards for aquaculture operations and specific limitations are set on an individual or local basis.

Public health and safety laws are also applicable to aquaculture operations. These are designed to protect both the consumer and the aquaculture industry. All products that are grown in culture operations must be pure and wholesome to eat, and must be produced under sanitary conditions, according to the federal Food, Drug, and Cosmetic Act, which is administered by the **Food and Drug Administration (FDA).** The FDA also regulates the packaging of foods and the use of drugs and feeds for cultured species.

The federal government regulates activities in navigable waters that might obstruct vessels or impair the flow or circulation of water. Permits from the Army Corps of Engineers would be needed for the installation of an aquaculture facility that might create obstructions, alter the course of navigable waters, or involve excavation, filling, or discharge of dredged or fill materials into navigable waters. If lights, buoys, or other signals are used in navigable waters to mark fixed or floating aquaculture facilities, approval must be granted by the Coast Guard.

The federal government also supports aquaculture research through the **United States Department of Agriculture (USDA)** as well as the Sea Grant College Program, which was established in 1966; there are currently 29 Sea Grant programs in various states. In the 1982 fiscal year, Sea Grant funded aquaculture projects composed over 10% of its available budget monies, $2.35 million in federal funds, which was matched by $1.35 million in nonfederal funds. Besides supporting research in the area of aquaculture, a Marine Advisory Service, part of the **National Oceanic and Atmospheric Administration (NOAA),** was formed to help fishermen and culturists, theoretically in the same way that the USDA Extension Service has helped the farmer.

In 1980, the federal government passed the **National Aquaculture Act (NAA)** to be implemented through a coordinating group, the Joint Subcommittee on Aquaculture (JSA) of the Federal Coordinating Council on Science. The act mandated the implementation of a National Aquaculture Development Plan put together by the secretaries of agriculture, commerce, and interior; the plan was released in August 1983. The act also authorized the awarding of contracts and grants to carry out the implementation and required the establishment of services that would collect and distribute information on technical, scientific, economic, and legal problems. Funding for these projects, which from 1981 to 1983 was $70 million, came from the Departments of Agriculture, Commerce, and Interior.

The National Agricultural Research, Extension, and Teaching Policy Act of 1977 was amended to support the 1980 NAA by establishing "aquaculture assistance programs" that could award grants for research and extension "to facilitate or expand promising advances in the pro-

duction and marketing of aquacultural food species and products." In addition, provisions in the amendments were made for the establishment of aquacultural research, development, and demonstration centers.

A series of federal **mitigation** laws also pertain to aquaculture. Mitigation, in the sense used in this text, represents compensation for damaged resources, and often takes the form of new hatcheries to supplement natural stocks that can no longer be self-sustaining because of habitat alterations. Federal legislation such as the 1934 Fish and Wildlife Coordination Act, passed because of the proposed transfer of the Jones Hole National Fish Hatchery, and the Mitchell Act, passed after dam construction in the 1930s blocked the salmon and steelhead migrations in the upper Columbia and Snake rivers, are examples of mitigation laws.

LOCAL LAWS

Provinces, states, counties, and towns may all have laws that in some way regulate the business of aquaculture. Some of these laws have been passed recently and pertain specifically to aquaculture, while other laws may be much older and only relate coincidentally. Fishery laws, designed originally to protect and conserve natural populations of fish and shellfish, may technically apply to aquaculture. For example,

1. Some fish and shellfish may not be sold in certain seasons, a regulation passed to prevent harvesting during the breeding season that would damage the structure of the wild population. However, culturists may also be prevented from selling their product at that time even though it would result in no harm to the natural stock.

2. Animals smaller than a specified size may not be caught and sold, a

law designed to be sure that a minimal number of animals would be able to mature and breed before being caught, but the culturist may not be able to sell small animals either, although they will never contribute to the wild stock.

A legislative distinction between aquatic species that are part of the natural population of an area and those that are cultured could be made by lawmakers (such as was done in Hawaii where farm-raised mullet may be sold during the closed fishing season). Failure by governments to make such a distinction results in considerable ambiguity, and only serves to hinder the aquaculture industry.

The "Lease" Concept

Land under nonnavigable streams and ponds are normally privately owned; in leasing them for the purpose of aquaculture, the fish farmer is generally subject to the rules governing the aquisition or leasing of private lands. Legal problems have sometimes arisen when public lands are leased. This includes all subtidal land, and in all but five states the intertidal land between mean high tide and low water also belongs to the public. Much of the legislation concerning aquaculture centers on the granting of leases. The **lease** is the basic document that gives the culturist the right to occupy public foreshore and subaquatic land. The lease can be considered a tool to ensure the wise use of public resources by way of the conditions contained in the lease. The laws governing the granting of the lease and the conditions in the lease vary, but often considered in the lease laws are the size of the leased area, the cost and length of the lease, who can apply for the lease, and the revocation of the lease.

The size of the leased area can be a function of the species and/or the type of culture facility. In California, the Fish and Game Commission must find that making the lease is in the public interest, and no

lessee can be granted a lease so large as to foster a monopoly. In New Jersey an individual may only apply for up to 2 acres of bottom land per lease application. The New York Fish and Wildlife Law, on the other hand, established a minimum area to be leased—50 acres for shellfish bottom culture and 5 acres for off-bottom culture. Maine's legislature, in 1987, passed an act that, among other things, limits leases to 100 acres while preventing any tenant from accumulating more than 150 acres, and requires that the lease applicant write an assessment of how the culture project will impact "significant" marine organisms and established fisheries in the area.

The rent to be paid to the leasing authority is established so that participation by aquaculturists is encouraged and yet the administrative costs to the lessor are not a burden to the taxpayer. The cost of the lease can be established by

1. Bidding, as is done in New York and California.
2. A base rental rate plus a percentage of the lessee's profit, which is done in Florida.
3. The authority setting the rent, as in New Jersey, which may be done as a function of the particular area, the species to be raised, and the size of the area to be leased. A legislative maximum and minimum rental charge is established in Louisiana and Massachusetts.

The duration of the lease must be such that it is in the public's best interests, but at the same time should be long enough so that the aquaculturist is not in danger of losing the initial capital investment. In addition, renewals must be available to the lessees. Leases of 5 years with two to four uncontested renewal options, also of 5 years each, might be a reasonable policy. In France, mussel and oyster leases may last up to 25 years, in Australia leases for oyster farmers can be 15 years, and in Spain the bay areas used for raft mussel culture can be leased for 10 years and are renewable.

In some cases, the lessee must be a resident of the area where the lease is held. In addition, the state may require a minimum performance standard or minimum qualifications of the lessee so that the leased land will not be tied up by unproductive operations. If a lessee fails to meet the minimum performance standards of the leasing authority, the lease can be terminated. Other reasons for termination of an aquaculture lease could be nonpayment of rents or violations of the lease agreement—such as failing to post signs or working outside the leased area—or by the authority of the leasing agency for a public purpose.

Canada

In the Canadian provinces, legislation has been enacted by the local governments and by the federal government for the individual provinces (for example, the federal Saskatchewan Fishery Regulation). Licenses are issued to culturists depending, usually, on whether they are using coastal waters or inland lakes and streams, and according to the type of organisms that are being grown or held. These licenses are granted by various local agencies. Legislation of aquaculture, in one form or another, exists in each of the provinces. Alberta's aquaculture is dominated by the rainbow trout, and the Department of Energy and Natural Resources, by way of the The Fish Marketing Act, controls this industry. The Manitoba Fishery Regulations serve the same purpose in that province that also produces mostly trout. Trout is virtually the only cultured fish in Ontario where the most important statute is the Game and Fish Act, and in Quebec the Quebec Fishery Regulations function in the same way. In Newfoundland it is the Oyster Fishery Act and the Trout Hatcheries and Nurseries Act, and in New Brunswick it is the Fish

and Wildlife Act. Prince Edward Island's shellfish industry has been directed by the Oyster Fisheries Act of 1906 and the Oyster Area Registry Act of 1913. Perhaps it is worth looking at one of the provinces, British Columbia (B.C.), in greater detail so we can appreciate how the Canadian aquaculturist must prepare before going into business.

British Columbia is one of Canada's leading aquaculture provinces because of its extensive sheltered coastline and clean freshwater. In addition, it has well-established trade and information ties with Japan and its neighboring American states, Washington and Alaska. Although oysters are the most important cultured organisms, trout and salmon are also popular, and small operations revolving around clams, carp, mussels, abalone, and seaweeds may be on the horizon. The province has a good working relationship with the federal Department of Fisheries and Oceans, the agency that has most of the responsibility for the regulation of aquaculture on the federal level. In fact, the provincial and federal governments have designated different lead agencies for different industries. The lead agency for the culture of salmon and most marine organisms is the federal Department of Fisheries and Oceans, while for oysters and trout it is the B.C. Ministry of Environment.

Control over the granting of aquaculture permits belongs to the federal government, but the responsibility is delegated to the B.C. Minister of the Environment, along with control of importation and transportation. But oysters, because of their importance, have received special attention. The federal government becomes directly involved in the processing of oysters, pollution of oyster beds, and theft. However, the B.C. shellfish industry is largely governed by the provincial Fisheries Act, in particular, Part III, "Culture and Harvesting of Shellfish." The culture of freshwater fish, but not salmon raised in freshwater, is governed by the "Fishfarms" regulation enacted in 1965, which concerns the issuing of licenses. British Columbia makes licenses available called **private fish pond permits,** which are for noncommercial purposes and apply only to artificially constructed ponds and pools, and the **commercial fish farm license.** Other permits are needed to transport live trout within the province, and one needs a **water license,** under the B.C. Water Act, if water is to be diverted from a lake or stream (this is not the common law position that states that the owner of land along a river or a stream is allowed reasonable use for his home or farm).

There are other laws of which the aquaculture community must be aware. The leasing of Crown land, including subaquatic lands, is authorized by the B.C. Land Act, under which an **oyster license** is issued if only bottom culture will be conducted, but a lease must be granted if structures are to be built (in which case, consent of the upland adjacent riparian owner is required). The B.C. Fisheries Act states that any processor of aquatic organisms must hold a processing license, and regulates the building of fish passes or fish ladders that might be constructed by salmon ranchers. The B.C. Oyster Board, given its power by the Natural Products Marketing Act, regulates marketing of oysters within the province. The Fish Inspection Act provides for appointed inspectors who may seize and detain products that are tainted or somehow unwholesome.

United States

Many individual states have laws that affect the culturist either directly or indirectly. Those states with a long history of aquaculture, such as Hawaii, Florida, and Virginia, have a considerable number of statutory provisions pertaining to the holding of aquatic organisms and their place in the public waters, while other states whose economies have been domi-

nated by industries other than aquaculture often do not clearly define the rights of the culturist. Similar laws in different states do not necessarily result in equal patterns of rights due to the complexities of some of the ownership statutes in particular states.

Several states have recently developed **aquaculture plans** in an effort to coordinate the activities in that state, to make suggestions to aquaculturists with reference to species and facilities they should consider, and to make recommendation to the legislature concerning regulations and support. The plans may be written by a panel appointed by the state government or by a cooperating federal agency such as NOAA or the USDA (see Figure 12.1).

The control of fisheries is part of the sovereign power of each state. In most instances, control of finfish fisheries is maintained on a state level, but the responsibilities for shellfish in the intertidal areas have often been delegated to the municipalities. Troubles can arise when the suitable subtidal land extends beyond municipal boundaries; in Connecticut this problem has been dealt with by allowing municipalities to jointly regulate such activities so that allocation of cultivation areas is regional rather than only local. There are also questions concerning the degree to which general fisheries laws, which vary considerably between states, pertain to aquaculture.

The rights of the aquaculture lessee, of course, varies from state to state, and even from species to species. In the Chesapeake Bay, Virginia will lease up to a 5000-acre maximum, but Maryland limits the cultivation of oysters to individuals with leases of no more than 500 acres. Another factor to consider is that most states will not lease areas for private cultivation that have traditionally been natural-growing areas. This is based on the historical dependence of the people for subsistence on a resource that is considered common property.

Figure 12.1 Aquaculture plans are designed to give the industry a focus and to alert governments to problems and solutions with reference to permits, laws, and licenses.

States often require that aquaculture leases be granted only to residents of that state, and in some cases municipalities may have similar requirements. Most states limit their shellfish resources to residents, but similar laws concerning free-swimming fish have not consistently held up in the courts. In Virginia, an applicant for a lease must be a state resident, and if the applicant is a corporation, it must be chartered under the laws of Virginia and 60% of the stock of that corporation must be owned by residents of the state.

In 1969, Florida became the first state to adopt legislation authorizing the use of the water column for aquaculture; California and Maine have since enacted similar laws. **Water column** is defined here as the "vertical extent of water, including the surface thereof, above a designated area of submerged bottom land," and although Florida had been leasing bottom land for oyster and clam culture since 1881, the leasing of the water column presented new problems. However, in principle the fundamental question did not change: How does this law, passed so that the state could develop and manage a par-ticular resource, conflict with the traditional and statutory rights of the citizens of the state? The law had several major points to make: first, the lessee was required to mark off the leased area with appropriate signs, and signs must be posted along the shoreline where the public is afforded access, and second, the culturing process should have no adverse effects on the water, and the original condition of the leased land must be able to be restored within one year after the termination of the lease (the Department of Natural Resources being required to survey the land before the lease is granted). In addition to this, the state leases must be approved by the local county commissions that hold veto power over the application.

Rather than lease the water column, fish such as salmon may be grown in the open sea by ranching. However, the release of fish may be as complex as confined culture from a legal standpoint. In Oregon, ocean ranching may mean a private game fishery license, a chum salmon hatchery permit, a permit from the state engineer to divert water from a stream or river, a wastewater discharge permit, construction permits, and approval from local zoning boards. In addition, private hatcheries in Oregon must be located close to the sea but may not be on a river system that has an established public hatchery; disease inspection is required before fish can be released, and the numbers released are fixed by the permits.

When more than one party is interested in leasing a particular plot, the state must have a method for choosing a lessee. In Maine, the statute recognizes those with traditional interests in the area such as riparian owners, adjacent landowners, and fishermen that have worked in that area. The Florida and New Jersey laws, on the other hand, are most concerned with the efficient use of the plot, and the lessee must demonstrate competency to cultivate the species of interest. In California the key question is revenue for the state, so competitive bidding for the lease is carried out with the lease being awarded to the highest responsible bidder.

Once a lease has been granted, and the lessee has posted signs, then the law protects the lessee from trespass and damage or theft by statutes for punishment of up to six months of imprisonment, depending on the state. These statutes are, in effect, creating a property interest, so the lessee should also be able to bring civil action against a violator.

Many states support aquaculture research, although a commonly voiced complaint is that the aquaculturist, who is the immediate benefactor, pays for only a small part of the costs. In Louisiana, however, House Bill 149, sponsored by the crawfish farmers, calls for a tax on crawfish bait that is then used to support research and marketing.

SUMMARY

In many places, the aquaculture industry is new, and old laws and regulations have been applied that hinder its development. Laws that may affect the industry can be divided into common, international, federal, and local laws. Common laws deal with water rights and culture of organisms in public waters. International laws pertain to the establishment of exclusive economic zones and fishing agreements between nations, and therefore are essentially restricted (from the aquaculture point of view) to fish or turtles that are released and will return to the culturist. Federal laws are largely concerned with trade, pollution of waterways, consumer health, development of coastal lands, and marketing. Federal governments often delegate responsibilities for such things as water discharge quality, and so on, to the local governments. Local laws (those of provinces, states, counties, and towns) normally apply more directly to the culturist and include leasing policy and building permits.

CHAPTER 13

ECONOMICS

Fish and shellfish are an important part of our economy. According to the National Marine Fisheries Service, the per capita consumption of seafood has increased dramatically in the past two decades. Traditional fisheries in the United States are not able to meet the demand for these products; thus, there is clearly a growing market for aquaculture products. The United States exported $4.7 billion of fishery products in 1989 but imported $9.6 billion (compared to $8.9 billion in 1988, and this trade deficit is growing). The reasons why fishing has not kept pace with consumption in the United States are complex, but are probably related to an increased demand, overfishing of natural stocks, and the cost of operating a fishing fleet. For example, many of the trawlers that collect shrimp in the Gulf of Mexico are variations of ships designed years ago when energy was cheap; but now those ships, which require about a gallon of diesel fuel for every pound of shrimp caught, are finding it harder to make a profit.

Clearly, the relative success of the fishing industry must affect the market. The prices and quality of fish change quickly as a result of the uncertainty of the catch. Aquaculture, however, is a means by which to produce a more uniform product on a steady basis, thereby increasing the marketability of fish and shellfish.

To make an impact on the economy, the culture of freshwater and marine organisms must be competitive with other types of food-producing industries, including farming and fishing. A critical mistake many speculators have made in the recent past is to become lured into aquaculture by the promise of substantial financial rewards based on small-scale biological pilot studies without due consideration to economics. Schemes to raise valuable fish or crustaceans on a cheap feed are often discussed but rarely seen. "Dramatic new discoveries" in aquatic culture, those that might potentially be the most profitable, are just that—new and untested. Some of these advances may be applicable to large-scale production; but others, sometimes for reasons unrelated to biological questions, are doomed to failure. In aquaculture it is important to follow the old adage, "If it sounds too good to be true, it probably is."

However, with careful planning, an entrepreneur or culturist who has both eyes open can invest in or start a farm and reap the financial rewards from that new fish or shellfish venture. The most important first step is to construct an economic model that will fairly predict the costs, production, and the profits. The more complete the model, the better one can judge the success of the project; this means that the person(s) building the model must be familiar with capital costs, operating costs, the size of the market and the marketing procedures, and the capacity of the system to produce the product. The importance of this model cannot be overemphasized; consultation with an aquacultural economist may be the best way to begin the farm.

ECONOMIC MODELS

Economic models may be extremely complex or very simple, but all strive toward the same end: predicting financial feasibility. This is done by estimating the cost of starting the aquaculture project (capital costs), the cost of running the project for a given period of time (operating costs), and the value of the resultant crop. These models differ from farm to farm; obviously, a small company that grows seaweed in an open bay will not be properly represented by a model built by a large catfish farmer who must consider pumping water into ponds and feeding the fish. The major items to consider for most culturists include the following.

Capital Costs

Capital costs refer to items that are purchased once, such as a parcel of land, or purchased very infrequently, such as a tractor or a computer, which can be depreciated over a long period of time. Capital costs normally can be broken into two categories, **construction costs** and **equipment.** Some other costs that are only seen at the beginning of the project, such as hiring an economic consultant to help build the first model, are also included in this category.

Construction Costs When a farm is built, the initial costs for construction and labor can be substantial. Engineers should be hired to design the facility properly; a qualified engineering firm can save the aquaculturist considerable money and trouble, since properly answered questions about the size and type of pumps or pipes to be used, the properties of the soil, the design of the buildings and ponds or raceways, are critical. It is far more economical to do the construction correctly the first time rather than fix it later.

Permits must also be obtained. These may be related to the culture operation directly or concern building regulations. The cost of the permits will vary according to the size of the operation, the location, and the organisms to be cultured, and will usually range from a few hundred to a few thousand dollars.

Pond construction is often a major expense. The costs depend on the design of the pond and the amount of earth that must be moved. Naturally, the cost will change according to what type of liner or sealing is needed to prevent excessive drainage, if this is a problem. If concrete or fiberglass are used, estimates should be obtained from several companies, since the costs for these items may vary considerably. Many preformed fiberglass tanks and raceways are commercially available, but molds can be made for essentially any shape that is needed. Greenhouses for algal culture or hatcheries are also sometimes major expenses.

Aeration systems, if needed, must be included in the calculations. Attention should be paid not only to the capital cost of the aerator, but also to the cost of its operation; in some cases it may be cheaper to have an aerator designed and built specifically for a farm, although in many instances a standard one will work just as well.

Construction of an office, warehouse, hatchery, or laboratory is often needed. It is usually impractical to try to separate these facilities physically from the farm; long distances result in constant telephone communication and a loss of appreciation of the different facets of the farm. A warehouse may be needed when feed or chemicals must be stored in large quantities; warehouses are generally inexpensive to build and allow the farmer to make cheaper bulk purchases and reduce the number of transactions made. A small laboratory, where animals can be examined for signs of disease, or where water quality can be tested, is required for most aquaculture projects. Separate buildings may not be necessary, so a single structure may serve as office, warehouse, and laboratory.

An electrical system must be installed. In some cases this means power only to the buildings, but in some instances power also must be available at the raceway or pond site to operate pumps or harvesting devices. Some farmers will probably find that bringing electrical lines to the ponds is not practical and will invest in electrical generators that can be moved to any location when needed.

A water system must be installed. This means either diverting water into the aquaculture facility, or digging wells. The cost of digging wells varies greatly from place to place according to the depth of the wells, the character of the soil, and the size and number of the wells. The deeper the well, the more powerful (and expensive) are the pumps needed to bring up the water. Once the water is up, it may have to be treated or passed through filters, and a delivery system of pipes must be designed. Besides the cost of the pumps and pipes, one must also consider the price of their installation.

Other costs—such as building service roads or putting up fences, among other things—will increase the cost of the farm's construction.

Equipment Costs Equipment requirements vary from one aquaculture facility to the next, depending on the type of crop and the size of the farm, but several types of equipment almost always will be needed.

Office equipment must be purchased. This may range from a couple of desks and chairs for small operations, to large offices with photocopiers and computers for big farms. Laboratory equipment may need to be bought, and again the quantity and kind can vary considerably, although at least a microscope and a water testing kit seem to be standard fare. Tools for general maintenance and repair of equipment are required; for many aquaculture operations, it is also a good idea to have on hand special tools related to plumbing and the care of pipes. Vehicles, perhaps a truck or tractor, are also needed, along with harvesting and feeding devices. These may include automatic feeders, boats, nets, traps, or large machinery.

Operating Costs

The cost of operating a farm on a regular basis, again, can be divided into two major categories: **fixed operating costs,** those that do not change as a function of the size of the harvest, and the **variable operating costs** that depend on production.

Fixed Costs These are costs that do not change as the efficiency of the culturing operation changes. While these "fixed" costs may in fact differ from one year to the next, the difference is a function of events outside of the culture operation. For example, land taxes may change from one year to the next because of the political structure of the area in which the farm is situated, and although these changes may impact the aquaculturist, they are not a result of the culture operation, so they are considered fixed costs.

Lease rent is considered a fixed cost because the amount that must be paid to use the land will not change (in most states) with changes in production. The cost of the lease may change if the location or the size of the leased area changes, but if these factors are unaltered, so is the amount paid for the land.

Depreciation on equipment is considered a fixed, predictable cost. Depreciation is an asset's loss of value over time as a result of age, obsolescence, and use. At any point in time, the depreciation is the difference in the cost of the asset less its salvage value at that time. Some items, such as old plastic pond liners, have little or no salvage value when they are finally replaced. Depreciation may be taken on items such as the plastic pond liners, or items with salvage value such as pumps, trucks, office equipment, and so on— anything for which the length of the useful life can be estimated. There are a number of ways to calculate the depreciation of an item, the most commonly used being the straight-line method, which is calculated by

$$\frac{\text{cost} - \text{salvage value}}{\text{estimated life in years}} = \text{annual depreciation}$$

Any interest to be paid on loans taken to pay for the capital costs or the operating costs must be considered too. The rate of interest will fluctuate with the availability of the capital that can be borrowed (capital is thus considered scarce in many developing countries) and with the risks that are involved. For commercial aquaculture, the risks tend to be high because it is a relatively new field. Furthermore, newer and untested culture methods are riskier than conventional ones. Business insurance costs are also fixed operating costs and are closely related to interest rates. The more risk there is, the higher the insurance rate. In many countries, however, such insurance is not available.

Legal fees, audit fees, and maintenance costs are also fixed operating costs. For some aquaculture businesses, marketing may also be part of the operation and represent a fixed cost. The salaries and benefits of some staff members may also be considered as fixed, but only those employees who are hired on an annual basis and remain on staff regardless of production. Such permanent employees may be the farm manager, technicians, mechanics, security, and office staff. In contrast, the wages and benefits of the laborers, whose numbers are a function of the size of the crop and the duration of time it takes to do the work (such as harvesting), are best considered as variable costs.

Variable Costs   These are expenditures that change from year to year with production and are therefore more difficult to predict. Probably the most important of the variable costs for most farms is associated with the feed or fertilizer, which can amount to half the operating budget. Variable costs exist because the size of the crop may change each year as a result of parasites and disease, availability of seed stock, differences in growth rates of the cultured species, pond management techniques, and other factors. In addition, feed costs change as the costs of the ingredients in the feed change.

The total feed expenditure depends on the **conversion ratio** of the organism and the cost of the feed:

$$\text{cost of feed per pound of organism} = \text{conversion ratio} \times \text{price per pound of feed}$$

(Conversion ratios relate the amount of *dry* feed required to produce *living* tissue; living tissue is often about 90% water.) So if a fish can convert 1.5 pounds of dry feed into 1.0 pound of meat, and the cost of one pound of the feed is $0.20, then the cost of the feed needed to make one pound of fish is $0.30. It is critical to use the right type of feed to give the best conversion ratio, and not to over- or underfeed. Transportation and storage of feed or fertilizers can also affect the farmer.

The yearly feed requirements, in many instances, can be calculated in advance. This requires that the conversion rate be established, and that the size of the ponds, the stocking density, the size of the animals at the beginning of the growout season, and the size at which they will be harvested, all be known. For example, suppose that a 100-acre fish farm starts its growing season in the spring by purchasing 2-inch fingerlings, each weighing about 0.05 pounds. The farmer knows that to grow the animals to the marketable size of 1 pound will take 30 weeks using a feed that is converted at a ratio of 1.3 : 1. To make the desired profit, the farmer must harvest 2200 pounds of fish per acre, so the total harvest should be 220,000 pounds or 220,000 fish. The farmer has found that on the average there is loss of 15% of the fish before the harvest, so the farmer must stock each acre with 1.15×2200 fingerlings, or about 2530 organisms, in the spring. This means that each acre will start out with about 126.5 (= 0.05×2530) pounds of fish, or 12,650 pounds for the entire farm. If the farmer starts with 12,650 pounds of fish and harvests 220,000 pounds 30 weeks later, there is a difference of 207,350 pounds. Since each pound of fish that is produced requires 1.3 pounds of feed, the culturist knows that 269,555 (=$207,350 \times 1.3$) pounds of feed will be used.

The purchase of seed organisms (if needed) is also a variable cost. A source of good-quality seed is critical to aquaculture operations; the source must be reliable since nothing can be used as a substitute and the farm cannot operate without its initial stock. The cost of operating a hatchery can be substantial because relatively highly skilled labor is required, and, of course, a hatchery means additional capital costs. For large facilities, the inclusion of a hatchery on the farm may be a wise investment, and excess seed can be sold to other farmers. For species not subject to artificial spawning, the costs and availability of the seed become especially hard to predict. The cost of young fish or shellfish that have been produced in a hatchery is less variable than the price of collected seed, but even for hatchery seed the price is difficult to predict and depends on the hatchery's production, the demand for the young animals, and the age of the animals that are purchased. The seed's age (size) is very important; for example, it is more expensive to buy shrimp PLs than newly hatched shrimp nauplii, but the nauplii have a higher mortality rate and require special feeding

until they can accept what the larger juvenile shrimp normally consume. In addition, the length of time that the culturist must keep the animals until they reach a marketable size will be shorter if the PLs are purchased.

Mechanization is not yet available to replace much of the manual labor in a typical aquaculture facility. This is especially true in many developing countries where cheap labor is far more abundant than the capital to mechanize. In addition, if unemployment is a problem in a particular area, unskilled labor can be used in lieu of machinery, and the general standard of living may be raised as a result. Labor utilization often becomes more efficient as the size of a farm increases; this is true with regard to harvesting or feeding and stocking.

Other variable costs include the cost of electricity to operate the farm and the office, and the cost of laboratory and office supplies.

Production

In addition to knowing the cost of operating the farm, the culturist must also be able to estimate the **production** of the farm and the value of the product. The farmer's net income per year (profit) per unit area (usually acres or hectares) may be computed when the cost of production per year per unit area (C), the price received for the product (P), and the actual production per unit area (Q), are known:

$$\text{profit per unit area} = QP - C$$

If a 100-hectare shrimp farm had an annual operating cost of $650,000, and it was able to sell the shrimp at $6.75/kg and grow 1200 kg of shrimp per hectare per year, then the profit for that year would be $160,000. The factors affecting the production of shrimp on that farm would be the shrimp stocking rate, survival of the shrimp until the harvest, and weight of the shrimp when they are harvested. If these can be increased, production will increase.

Increasing the survival and growth of the fish in the ponds means using good culture management techniques: using proper feeds or fertilizers, maintaining good water quality, monitoring the ponds for signs of disease or parasites, and avoiding overcrowding.

The rate of stocking is related to the number of individuals that can be supported by the pond under natural conditions, which is also called the **maximum standing crop.** This can be increased by adding feed or fertilizer to the ponds, aerating the ponds, and using polyculture and stock manipulation techniques. Feeds and fertilizer obviously increase the cost of the operation, but the production increases may compensate for the costs. There is normally an increase in profits when feed or fertilizer is added to ponds of fish and shellfish, although this becomes less effective as increased labor is needed for harvesting and adding the feed, and the organisms become less efficient at converting feed to meat. Increasing the amount of feed increases the demand for oxygen, and this oxygen demand increases again when more animals are added to the pond. Aeration must then be increased. If the costs of extra aeration (capital and operating costs), feeding, and labor are greater than the profit made by adding more animals, then another way must be found to increase profit.

Special models can be built to estimate production in a variety of circumstances (variable stocking, feeding, temperature, and so on). These models are based on what is known about the physiology and growth of the organisms in question; the model builder's experience plus the published literature on the subject are used in the construction of the model. Once it is built, the model can be used to see how changing the culture parameters will change the final yield. For example, a model was built for the catfish, *Clarias,*

and then tested against actual growout experiments; it was discovered that doubling the feeding rate could increase the biomass yield by over 500%, while the size of the fish that were originally stocked was relatively unimportant. Models of complex biological systems are necessarily greatly simplified, and their accuracy is far from absolute; in open and semiclosed systems, parameters like temperature, and especially disease, cannot be appreciably controlled or predicted and therefore make the models gross approximations, at best.

Polyculture allows a greater stocking rate because more efficient use of space and feed can be realized when animals or plants that utilize different environments and resources in the pond can be kept together. Normally a secondary organism is added to increase the income derived from a given pond, and stocking rates are developed over time for the two (or more) species so that the maximum profit is obtained. Since one crop organism will probably be more valuable than the other(s) in any polyculture system, keep in mind that the maximum harvest (weight) may not be the same as the harvest with the highest value.

The stocking rate can also be increased by manipulating the stocks so that the pond area is used more efficiently. There are several methods for doing this. Stocking animals of the same species but of different ages (sizes) is one technique; young animals can be stocked at a high density because they are small, while adults stocked at the same density would be crowded and unable to grow well. Stocking the young animals at the same density as adults would mean inefficient use of most of the pond space. However, if the largest animals are periodically removed from a pond and are replaced with young animals, the space is used more efficiently. This technique has increased the harvest of milkfish in some areas by 100%. In regions where there are significant seasonal climatic differences, a pond may be used effectively for one species for part of the year and another species for another part of the year. That is, use the species that is best adapted for the temperature and light conditions of that particular season.

The culturist must also consider **risk** while building the model. There are ways to increase profits, but there is also usually a chance that profits will decrease when a new strategy is initiated. For example, heating the water may make the organisms grow more quickly and therefore result in larger fish or shellfish, but there is also an increased chance that disease at higher temperatures will destroy the entire crop! "Less experienced or less financially secure farmers may feel that firm survival is a crucial concern and may decide to pursue a more conservative strategy in which potential for losses is considered important. More experienced or financially secure farmers may not doubt their survival and be willing to face the higher risks that will bring higher expected income. Several poor seasons could ruin the inexperienced, poorly financed producer, whereas experienced, well-financed producers can survive poor results better." (see Hatch, Sindelar, Rouse, and Perez, 1987.)

Besides increasing the production of the aquaculture operation, the profit margin can be increased by producing a more valuable harvest of organisms. This is often the more realistic approach to stable profits. Ways to improve the quality of a harvest include better preservation and transportation techniques, careful handling during harvesting, and **grading** the organisms, because fish and shellfish of a uniform size are preferred to animals of assorted sizes (see Figure 13.1). (This increases their value because purchasers normally want animals for a specific purpose, and only animals in a particular size range are acceptable. For example, restaurants want fish filets all about the same size, and the same holds true for shrimp, clams, and oysters.) The culturist can in-

Figure 13.1 Workers use a grading device to sort shrimp of different sizes. (Photo courtesy of Sort-Rite International, Inc., Texas.)

crease the value of the product by treating it in some way, such as smoking or dressing it. Also, if the product can be produced during a season when demand is highest, or when supplies are the lowest, the value of the harvest will be greater.

Before an aquaculture operation can begin, the culturist must have some idea about who will buy the product after it is harvested. Retail markets, such as restaurants and fish markets, are only one of several possible outlets the culturist must consider. There are at least three others: fee fishing, live hauling, and processors. Fee fishing is extremely important to many catfish farmers; in such operations, the farmer allows paying fisherman access to the ponds at certain times of the year (see Figure 13.2). Fee fishing profits can be extended by offering products (bait, for example) for sale to the fisherman, as well as services (cleaning or even taxidermy). Live haulers buy fish from the culturist and deliver them to processors,

Figure 13.2 Fee fishing is a popular method of marketing catfish in the southeastern United States. (Photo from Lee, 1981. Copyright © The Interstate Printers & Publishers, Inc., Illinois. Reprinted by permission.)

distant markets, and other pond owners or fee fishing operations. Processors are only an outlet for culturists who are in a area where farming is very common. The processor buys the products from the culturist (and in many instances, also does the harvesting) and cleans, packages, and distributes the fish to markets that may be unavailable to the farmer. Culturists receive less money from processors than they would from fee fishing or retailers, so these markets are saturated first before the processor is called.

To estimate the value of a product when one is building an economic model, more must be known than the current wholesale and retail prices. The culturist must also understand what the demand is for the product at the moment, how that demand may change over time, and from what areas competition will come. Competition can come from other culturists as well as from natural catches, attention therefore should be paid to the status of the fishery of interest. Certain species become more popular as the average income of the consumers increases, and in a growing economy these are the animals with the greatest potential for development. In some cases, the elasticity of the product must also be known: How will the demand change as prices change? as the purchasing power of the consumers changes? as the amount of product on the market changes? Substitute products affect the market and price stability. For example, freshwater prawns can be substituted, to some extent, for saltwater shrimp. Recently "artificial crab," which is made by blending, treating, and cooking cheap blocks of fish, such as pollack, has made a considerable impact on the fish market. The treated fish, called **surimi,** is shaped and colored to appear as crab, lobster, shrimp, or many types of fish.

Two examples of economic models follow.

Model 1 A small facility in Colorado grows the tropical freshwater prawn, *Macrobrachium rosenbergii*, using heated (geothermal) ponds. The facility is owned by a family whose members participated in its construction and also do most of the work on the farm. The land belongs to the family who used it for traditional agricultural purposes until recently, so there is a source of water on the land as well as buildings, roads, storage, excavation equipment, pipes, pumps, a shop, trucks, and a small boat. Manure is used to supplement the purchased feeds. Such a facility probably would be unprofitable if it were not family built and run, but in this case the annual net profit for the 10-acre farm would be almost $15,000 (see Table 13.1).

Model 2 Biologists wanted to know under what circumstances a shrimp farm in Florida would be profitable. What size should the farm be? Should it produce shrimp for eating or bait shrimp for fishermen? What would be the effect of the price of the land on ability to make a profit?

To examine these questions, they built a model comparing six different farms: 100, 500, or 1000 acres of ponds for either bait shrimp or food shrimp. The actual total size of the farms was estimated to be 120% of the area of the ponds alone. The capital costs for bait and food farms were assumed to be the same, except the bait farm would require live holding tanks. The cost of land was estimated to be between $15,000 and $250 per acre. (This model was published in 1971; although costs have risen, the model is still a useful tool and requires only

TABLE 13.1 Economic Model for a small family-owned *Macrobrachium* Farm*

Capital costs

1. Land	Available
2. Water	Available
3. Pond construction	
a. Earth moving	1200
b. Clay seal	4500
c. Piping, valves, etc.	500
d. Wells	Available
e. Buildings	Available
f. Roads	Available
4. Equipment	
a. Feed storage bin	Available
b. Pumps and assembly	1000
c. Aerators	400
d. Lab and instruments	500
e. Row boat	Available
f. Harvesting and packing	1000
g. Shop and maintenance	Available
h. Auxiliary generator	None
i. Vehicle	Available
5. Heat loss control facilities	None
Total capital costs	9100

Operating Costs

1. Feed	225
2. Imported stock	500
3. Labor ($4.50h^{-1})	
a. Stocking	None
b. Feeding	1000
c. Pond management	None
d. Harvesting and packing	None
e. Transport to market	100
4. Marketing	None
5. Communication	
a. Phone	150
b. Mail	50
6. Power	
a. Electric	75
b. Gas	None
7. Taxes	220
8. Amortization (12% capital cost)	1100
9. Bookkeeping and administration	100
10. Pond cover replacement	None
Total operating costs	3520

Return

1. Production	1000 lbs
2. Sale of byproducts (S)	None
3. Market values ($5.00 lb^{-1})	5000
Profit	1480

* Costs per acre for a 10-acre farm.
Source: Klemetson and Rogers, 1984. Copyright © Elsevier Applied Science Publishers. Reprinted by permission.

some price adjustments.) Pond construction would cost $850 per acre and maintenance would be 10% of this again after 10 years. The cost of the hatchery, regardless of the size of the farm, would be about $82,500, plus $1,000 per year for utilities and $5000 every three years for repairs. After 16 years, the land would be sold for 80% of the original cost.

The value of the shrimp to those who caught them (the **ex-vessel price**) in 1970 was $0.72 per pound, which means that if the shrimp farm produced 1000 pounds of large shrimp per acre, with one crop per year, the gross revenue would be $720 per acre per year. The bait shrimp farm could produce four crops per year of small shrimp, at 30,000 shrimp per acre and a price of $15 per thousand, giving an estimated gross revenue of $1800 per acre per year. The price of the feed for the food shrimp farm was estimated to be $189 per acre per 180-day growing season (grown only during the warmest months of the year), and for the bait shrimp it was $118 per acre per 90-day growing season (bait shrimp can be grown throughout the year).

Gravid female shrimp would be caught to supply the farm with seed stock for the ponds. The cost of ship time would vary with the season, and it was assumed that it would cost four times as much per year to supply a bait shrimp farm as a food shrimp farm of the same size.

The cost of the labor would vary with the size and type of farm. The labor expenses to run the shrimp hatchery would be the same for either type of farm, being $147,500, $123,000, and $106,000 for the 1000-, 500-, and 100-acre farms, respectively. The labor expenses for pond operations in the bait shrimp farm would be 135% to 150% that of the food shrimp farm.

Based on this model, it was shown that the 500- and 1000-acre bait shrimp farms probably could be profitable, while other farms had negative cash flows (see Table 13.2). Even if the food shrimp farm could produce two harvests per year, it would be unable to support itself. Many aspects of shrimp farming have changed since 1970, but the principles used to develop this model are still valid.

MANAGEMENT

As an economic entity, aquaculture has had its share of past failures, and as in the case of most businesses, many of the reasons for the failures are clear in hindsight. No small number of the failures stem from poor management practices; but improved management techniques that would better focus the direction of a farm, as well as coordinate operations and production, have not been forthcoming. **Management,** as the term is used here, includes the skills of business planning, organization, and monitoring the operation so that changes can be made as the economic climate evolves.

Planning, for the owner of a small seasonal fish farm, would include deciding what organisms to grow, how long a growout period will be, how the fish will be harvested, to whom to sell the fish, and how to raise any capital needed at the start of the season. Planning also means developing some sort of annual economic projection that allows the owner to predict his or her expenses and revenues (this can be very comprehensive, or short and informally done). Organization means hiring help for feeding and harvesting, and purchasing the feed and chemicals needed for the growing season and arranging for their delivery. Monitoring includes supervising the hired help, and keeping track of the actual performance of the farm and comparing this with the economic/production projection that had been developed at the start of the season; if the actual performance does not match the expectations developed in the projection, then changes can be made to improve progress. The owner of the fish farm may not have had formal training as a manager and probably does not think of himself or herself as "management." Regardless of this, the difference between a profitable season and a failed farm is *good management.*

Large aquaculture facilities tend to be the most profitable, but as the size of the operation grows, so does the need for increased management. Levels of management tend to develop, and as our theoretical fish farm gets larger the owner finds that she has developed a management structure under her. She has hired a number of field foremen to oversee the labor. Above the foremen are supervisors, considered **lower management,** who are responsible for staffing and assist in organizing the feeding and harvesting. Above the supervisors are members of **middle management** who are responsible for short-term planning and controlling the operations. There is also the **top management,** which includes the board of directors and the president, who are responsible for the long-term planning that molds the future of the company. Within any of the management levels, there should be assigned jobs such as purchasing and accounting, production, and marketing. Most aquaculture operations do not have such an extensive management pyramid as this, but even for much smaller companies, the management duties of all employees must be clearly understood.

MARKETING

Marketing brings a desired product to the consumer. Therefore, one who markets a particular type of fish must identify where there is a demand for it, when the demand (price) is greatest (see Figure 13.3), what form (whole, filet, dried, smoked, and so on) is desired (see Figure 13.4), what sizes bring the best prices, what is the best way to transport the product to different destinations, and whether advertisement can affect sales.

With such a new industry as aquaculture, these questions may be difficult to answer, since consumption patterns for particular cultured fish and shellfish may not be established. This is especially true when the fish being cultured is unfamiliar to the consuming public. In these cases, a market must be developed by such tech-

TABLE 13.2 Economic Model of a Florida Shrimp Farm in 1970

Labor Costs for Hatchery and Management

Personnel	Unit Cost	1000 Acres		500 Acres		100 Acres	
		Number	Cost	Number	Cost	Number	Cost
Hatchery technicians	$ 7500	3	$ 22,500	2	$ 15,000	2	$ 15,000
Supervisors	18,000	1	18,000	1	18,000	1	18,000
Mechanics	11,000	2	22,000	2	22,000	2	22,000
Maintenance men	7500	1	7500	1	7500	1	7500
Labor Subtotal			70,000		62,500		62,500
Secretaries	6500	1	6500	1	6500	1	6500
	5000	2	10,000	1	5000		
Sr. manager	25,000	1	25,000	1	25,000	1	25,000
Jr. manager	12,000	3	36,000	2	24,000	1	12,000
Management Subtotal			77,500		60,500		43,500
Total costs			$147,500		$123,000		$106,000

Labor Costs for the Operation of Ponds in Food Shrimp Farm

Personnel	Unit Cost	1000 Acres		500 Acres		100 Acres	
		Number	Cost	Number	Cost	Number	Cost
Permanent skilled	$ 7500/yr.	20	$150,000	15	$112,500	15	$112,500
Seasonal (6 mos.)	$ 350/mo.	70	147,000	45	96,600	27	56,700
Seasonal foreman (6 mos.)	$ 750/mo.	5	22,500	4	18,000	3	13,500
Supervisor	$18,000/yr.	1	18,000	1	18,000	1	18,000
Total costs			$337,500		$245,100		$200,700

Labor Costs for the Operation of Ponds in the Bait Shrimp Farm

Personnel	Unit Cost	1000 Acres		500 Acres		100 Acres	
		Number	Cost	Number	Cost	Number	Cost
Permanent skilled	$ 7500	20	$150,000	15	$112,500	15	$112,500
Permanent unskilled	$ 4200	70	294,000	46	193,200	27	113,400
Foreman	$ 9000	5	45,000	4	36,000	3	27,000
Supervisor	$18,000/yr.	1	18,000	1	18,000	1	18,000
Total costs			$507,000		$359,700		$270,900

A 16-year Balance Sheet for the 500-acre Bait Shrimp Farm

	Capital				Operating Costs					
Year	Live Holding Tank	Ponds	Hatchery	Land	Labor for Ponds	Labor for Hatchery and Management	Food	Egg Acquisition	Total Revenue	Net Cash Flow
1	$30,000	$425,000	$82,500	Range from $9,000,000 to $150,000	$359,700	$123,000	$236,000	$10,920	$900,000	Range from − $9,537,500 to − $687,500
2		250	1000							$169,230
3		250	1000							$169,230
4		250	6000							$164,230
5		250	1000							$169,230
6		250	1000							$169,230
7		250	6000							$164,230
8		250	1000							$169,230
9		250	1000							$169,230
10		42,250	6000							$121,230
11		250	1000							$169,230
12		250	1000							$169,230
13		250	6000							$164,230
14		250	1000							$169,230
15		250	1000							$169,230
16		250	1000							Range from $7,200,000 to $120,000

Source: Anderson and Tabb, 1971. Copyright © The Gulf and Caribbean Fisheries Institute, Inc. Reprinted by permission.

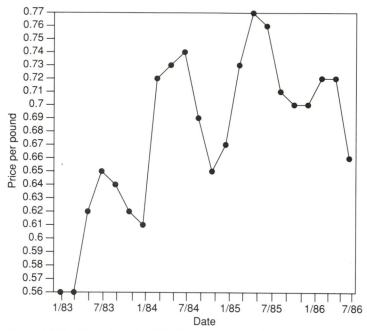

Figure 13.3 The price of catfish, like many other foods, varies with the time of the year. Price is related to the amount of the product that is on the market and the seasonal demand.

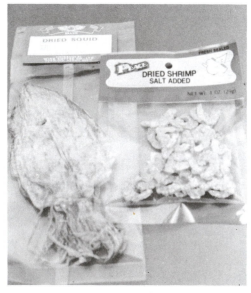

Figure 13.4 Products must be marketed in a form that consumers want to purchase. In Japan, dried squid and shrimp are popular; these products have storage and shipping advantages over fresh and frozen seafoods.

niques as promotions and education, which may include the distribution of cooking suggestions or demonstrations of preparation in supermarkets. Increasing the market for a new product sometimes means processing the fish into an easily used form, such as a breaded fishcake with cooking instructions on the package.

Socioeconomic surveys are one of the most important sources of local data on the potential for aquaculture products. These surveys, which take the form of personal interviews or mail questionnaires, aim to cull basic information about consumer habits and opinions. (Consumers may constitute restaurants, households, or institutions.) Marketing strategy may be developed after the data have been analyzed and patterns are established; such surveys include questions about location, size of the consumer's budget, race or ethnic background, income, age, or sex, among other things.

A potential market can also be estimated by small-scale trial experiments; in these, test stores or areas are chosen where the product is sold. The consumer's acceptance of the new product is recorded as a function of sales. Variables such as the effects of advertising, price, and the form and packaging can be introduced.

Some types of fish are well known to the consuming public, but for new fish or shellfish, a market must be developed. Part of developing a market includes calling the product by a name that appeals to the consumer: "mushroom," "caviar," and "tripe" appeal more to the palate than "fungus," "salted fish eggs," and "cow stomach." Some economists feel that the failure of horse meat to establish itself in the United States during the 1970s, despite the marketing attention it received, was because no good alternative name was used—people didn't like the sound of "horse meat." In some areas, dogfish is now marketed as "salmon shark," blowfish is called "sea squab," and monkfish is called by its French name, "lotte."

Many aquaculture groups, such as the United States Trout Farmers Association, the Louisiana Crawfish Promotion Board, and the Catfish Farmers of America, assist the farmer in developing new markets. The United States Department of Agriculture funds the Foreign Agriculture Service (FAS), which looks for potential new markets overseas for agriculture and aquaculture products.

FINANCING

Because of the considerable capital investment that often must be made before a commercial aquaculture project can begin, many aquaculturists must borrow money. In the past, many lenders have been uncomfortable about lending money for aquaculture because

1. They believed the risks were too great.
2. In many areas, it was a new indus-

try, and therefore the rules about business decisions were not established.

3. Much of the capital goes toward purchasing specialized equipment and land that have very limited resale potential and, therefore, are of little value as collateral.

4. Even the laws governing the rights of the aquaculturist are unclear. For example, with reference to the Canadian Bank Act, it is unclear whether banks can take security on fish in the water the same way they can with cattle on land.

Besides these problems, banking procedures often require that working capital be repaid after only a short period (one to three months), which is often not possible when fish, seaweed, or invertebrates are being grown to a marketable size. However, increasingly more commercial and development banks, as well as venture capital companies, are beginning to understand what is involved in aquaculture and are willing to lend money for good aquaculture projects.

The financial institution that is willing to consider an aquaculture venture will begin by comparing the project to nonaquaculture opportunities. Typical questions to be raised might include the following:

1. Is the project large enough to warrant the time needed to study the proposal?

2. If the project is to be in a foreign country, what is the stability of that country's economy and what is the political and social climate?

3. Who will be running and managing the project? Do they have aquaculture and business experience? If so, what do their past performances say about the potential success or failure of this particular venture?

4. Do the assumptions made by the

aquaculturist about this project seem realistic, are the goals achievable, and does the submitted feasibility study give all the necessary details?

5. What does the market analysis of the product indicate, and are the prices stable, declining, or rising?

In the past, some aquaculture ventures were supported by tax shelter investors. In these cases, the goals of the investors were different and included getting an immediate deduction for federal income tax purposes, and receiving future **ordinary income** (profits from the farm's activities) and/or **capital gains** (money made by selling the investor's share of the business). The United States' tax reforms of 1987 have eliminated most of the tax shelter opportunities available to business.

In some countries, an aquaculture project can be partially or completely financed by the government. In these cases, the project must not only demonstrate developmental potential, but must also conform to the objectives of that government's economic plan. The questions asked by the government will probably reflect not only a concern about the culturist's ability to produce a product, but will explore questions relating to increasing the domestic consumption of the product, how the product will affect foreign exchange (in particular imports of a similar nature), how the local fisheries industries will be affected, and how the project will affect the employment opportunities in that country. In many cases, the answers to these questions are not entirely clear; for example, what would be the interaction between salmon ranching and salmon fisheries in an area? It has been suggested that salmon ranching will stimulate overfishing, and if the price of salmon is high enough, the natural stocks will be fished to near extinction. The range of prices under which natural stocks and ranched salmon can probably

coexist can be maintained by proper management techniques.

INSURANCE

Insurance is a way to spread the risks so that the culturist does not bear the burden alone if something unforeseen should occur. It is also critical for many banks and investors who insist that a risky project must be insured. There are a number of types of insurance that may be needed by the owner of a commercial aquaculture facility. Besides standard types such as health insurance for the employees and fire insurance for the building, other types are available as well.

Mortality insurance provides against losses of the stock due to identifiable agents or within a stated period of time. Mortality insurance may protect against predation, or failure of some piece of equipment, or disease. Mortality insurance can be purchased as a blanket type of coverage protecting the stocks in the event of any kind of loss, and in this case it is termed **all-risk** coverage. The cost of mortality insurance, if available, is a function of an assessment of the risks that are involved; for European trout farmers, the cost may vary from 1% to 20% of the maximum value of the stock. Some of the things that an insurance company will want to know before a policy is issued include:

1. Is there an alarm system?
2. What is the water quality?
3. How long has the farm been in business?
4. How big is the facility, who built it, and when was it built?
5. Is the water treated?
6. Have there ever been any disease or pollution problems?

Aquaculturists who use cages or rafts may also want some sort of liability insurance to cover accidents that may occur with boats or swimmers. Product liability may be desired and protects the culturist in the event that they should be at fault because their activities resulted in the spread of a disease or parasite to stocks of fish that do not belong to them, or because a product from their farm resulted in someone's injury or illness.

ECONOMIC EFFICIENCY VERSUS BIOLOGICAL EFFICIENCY

From time to time, biologists who are unfamiliar with aquaculture have commented that aquaculture is impractical because it is an "inefficient" system. What inefficiency means, in biological terms, is that the system requires more of a particular resource, such as protein, than it is supplying. Certainly this is true of some aquaculture systems. Trout, for example, have been raised on a diet of ground carp and salmon cannery wastes, and although this means that more fish protein is going into the system in the form of carp and salmon than is coming out in the form of trout, many people continue to use similar feeding regimens. This may be biologically inefficient, but since the farmer is still able to sell the trout for considerably more than the cost of the ground carp and the cannery wastes, he is still making a profit. Thus, economic and biological efficiency are clearly different concepts.

SUMMARY

Aquaculture is a business, and is therefore governed by the laws of economics. Perhaps the most valuable tool that the culturist and business person have is the economic model. The model allows them to see quickly how much capital is needed, what the operating costs are, and how costs will change over time under variable conditions; in addition, production (value) can be predicted. Thus the model allows the culturist and entrepreneur to see whether a project is worth investing their time and money in; and once the project is started, the investors can see how the money is flowing in and out of the business.

It is management's role to organize the business by developing policy, setting goals, and monitoring the daily and long-range activities. Since profit is the key motive of the business, it is important to bring the product to the consumers in a form they find acceptable, at a time when the demand is greatest, and to create an interest in the product if one does not already exist; this is called marketing. Because aquaculture is a relatively new (and therefore risky) business, financing is sometimes difficult to obtain, although this has begun to change recently. Insurance may be sought to reduce the risk to the culturist.

APPENDIX 1

METHODS OF WATER QUALITY ANALYSIS

Perhaps the most underestimated problem for the inexperienced aquaculturist concerns water chemistry. Poor water means poor growth and disease. The logical question, then, is: "What makes water poor?" This question is harder to answer than it might seem at first because what is acceptable for a trout farmer might be very different from what is acceptable for someone growing seaweeds. There is a considerable body of literature on environmental parameters for different species of plants and animals, in terms of both an "ideal" environment and physiological tolerances. These parameters are usually discussed in terms of

1. Temperature
2. pH
3. Turbidity
4. Standard chemical measurements, which include
   a. Salinity for saltwater systems
   b. Hardness for freshwater systems
   c. Dissolved oxygen
   d. Carbonates
   e. Phosphorus

f. Several nitrogen compounds
g. Occasionally a few minerals such as silicon

STANDARD CHEMICAL ANALYSIS OF WATER

A number of kits are now sold that make much of the difficult water chemistry of the past a simpler and quicker chore (see Figure 1a.1). The accuracy of these kits is certainly less than that achieved by the older analytical methods, but such accuracy is usually not of great concern to the culturist who only wants to maintain safe minimum and maximum levels in a system. To put it another way, a thermometer that can measure to an accuracy of $\pm 1°C$ or $\pm 2°F$ is probably acceptable for almost all commercial culture work, and buying one that could measure to 1/1000th of a degree would not make that person a better fish farmer.

In some instances, greater accuracy than can be achieved using a kit *is* desired, and sometimes older standard methods may be cheaper to perform if there are many such tests to be run daily. In this

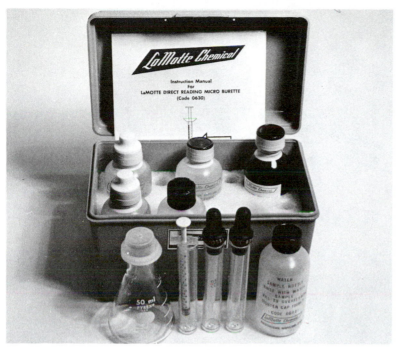

Figure 1a.1 A kit used for measuring dissolved oxygen in seawater. (From Spotte, 1973.)

appendix, the chemical methods used for detection will be discussed, because not only can they be used "as is," but they are also the basis of most of the simple kits.

Several companies now produce relatively accurate and simple-to-use electrical devices that measure salinity, pH, temperature, dissolved oxygen, alkalinity, and even specific ions such as ammonium (see Figure 1a.2). There are greater capital costs associated with the purchase of such pieces of equipment, but they are usually well worth the expense if a significant number of water samples are tested on a regular basis.

Whenever water samples are taken, the sampler must exercise some common sense. If a pond has poor circulation and is therefore poorly mixed, replicate samples should be taken at several locations and at different depths in the pond in order to better measure the true condition of the entire pond. The very surface of the water is not representative of most

of the water in a pond, and generally should not be used. All sampling containers should be clean, and it is good practice to rinse them once with the water to be sampled. Likewise, all laboratory glassware should be clean (a little soap film left in a beaker can ruin an experiment) and covered when not in use to prevent dust from entering. When sampling water for O_2, pouring must be avoided. Pouring causes air bubbles to form (this will increase the O_2 level in oxygen-poor water, and water that is supersaturated with oxygen, as can occur during periods of active photosynthesis, will release O_2, so the analysis will show merely saturation). These and other logical procedures must be implemented if any subsequent analysis is to be valid.

Salinity

The amount of salt in a seawater sample can be calculated by measuring the amount of chlorine in the water. This

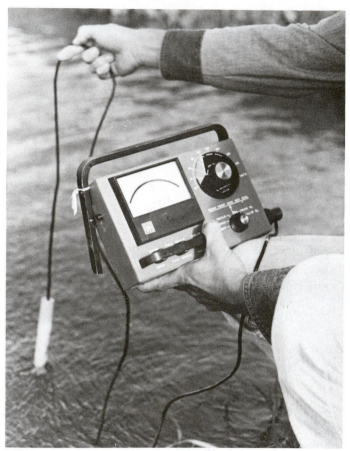

Figure 1a.2 An electronic meter that can be used to measure dissolved oxygen in water by using a selective probe. (From Stickney, 1979.)

can be done because the ratios of all the major elements in the oceans, including chlorine, are constant. As discussed in Chapter 2, these are the conservative elements. Thus, if you know how much chlorine is in a seawater sample, you will also have an accurate measure of the amount of sodium, potassium, sulfate, bromide, calcium, and so on. Just as **salinity** is the weight of the dissolved salt in 1 kilogram of water, **chlorinity** is the amount of chlorine in 1 kilogram of water. The relationship of the terms is

$$salinity = 1.807 \times chlorinity$$

Into a known volume of seawater containing a small amount of potassium chro-mate, which serves as a color indicator, silver nitrate is titrated. The silver combines with the chlorine; when all the chlorine has been bound, the silver combines with the indicator to give a red color, at which point the titration is halted. Since the amount of silver used to bind up all the chlorine is known, the amount of chlorine in the sample is therefore also known. The chemical reactions are

$$Cl^- + Ag + \rightarrow AgCl$$
$$2\,Ag^+ + CrO_4{}^{2-} \rightarrow Ag_2CrO_4 \text{ (red)}$$

A nonchemical method for the measurement of salinity, based on water's refractive index, is also popular. As salt is dissolved in water, the water "bends" light to a greater extent. Small hand-held **opti-cal salinometers,** which measure the re-fractive index of water, are easy to use and are generally accurate enough for the cul-turist's needs; they are therefore a popu-lar alternative to chemical or electrical measurements of salinity.

Hardness

Hardness is a measure of the amount of dissolved calcium and magnesium in a sample of freshwater. A water sample of a known volume is adjusted to pH 10, and an indicator dye, Eriochrome Black T, is added to the sample. This is titrated with ethylenediamine tetraacetic acid (EDTA), a chemical that complexes with metal ions. The water sample starts out wine-red color while the EDTA and the ions are being complexed, and the solution turns blue when there are no longer any free calcium and magnesium ions. The amount of EDTA used to bind the ions is therefore a measure of the amount of dis-solved calcium and magnesium.

Oxygen

Poor **oxygen** can not only potentially damage the organisms that are being cul-tured, but can also result in biological fil-ters not working to full capacity with the consequent buildup of organic contami-nants and ammonia.

A sampling bottle is completely filled up with water (there must be no air left on top); the bottle is then stoppered until testing can begin. Normally these samples are processed as soon as possible, but if they must be stored for any length of time, it is best to chill them to slow down any respiration or photosynthesis that may take place as a result of small animals, plants, or bacteria in the water.

A manganous sulfate solution is added to the water sample, followed by the addi-tion of a strong alkaline iodine mixture made from sodium hydroxide and po-tassium iodide. Any oxygen in the water sample quickly oxidizes the manganese hydroxide ions and a precipitate forms.

The reactions are

$$Mn^{2+} + 2\,OH^- \rightarrow Mn(OH)_2$$
$$2\,Mn\,(OH)_2 + 0.5\,O_2 + H_2O \rightarrow 2\,Mn(OH)_3$$
$$\text{precipitate}$$

At this point, samples need only be stored in airtight bottles. When analysis begins, the samples are acidified with sulfuric acid and the precipitate dissolves. The re-actions are

$$2\,Mn(OH)_3 + 2\,I^- + 6\,H^+ \rightarrow$$
$$2\,Mn^{2+} + I_2 + 6\,H_2O$$
$$I_2 + I^- \rightarrow I_3^-$$

The sample is then titrated with sodium thiosulfate until a light yellow color re-mains. At this point, an indicator (starch) is added, turning the sample dark blue. Titration is continued until the disap-pearance of the blue color caused by the complete reduction of the iodine mole-cules by the thiosulfate. The reaction is

$$I_3^- + 2\,S_2O_3^{2-} \rightarrow 3\,I^- + S_4O_6^{2-}$$

Thus, this method is indirect: oxygen lev-els affect the manganese, which in turn affects the iodine, which can complex with the starch, and it is the iodine com-plexed to the starch that is actually mea-sured.

Carbonates

Carbonates (CO_2, H_2CO_3, HCO_3^-, and CO^{2-}) are important because they serve as nutrients for plants as well as buf-fering the water, preventing extreme pH shifts. The pH of the water is the basis of the most popular chemical method for carbonate determinations.

By adding a known amount of acid to a water sample, and then reading the pH, the total carbonate content of the water at a specific temperature can be calculated. The key to the success of this method is that the concentration of the acid must be precisely known and the volumes of acid and water used are measured very carefully.

Phosphorus

There are several tests for inorganic **phosphate** in water. One of the most widely used tests is carried out by adding a small, measured volume of water to a clean test tube along with an indicator, phenolphthalein. To this is added a reagent containing sulfuric acid, potassium antimonyl tartrate, ammonium molybdate, and ascorbic acid. Time is allowed for the reagent and phosphate to react; then the absorbance of the mixture is measured at 880 nm using a **spectrophotometer.** A series of clean test tubes is prepared containing serial dilutions of a phosphoric acid standard. The same procedure is performed on the series of standards as was carried out on the water sample; then a comparison is made between the unknown water sample's absorbance and those having a known concentration of phosphate. This is known as constructing a **standard curve** and is one of the most useful and commonly used procedures in analytical water chemistry (see Figure 1a.3).

The principle of this test involves the formation of phosphomolybdic acid. The amount of this acid formed is a function of the amount of phosphate in the water. The ascorbic acid then reacts with the phosphomolybdic acid to give a blue color. Small amounts of phosphate that may be present on pieces of improperly washed glassware give elevated results.

Nitrogen

Practically all the **nitrogen** in water is in the form of N_2, which does not affect most biological systems. N_2 is almost never measured. Nitrite and nitrate are

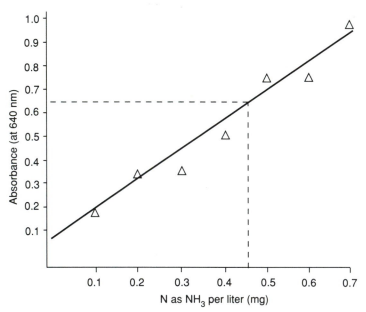

Figure 1a.3 A standard curve. A series of water sample standards having known concentrations of ammonia were analyzed using Bower and Holm-Hansen's method (1980). The absorbance of each of the known standards (0.1, 0.2, 0.3, 0.4, 0.5, 0.6, and 0.7 mg of N as NH_3 per liter) is graphed and a straight line is fit to the points. This is the standard "curve." An unknown water sample is also analyzed and found to have an absorbance of 0.640. Using the standard curve, the concentration of the N as NH_3 in the unknown sample is estimated to be about 0.46 mg per liter.

often measured, but clearly because of its toxicity, the most commonly monitored nitrogen compound is ammonia.

Nitrite is measured by adding to the water sample an acidic sulfanilamide solution that results in a diazo compound. To this is added a naphthylethylenediamine reagent, resulting in the formation of an azo dye that is measured 543 nm. Again, as in the case of the phosphates, a standard curve is constructed using solutions of known nitrite concentrations.

Analysis of nitrate is carried out by passing the water sample through a glass column that contains copper-coated cadmium filings. This reduces all the nitrate to nitrite. The water that has passed through the column contains both the original nitrite and nitrite that was made from nitrate. This water is then analyzed as described earlier for nitrite. If the original concentration of nitrite is known, and the combined nitrite plus nitrate value is known, then the nitrogen in the form of nitrate can be calculated.

Probably the most popular method for analysis of ammonia in water is the phenol-hypochlorite technique. A phenol-alcohol solution is added to a water sample. This is followed by the addition of a solution of sodium nitroprusside, which acts as a catalyst. An oxidizing reagent, an alkaline mixture of sodium citrate in sodium hydroxide plus sodium hypochlorite, is finally added to the sample. This results in a blue indophenol being formed that is measured at 640 nm. Again, a series of standards must be used to compare against the water sample. In recent years there has been an effort to remove phenol, a carcinogen, from the laboratory. Therefore, alternative procedures have been developed and their use will hopefully become more widespread. Sodium salicylate has been used in place of phenol, and the use of gas-permeable membranes and electrodes (in freshwater systems especially) is becoming more common.

Silicon

Silicon can be the limiting nutrient in cultures of diatoms. Its concentration in water therefore may have some limited interest for the aquaculture community.

The molybdosilicate method for silicon determinations is commonly performed. In this procedure, the water is first "digested" by heating the sample with sodium bicarbonate and sulfuric acid; this procedure converts all the "unreactive" silica to the "reactive" form. To this is added hydrochloric acid, followed by a solution of ammonium molybdate, and finally a solution of oxalic acid. Under acid conditions, the molybdate complexes with the silica to give molybdosilicic acid, and with phosphates to form molybdophosphoric acid (which is then destroyed by the oxalic acid to prevent interference). The absorbance of the yellow color is read at 410 nm using a spectrophotometer, and it is compared to standard silica solutions developed along with the unknown water sample.

POLLUTANTS

When initially considering a water source for a commercial farm, the aquaculturist is advised to consult a water chemistry laboratory and have a more complete analysis of the water carried out. Many pollutants that could seriously damage the crop (or the consumer) are not easily analyzed or require sophisticated equipment. Heavy metals, such as mercury or cadmium, which could be concentrated by growing organisms, are not detectable in water samples using simple analytical methods. Organic pollutants, including pesticides, PCBs, and petroleum products, not only are difficult to separate and detect, but are also difficult to identify even when "something" has been isolated by an experienced chemist. This book's purpose is not to discuss the analysis of all possible chemicals and compounds that *could* be present in a water system.

Even after the water has been analyzed in a water chemistry laboratory for an assortment of possible contaminants, the culturist must decide what the data mean on a case-by-case basis. For example, most water samples will contain some mercury, but how much is too much for aquaculture? Toxicological tests are normally run in a laboratory for 24 or 48 hours, during which time deaths or other obvious effects are monitored and recorded. After one of these **bioassay tests** is completed, a "safe level," that is, a level where there is "no effect," may be proposed, and these "safe levels" are often the basis for government regulations. There are two unfortunate problems with these tests as they are normally carried out. First, compounds that look as if they have no effect after one or two days of exposure may have a pronounced effect on growth and reproduc-

tive capacity over longer periods of time. Second, many of the tests are run on adults of **standard species,** which are practically always the most tolerant and hardiest (that's *why* they are favorites of laboratory personnel—they are easy to maintain!). As an example, in a test comparing minnows and cultured snook, in which both fish were exposed to EDB and aldicarb, the snook were found to be more sensitive than the "standard" minnows, and embryonic fish were more sensitive than larval fish.

In terms of pollutants, therefore, it is difficult to say what is "safe." Low concentrations of pollutant "X" may not cause massive fish kills, but may cause depressed growth and reproduction. It is advisable to consult others who are growing the same or similar organisms and find out what they have established as acceptable.

APPENDIX 2

FLUID MECHANICS PROBLEMS AND APPLICATIONS

The study of fluid mechanics can be divided into statics (fluids at rest) and dynamics (fluids in motion). A detailed understanding of the principles of fluid mechanics is probably not essential for the daily undertakings of the aquaculturist. However, when a farm is being built, or when new machinery or pipes are being purchased or installed, fluid mechanics must often be considered.

Many of the calculations necessary to answer even the most commonly asked questions faced by the culturist are complex and require not only a working knowledge of trigonometry, algebra, and geometry, but calculus and physics as well. Since they are used so sporadically, few culturists are completely comfortable with the required mathematics. Even a brief review of these skills is clearly beyond the scope and purpose of this appendix. Therefore, it is strongly advised that an engineer be consulted before beginning major construction. The cost of their consulting fee is almost always less than the consequences of guesswork, which can lead, for example, to the collapse of a tank wall.

Some common terms and conversions are given in Table 2a.1. Engineering texts do not always agree on notation; the terms used in this book are those adopted by Wheaton in *Aquacultural Engineering*.

FLUID STATICS

The culturist will face a great many problems relating to the force that water exerts on a surface. Will the side of a dam or tank collapse when water is pumped in? What is the pressure exerted on a valve controlling the flow of water out of a holding tank? How can the buoyancy of a raft be determined while it is being built?

An example of a common statics problem involves burst pressures of pipes. If a pipe bursts, it is because the outward force of the water on the internal surface of the pipe was greater than the inward forces exerted by the pipe. How can we determine which pipes are safe to use? This type of problem can be solved using the equation

$$\sigma_w = \frac{DP}{2t_t} \qquad \textbf{(Eq. 1)}$$

TABLE 2a.1 Fluid Mechanics Terms and Units

Area (A)	m^2
Velocity (v)	m/s
Density (rho, ρ)	Mass per unit volume (kg/m^3)
Newtons (N)	Force needed to give an acceleration of 1 m/s^2 to 1 kg
Gravitational constant (g)	Acceleration of a falling object on the earth, that is, 9.8 m/s^2
Specific weight (gamma, γ)	Weight per unit volume (N/m^3) = (ρ) (g); the specific weight of water is 9800 N/m^3
Absolute viscosity (mu, μ)	Resistance to shear stress (Pa s)
Pascals (Pa)	A unit of pressure (N/m^2)
Kinematic viscosity (nu, ν)	μ/ρ
Pressure (P)	Force per unit area (N/m^2)

Where

σ_w = The allowable stress, measured as Pascals (Pa), for the material out of which the pipe is constructed. (These stress values are available from the manufacturer or from engineering handbooks.)

D = pipe diameter, measured in meters (m).

P = Pressure, as Pa, exerted by the fluid. This can be calculated or measured using a manometer or a Bourdon gage.

t_t = Thickness of the pipe, as m.

EXAMPLE 1. What thickness must the walls of a PVC pipe be to resist bursting if the internal force exerted by water is 0.15 M (mega = 10^6) Pa, the diameter of the pipe is 1.5 m, and the allowable stress for PVC is 48 MPa?

$$2t_t = \frac{DP}{\sigma_w}$$

$$t_t = \frac{(0.5)\,(1.5\text{ m})\,(0.15 \times 10^6\text{ Pa})}{(48 \times 10^6\text{ Pa})}$$

$$t_t = 0.0023\text{ m} \quad \text{or} \quad 2.3\text{ mm} \; \approx$$

Normally, a safety factor is built into any designed system. For pipes, the safety factor is usually 1.5 to 2.5 times what would be needed to withstand the water pressure, so the actual thickness of the pipe used in this example might be about 5 mm.

FLUID DYNAMICS

Fluid statics problems concern the weight of fluids at rest. Fluid flow tends to be more complex, and often is not subject to exact calculations. The types of problems one may deal with are varied and could include drag, flow of compressible fluids, and flow through pipes, nozzles, open channels, and other structures.

The law of **conservation of mass** is one of the important principles of fluid dynamics. It states that in a closed system, the mass flowing past one point is the same as that flowing past another point per unit time, regardless of the diameters of the pipes at those points. Conservation of mass problems often involve the **continuity equation,** which has several forms, including

$$Q_1 = (\gamma_1)(v_1)(A_1) \qquad \textbf{(Eq. 2)}$$

where γ is the specific weight of the fluid, v is the velocity of the fluid, and the cross-sectional area of the pipe (A) is 3.14 times the square of the pipe radius, and

$$Q_1 = Q_2 \qquad \textbf{(Eq. 3)}$$

Therefore,

$$(\gamma_1)(v_1)(A_1) = (\gamma_2)(v_2)(A_2) \qquad \textbf{(Eq. 4)}$$

But for fluids that are essentially noncompressible, like water,

$$\gamma_1 = \gamma_2$$

It is therefore true that

$$(v_1)(A_2) = (v_2)(A_2) \qquad \textbf{(Eq. 5)}$$

EXAMPLE 2. In the bifurcated pipe in this example, the flow past point A is 2 m^3/s, and the velocity of the water is 0.5 m/s (see Figure 2a.1). At point B the diameter is 60 cm. At point C the flow is 0.75 m^3/s, and at point D the velocity is 10 m/s. What is the velocity at point B? What is the radius of the pipe at point D?

$$Q_A = Q_B = 2 \text{ m}^3/\text{s} = (A_B)(v_B)$$
$$\frac{(2 \text{ m}^3/\text{s})}{(3.14)(0.3\text{m})^2} = v_B$$
$$7.08 \text{ m/s} = v_B$$
$$Q_D = Q_A - Q_C$$
$$Q_D = 2 \text{ m}^3/\text{s} - 0.75 \text{ m}^3/\text{s} = 1.25 \text{ m}^3/\text{s}$$
$$Q_D = (A_D)(v_D)$$
$$1.25 \text{ m}^3/\text{s} = (3.14)(r_D{}^2)(10 \text{ m/s})$$
$$0.04 \text{ m}^2 = r_D{}^2$$
$$0.2 \text{ m} = r_D \quad \text{≋}$$

The law of **conservation of energy** is also an important concept. Fluids have three energy components:

1. **Potential energy due to elevation, PE$_e$:** a raindrop that is forming in a cloud has potential energy before it falls.

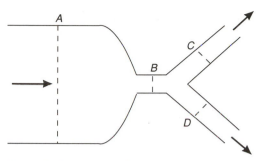

Figure 2a.1 Bifurcated pipe.

2. **Potential energy due to pressure, PE$_p$:** the energy of water in a fire hydrant before it is opened is potential.

3. **Kinetic energy, KE,** of fluids in motion.

According to the law of conservation of energy, in a frictionless system with two points, A and B,

$$PE_{eA} + PE_{pA} + KE_A = PE_{eB} + PE_{pB} + KE_B \qquad \textbf{(Eq. 6)}$$

However, this is an ideal system. In practice, energy can be added between points A and B, for example, if a pump is placed between them. Energy can also be lost. Lost energy can be in the form of **minor losses** and **pipe friction losses.** Minor losses are calculated by

$$\text{minor losses} = K\left(\frac{v^2}{2g}\right) \qquad \textbf{(Eq. 7)}$$

where K is a constant resulting from internal friction such as is generated when the fluid passes from one size pipe to another, or it changes direction, or the internal configuration of the pipe changes in any way because of a bend in the pipe as might be seen at T-junctions, valves, nozzles, and so on. K values have been generated experimentally and can be found in engineering handbooks (Table 2a.2).

Pipe friction losses depend on a number of factors, including the internal diameter of the pipe, roughness of the in-

TABLE 2a.2   Minor Loss Coefficients for Several Common Fittings

Fitting	K
Globe valve (fully open)	10.0
Angle valve (fully open)	5.0
Swing check valve (fully open)	2.5
Gate valve (fully open)	0.19
Close return bend	2.2
Standard tee	1.8
Standard elbow	0.9
Medium sweep elbow	0.75
Long sweep elbow	0.60
Entrance (square edge)	0.5
Entrance (slightly rounded)	0.2
Entrance (inward projecting)	0.8

Source: Wheaton, 1977.

ternal surface of the pipe, the velocity at which the water is traveling, and the salinity and temperature of the water. These factors contribute to two concepts. The first is **relative roughness,** which is the ratio of the absolute roughness (that is, the size of the small bumps and pits inside the pipe) to the diameter of the pipe. In-

tuitively we can see why bumps, which are 0.01 cm high, would affect flow through a 1-cm-diameter pipe more than flow through a 10-m-diameter pipe. The second concept is the **Reynolds number (RE),** which is the ratio of the inertial force of the water flowing through the pipe (density, velocity, and pipe diameter) to the viscosity of the water resisting the flow. That is,

$$\text{RE} = \frac{(\rho)\,(v)\,(D)}{\mu} = \frac{(v)\,(D)}{\nu} \qquad \textbf{(Eq. 8)}$$

(See Table 2a.1 for viscosity terms; viscosity values can be obtained from Fig. 2a.2.) Once both the relative roughness and the Reynolds number are known, the friction losses, f_f, can be found using **Moody's diagram** (Figure 2a.3). To do this, the curve of the relative roughness line is followed to where it intersects with the Reynolds number, then a horizontal line is extended to the f_f value.

The head lost (h) because of pipe friction is calculated by

$$h = (f_f) \left(\frac{L}{D}\right) \left(\frac{v^2}{2g}\right) \qquad \textbf{(Eq. 9)}$$

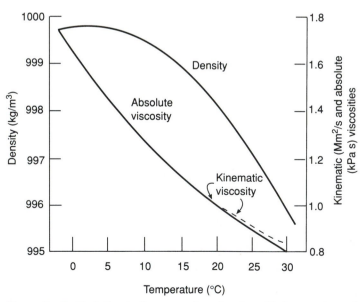

Figure 2a.2   Variation in the properties of water with temperature at atmospheric pressure. (From Wheaton, 1977.)

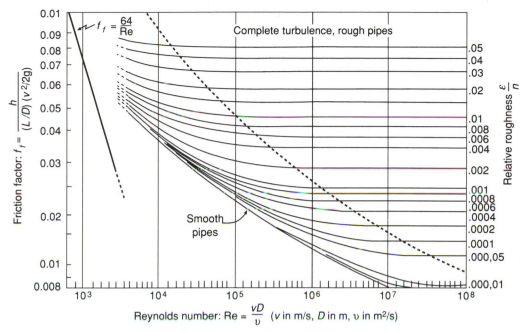

Figure 2a.3 Moody's diagram. (From Moody, 1944. Friction factors for pipe flow. Transactions of the American Society of Engineers 66:671−678. Copyright © ASME. Reprinted by permission.)

where L is the length of the pipe and g is the gravitational constant.

$$\text{minor losses} = K\left(\frac{v^2}{2g}\right) \qquad \textbf{(Eq. 7)}$$

EXAMPLE 3. Calculate the head lost when water at 10°C is passed through a steel pipe that has a 2 cm internal diameter and is 200 m long. The pipe has a swing check valve, and the water is traveling at a velocity of 1.5 m/s.

From Table 2a.2 we see that the K value for a swing check valve is 2.5, so

$$\text{minor losses} = 2.5\frac{(1.5 \text{ m/s})^2}{(2 \times 9.8 \text{ m/s}^2)} = 0.29 \text{ m}$$

TABLE 2a.3 Absolute Roughness for Several Types of Pipe

Pipe Material	Absolute Roughness (cm)
Riveted steel	0.091–0.91
Concrete	0.03–0.31
Wood stave	0.018–0.091
Cast iron	0.026
Galvanized iron	0.015
Asphalted cast iron	0.012
Commercial steel or wrought iron	0.0046
Drawn tubing	0.00015
PVC pipe	Approx. 0.000005

Source: Moody, 1944. Friction Factors for pipe flow. Transactions of the American Society of Engineers 66:671–678. Copyright © ASME. Reprinted by permission.

To calculate f_f we must find the relative roughness and the Reynolds number; therefore,

$$\text{relative roughness} = \frac{\text{absolute roughness}}{D}$$

and we can see from Table 2a.3 that for steel pipe the absolute roughness is 0.0046 cm, so

$$\text{relative roughness} = \frac{0.000046 \text{ m}}{0.02 \text{ m}} = 0.0023$$

$$\text{RE} = (v)\frac{(D)}{\nu} \qquad \textbf{(Eq. 8)}$$

and from Figure 2a.2 we see that saltwater at 10°C has a kinematic viscosity of about 1.3 Mm²/s, so

$$\text{RE} = \frac{(1.5 \text{ m/s}) (0.02 \text{ m})}{(1.3 \times 10^{-6} \text{ m}^2/\text{s})} = 2.3 \times 10^4$$

Using Moody's diagram, we see that the f_f is about 0.031; therefore,

$$h = f_f \left(\frac{L}{D}\right) \left(\frac{v^2}{2g}\right) \qquad \textbf{(Eq. 9)}$$

$$h = (0.031) \left(\frac{200 \text{ m}}{0.02 \text{ m}}\right) \frac{(1.5 \text{ m/s})^2}{(2 \times 9.8 \text{ m/s}^2)} = 35.6 \text{ m}$$

The total head loss is loss due to friction plus minor losses, or

$$3.56 \text{ m} + 0.29 \text{ m} = 35.88 \text{ m}$$

APPENDIX 3

MOLECULAR BIOLOGY IN AQUACULTURE

During the past two decades, certain branches of biology have made unprecedented jumps in technology—particularly the disciplines of molecular genetics, immunology, and cell and tissue culture. Synthesis and sequencing of nucleic acids (DNA and RNA) and proteins are now commonplace. Along with the acquisition of a great deal of basic knowledge on a fundamental level, an array of new laboratory techniques have been developed. So far, this has had little impact on the aquaculture industry, but this seems to be changing. Some researchers are beginning to see applications for molecular biological techniques in culture operations. While it is impossible to give a thorough introduction to molecular biology here, a few basic concepts illustrate its potential importance to the aquaculture community.

VACCINES

Vaccines are preparations that stimulate the production of **antibodies** to particular **antigens.** Antigens are any foreign substance that is recognized by the body's immune system. Antibodies are proteins that specifically recognize each foreign substance, and are generally most effective in combating the extracellular phases of bacterial and viral infections. [An antibody is technically referred to as an immunoglobin (Ig); there are five classes of antibodies (designated M, G, A, D, and E) that have been identified in mammalian serum.] Antibodies work by binding to the antigens, a reaction that renders both inactive. They are produced in special lymphocytes called **B cells,** while **T cell** lymphocytes are responsible for defense against other types of antigens (such as protozoans, or intracellular bacterial and viral infections).

Many contagious diseases are the products of microbial antigens. If, as a result of a vaccine, antibodies are present in a host that will react with newly introduced antigens, then those antigens often will be unable to multiply to the point where the disease symptoms are detectable. The production of vaccines is not a new notion; they have been used for decades to help prevent the spread of disease among people and farm animals. Recently, vaccines have been produced to fight diseases to which cultured fish are subject. Vaccines contain microbes (or parts of them) that cause the disease for which resistance is sought. Therefore, the vaccines themselves must be rendered harmless before they are used; this is done in one of two ways:

1. The microbes are killed (e.g., as in the Salk polio vaccine).
2. The microbes are treated to make them harmless although they are still alive (e.g., as in the Sabin polio vaccine).

The antigen in the vaccine is introduced into the host in either a simple saline solution or in an **adjuvant.** Adjuvants are oily substances that are mixed with the antigen before injection; the adjuvant serves both as a depot for the slow release of the antigen and as a nonspecific activator of the lymphoid system, thus enhancing the immune response.

For fish, the most popular vaccination methods are by

1. **Dip (immersion),** which requires a short exposure to a high vaccine concentration.
2. The **bath method,** which requires longer exposure to lower concentrations.

Injections of vaccines are also used for fish, but are less common. **Oral vaccines,** theoretically, are very attractive because they would allow the farmer to immunize large numbers of fish with no increase in labor, but experimental trials have hinted at several problems in design, most important, the vaccine's low or inconsistent potency.

Commercial fish vaccines for protection against vibriosis, furunculosis, and redmouth are available. Researchers are also developing vaccines for protection against *Edwardsiella,* channel catfish virus, viral haemorrhagic septicaemia, and others. The length of time that a vaccine offers protection is variable, depending on the vaccine and the age and health of the immunized animal.

MONOCLONAL ANTIBODIES AND *ELISA* ASSAYS

Highly specific antibody preparations, called **monoclonal antibodies,** have found a variety of applications in molecular biology. The production method of monoclonal antibodies can be summarized as follows:

1. Mice are inoculated with the desired antigen.
2. Spleens are removed from the mice whose serum tests positively for the presence of the antibody that is sought.
3. The spleen cells are mixed with a commercially available culture of mouse myeloma (tumor) cells; these are transformed B lymphocytes similar to the spleen's B cells. The mixing of the two cell types, in the presence of polyethylene glycol, leads to their fusion.
4. Excess and unfused myeloma cells are killed.
5. The fused cells are diluted with cell culture medium and grown in wells in a plastic multiple-well plate.
6. Cell cultures are tested for antibody production.
7. Cell cultures that produce the antibody of interest are diluted so that there are only single cells in each culture.
8. These cultures that start out with a single cell (hence the name, *monoclonal*) will produce specific antibodies (see Figure 3a.1).

Testing for the production of low levels of specific antibodies is the most difficult step in this procedure. It can be done with functional tests, or a **radioimmunoassay**

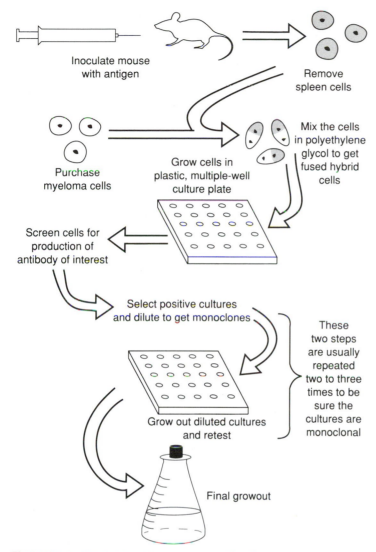

Figure 3a.1 Production of monoclonal antibodies.

(RIA), but the most common technique is to use an **enzyme-linked immunosorbent assay (ELISA).** This procedure is based on

1. Binding the antigen of interest to the walls of a multiple-well plastic plate, followed by a washing-off of excess material.

2. Adding a serial dilution of the serums from the inoculated mice that contains the antibodies that will bind to the antigens, then again washing the well to remove the unbound material from the serum.

3. Adding a conjugate of an enzyme, alkaline phosphatase, that has been bound to an antibody (for example, goat antimouse IgG) that will form a complex with the mouse proteins (that are themselves antigens to antibodies produced by the goat).

4. Adding a chromatic substrate, p-nitrophenyl phosphate, to the well. A yellow color will develop if the antibody to the original antigen

is present. The intensity of the color, measured spectrophotometrically, is used to determine the amount of antibody–antigen reactions, and therefore the amount of antibody in the serum.

How can these techniques be used in aquaculture? Here are several examples:

1. Effective treatment of *Edwardsiella* infections in catfish ponds is based, to a large extent, on an early diagnosis. The traditional method used, bacteriological confirmation, is slow, requiring four to five days. Monoclonal antibodies, made by injecting mice with heat-killed bacteria, can be used to detect *Edwardsiella* in three to four hours.

2. *Renibacterium* is responsible for bacterial kidney disease. The detection of a specific extracellular product of this organism, rather than the organism itself, can be carried out with an ELISA that can detect the antigen at levels as low as 2–20 ng.

3. "Dip stick ELISAs," for redmouth and furunculosis, have been developed so that tests can be made in the field by the fish culturist, rather than in a laboratory. In place of the multiple-well plate, scientists testing this technique used a modified plastic knife to probe the fish's kidney; assays were run in 25 minutes.

4. Monoclonal antibodies may also be used for purposes other than detection of pathogens. Researchers in Europe have injected mice with carp spermatazoa and used the spleens to make mouse anticarp-sperm monoclonal antibodies; it is hoped that sterile populations of fish can be produced by treating the fry with similar antibodies that would selectively attack the gonadal tissue to prevent maturation.

RECOMBINANT DNA AND GENETIC ENGINEERING

Much has been written in the scientific and popular presses about the use of these techniques. The basic idea is rather simple, and the possibilities for application are nearly endless.

With most of the details omitted, recombinant DNA may be generated as follows:

1. Isolate the DNA from **plasmid**-containing bacteria or yeast (plasmids are extrachromosomal circular DNA molecules). Then, isolate the plasmids from the rest of the DNA. The plasmids will act as the carriers (**vectors** or **cloning vehicles**) for the gene of interest. Other vectors, including bacteriophages or other viruses, are sometimes used.

2. Strands of chromosomal DNA containing the gene of interest can be "cut" from the chromosomes, and the closed circles of plasmid DNA can be opened at specific locations with the use of **restriction enzymes.** Restriction enzymes are endonucleases that have been isolated from a variety of organisms; they recognize specific sequences on the DNA, and therefore the number of cuts made by the enzyme is limited to these few sites. The sites of the cuts on the double helix strands of the DNA are often not directly opposite to each other.

3. Chromosomal DNA fragments are mixed with opened plasmids. The fragments are inserted into the plasmids and are enzymatically combined, often with enzymes called **ligases,** to form a new circular strands of DNA. When vectors are cleaved and mixed with the chromosomal products of a particu-

lar organism, there are a great many (sometimes millions) different molecules formed. Because of this, researchers have developed some methods for greatly reducing the number of different DNA fragments to be inserted into the vectors; these are techniques involving separation of the fragments by elec- trophoresis, or the isolation of cytoplasmic mRNA from specialized cells that produce only the desired protein (and perhaps a few others) and then producing a double-stranded **complementary DNA (cDNA)** strand from the mRNA single strand by using an enzyme called **reverse transcriptase.**

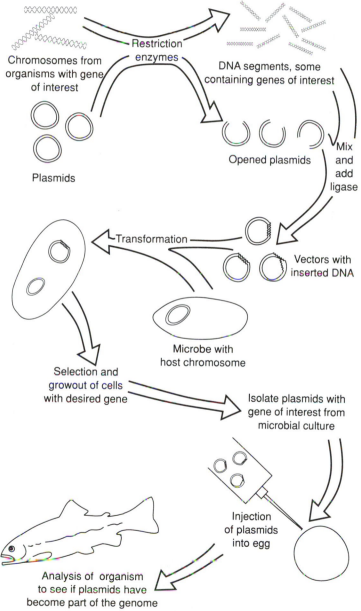

Figure 3a.2   Generation of recombinant DNA and transgenic animals.

4. The new hybrid plasmid is inserted into the recipient cells by mixing the two in a cold $CaCl_2$ solution, a procedure called **$CaCl_2$ transformation.** The recipient cell must have the replication enzymes that are active on the plasmid (which must be the case if the donor and recipient cells are of the same species).

5. After the $CaCl_2$ transformation, the cells with the plasmids of interest should be separated from the cells that did not receive a plasmid, and from those that contain the plasmid but with the wrong segment, or no segment, inserted. If the plasmid used contains a gene for resistance to an antibiotic, and if the cells are grown in a medium with that antibiotic present, then only the cells with the inserted plasmids will survive. There are several techniques that distinguish original plasmids that may have been inserted from the hybridized plasmids. Colonies of cells containing the inserted gene of interest are normally identified by a technique called *in situ* **hybridization,** which can be used for any gene for which radioactive mRNA is available. If the gene product is a protein, immunological tools may be used to identify the colonies of cultured cells that contain the gene of interest (see Figure 3a.2).

Genetic engineering has allowed us to make large quantities of rare molecules (such as hormones), some of which may be used on a regular basis by culturists in the future. Bovine growth hormone, **somatotropin,** has been produced in this manner; this hormone has recently been shown to double the growth rate of coho salmon. Likewise, the gene for **tuna growth hormone (TGH)** has been inserted into the bacterium *Escherichia coli*, after which the hormone has been synthesized by the bacteria; TGH has been shown to stimulate the growth of the snapper, *Pagrus*. Another example involves the growth of abalone: Experiments have revealed that insulin and mammalian growth hormone, added to seawater, increase the growth rate by raising the efficiency of nutrient assimilation and utilization in these mollusks; therefore, efforts have been made to construct a recombinant DNA gene bank of abalone DNA (isolated from sperm and cloned in plasmids) for introduction into microbes for mass production of the natural growth-controlling hormones that may be more effective than mammalian hormones and that someday may be used in culture.

Recombinant DNA, isolated from microbes in culture, has also been directly injected into fertilized and unfertilized eggs to produce **transgenic** animals. A vector containing the structural gene coding for human growth hormone was injected through the embryonic blastodisc of newly fertilized tilapia eggs; the tilapia were analyzed after 90 days, and it was found that some of the fish had integrated the injected DNA into their genome. The possibility of developing transgenic fish for culture is extremely exciting.

APPENDIX 4

FEED FORMULAS

There exist a surprising number of published feed formulas, but there are also a considerable number used in commercial culture operations that have never been published. If farmers or feed manufacturers spend their time and money developing a new diet, it is likely they will not want to share it with business competitors.

No perfect diet for any particular species exists, because the requirements change as a function of the water quality, environment, and other flora and fauna present in the culture system. However, some diets have been shown to be generally successful, and their compositions may be used as guides to new formulas. Some examples of these experimental formulated feeds are given in this appendix.

Diet 5 for Channel Catfish *Ictalurus punctatus*

Ingredient	Percent
Soybean meal	25.0
Wheat flour	15.0
Menhaden fish meal	10.0
Cottonseed flour (defatted glandless)	24.0

Vitamin mix[a]	0.5
Mineral mix[b]	5.86
Carboxymethyl cellulose	2.0
Lysine HCl	0.143
Soybean meal	7.31
Ethoxyquin	0.0125
Cellulose	10.17

[a] Vitamin mix (per kilogram of diet): vitamin A, 5500 IU; vitamin D_3, 100,000 IU; vitamin E, 5000 IU; thiamin, 2 g; riboflavin, 2 g; pyridoxine, 2 g; folacin, 500 mg; ascorbic acid, 40 g; d-calcium pantothenate, 5 g; biotin, 10 mg; choline, 55 g; niacin, 10 g; vitamin B_{12}, 2 g; vitamin K, 1 g; inositol, 10 g; ethoxyquin, 0.15 g.

[b] Mineral mix (as a percentage of feed): defluorinated $CaPO_4$, 2.1%; NaCL, 0.6%; pulverized oyster shell 1.1%; KH_2PO_4, 1%; KCl, 0.1%; $MgSO_4$, 0.3%; (per kilogram of feed): $FeSO_4$, 500 mg; $MnSO_4$, 350 mg; ZnO_3, 150 mg; $CuSO_4$, 30 mg; KIO_3, 10 mg; $CoCl_3$, 1.7 mg; Na_2MoO_4, 8.3 mg; $NaSeO_3$, 0.2 mg.

Source: E. H. Robinson, S. D. Rawles, and R. R. Stickney. Evaluation of glanded and glandless cottonseed products in catfish diets, *Progressive Fish-Culturist,* 46 (1984), pp. 92–97.

Diet 4 for Rainbow Trout *Salmo gairdneri*

Ingredient	Percent
Menhaden meal	25.0

325

Soybean meal	25.0
Wheat middlings	22.7
Wheat feed flour	5.0
Blood meal, ring dried	10.0
Fish oil	9.5
Vitamin premix[a]	0.4
Mineral premix[b]	0.1
Ascorbic acid	0.075
Choline chloride, 50%	0.176
Pellet binder	2.0

[a] Vitamin premix (per kilogram of feed): d-calcium pantothenate, 105.8 mg; pyridoxine, 30.9 mg; riboflavin, 52.9 mg; niacinamide, 220.5 mg; folacin, 8.8 mg; thiamine, 35.3 mg; vitamin B_{12}, 0.02 mg; menadione sodium bisulfite complex, 11 mg; vitamin E, 353 IU; vitamin D_3, 441 IU; vitamin A, 6613 IU.

[b] Mineral premix (as a percentage of mix): $ZnSO_4$–H_2O, 41.2%; $MnSO_4$–H_2O, 12.3%; KIO_3, 0.17%; $CuSO_4$, 0.77%; inert carrier, 45.56%.

Source: G. Reinitz. Performance of rainbow trout as affected by amount of dietary protein and feeding rate, *Progressive Fish-Culturist,* 49 (1987), pp. 81–86.

Diet 23 for the Milkfish, *Chanos chanos*

Ingredient	Percent
Soybean meal	50.26
Corn starch	39.49
Corn oil	2.00
Alphacel	2.00
Cyphos	2.00
Pellet binder (Permapel)	1.25
Salt	0.70
Mineral mix[a]	0.50
Vitamin mix[b]	1.80

[a] Mineral mix in corn starch: $CuSO_4$, 20 ppm; $MnSO_4$, 95 ppm; $ZnSO_4$, 135 ppm; KIO_3, 10 ppm.

[b] Vitamin mix in corn starch: vitamin C, 65 ppm; biotin 0.1, ppm; folic acid, 5 ppm; vitamin A, 5500 IU/kg; pyridoxine, 20 ppm; riboflavine, 20 ppm; thiamin, 20 ppm; vitamin E, 50 IU/kg; vitamin K, 10 ppm; vitamin B_{12}, 0.02 ppm; vitamin D_3, 1000 IU/kg.

Source: B. S. Shigemoto, E. Ross, R. W. Stanley, J. R. Carpenter, and R. A. Shleser. The effect of protein levels on growth of the milkfish *Chanos chanos, Proceedings of the Warmwater Fish Culture Workshop,* Special Publication No. 3, World Mariculture Society, Baton Rouge, LA, 1983, pp. 166–171. Copyright © World Aquaculture Society. Reprinted by permission.

Diet AA for Tilapia

Ingredient	Percent
Brown fishmeal (herring type)	57.0
Corn starch	10.0
Dextrin	10.0
α-cellulose	6.375
Binder (Na-carboxymethyl cellulose)	2.0
Chromic oxide	0.5
Cod-liver oil	4.0
L-ascorbic acid	0.125
Mineral mix[a]	2.0
Vitamin mix[b]	2.0

[a] Mineral mix (as a percentage of mix): $CaH_4(PO_4)_2$–H_2O, 72.78%; $MgSO_4$–$7H_2O$, 12.75%; NaCl, 6%; $FeSO_4$–$7H_2O$, 2.5%; $ZnSO_4$–$7H_2O$, 0.5%; $MnSO_4$–$4H_2O$, 0.25%; $CuSO_4$–$5H_2O$, 0.08%; $CoSO_4$–$7H_2O$, 0.05%; $CaIO_3$–$6H_2O$, 0.03%; $CrCl_3$–$6H_2O$, 0.01%.

[b] Vitamin mix (per 100 g diet): vitamin A acetate, 2000 IU; cholecaliferol, 1000 IU; α-tocopherol acetate, 10 mg; thiamine hydrochloride, 5 mg; riboflavin, 5 mg; pyridoxine hydrochloride, 4 mg; vitamin B_{12}, 0.01 mg; calcium pantothenate, 10 mg; folic acid, 1.5 mg; menadione, 4 mg; p-aminobenzoic acid, 5 mg; inositol, 200 mg; niacin, 20 mg; biotin, 0.6 mg; choline chloride, 400 mg. Alpha-cellulose was the carrier.

Source: A. K. Soliman, K. Jauncey, and R. J. Roberts. The effects of varying forms of dietary ascorbic acid on the nutrition of juvenile tilapias (*Oreochromis niloticus*), *Aquaculture*, 52 (1986), pp. 1–10. Reprinted by persmission of Elsevier Science Publishers.

Diet for *Cancer irroratus*

Ingredient	Percent Dry Weight
Fish meal	25.0
Ground yellow corn meal	25.0
Soybean meal	15.0
Wheat bran	5.0
Wheat middlings	5.0
Corn fermentation solubles	5.0
Crab meal	4.5
Alfalfa	3.0
Bone meal	1.5
Vitamin premix[a]	1.0
Menhaden fish oil	5.0
Gelatin	5.0

[a] Vitamin premix (per kilogram of final feed): vitamin B_{12}, 5.4 mg; vitamin K, 1.8 g; riboflavin, 2.7 g; niacin, 22.7 g; pantothenic acid, 4.5 g; pyridoxine, 910 mg; folic acid, 360 mg; selenium, 41.2 mg; vitamin A, 2300 IU; vitamin D_3, 900,000 IU; vitamin E, 2300 IU.

Source: S. Rebach. Pelletized diet for rock crabs, *Progressive Fish-Culturist*, 43 (1981), pp. 148–150.

Dry Salmon Diet

Ingredient	Percent
Fish carcass meal	44.5
Dried whey product	17.0
Wheat germ meal	16.5
Cottonseed meal	15.0
Soybean meal	6.0
Vitamin supplement[a]	1.0

[a] Vitamin supplement (per 453.60 g): thiamine mononitrate, 0.15 g; riboflavin, 0.69 g; pyridoxine hydrochloride, 0.30 g; niacin, 4.77 g; d-pantothenic acid, 0.68 g; inositol, 13.65 g; biotin, 0.03 g; folic acid, 0.10 g; DL-α-tocopherol acetate, 10.50 g (10,500 IU); ascorbic acid, 25.50 g; carrier (wheat middlings or cottonseed meal, passed through a U.E. sieve No. 30), 397.23 g.

Source: L. G. Fowler, and R. E. Burrows, The Abernathy salmon diet, *Progressive Fish-Culturist*, 33, (1971), pp. 67–75.

Diet 4 for Young Shrimp

Ingredient	Percent
Glucose	5.6
Sucrose	10.0
Starch	4.0
Chitin	4.0
Glucosamine	1.5
Powdered cellulose	4.0
Purified soybean protein	50.0
Methionine	1.0
Tryptophan	0.2
Glutamic acid	0.2
Glycine	0.1
Citric acid	0.3
Succinic acid	0.3
Refined soybean oil	8.0
Salt mixture[a]	7.7
Vitamin mixture[b]	2.6
Cholesterol	0.5

Diets were added to agar and water for feeding.

[a] Salt mixture (percentage): K_2HPO_4, 30.0%; KCl, 9.4%; $MgSO_4$, 14.8%; $FeSO_4-7H_2O$, 1.4%; $Ca_3(PO_4)_2$, 27.4%; $MnSO_4-7H_2O$, 0.2%; $CaCO_3$, 16.8%.

[b] Vitamin mixture (mg/kg dry diet): p-aminobenzoic acid, 50; biotin, 2; inositol, 2000; nicotinic acid, 200; Ca-pantothenate, 300; pyridoxine HCl, 60; riboflavin, 40; thiamine HCl, 20; menadione, 20; β-carotene, 48; α-tocopherol, 100; vitamin B_{12}, 0.4; calciferol, 6; ascorbic acid, 20,000; folic acid, 4; choline chloride, 3000.

Source: A. Kanazawa, M. Shimaya, M. Kawasaki, and K. Kashiwada, Nutritional requirements of prawn— I. Feeding on artificial diet, *Bulletin of the Japanese Society of Scientific Fisheries*, 36, (1970) pp. 949–954. Reprinted by permission.

Diet CD2 for Crayfish

Ingredient	Percent
Shrimp meal	25.0
Full fat soybean meal	18.0
Fish meal	12.0
Distillers dried grains	8.0
Wheat gluten	8.0
Rice bran	6.0
Whole wheat flour	5.0
Casein (high nitrogen)	5.0
Cod liver oil with astaxanthin	5.0
Cellulose	3.5
Refined soy lecithin	2.0
Phytosterol mix	1.5
Vitamin mix BML-2[a]	1.0

[a] Vitamin mix BML-2 (percent): thiamin mononitrate, 0.5%; riboflavin, 0.8%; nicotinic acid, 2.6%; Ca-pantothenate, 1.5%; pyridoxine HCl, 0.3%; vitamin B_{12}, 0.1%; folacin, 0.5%; biotin, 0.1%; inositol, 18.0%; ascorbic acid, 12.5%; para-aminobenzoic acid, 3.0%; cellulose, 60.0%; BHA, 0.1%.

Source: L. R. D'Abramo, J. S. Wright, K. H. Wright, C. E. Bordner, and D. E. Conklin, Sterol requirements of cultured juvenile crayfish, *Pacifastacus leniusculus, Aquaculture*, 49, (1985), pp. 245–255. Reprinted by permission of Elsevier Science Publishers.

Diet 18 for Aquarium Fish

Ingredient	Percent
Shrimp meal	15.0
Fish meal	15.0
Soybean meal	44.8
Rice polishings	10.3
Wheat bran	10.3
Fish oil	4.0
Vitamin premix[a]	0.5
Marigold petal meal	0.13
Ethoxyquin antioxidant	0.03

[a] Vitamin premix (milligrams per kilogram): α-tocopherol acetate, 55.0; ascorbic acid, 440.0; inositol, 110.0; choline chloride, 550.0; menadione Na-bisulfite, 11.0; niacin, 148.5; riboflavin, 19.5; pyridoxine HCl, 22.0; thiamine HCl, 22.0; Ca-pantothenate, 39.6; biotin, 0.44; folic acid, 4.95; vitamin B_{12}, 0.02; vitamin A palmitate, 5550 IU/kg; vitamin D_3, 1100 IU/kg.

Source: M. Boonyaratpalin and R. T. Lovell, Diet preparation for aquarium fish, *Aquaculture*, 12 (1977), pp. 53–62. Reprinted by permission of Elsevier Science Publishers.

APPENDIX 5

FEED ORGANISMS

Although much progress has been made in the development of artificial feeds for aquatic organisms, many times it is necessary to use live feeds. This is the case for filter feeding organisms such as bivalve mollusks, and for larval fish and shellfish. Unfortunately, the physiology and culture of these **feed organisms** have received less attention than they deserve, which has given rise to variable results reported in the scientific literature with respect to organisms' value as feed.

Initial stocks of feed organisms should be obtained from a reliable source, as most culturists would find it extremely difficult to spot misidentified microscopic organisms. For example, much time and effort would be wasted trying to grow *Dunaliella salina* in enriched seawater with a low salinity if a culturist thought she had actually purchased and was growing *Dunaliella tertiolecta*.

The choice of live feed is critical—the wrong selection may result in costly reductions in productivity and survival of the crop. Feed organisms should be selected based on their size, nutritional value, and ease of culture. In many instances, the culture conditions in which the feed organism is grown will determine its nutritional value to the primary crop organism.

MICROALGAE

Microalgae are single-celled or form small colonies of similar cells; the seaweeds are referred to as macroalgae because they form large colonies with specialized cells.

The blue-green algae, of the phylum **Cyanobacteria** or Cyanophyta, are not really an "algae" since they have no organized nucleus and are therefore considered members of the kingdom Monera. The Cyanobacteria are more properly thought of as an evolutionary bridge between algae and bacteria. One type, *Spirulina*, is cultured for direct consumption, not as a feed organism; therefore, *Spirulina* was discussed in Chapter 7. All the other "algae" discussed in this chapter are members of the kingdom Protista (that is, they have an organized nucleus).

Many phyla of algae are of limited interest to culturists since they are rarely, if ever, grown. Sometimes, however, they take on significance as contaminants in cultures of desired species. These include members of the phyla Xanthophyta, Chrysophyta (sometimes called golden-brown algae), Euglenophyta, Charophyta, Prasinophyta, and Cryptophyta. Members of one phylum in particular, the

Dinophyta or Pyrrophyta, are of particular interest as pests. These **dinoflagellates** are well known in many coastal areas where certain species cause **red tides.** Dinoflagellates are almost always single-celled and have two flagella and an eyespot(s). There are soil and freshwater species, but most are found in brackish or marine waters. Dinoflagellates can cause severe problems in algal cultures because they have the capacity to grow very quickly and therefore rob the desired species of space, nutrients, and sunlight; in addition, some dinoflagellates can secrete toxins, and others are actually predators and graze down the other types of algae.

Three phyla include essentially all the microalgae used as feed in aquaculture. The first of these, the phylum **Haptophyta,** is a small group whose members have two flagella and a small structure called the **haptonema.** The important genera of this phylum for the aquaculturist are *Isochrysis, Monochrysis,* and *Pavlova.* A tropical strain of *Isochrysis galbana,* known as clone **T-ISO,** has become popular recently because it grows very quickly over a wide range of environmental conditions.

The green algae (phylum **Chlorophyta**) are also important. Some members of this group are unicellular, others form small colonies, and some seaweeds are included. This phylum varies greatly with regard to reproduction, structure, and ecology. Three of the commonly reared genera are *Dunaliella, Tetraselmis,* and *Chlorella.*

Probably the most important microalgae for the culturist are members of the phylum **Bacillariophyta,** commonly called the **diatoms,** that have a distinctive cell covering composed of silica (see Figure 5a.1). As was the case for the green algae, this group has made many adaptations resulting in a number of reproductive and ecological variants. The diatoms are subdivided into two groups: those that are elongate, the **pennate diatoms,** and **centric diatoms,** those with cells that,

when viewed from the top, are circular, oval, triangular, or polygonal. Some of the cultured pennates are *Phaeodactylum, Navicula, Amphora,* and *Nitzschia;* centrics include *Cyclotella, Chaetoceros,* and *Skeletonema.*

Characteristics of Algal Growth in Culture

Algal cells are generally cultured in one of two ways. They are either grown in **continuous cultures** where cells are constantly being removed and replaced with fresh media, or they are grown in **batch cultures.** In batch culture, the medium is inoculated with algal cells, and the culture is harvested all at once after a certain amount of cell reproduction has taken place. While the dynamics of continuous cultures are often complex, the batch culture's growth phases are simple and well documented. They consist of five stages (see Figure 5a.2):

1. **Lag phase,** immediately after inoculation, during which there is little or no apparent change in the number of cells in the culture.
2. **Exponential** or **log phase,** during which there is a rapid increase in the number of cells in culture.
3. **A declining growth phase.**
4. **Stationary phase,** during which, again, there is no apparent change in the number of cells.
5. **Death phase,** when the number of cells declines.

The lag phase may be real in some cases, the cells being unable to divide because of the shock of the change in the medium conditions or the physiological state of the algae. However, the lag phase is sometimes only an observational phenomenon during which time the cells are dividing but the initial number of algal cells was so small that their change in numbers is not easily seen.

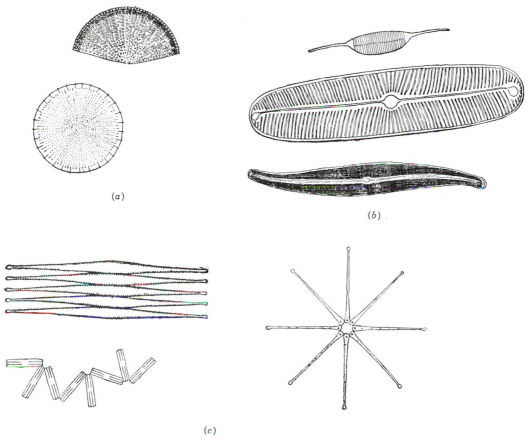

Figure 5a.1 Examples of diatoms. (*a*) Centric, (*b*) pennate, and (*c*) chain-forming diatoms. (From Ward and Whipple, 1918. After Schröter, Smith, and Wolle.)

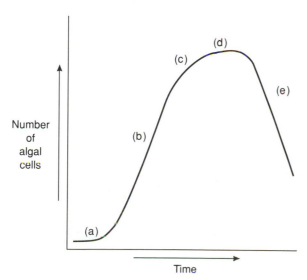

Figure 5a.2 Phases of an algal culture. (*a*) Lag, (*b*) log, (*c*) slowing growth, (*d*) stationary, and (*e*) death.

Growth during the exponential phase that is not limited by light or nutrients can be expressed by the equation

$$W = W_o e^{kt} \qquad \textbf{(Eq. 1)}$$

in which W_o is the algal biomass present in the culture medium at the time of the inoculation, t is the amount of time that the culture has been growing since inoculation, e is the base of the natural logarithms (2.718), k is the relative growth constant, which is a measure of the efficiency of that species in that particular medium under a fixed set of conditions, and W is the algal biomass present in the medium after time t. In large populations of cells such as is normally seen in aquaculture operations, we can express the increase in the number of cells (N) as

$$N = N_o e^{kt} \qquad \textbf{(Eq. 2)}$$

Another often-used expression of algal growth is the **doubling time, G,** which is the average amount of time that an algal cell in the culture medium will take to divide after the previous division. For example, if we inoculate a sterile medium with 1 million cells and there are 2 million cells 12 hours later, then G is 12 hours. This can be calculated using the formula

$$G = \frac{0.301}{k'} \qquad \textbf{(Eq. 3)}$$

and in turn, k' is calculated by

$$k' = \frac{\log N - \log N_o}{t} \qquad \textbf{(Eq. 4)}$$

The value of k' is a function of the physiology and morphology of the algal species in question. Generally, when conditions are optimal for a species, k' will be a function of the size of the cell. Large cells have a reduced surface/volume ratio, so smaller cells are able to take up more

nutrients through the cell surface in a given period of time in relation to their weight. Thus, small species often grow more quickly than large ones. When cells form colonies, they further reduce their surface area relative to the volume.

The culturist should remember that for two very different species of algae, the optimal growing conditions are probably also very different with respect to the types and amounts of nutrients added, as well as temperature and other factors. Growth is also strongly correlated to light, as might be expected for photosynthetic organisms. When light is limiting, growth proportionally declines with light intensity; when light is saturating, the growth of algae is independent of light. Some species can adapt quickly to high or low light levels by changing the amount of chlorophyll in the cells.

Following the exponential growth phase of number, of cells in the culture may continue to rise, but at a diminished rate. Primary reasons for the end of the exponential phase may include

1. Changes in the concentrations of nutrients.
2. Self-shading, which reduces the amount of light to which cells in the culture are exposed.
3. Changes in the medium that are the result of the culture itself, including changes in pH, or the build up of metabolic waste products, or substances called **autoinhibitors** that are secreted by some species.

The stationary phase of the culture may last for a long time or may be essentially nonexistent, and the culture will quickly decline. This varies with the species and the culture conditions.

Batch culture works for some aquaculture operations, but the use of continuous cultures is often more practical and desirable. The physiological conditions of the cells in batch cultures change as they leave one phase and enter another, but when a continuous culture is maintained, the cells always can be harvested during the growth phase and the culture retains its physiological status over an extended period of time. Continuous culture is usually carried out using devices called **chemostats** and **turbidostats** (see Figure 5a.3). Using some sort of overflow device, fresh medium enters the algal culture and displaces some of the old medium containing the algal cells. The chemostat has a pump that adds new medium at a continuous, fixed rate, while the turbidostat has a photoelectric device that controls the flow of fresh medium as a function of the density (=turbidity) of the culture so the flow is not continuous.

To understand the dynamics of a chemostat, let us first consider the imaginary case of particles, which unlike algal cells, are not growing and reproducing. For a set amount of time, t, the number of particles in the chemostat that are carried out, N, will be a function of the dilution rate, D. In turn, D is the ratio of the flow rate, f, of the new media going into the vessel to the volume, v, of the vessel, that is, $D = f/v$. Therefore, the rate of change is

$$\frac{dN}{dt} = N\left(\frac{f}{v}\right) = ND \qquad \textbf{(Eq. 5)}$$

Now suppose that in place of the particles we use algal cells that are multiplying at a

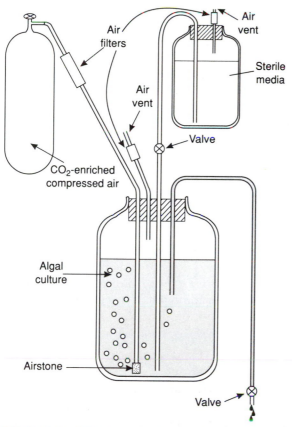

Figure 5a.3 A simple system for continuous production of algae. Portions of the culture are removed regularly by opening the lower valve; after some of the old medium with algae is removed, the lower valve is closed and the upper valve is opened to replace the medium. All air entering or leaving the system is filtered.

constant rate, g. The rate of increase of the cells at any instant is Ng. But since cells are being washed away as well as being produced, the rate of change in the chemostat is

$$\frac{dN}{dt} = Ng - ND \qquad \textbf{(Eq. 6)}$$

We may also express this in terms of biomass, x, in which case Eq. 6 becomes

$$\frac{dx}{dt} = xg - xD \qquad \textbf{(Eq. 7)}$$

If the growth rate is equal to the dilution rate ($g = D$), then there is no net change in the biomass of algal cells in the chemostat during the time period t. That is, $dx/dt = 0$. This is considered a steady state, and that is the condition normally sought.

Equations have been generated that show, theoretically, how to get the maximum production from a system in which the algae is continuously harvested. These are based on the optimum dilution rate and the nutrient concentration in the incoming media. In practice, these equations are rarely used, and culture manipulations are determined empirically after a series of repeated trials.

Algal Media

The growth of algae is a function of the medium in which it is grown. The ideal medium gives the culturist maximum growth in a relatively short period of time for the desired species, inhibits the growth of possible contaminants (although this is often very difficult), and is cheap and easy to prepare. It would serve the culturist well to establish the best species of algae for her needs and then try to find the best medium for that particular species; one need not use only published media, and it is certainly easier to conduct initial small-scale culture experiments early on to develop the best medium under a given set of conditions than it is to switch later after mass production has begun.

Media also affects the composition of the algae. Experimental evidence suggests that this happens in two ways. First, algae can take up compounds and ions in the water and carry them to the primary culture species. This may be the case with media containing certain trace metals; it was shown experimentally that such media will make *Isochysis* and *Dunaliella* significantly more nutritious for brine shrimp, although those metals did not alter appreciably the algae's growth rate. Second, and perhaps more commonly, the medium can alter the basic metabolism of the algae so its cell composition will change. For example, biologists showed that a relatively nutrient-poor medium resulted in cells of *Dunaliella* and *Tetraselmis* that were lower in protein and higher in carbohydrate than those grown in a standard medium; interestingly, these high-carbohydrate cells were better able to sustain growth of juvenile oysters. Despite the importance of media composition on algal chemistry, it is often overlooked by those growing algae. A similar phenomenon may also occur as temperature and light conditions change. It is known that the temperature at which organisms are grown affects their fatty acid composition (as temperatures decrease, so does the degree of fatty acid saturation).

There are two major considerations to keep in mind when designing a medium for a specific alga:

1. pH and the ionic composition of the medium
2. Nutrients that the cells require

In the laboratory, a culture medium can be made from distilled water and pure chemicals from stockroom bottles (see Table 5a.1), but the culturist working on a large scale will have to use the water

TABLE 5a.1 Composition of an Enrichment Medium Used to Grow *Tetraselmis* to Be Fed to Barnacle Nauplii[a]

	Weight
Macronutrients	
$NaNO_3$	200 mg
Na glycerophosphate	100 mg
Liver extract	5 mg
Yeast extract	25 mg
Trace Metals	
Fe:NaEDTA	1 mg
$FeCl_3–6H_2O$	3 mg
$CuSO_4–5H_2O$	0.01 mg
$ZnSO_4–7H_2O$	0.02 mg
$CoCl_2–6H_2O$	0.01 mg
$MnCl_2–4H_2O$	0.2 mg
$Na_2MoO_4–2H_2O$	0.01 mg
Vitamins	
Thiamin–HCl	10 mg
Ca pantothenate	20 mg
Biotin	1 μg
B_{12}	0.5 μg
Seawater	to 1 liter

[a] Adjust pH to 8.0.

available at the culture facility. It is wise to have a laboratory run some standard tests on that water; these would include salinity or hardness, nutrients, pH, perhaps an examination of some of the possible inorganic and organic contaminants that have been reported in the area, as well as an identification of the native algae species.

In most cases, the nutrients that limit growth of algae are nitrogen and phosphorus. While ratios vary for different species under different conditions, in most cases microalgae will use 10 to 20 times as many nitrogen molecules as phosphorus molecules. Nitrogen and phosphorus (along with silica when the species to be grown is a diatom) are considered the **macronutrients. Micronutrients** are metals, such as iron or manganese, or vitamins. There are a number of well-designed general media that can be used for a variety of species.

Normally, a laboratory keeps several test tubes of each species of alga that is of interest; often these **prime cultures,** or **stock cultures,** will be in a medium containing agar. Prime cultures are unialgal and whenever possible should also be bacteria free, or **axenic.** Duplicates of the prime cultures should be kept in several places in case one is destroyed accidentally. To begin a mass culture, the cells from one of the prime cultures are transferred to a larger vessel, which eventually is used to start a still larger culture, and so on until the largest vessels have received their initial algal inoculum. These transfers of cells are done when the population is in the exponential growth phase.

Water for laboratory cultures is often filtered to remove wild algal species, as well as zooplankton that might graze down the cells. Water for prime cultures may be autoclaved, and is sometimes pretreated by aging in the dark or by mixing with small amounts of powdered charcoal. For large-scale cultures, the water is pretreated only by filtration and/or UV radiation. The macronutrients, trace metals, and vitamins are added to the seawater just before the algal inoculation.

Culture Facilities

Small laboratory cultures of algae are best kept in flasks or tubes that are easily cleaned and autoclaved, and can be partially closed to reduce bacterial contamination. Screw-top tubes, or Erlenmeyer flasks with caps or cotton plugs, are most common. Fluorescent lights are normally used as a light source, and although special lights have been designed to give spectral emissions similar to "daylight," they are often very expensive, and there is little evidence that they yield superior growth of algae. The amount of light needed varies as a function of

1. Species
2. Size of the culture

3. Glassware used
4. Density of the cell cultures
5. Duration of the illumination

Illumination of laboratory cultures typically varies from 1000 to 10,000 lux. The photoperiod is normally controlled by electric timers and extends from 12 to 24 hours per day in most laboratories; not all species of algae can live in continuous illumination, although many of the commonly cultured species can (see Figure 5a.4).

Temperature must be controlled in the laboratory just as illumination is. Some species grow best in cool conditions, and virtually no algae exist at temperatures above 35°C. Most of the commercially cultured species grow best at 18°–22°C. Indoor algae cultures may be kept in incubators or rooms that are temperature controlled with fans or air conditioning; some culturists use water baths to regulate temperatures, but this is generally too expensive and uses too much space.

Laboratory cultures are often aerated. This has a dual function. First, the aeration keeps the cells suspended so that the light is used more efficiently, and second, it supplies the carbon, in the form of CO_2, to the plants. Compressed air may be used, but since it is low in CO_2 (0.03%), a mixture of CO_2 (3% to 6%) in air is often used for dense cultures, but the pH must be monitored to be sure that it does not decline below acceptable levels for survival. Aeration should be gentle immediately after the culture receives an inoculation and can be increased as the culture grows.

The mass culture of algae for aquaculture is usually done by continuous culture rather than batch culture (although batch method is used to produce the inoculums for continuous culture) (see Figure 5a.5). Chemostats that are made of 20 liter clear cylinders are one popular way to produce large numbers of algal cells in continuous culture. A variation on this theme includes the use of large (50 liters) transparent plastic bags that are easy to clean, cheap, and less heavy than cylinders.

Larger cultures must be grown outside in ponds, pools, flumes, greenhouses, or

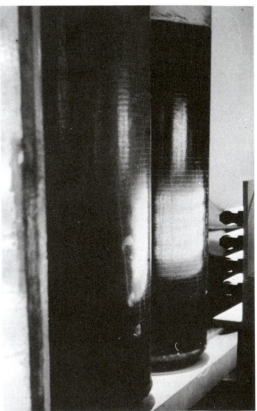

Figure 5a.5 Large plastic tubes used to grow algae that eventually will be used to inoculate a larger outdoor culture. (Photo courtesy of The Oceanic Institute, Hawaii.)

Figure 5a.4 Cultures of algae kept in a temperature- and light-controlled room.

raceways (see Figure 5a.6). Such outdoor facilities are sometimes difficult to maintain because of outside contamination, circulation problems, and variable temperature and light conditions. Cultures should be kept relatively shallow so that light can penetrate to the algae near the bottom. However, if a culture is too shallow, the culturist is not making efficient use of the land; a reduction in depth also means that the volume is reduced, which may result in temperatures that vary too much with ambient conditions. Circulation is critical to maintaining a suspension of the cells (to make the best use of the nutrients and light). This can be accomplished by aeration that moves the water, or the use of airlifts, foils, paddlewheels, or other devices.

Microalgae Products

Recently, there has been great interest in the production of certain strains of microalgae because of the natural products that are extractable. *Dunaliella salina* (see Figure 5a.7) is being grown because certain strains produce significant amounts of **β-carotene;** in fact, this molecule may make up as much as one-fifth of the algae's dry weight. Beta-carotene is used as a food coloring and as a vitamin supplement, and may have anticancer properties.

Several long-chain polyunsaturated fatty acids are also made in large quanti-

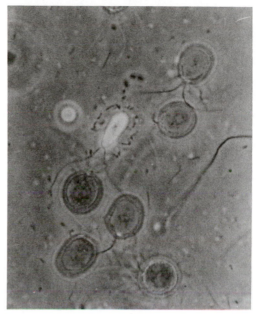

Figure 5a.7 Beta-carotene-rich *Dunaliella*. (From Sommer, 1988. Copyright © Archill River Corporation. Reprinted by permission.)

ties by some algae. Some of these may affect cholesterol levels in the blood. Some vitamins, polysaccharides, and coloring agents may be economically removed, although much work is yet to be done.

Researchers in India have suggested that nitrogen-fixing cyanobacteria cultures can be dried into flakes and used as a biofertilizer in rice fields (see Figure 5a.8).

Figure 5a.6 Algal plant in Czechoslovakia made of glass plates. The three plates shown have an area of 1000 m². (Photo from Richmond, 1986. Courtesy of Dr. A. Richmond, Israel. Copyright © CRC Press, Florida. Reprinted by permission.)

Figure 5a.8 Dried flakes of cyanobacteria that can be used in rice fields. (Photo from Richmond, 1986. Courtesy of Dr. L. V. Venkataraman, India. Copyright © CRC Press, Florida. Reprinted by permission.)

The cyanobacteria establish themselves in the rice field and improve the quality of the soil and, therefore, the size of the crop.

ROTIFERS

Rotifers, or "wheel animals," are a phylum of small organisms that are primarily found in freshwater, although species from brackish and marine environments are not uncommon (see Figure 5a.9). For the aquaculturist, rotifers offer several advantages as feed organisms

1. They reproduce quickly; it is estimated that populations under favorable conditions can double every one to five days. (One species, in intensive laboratory culture, has recently been reported to have a doubling rate of less than nine hours!)

2. Rotifers are small and therefore will be accepted by some organisms that cannot injest larger zooplankton; thus they are an important "first food" for many fish and crustaceans.

3. They are nutritious, and their actual nutritional value can be improved, as it can for other zooplankton, by **packing** the rotifer with specific strains of algae or other feed.

Problems do exist for the rotifer culturist. In particular, it is sometimes difficult to synchronize rotifer production and demand. One thing that can be done about this is to be alerted when problems first appear in the culture; if a sample of the culture is examined under a microscope and either the rotifers are swimming more slowly than usual or there are few eggs on the females, then there is probably a water-quality problem. It has been suggested that since rotifers, like brine shrimp, produce resting stage eggs, it may be possible to mass produce these resting eggs for times when the laboratory population is unable to keep up with the demand. However, unlike brine shrimp, rotifer eggs hatch into females that begin to reproduce quickly; therefore the eggs are best used to inoculate new cultures rather than used directly as feed. The production of resting stage eggs can be con-

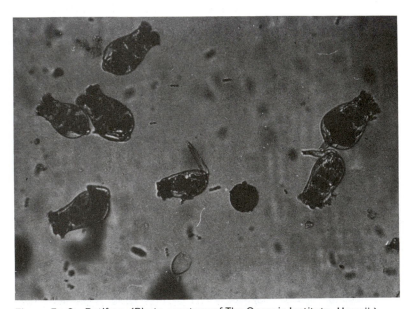

Figure 5a.9 Rotifers. (Photo courtesy of The Oceanic Institute, Hawaii.)

trolled to some extent by manipulations of the temperature and the diet.

Brachionus plicatilis is probably the most commonly grown of the marine species. At least two genetic strains are being cultured that differ in size; the most commonly cultured of the strains is larger with only 1% to 2% of the animals smaller than 150 microns, while of the smaller strain, originally isolated from salt ponds in Israel, 80% are less than 150 microns.

Several systems have been devised for the laboratory rearing of *Brachionus*. One system uses fiberglass tanks, kept in open sheds and filled with seawater, that are stocked initially with rotifers at a density of 1/ml. The rotifers are fed a suspension of torulose yeast that is thoroughly mixed in water before being added. Several hours after feeding, the rotifers rise to the surface at which time they can be skimmed off. If a very dense collection of rotifers is needed, the skimmed water containing the animals can be diluted 1:1 with deionized water, causing the rotifers to sink so they can be removed in a more concentrated form. (Interestingly, not all the rotifers sink under these conditions; those bearing eggs eventually float to the top and can be added back into the fiberglass tanks.)

Culturists in Italy have developed a system that consists of the co-culture of *Brachionus* with algae in plastic bags. Yeast is added after the rotifer population begins to grow quickly. Of the rotifers produced in this system, 88% are grown on yeast and only 12% on the more expensive live algae. Using yeast solely, however, causes a decline in the rotifer's reproductive rate.

It has been shown that the species of algae that is fed to the rotifers is important, since it will affect the rate at which the rotifer population grows as well as the chemical composition of the animals. The most commonly used types of feeds are yeast, dried algae, and fresh algae. Some of the algae cultures used successfully are *Dunaliella, Monochrysis, Nannochloris, Exu-*

viella, Isochrysis, and *Phaeodactylum.* In some cases, the amount of available food seems to affect the size of the rotifers and their eggs. Temperature and time of feeding are also important for rotifers and other feed organisms. Under cool conditions, rotifers will have higher levels of fats and carbohydrates than when grown in warm water, and rotifers that are not fed for a period of time before they themselves are used as food will contain significantly lower levels of fat and carbohydrates, although only slightly less protein when compared to the rotifers that had just received their feed.

DAPHNIA

Daphnia, also called the "water flea," live in freshwater and belong to a group of crustaceans called the **cladocerans** (see Figure 5a.10). *Daphnia magna* is the most commonly cultured and also the largest, reaching up to 2.5 mm; *Daphnia pulex* is also frequently grown. For the culturist they serve the same function the brine shrimp does, namely, being an active live feed for those organisms that are too large to be fed rotifers.

Daphnia have long been favorites of biologists because they have the potential to reproduce quickly and are easy to maintain, often being kept in a mixture of pond water, horse manure, and garden soil. Cottonseed oil, wheat bran, and boiled lettuce leaves have also been added to their cultures. The facilities used to grow *Daphnia* are similar to those used for

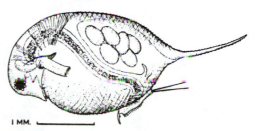

Figure 5a.10 *Daphnia pulex.* (From Ward and Whipple, 1918.)

rotifer culture; these animals can be reared in tanks, cylinders, pools, plastic bags, and so on. Although maintaining a culture of *Daphnia* is fairly simple, it may be difficult to produce consistently high yields, and much effort has gone into the production of the ideal media for their growth. Because they are grown in high densities, and are often co-cultured with algae, it is important to maintain a close watch on the water conditions; even though the animals survive in high levels of ammonia and withstand large changes in pH, their growth and reproductive rates may fall below acceptable levels in poor water. A slow replacement of the culture water over time is probably advisable.

In all likelihood, algae is the ideal feed for *Daphnia. Ankistrodesmus* is known to be a good food species, whereas *Chlorella* and *Chlamydomonas* may be inadequate. Inert feeds can also be used. Micronized defatted rice bran, a very cheap commodity, was found to support reproductive populations, which in six weeks increased in size from 100 animals/liter to 12,000 animals/liter, with a food conversion rate of 1.0.

COPEPODS

Copepods are a very diverse group of crustaceans. Although they exist in all aquatic environments, most of the cultured species are marine (see Figure 5a.11). Planktonic copepods can be collected easily in great numbers with a plankton net in the spring; several collected gravid females placed in a test tube can quickly produce a breeding population.

Numerous laboratory studies on rearing copepods suggest that many species may be grown on a large scale with little difficulty. The Danish Institute for Fisheries and Marine Research has developed a particularly good system for rearing *Acartia tonsa;* many generations of copepods have been grown there in 200 450-liter tanks. In this system, the copepods are kept in concentrations of 50–100/liter, and eggs are collected daily from

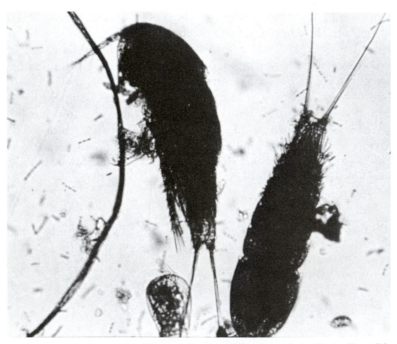

Figure 5a.11 Copepods. (Photo courtesy of The Oceanic Institute, Hawaii.)

gravid females. The myriad other culture methods reported in the scientific literature reflect the diversity of their natural habitats.

One of the most easily cultured copepods is the benthic *Tisbe*. Several species of *Tisbe* have been grown using an assortment of unicellular algae species as food. In addition, *Tisbe* and other benthic copepods, such as *Amphiascella, Nitrocra, Schizopera*, and others, have been cultured on diets of macroalgae and even vegetables (one study shows that they are supported best by leafy species, such as spinach and lettuce, rather than those like peas and potatoes that have a higher caloric content). Populations of copepods have even been kept for up to three years in cornmeal agar. Another commonly grown copepod is *Tigriopus*, which has been fed diatoms, *Dunaliella*, and *Tetraselmis*. *Euterpina* has been grown on a mixture of algae including diatoms, dinoflagellates, and *Tetraselmis*.

Although pelagic species are generally more fragile than benthic species, a number of these have been reared with great success. Some of these include

1. *Centropages*, which has been grown on a diet of *Thalassiosira* and *Prorocentrum*.
2. *Acartia*, fed *Thalassiosira, Chroomonas, Rhodomonas*, and *Isochrysis*.
3. *Labidocera*, reared on the dinoflagellate *Gymnodinium*.
4. *Calanus*, populations of which have been supported by *Thalassiosira*.
5. *Rhincalanus*, a predatory copepod, that has been grown on a diet of mixed diatoms and brine shrimp nauplii.

ARTEMIA, THE BRINE SHRIMP

The most commonly cultured animal to feed to larval fish and crustaceans is the **brine shrimp.** It has been cultured for a long time, although its importance to the aquaculture industry has only been recognized during the last 50 years. There currently exists some confusion among scientists concerning the number of true species of **Artemia;** for our purposes, we will simply refer to each geographical strain as *Artemia* and will not assign a species name. The most commonly used strains come from Canada, the United States (Great Salt Lake and San Francisco Bay), Brazil, Colombia, China, Italy, and Australia. *Artemia* are found in isolated hypersaline bays and lakes over five continents, but they are not found in the oceans. Brine shrimp have been introduced by humans into many areas where they have established reproductive populations. In general, although variations among geographical strains exist, they are very adaptable creatures, capable of living in a wide range of temperatures, in low oxygen environments, and in salinities of 20–340 ppt (although they are usually found only in water with salinities above 45 ppt). Because they are such a hardy organism, they are a very popular animal to rear.

Probably the most important feature to the culturist is their production of **cysts,** or **winter eggs.** When conditions are unfavorable, particularly when oxygen levels are low, *Artemia* do not produce live nauplii; instead they produce cysts. The cysts can be collected in great numbers, dried, and kept for very long periods.

Brine shrimp are often used in the commercial production of salt. Solar saltworks are a series of ponds, each with its increasing salinity and its particular ecosystem. Saltwater is pumped into the first set of ponds where evaporation begins; these ponds are normally 3.5% to 9% salt and contain a rich diversity of bacteria, algae, protozoans, mollusks, crustacea, and some fish and vascular plants. Water is then moved from the low-salinity ponds to intermediate ponds with salinities of 8% to 18%, characterized by low species diversity and high levels of dissolved organic compounds. The most important organisms in these ponds are a cyanobacterium, *Coccochloris*, and brine shrimp. High levels of the *Coccochloris* reduce evaporation and result in a viscous mucilage; a healthy brine shrimp population is needed to graze down this cyanobacteria. Water is finally moved to the high-salinity ponds, 19% to 29% salt, where dissolved organic matter levels are very high and the chief organism in the water column is a red bacteria of the genus *Halobacterium*, often turning the water an orange-red color. *Artemia* die when they pass into this final stage, probably because of a lack of adequate feed algae, and the dissolved material from their bodies seems to stimulate the bacteria's uptake of other organic molecules. The introduction of brine shrimp into intermediate ponds of saltworks has increased the production and quality of the salt.

The various geographical strains of newly hatched *Artemia* nauplii have yielded contradicting results in culture experiments. Although much emphasis has been placed on these differences in strains, the extent of these differences varies from year to year, suggesting that they may be a function of the physiological state of the parent, rather than completely a function of the genetics of the particular strain. Perhaps the most important differences between strains are the fatty acid compositions of the newly hatched nauplii; some strains contain high levels of an acid that is important for freshwater fish but can be detrimental to marine fish, and other strains are high in an acid that is essential for marine fish.

It should be understood that the chemical composition of the nauplii varies as a function of its treatment prior to being given to the primary culture organism. Newly hatched nauplii have a high energy content, but 48 hours after hatching, a large portion of their lipids have been used and their caloric content is lowered. Alternatively, the nauplii can be fed before they are given to the primary culture organism; this not only will alter the composition of the *Artemia*'s tissues, but their gut will be packed with additional nutrients in the form of algae or yeast.

Hatching the cysts is a simple process. The **hatching efficiency,** that is, the percentage that will hatch, varies from dealer to dealer, strain to strain, and year to year. The cysts can be purchased in amounts from a few grams to kilogram quantities. These cysts are very small, 200,000 to 350,000/g; they are usually packaged in airtight containers which, when opened, should be shut again and stored in the cold. The cysts are hatched in gently bubbling seawater that has been diluted to 15–25 ppt. (They will hatch at higher salinities, but do so more quickly in water with a lower salt content.) Hatching begins in 24–72 hours and continues for several days. The newly hatched nauplii can be collected by turning off the air supply, causing the unhatched cysts to sink while the actively swimming nauplii can be siphoned out of the water column (see Figure 5a.12). A weak light directed at the side of the culture vessel can be used to aid in their collection, since the newly hatched animals are positively phototactic. A number of systems have been designed that produce nauplii automatically; these systems may be of great benefit if nauplii are needed at a constant daily rate.

Another technique often used during hatching is **decapsulation,** the removal of the hard shell protecting the cyst. Decapsulation tends to improve the hatching efficiency, speeds the hatching, and dissolves the empty cysts that sometimes get caught in the feeding apparatus of the primary culture organism. The cysts are hydrated in water for about an hour, at which point a dilute hyperchlorite solution is added; after a few minutes, the eggs are rinsed thoroughly and can be either hatched immediately or stored at low temperatures in a saturated salt solution. In situations where feed is needed immediately, the decapsulated cysts themselves can be used.

Figure 5a.12   Brine shrimp nauplii. (From Spotte, 1973.)

Although the brine shrimp are primarily used immediately after hatching for feeding juvenile fish and crustacea, they can be packed with a feed for a brief period to improve their nutritional quality, as mentioned earlier, or they can be grown out to adult animals. Packing may be done with any species of algae that the nauplii can ingest and that has the desired composition; sometimes yeast is also used. Microencapsulated diets, with specific nutrients added, also may be tried.

Research has been conducted on the nutritional requirements of *Artemia*, but the results have been contradictory, perhaps because of bacteria contaminations. Some of the inert products that seem to support good growth are dried *Spirulina*, rice bran, soybean meal, and whey powder.

The choice of an alga to feed brine shrimp is critical since some species may be too large for the young nauplii to ingest. Some of algae species that have consistently yielded acceptable growth are *Dunaliella, Tetraselmis, Nanochloris, Isochrysis, Rhodomonas, Amphora, Cyclotella,* and *Nitzschia.* The exact nutritional requirements of brine shrimp change as they develop, and also seem to be a function of environmental conditions such as salinity.

The mass culture of adult brine shrimp has been accomplished in a number of ways. Several hypersaline bays and lakes have been used for rearing *Artemia.* That body of water must have a suitable species of algae naturally present. The introduction of a particular algal species by the culturist to support the brine shrimp is often difficult, since competition with the native algae is likely to be severe. Nutrients, in some form, are added to the water to encourage the growth of the algae, and the brine shrimp grow and reproduce, and are harvested. In a number of cases, when all the conditions are near optimal, these operations can be highly successful with limited management. More often, the brine shrimp must be raised in raceways or ponds; algae either is cultured together with the *Artemia* in the same facility or is grown separately and pumped into the water containing the brine shrimp.

The use of more complex techniques for high-density cultures has been worked out on a laboratory scale, and reports exist of converting 10 g of cysts into 2 kg of *Artemia* in two weeks, in only 1 m^3 of tank space.

Systems showing particular promise for growout of *Artemia* have recently been demonstrated by biologists at the State University of Ghent, in Belgium, the center for *Artemia* culture research for the last decade.

OTHER LIVE FEEDS

Although algae, *Daphnia*, rotifers, copepods, and *Artemia* are the most commonly cultured live feeds, others have been used. In certain cases, when none of the standard alternatives is satisfactory, new avenues must be explored. These include organisms that are closely related to those mentioned earlier, as well as those that are considerably different.

Among those that are similar are *Moina*, a cladoceran like *Daphnia*. These sometimes can be collected in large numbers, and have been cultivated at times. Some fairy shrimp related to *Artemia* have been reared in freshwater environments and show some promise.

Protozoans have some potential as a small live food since they can reproduce very quickly and do not have the chitin exoskeletons associated with the Crustacea. *Fabrea salina*, a ciliate of 50 to 500 microns, has been grown on diets of *Tetraselmis* and yeast; in raceways, it may reach a density of more than 180/ml and maintain a doubling time of about 13 hours. Other large ciliates can also be used.

Frog tadpoles have been used for fish culture. They are easy to obtain, but some fish do not accept them readily.

The larvae of some fish can be used as feed for other fish, and carp eggs have been acceptable to such fish as the largemouth bass.

The larvae of invertebrates can also be used. These may be collected with a plankton net, or gravid adults are brought into the laboratory where their eggs hatch. Many species of barnacles can be used, since large numbers of the adults often can be collected easily in the same place. Clusters of fertilized barnacle eggs are found in the mantle cavity during the breeding season, and these sacs can be removed merely by breaking the barnacle open. The barnacle nauplii will hatch immediately after being removed if they are far enough along in their development, or otherwise will hatch in a short period of time if kept in clean seawater.

The mysid crustacean, *Mysidopsis*, has been grown in tanks on a diet of newly hatched *Artemia*. Amphipod culture has had some success, especially *Gammarus; Corophium* has been grown using *Rhodomonas* and *Isochrysis*.

Several small **oligochaete worms** are raised as feed for fish. *Tubifex* has been observed to go through its entire life cycle on a diet of tainted lettuce and reaches very high densities in culture. These animals are commonly associated with slow-moving water that has a high organic load, although this is not always the case; recently, very good production has been achieved using quickly-moving water. A similar oligochaete, *Branchiura*, grows well in a mixture of soil and activated sludge. Both worms grow best in warm water.

APPENDIX 6

OTHER CULTURED MOLLUSKS

SCALLOPS

Among the most valuable of the mollusks are the **scallops.** Different species found around the world have been commercially cultured, although nowhere is the industry very large, except in Japan and China.

In Japan, the culture of the **sea scallop, *Patinopecten yessoensis,*** is an outstanding example of how aquaculture can make a rare delicacy available to the general public. In the mid-1960s, a scallop seed collector was developed that allowed the industry to expand rapidly; there are about 10 times the number of scallops available for consumption in Japan now as there were before a culture technique was developed.

P. yessoensis is found on cold water seabeds of mud or gravel in waters from 10 to 30 m deep off the northern coast of Japan. The scallops dig depressions for themselves and are capable of swimming short distances. When water temperatures rise above 20°C in the summer, growth stops; spawning takes place in March and April after they reach two years of age and are about 10 cm in length.

The planktonic larvae of the sea scallops remain in the water for about 40 days; collection of the seed therefore is done in the spring and early summer. The seed collectors are onion bags filled with either worn-out synthetic fishing nets or cedar twigs. The scallops will grow in the collectors to 7–10 mm by mid- to late summer, when they can be harvested and transferred to intermediate culture areas where they are either sown or hung in pearl nets.

When the *Patinopecten* have reached 3–5 cm, they can be moved to a growout area. Here again they may be either sown or hung. Sowing is done in water at least 10 m deep to avoid wave action, and the seabed is firm but not rocky (which would make harvesting by a dredge difficult); predators such as starfish are removed. Several types of hanging culture systems are used, of which the most popular are cylindrical and pocket nets (see Figure 6a.1). The greatest production occurs with cylindrical nets. In Japan, most scallops are boiled before being shipped; few are marketed unprocessed.

The **purple-hinge rock scallop, *Hinnites multirugosus,*** is currently considered a good aquaculture prospect. Its chief ad-

Figure 6a.1 Scallops being cultured in cylindrical nets. (Photo courtesy of Liang Hongwu, Bureau of Aquatic Products, People's Republic of China.)

vantage is the fact that the adductor muscle (often the only part of the scallop that is eaten) represents 40% to 50% of the animal's total weight. *Hinnites* can be spawned almost year-round, and the larvae can be reared by maintaining a temperature of 14°–18°C and feeding them a diet of mixed unicellular algae. This scallop's long larval period is a potential problem for culturists, but wild spat may be collected from bed areas so larval culture may not be required (however, collection of wild spat is usually much less reliable). Unlike other scallops, this species secretes a cement much like an oyster does when it is 20–40 mm in length, and remains in that fixed position for the rest of its life. The attached scallops can be grown in open culture systems and reach a marketable size (100–130 mm) in three years or less.

Another species, currently cultured in Peru, is *Chlamys purpurata*. The center of the industry is the Bay of Paracas, where farmers lease land from the government and use nets, with attached floats on top and rocks on the bottom, to delineate the leased area and to prevent the scallops from migrating. The growout takes place on the bay bottom (or sometimes in nets suspended in the water column). The areas are seeded with small scallops (20–30 mm) collected nearby. There are few natural predators of the scallops near the Bay of Paracas, so these seed animals are collected quite easily in local shallow waters. Should the natural production of seeds decline, this species can be reared in

the laboratory and taken to culture beds at about 10 mm with excellent chances of survival. These scallops grow to 75 mm in about half a year and can then be harvested.

Similar species, *C. farreri* and *C. nobilis*, are grown in China using floating ropes (like those used in seaweed culture) and screen cages. Spat production is artificial, but good results have also been obtained collecting wild seed scallops.

The **bay scallop**, *Arogopectin irradians*, is found off the East Coast of the United States. This is common in many protected waterways, over mud flats, and in eel grass. A hermaphrodite that spawns first as a male and then as a female (although sometimes both gametes are shed together), it can be ripened throughout the year by adjusting the temperature and diet. Despite much interest in this species, growout experiments in open culture systems have yielded mixed results, and there are no major commercial bay scallop culturists.

COCKLES AND "OTHER CLAMS" AND OYSTERS

Cockles are clamlike animals belonging to the molluskan family Arcidae, which includes the often cultured *Anadara*. (The English "cockle," *Cardium*, a member of the Cardiidae family, is not an important culture organism.) *Anadara* culture is carried out largely in Southeast Asia. While *Anadara* is normally collected as seed from the environment, it is possible to induce spawning of ripe cockles by thermal manipulation. The larvae can be reared on a diet of diatoms and *Monochrysis*.

A. granosa, sometimes called the **blood clam,** is found on coastal mudflats; because they have no siphon, they must live in shallow sediments, so they are not as well protected from predators and rapid changes in temperature as are burrowing bivalves. Their preferred habitat is a low-energy tidal flat that is up to 90% silt. *Anadara* can withstand a certain amount of variability in the salinity, but grows best in environments that are 25–30 ppt. This species spawns throughout the year, but there is a peak reproductive season that varies with the local conditions and latitude. Culture of the cockle is carried out in China, India, Malaysia (where it is the most important of the cultured bivalves), the Philippines, and Thailand.

Anadara culture consists of collecting small seed organisms about four months after they settle. The seed cockles are sold to culturists who distribute them in intertidal flats where growth is known to be good. In India and Thailand, the flats are sometimes protected with split bamboo screens. The cockles are harvested in six months to one year, and weigh 7–35 g. Initially the density of the seed cockles may be as high as $2000/m^2$, but they are thinned as they grow to $200—600/m^2$.

In China, the culture method is slightly different; seed cockles are raised from collected spat, then transferred to enclosed pools, and finally moved out to the tidal flats for growout. It takes about a year for the seed to reach the young cockle stage, and another two to three years before they can be marketed at 8–9 g. In Japan, *A. subcrenata* are collected as spat using a device resembling a giant bottle brush made of palm fiber; spat are moved to nursery beds, and are finally sown in a growout area. *A. broughtoni* is grown, to a limited extent, in Korea and Japan.

In recent years, some research has been done on the culture of the **giant clams** of the Indo-Pacific, *Tridacna* and *Hippopus* (see Figure 6a.2). These clams have long been eaten by the people of Southeast Asia and the Pacific, but little was known about their use as food outside this region until recently (although small, dried whole clams or adductor muscles were sometimes exported). All the tissue, except the kidney, is edible. In recent years, for a variety of reasons, many of the

Figure 6a.2 An ocean-based giant clam farm. (Photo from Heslinga and Fitt, 1987. Courtesy of Gerald Heslinga. Copyright © The American Institute of Biological Sciences. Reprinted by permission.)

reef-dwelling stocks have disappeared, which has sparked interest in culture.

The giant clams live in clear waters and house in their mantle tissues one-celled dinoflagellates, *Symbiodinium*, called **zooxanthellae.** The zooxanthellae are symbionts that supply the clam with its food, in return for protection and a source of raw materials. Considering that these clams do not feed (or if they do feed, it is a secondary energy acquiring mechanism), their growth is remarkable. After spending the first one and a half years in a hatchery, *T. gigas* (the biggest of these clams and one of the species with the greatest potential for culture) can reach 6 kg in as little as five years. Cultured *T. derasa*'s meat weight increases fourfold between years 3 and 5, while the shell weight increases seven times. Growth slows after year 5 when the animals become reproductive. This means that if harvesting were to take place after five years, the culturist could produce 22 metric tons of meat and 140 metric tons of shells/hectare/year.

All the giant clams have been spawned in the laboratory using conventional methods; serotonin has also been injected to induce spawning. These clams are hermaphrodites, first releasing sperm and then releasing eggs (15 minutes to over an hour later). The larvae are easily raised on *Isochrysis*.

Increasing the production of the giant

clams is being investigated. Two possible paths to follow are to

1. Add nutrients to the water to increase the growth rate of the zooxanthellae.
2. Select the best strains of zooxanthellae for a species under a particular set of conditions and then inoculate the young clams with that particular dinoflagellate (the clams host the dinoflagellates only after they have begun their benthic existence).

On the Bahama Islands, the culture of two clams, *Asaphis deflorata* and *Codakia orbicularis,* has been suggested. Both these species have been spawned in the laboratory and their larvae reared to the juvenile stage. What makes these animals especially interesting is that they both may get some of their required nutrients as deposit feeders (that is, from the sediment) rather than exclusively as filter feeders. In addition, *C. orbicularis* is found in highly stressed environments (low salinity and dissolved oxygen, high levels of hydrogen sulfide) where culture normally cannot take place; this may be because of the presence of symbiotic bacteria in the gills of this clam that can use the hydrogen sulfide as an energy source. As in the case of the giant clams, symbiotic microorganisms may make this animal profitable to grow in areas were the culture of other bivalves would not be considered.

The **razor clam, *Sinonovacula constricta,*** is the most important of the cultured bivalves in parts of China. Fisheries scientists predict the best time to collect the seed clams, which are then sown into rearing beds. The marketable clams are harvested six months to a year later. While yields of 15 to 30 metric tons per hectare are normal, this may reach up to 80 metric tons.

In Portugal, the **little-neck clam** (sometimes called the **carpet shell clam**), *Venerupis decussata,* is farmed in the southern waters off the coast of Algarve. The juvenile clams are collected in the intertidal zone and are transplanted to beds where they are allowed to grow for several years until they reach a marketable size. The farmer's clam beds are cleared of rocks and smoothed; predators are removed when possible. This transfer to the farmer's beds is done not because the beds are better suited for growth (in fact, they are intertidal just like the collection area), but because the beds are located in an area over which the farmer has control, so no one else can take the clams. The beds are also densely stocked compared to the natural intertidal areas, so harvesting is more efficient. There has been some experimental spawning and larval rearing of this animal. Spawning takes place after a long conditioning period when gametes are added to the water and the temperature is raised. Larvae can be reared on *Isochrysis;* this technique has allowed researchers to produce 5 mm seed clams after three months.

Although it is distributed worldwide, the freshwater clam *Corbicula* is cultured only in Asia where it is used not only as a food, but also as a treatment for liver disease. In Taiwan, seed clams are purchased by farmers who grow them in clean, well-aerated freshwater ponds, water reservoirs, and drainage canals. The seed clams weigh about 0.1 g and are harvested at 5 to 6 g with iron rakes after seven or eight months.

Tapes philippinarum, the short-necked clam, is found throughout Southeast Asia on intertidal and subtidal sand/mud beaches. Farmers in China loosen and smooth juvenile culture beds to encourage the wild larvae to settle; spat can also be produced in a hatchery and distributed over the beds when they reach the "seed size." The spat are allowed to grow for about a year, after which they are collected and transplanted to growout beds where they remain for about one more year. *T. japonica,* the **asari** or **Manila clam,** is the most commonly cultured clam in Japan (see Figure 6a.3). Seed clams are

Figure 6a.3 *Tapes japonica.* (From Conrad, 1985. Copyright © Archill River Corporation. Reprinted by permission.)

collected and transplanted to uncrowded, shallow, sand beds where growth is unhindered. In the spring and fall the seed clams are sown during slack water, so the currents do not carry them off before they settle. The clams are harvested when they reach about 10 g; this takes one to two years. There is also limited culture of this species on the west coast of the United States.

The clams of the genus *Meretrix* are cultured in several parts of Asia. In Taiwan, seed clams weighing about 1 g are collected throughout the year, but particularly in the summer. Since about 1971,

some culturists have begun to farm **subseed clams,** very small clams that are sold to "seed farmers" as a mixture of sand and clams; the sand/clam mixtures are grown in brackish water ponds, sometime fertilized, until the clams reach seed size and are then sold to the "growout farmers" who will produce the marketable product. The seed clams are grown on sandy tidal flats and in old milkfish ponds. The beds are surrounded by bamboo poles and nets that keep the clams from escaping to deeper water with the tide (these clams sometimes float with the aid of a mucilaginous thread that they secrete). The clams are stocked at 2–5 metric tons per hectare and are harvested about one and a half years later at 25–30 g each.

Many other species of clams are cultured or have been suggested as possible candidates. In the United States these include the soft clam (*Mya*), razor clams (*Ensis* and *Siliqua*) (see Figure 6a.4), surf clam (*Spisula*), butter clam (*Saxidomus*), and "littleneck" (*Protothaca*).

In the Philippines, the **windowpane oyster, *Placuna placenta,*** which grows up to 16 cm, is found on muddy bottoms at depths of 0.5–100 m. These mollusks have delicate, translucent shells that are

Figure 6a.4   The razor clam, *Siligua.* (From Weymouth, McMillin, and Holmes, 1925.)

used for window glazing and shell craft. Divers collect seed *Placuna* (25–40 mm) and sell them to farmers who maintain mudflat beds in sheltered areas protected by bamboo poles and barbed wire. The windowpane oysters are harvested in less than a year when they reach about 90 mm; larger shells become opaque and cannot be used.

PEARL CULTURE

Any mollusk that secretes large amounts of iridescent material **(nacre)** can potentially be used for pearl culture, but most cultured pearls are produced by only a few species. In Japan, the center of the pearl industry, these include four marine oysters and a freshwater "mussel" (which is not very closely related to the true mussels). The oysters are the following

1. *Pinctada fucata* is used in China, Japan, and Korea (see Figure 6a.5). It is the major pearl-producing species in Japan and the one that was used to develop the technology for the production of spherical pearls.
2. *P. maxima* is used in Australia,

Burma, Indonesia, and the Philippines. It produces the largest round pearls (to 18 mm in diameter).
3. *P. margaritifera*, known as the **"blacklip oyster,"** is the most suitable for producing the steel-black pearls. It is cultured in Japan, Tahiti, and Fiji.
4. *Pteria penguin*, the **Mabe** in Japan, is also grown in China and Korea. This species is used for the production of large half-round pearls.

For *P. fucata*, the best pearl formation occurs when the pearl's nucleus is inserted in the gonads immediately after the oyster has spawned. Farmers therefore may wish to be able to control the reproductive period; to inhibit gonad maturation, the animals can be placed in cold water, or the spawning can be accelerated so that the gametes are shed early by putting the oysters in a crowded cage placed in warm seawater. Research on the production of sterile triploid pearl oysters, which would eliminate these manipulations, has begun.

To begin pearl culture, the oyster is placed in a shallow tray of seawater. When the oyster begins to open, a wedge is placed between the valves. A piece of undamaged mantle tissue from another oyster is cut into a small square, and is inserted into the gonad along with a piece of shell, the **nucleus,** which has been machined to a rough sphere. The size and shape of the nucleus determines the size and shape of the pearl (the color is a function of the oyster species and the chemistry of the water). Depending on the size of the nuclei, from one to five are used per oyster. If a "half pearl" is desired, the nuclei are placed between the mantle and the shell valve.

After this operation, the oysters are placed in a cage covered by a wire frame; cages are suspended from rafts in calm waters while the oysters recover from the operation. After about six weeks they are

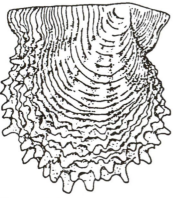

Figure 6a.5 The pearl oyster, *Pinctada fucata*. (From Kafuku and Ikenoue, 1983. Copyright © Elsevier Scientific Publishing Company, New York. Reprinted by permission.)

inspected and the dead oysters removed. The oysters are finally suspended from rafts or long lines for three to four years, being raised periodically to be cleaned of encrusting organisms. The waters that produce the best pearls are often not good for growth; therefore, it is common to start the culture out in a place where the oysters grow well, then to transfer them to an area where the pearl quality will improve. In successful operations, the graft tissue grows over the nuclei and forms **pearl sacs.** The epithelial tissue secretes a **nacreous** (pearly) substance that is deposited on the nuclei.

Hyriopsis schlegeli is cultured in freshwater in Japan. This "mussel" produces a pearl that is variable in shape and size, often of a pink color. *H. schlegeli* is found in some of the lakes of central Japan. It is an interesting species, which may live up to 40 years, and has a life cycle very different from the pearl-producing oysters. The male releases its sperm into

the water; it is taken up by the female, where fertilization and brooding take place. The embryos develop into **glochidum** larvae that are released into the water, where they attach to the gills of fish. In two to three weeks, the mussels release their hold on the fish, drop to the bottom, and begin a free-living existence (see Figure 6a.6).

The *H. schlegeli* are collected and divided into two groups: those that will be operated on and make pearls and those that will be sacrificed (called the **cell mussels**). The mantle of a cell mussel is cut into small pieces and is inserted into holes carefully made in the mantle of the other mussels. Sometimes a nucleus is added along with the mantle from the cell mussel. A pearl sac forms and a pearl begins to be secreted. The operated on mussels are taken to pearl farms where they are hung in a net or basket in a lake for two to three years. The pearls are harvested either by crushing the mussel or removing the

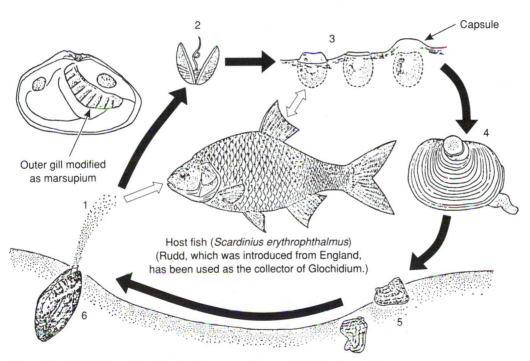

Figure 6a.6   The life cycle of the freshwater pearl mussel. (1) Discharge of glochidium larvae, (2) glochidium, (3) glochidium in the tissue of the fish, (4) the young mussel, (5) young mussels on the bottom, and (6) an adult mussel. (From Kafuku and Ikenoue, Copyright © 1983. Elsevier Scientific Publishing Company, New York. Reprinted by permission.)

pearl carefully and allowing the mussel to live. Because there is a dwindling number of pearl mussels available, the mussels are usually kept alive and may be used again to make pearls after a few years.

In the Mississippi River Valley of the United States, the pearly mussel *Lampsilis claibornensis* (and its relatives) have been harvested as a food and for the production of buttons and small trinkets. Pieces of the shells of these animals are exported to Japan, where they make excellent nuclei for pearl-producing mollusks. Its life cycle is similar to that of *H. schlegeli,* the young being hosted by freshwater fish.

OCTOPUSES AND SQUIDS

The cephalopod mollusks, which include **octopuses** (see Figure 6a.7) and **squids** (along with cuttlefishes and the chambered nautilus), have complex nervous systems that are used by scientists as models for studies in neurophysiology and learning; for this reason, there is a demand for these animals by laboratories. At the moment, however, there is little interest in producing these cephalopods in large culture systems for food (but this may not always be the case). The best culture systems developed so far are those for the laboratory model animals; in all likelihood, the species used as neurophysiology models are not those that would be cultured on a grander scale for food. A number of species are now being reared very successfully at the University of

Figure 6a.7 An octopus. (From Berry, 1912.)

Texas Medical Branch, including *Octopus joubini, O. bimaculoides,* and *Loligo opalescens.*

Regardless of the species, cephalopods share certain common characteristics, two of which have given them the reputation (unjustifiably?) of being difficult to raise

1. Poor or fluctuating water conditions must be avoided, as these organisms are sensitive to temperature changes, pH (especially squid), salinity, nitrogen, DO, turbidity, and many pollutants. For these reasons, closed culture systems may be more practical than open systems (see Figure 6a.8). However, octopus and cuttlefish, at least certain species, seem to be more tolerant than was generally believed in the past.

2. Many cephalopods are particular about the food they eat, often taking only moving fish or crustaceans. An important exception to this are cuttlefish, which can be quickly trained to take frozen fish and shrimp.

O. joubini, the pygmy octopus, is typical of the cephalopods with a benthic lifestyle and large (**telolecithal**) eggs from which come small adultlike hatchlings, rather than larvae that must metamorphose into octopuses. Because it grows quickly under laboratory conditions and is small when it reaches sexual maturity, two generations per year can be raised (this is up to twice as fast as it grows in the wild). The animals should be kept at about 25°C in full seawater, in tanks with shells or pipes that can be used as shelters. One or two small, live crabs or shrimp are fed to each individual per day. When the sexes are mixed, they mate readily at night; the eggs, which are attached to a substrate in the octopus's den, are cared for by the female. Hatching takes place in the evening 30 to 40 days after the eggs are laid. The hatchlings look and behave like small adults; they should be offered

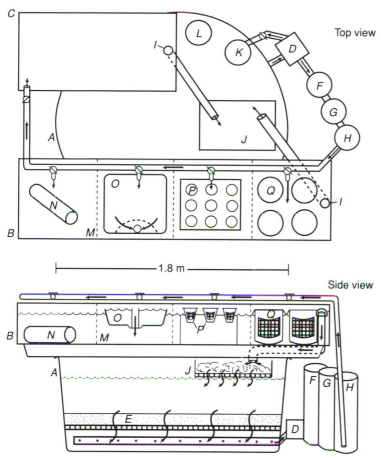

Top view

Side view

|———— 1.8 m ————|

Figure 6a.8   A closed octopus-rearing facility. (A) Water conditioning tank, (B, C) culture trays over the water conditioning tank, (D) pump, (E) biological filter bed, (F, J) mechanical particle filter, (G) activated carbon filter, (H) UV sterilizer, (I) screened standpipes, (K) protein skimmer, (L) cooling and heating unit, (M) screen dividers, (N) dens for octopuses, (O) small trays for hatchings, and (P, Q) individual rearing compartments for older octopuses. (Drawing from Hanlon and Forsythe, 1985. Copyright © American Association for Laboratory Animal Science. Reprinted by permission.)

live food immediately, although they may not accept it for a few days until the yolk has been completely absorbed. *O. joubini* are relatively easy to raise as long as the water conditions remain good, and there is enough live food. They can probably be produced economically as laboratory specimens and perhaps for the saltwater aquarium trade.

  O. bimaculoides is a larger species with an extended life cycle. Hatchlings can be stocked in water at densities of 300 to 700 per square meter and are fed mysid shrimp; as they grow, their diet changes to large organisms, first palaemonid shrimp and then penaeid shrimp. The growth and the life span of *O. bimaculoides* are very much affected by the temperature. When kept at 23°C, they grow more quickly initially, starting at an average weight of 0.07 g and reaching about 0.6 kg in 370 days, but growth of this species at 18°C surpasses that at 23°C as the culture extends into the second year.

  If, as in the case of the bullfrog, methods can be developed to induce the octopus to take nonliving food readily, the culture of these animals will quickly become more widespread. (Larger adults take some frozen and dead food, but this is only done to maintain the octopuses when live food is unavailable). Investigators have shown that some terrestrial and freshwater organisms (crawfish, bivalves, sunfish, and salamanders) can be substituted for marine organisms as feed.

Culture of the squid, *L. opalescens*, starts with capturing healthy, undamaged wild specimens to be transported to the culture facility (see Figure 6a.9). These are placed in large tanks or pens that allow this actively swimming and schooling animal enough room to move easily. The water is kept at 12°–18°C and conditions should be oceanic (salinity 35 ppt, pH 8). They feed on swimming shrimp and fish.

L. opalescens has been reared from egg to sexual maturity in seven months. The sexes are separated before spawning to prevent fighting. To induce spawning, a pair of squid are isolated and an artificial cluster of egg capsules is introduced in the water—this usually elicits mating behavior and egg laying. The eggs are then moved to a basket in a smaller rearing tank, and hatch about three to four weeks later in water of 16°C. The eggs are small, but the squid at hatching appear as fully developed miniature adults. The period of greatest mortality occurs when the young *Loligo* switch from getting their nourishment from the yolk sac to capturing live food. Copepods and then mysids are the chosen foods during early development (unfortunately, brine shrimp are not readily taken by the squid). In contrast to *L. opalescens*, another squid, *Sepioteuthis*, produces large eggs and quickly growing offspring that do not require the complex dietary changes described; recent results have shown that *Sepioteuthis* can be reared in the laboratory from the egg to 2.2 kg in four and a half months.

Unlike the octopuses, squids cannot be sold to most home aquarists, but *Sepioteuthis* has some potential for mass culture for laboratory specimens or even as food. Cuttlefish, which do relatively well in aquaria and are easy to breed, feed, and care for, may become common in home saltwater aquaria in the future.

OPISTHOBRANCHS

The **opisthobranchs** are gastropod mollusks that display various degrees of shell and mantle cavity reduction. Like the cephalopods, these animals are used primarily in laboratory studies of the ner-

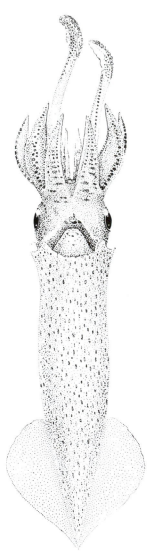

Figure 6a.9 The squid, *Loligo opalescens.* (From Berry, 1912.)

vous system. They have little, if any, chance of being eaten (indeed, for years *Aplysia* was believed to be highly toxic) or even of becoming popular as aquarium pets (see Figure 6a.10). Many different opisthobranchs have been raised in the laboratory, including *Hermissenda, Bursatella, Bulla, Elysia*, and others, but true mass culture has only been achieved for *A. californica*, the **sea hare.**

Mating pairs of *Aplysia*, 50 to 1000 g each, will produce viable embryos year-round when kept at 15°–24°C in seawater (30–33 ppt) and fed seaweed (*Ulva, Agardhiella*, and/or *Porphyra*). The egg masses are removed from the spawning tank, cleaned, and placed in aerated flasks where development continues. Just prior to hatching, the eggs are again washed and placed in a hatching medium along with microalgae (*Isochrysis* and *Chroomonas*), which serves as food for the larvae; the larvae hatch and are transferred, every three to four days, to fresh culture flasks until they are determined (by their pigmentation) to be competent to metamorphose. The premetamorphic larvae are then transferred to a flask containing the red algae *Agardhiella subulata*, which induces the metamorphosis to the juve-

nile form. They are then grown out, in a flow-through system, feeding on *Ulva* and *Agardhiella*. The *Aplysia* normally are used by scientists when they reach about 100 g.

CONCH

A large marine gastropod, the **queen conch *(Strombus gigas)*** is considered to have the potential for being commercially cultured in the Caribbean and the Netherland Antilles. The **milk conch, *S. Costatus,*** is much smaller than the queen conch, but has also been mentioned as an aquaculture candidate. The meat of the conch is exported from the islands to the United States, Europe, and South America, and the shells are prized by collectors around the world. As a result of overfishing, their natural stocks have been depleted, and an interest in conch culture has therefore been generated (see Figure 6a.11).

The females spawn from spring to fall, laying one to four egg masses (up to 3×10^5 eggs per mass) during that season. Eggs are collected from natural waters or from large breeding pens that are constructed to house, in low densities, reproductively mature individuals (sex ratio

Figure 6a.10 The sea hare, *Aplysia*. (Photo courtesy Tom Capo, Howard Hughes Medical Institute, Florida.)

Figure 6a.11 Overfishing has damaged the natural populations of the queen conch. (Photo courtesy of LeRoy Creswell, Harbor Branch Oceanographic Institution, Florida.)

TOP SHELL

The tropical **top shell,** *Trochus niloticus,* is a gastropod mollusk that traditionally has been harvested in the South Pacific as food and for the shell that is used in the production of buttons. The meat can be prepared by salting and/or smoking. As in the case of so many organisms, the natural stock has been shrinking because of overfishing and alteration of its natural habitat, the coral reef.

Broodstock are kept in large concrete tanks with flowing seawater. The *Trochus* graze on the diatoms that grow on the interior surface of the tank. Spawning takes place monthly in the early evening, at (or immediately after) the new moon. Trochophore larvae can be seen in the surface water of the tank in the morning, and complete veligers are present within 24 hours of the spawn. There is no feeding during the larval stages until the veligers settle, at 50–60 hours, when they begin to graze on the benthic algae in the tank.

1 : 1). These eggs are cleaned and placed in sterilized seawater. After a few days, the eggs hatch and the planktonic veliger larvae emerge. The larvae are fed motile unicellular algae for four weeks until they complete their metamorphosis, which like the abalone, can be triggered by chemicals produced by red algae. After metamorphosis and settlement, the young conch may be transferred to a tank or raceway containing filamentous algae on which they can graze (see Figure 6a.12); marine angiosperms, coarse macroalgae (such as *Ulva* and *Dictyota*), and commercial fish pellets are not ingested by the juveniles and therefore do not result in growth. The conch can be released to support the natural population when they have reached several centimeters.

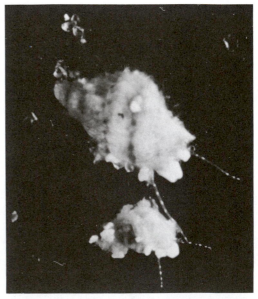

Figure 6a.12 Young conch in culture. (Photo courtesy of LeRoy Cresswell, Harbor Branch Oceanographic Institution, Florida.)

The growth rate of the young top shells is related to their density and the surface area of the sides of the tank (larger tanks have *less* surface area per unit volume for the animals to graze upon). Feeding takes place in the evening. Depending on the growth of the juveniles, the animals can be transferred to a reef for growout in one to several months. The best size top shell for this transfer has about a 1 cm shell diameter. The animals may be harvested from the reef after a growout period of two to three years. The use of this animal for restocking reefs is still experimental.

APPENDIX 7

OTHER CULTURED CRUSTACEANS

AUSTRALIAN CRAWFISHES

In recent years there has been some excitement, mostly as a result of work carried out by N. M. Morrissy, about the possibility of the culture of the Australian crawfishes of the genus *Cherax*. Until a few years ago, the species most often mentioned was the **marron, *C. tenuimanus*.** Since the summer of 1986–1987, when a hot spell killed a large portion of the marron in culture, attention has shifted to the **red claw** or **Queensland marron, *C. quadricarinatus*.** (see Figure 7a.1).

The marron is a native of southwestern Australia, where it is common in the rivers. Because it supports a large sport fishery, there are laws preventing its harvest from these rivers on a commercial scale; however, it can be raised legally in culture. The red claw is found in the Gulf of Carpentaria river system of Queensland and the Northern Territory, as well as parts of the Cape York Peninsula. There are a number of different strains of the red claw, some much more suitable for culture than others.

Marron and red claws are grown in ponds with a firm clay or gravel bottom. Unlike other types of crawfish, marron and red claw do not burrow into the sides of the ponds, so no damage is done to the structure and harvesting is simplified. Crawfish are susceptible to predation by water insects, catfish, eels, birds, and rats. If predation is a problem, fencing and wiring can be used to protect the crop, water can be filtered before it enters the ponds, ponds can be dried after harvesting, and shelters may be placed in the ponds. Shelters may consist of tree branches or pieces of plastic pipe, but these make harvesting more difficult.

C. tenuimanus and *C. quadricarinatus* are warm water animals, but temperatures should not exceed 28°C unless there is some sort of aeration device; the red claw is able to withstand slightly higher temperatures than the marron. Growth will stop below about 13°–14°C. These species require water with at least a 50% oxygen saturation (therefore aeration is suggested) and pH of 6.5 to 8.5, and the calcium content should be at least 60 mg/liter. Although live sales of *C. quadricarinatus* from Australia are restricted, postmetamorphic juvenile *C. tenuimanus* can be purchased legally from a dealer to start a culture. A reproductive population of marron should eventually establish it-

Figure 7a.1 An Australian red claw. (Photo courtesy of Robin Hutchins, F.A.C.T.)

self in a pond (assuming that the absolute size of the population is large enough and the environmental conditions are satisfactory). While high stocking densities are possible, one to four animals per square meter are commonly used.

The marron reproduces once a year, during August in Australia, as a response to the increase in day length and rise in water temperature with the approach of spring. This species becomes reproductive when it is 2 to 3 years old. After mating, the female carries the male spermatophore, using it later to fertilize the eggs that she extrudes. The fertilized eggs are carried under the curled tail and the female begins to hunt for a hiding place in which she will stay for several days until the eggs have become firmly attached to the pleopods. When the eggs are attached, she leaves her hiding place and begins to fan the water with her tail to ensure there is enough O_2 entering the egg's outer membrane. The number of eggs produced is a function of the size of the female, ranging from about 100 to 900. The red claw also becomes reproductive after one year, but there are multiple spawnings (three to five per season), and they will lay up to 1000 eggs, depending on the size of the female.

Cherax are detritivores, and in low-density culture situations, they require very little feeding. In higher densities, trout pellets can be used and are supple-

mented with alfalfa pellets, dried clover, composted hay or vegetables, or other plant materials; manures are sometimes used to increase the productivity of the ponds; however, the DO and general quality of the water should be carefully monitored following manuring. The greatest danger in Australian crawfish farming is overfeeding and the resultant O_2 problems. The amount of food given in the winter should be about 40% of that used in the summer.

These crawfish will grow to over 2 kg, but this takes years to do. It is best to harvest most of the crawfish when they are smaller, about 60–120 g (after one to two years), returning a few to the pond for the production of a new stock. Harvesting can be done with a seine net since both the marron and red claw are non-burrowing species, but traps and drop nets, and pond draining, may also be used.

A species related to the marron and the red claw, *C. destructor*, the **yabbie,** has also been considered as a possible aquaculture candidate because of its rapid growth, extended reproductive activities after a quick maturation, and ability to live in high densities and in waters with a low DO content. However, even with all these advantages, it is still not considered as good a prospect as the marron or red claw because of its burrowing habits, which result in pond damage and difficulties in harvesting, and because their size at maturity is significantly smaller than that of the other species.

CRABS

Crabs are decapod crustaceans like shrimp, crawfish, and lobsters. A number of species of crabs have been considered for aquaculture because of the high prices they bring at the market and the unpredictability of the crab fisheries catch in recent years. Despite the great interest in the culture of these organisms, so far they have generally resisted becoming major aquaculture species.

The **blue crab, Callinectes sapidus,** is commonly found along the East Coast of North America south to northern Argentina. It is found in brackish water and full seawater, generally in depths of less than 35 m. The blue crab lives for two to three years; it is an omnivore, taking in plants as well as benthic invertebrates and dead fish. It is a highly prized food supporting a large fishery.

The male blue crab will mate at any time during its last three intermolts, but the female will probably mate only once in her life, in estuarine waters during the molt between the juvenile and adult stages. She spawns 1 to 10 months after mating in the spring or summer, often going into deeper waters outside the estuary. The eggs, up to 2×10^6, are carried for a week to 10 days under the abdominal apron, attached to the swimmerets, before hatching.

The eggs hatch into the planktonic **zoeae** stage, which remain in the water column of the coastal ocean for four to seven weeks, the length of time being related to the amount of food available, temperature, and salinity. This long larval period, with its high mortalities, is the bottleneck in commercial culture. The last zoeal stage molts into the **megalops** stage, which is partially benthic and partially planktonic; at this stage, the organism begins to leave the full seawater environment and moves toward the estuary. The megalops finally molts into a juvenile crab that develops in estuarine waters, feeding on the abundant benthos and hiding in the tall grasses.

While the intermolt blue crab is normally considered a valuable catch, the soft-shell form of this crab (which exists for a short time after ecdysis, but before the new shell has a chance to harden) is especially prized, selling for five to six times as much as the hard-shell crabs.

Soft-shell crabs are produced in several ways. The idea behind all of the methods

is to collect blue crabs that are just about to molt **(peelers)**, hold them in some manner so they are protected when they are soft and vulnerable, and then quickly market them. To be economical, the time that the crabs are held should be as short as possible, and there should be no feeding (crabs do not feed just prior to molting). This process starts with the collection of the peelers using special devices, including drags that are pulled through grassy areas where the premolt crabs congregate and pots that are "baited" with a male crab to attract the unmated females. The peelers are selected by the fisherman, who can recognize their condition by the color of the shell and the "line" around the last two segments of the paddled swimming leg (a week or two before ecdysis, the line turns white, then pink, and finally red just before the animal leaves the old carapace). The captured crabs are placed in **floats,** wooden boxes that allow water to move through openings, or are brought indoors and placed on shedding tables. The blue crabs are observed often, particularly when they get to the red line stage. After a split occurs along the back of the crab's shell (the **buster** stage), ecdysis takes place in a few hours (see Figure 7a.2); the crab should be removed from the water within four or five hours of ecdysis or the shell will begin to harden. There are many ways to get crabs to molt more quickly in a laboratory than they would naturally, but these methods have not been used by most fishermen because they are impractical or result in a high percentage of mortalities during the molt; these include removing one or both eyestalks, removing most of the walking legs, or injecting the crab with molting hormones (ecdysterones).

Another swimming crab, the **serrated crab, Scylla serrata,** is cultured in Southeast Asia, where the meat and the ripe gonads are eaten. These crabs are grown in Taiwan in polyculture with milkfish and *Gracilaria,* and stocked with fish and shrimp in rice paddies. Monocul-

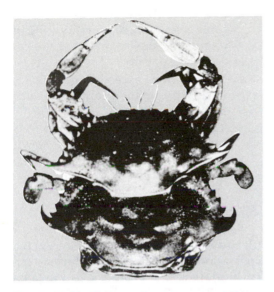

Figure 7a.2 Molting of a blue crab. (From Churchill, 1919.)

ture of this crab aims at producing mature females with ripe gonads so they can be sold for a higher price.

Estuarine ponds with sandy bottoms are used for the monoculture of *Scylla*. The pond walls are made of bricks or concrete and have a lip to prevent escape; a large square pond may be broken into four smaller ponds by these walls. The ponds are stocked with females that have just mated, and these are held for one or two months until the gonads have fully developed; weight gain during culture is not a consideration. Other ponds may be used to hold premated crabs that will later copulate and can be moved into the main pond. The crabs are fed trash fish and freshwater snails (which appear to be important to the maturation process). Crabs receive about 5% of their weight in feed each day, usually in the late afternoon, since feeding starts after dark. The serrated crab is harvested by nets or baited traps, particularly in the area of the water inlet.

Other crabs have been cultured in the laboratory or for short periods of time in pilot studies, but have not yet made any impact on the aquaculture industry. Three of these are

1. *Cancer*. This genus includes several species of crabs that are important to fisheries on the east and west coasts of the United States, and therefore established markets already exist for *C. magister* (**Dungeness crab**), *C. irroratus* (**rock crab**), and *C. borealis* (**Jonah crab**). Their life histories are well known and they have been raised in the laboratory through their larval stages.

2. *Menippe mercenaria*. The **stone crab** has been induced to spawn experimentally out of season in the laboratory by manipulating the temperature and feeding the crabs a diet of fish and mollusks. The eggs produced were viable and healthy, with 52% becoming megalopa and 36% reaching the juvenile crab stage. When reared in separate indoor tanks, these crabs may grow quickly, reaching (in the best case) over 230 g only seven months after hatching.

3. *Mithrax spinosissimus*. The giant **West Indian spider crab** is a herbivore that grazes on algae in the coastal environment. Like the species just described, there already exists a market for cultured animals because of an established fishery. To culture *Mithrax*, two systems have been used experimentally, one in which algae is grown and then brought to the crab, and another in which the crab and algae are cultured together. Some protein source added to the diet may increase growth. These animals spawn in captivity or eggs can be collected from gravid wild crabs; the larvae are easily reared and molt into the **first crab** stage in four to eight days without feeding. The crabs are then presented a diet of cul-

tured phytoplankton (brine shrimp nauplii have also been used). The early stage crabs can be stocked at very high densities, over 2.25×10^4 per square meter, if there is good water exchange.

"OTHER" SHRIMPS AND PRAWNS

In addition to the penaeids and *Macrobrachium*, other shrimps and prawns have been considered for commercial culture, and some have already been grown in the laboratory to be used as "model" crustaceans. Like *Macrobrachium*, they are caridean shrimp, but they belong to different families and are typically fisheries products in the northern hemisphere. Those most commonly considered are *Palaemon* and *Pandalus*.

Palaemon serratus is known as the **European common prawn.** This species breeds readily in captivity, although there are fewer eggs produced by this organism than by many other shrimps; this low fecundity may mean, unfortunately, that a large brood population must be maintained for commercial culture. Growth is best at about 22°C; lower temperatures have an especially adverse effect on younger shrimp, while warmer temperatures seem to inhibit growth of older shrimp more than the smaller *P. serratus.* Salinity and diet may also affect the growth. In a related species, *P. adspersus,* it was shown experimentally that there was relatively good growth when the shrimp were fed *Artemia* and the eggs of the shrimp *Crangon,* but growth was rather poor when mussel tissue and *Tubifex* worms were used.

Several other problems must be overcome before large amounts of capital can be invested in *Palaemon* culture

1. It grows slowly, taking about a year to reach maximum size even under the best laboratory conditions, but

P. serratus dies quickly at temperatures below 4°C, so overwintering in temperate environments requires an indoor facility.

2. Males rarely exceed 5 g (females reach 8 g) and therefore may be difficult to market.

3. The animals mature when they are small and therefore convert much of the feed given to them to gonadal material rather than to edible muscle.

Pandalus, including **P. borealis (deepwater prawn),** *P. platyceros* **(Pacific spot prawn),** and *P. prensor* **(far eastern shrimp),** have been cultured in the laboratory. Probably the most striking problem with this genus, for the purposes of commercial farming, is that these shrimps do not readily breed in captivity. Other problems to be overcome are a low fecundity even for captured females, and a growth rate slower than the penaeids. High temperatures cannot be used to speed growth because this results in high mortality rates. *P. kessleri,* which is found off Japan, has been considered to be a possible candidate for culture because it has a short larval life and hatches in a fairly advanced state of development; the problem with this species is that it has a narrow salinity tolerance, making it difficult to grow in a coastal environment.

LOBSTER

There is currently no commercial culture of **lobster.** However, these animals have been reared for a number of years in pilot scale projects and have received a considerable amount of attention from culturists. The reason for the attention is clear; lobster is a very popular seafood with a high market value, it is simple to prepare, and some species are easily shipped alive.

There are many types of lobster, including the slipper lobster and the coral lobster, but only one group of lobsters has been seriously considered as having possible culture candidates: those of the family Nephropidae. **Nephropid lobsters** have large claws (chelipeds) and are usually found in the higher latitudes. **Spiny lobsters** (without the enlarged claws, family Palinuridae) are normally found in the tropics and subtropics, and include the genera *Jasus* and *Panulirus*. Spiny lobsters grow quickly once they have achieved the juvenile stage, but unlike *Homarus* they have a long and complex larval cycle, including the characteristic **phyllosoma** stage, and therefore are probably not well suited for farming.

The species of interest to the culturist are **Homarus americanus**, the **American lobster,** and **H. gammarus** (=*H. vulgaris*), the **European lobster;** other lobsters similar to these, those of the genus *Nephrops,* have also been briefly considered for aquaculture.

Reproduction

The breeding cycle of *Homarus* is well known. In the summer months, the female molts and releases a sex pheromone that attracts the male. There is a "courting dance" followed by copulation, and the male leaves the female with sperm that can be stored by her for over a year until the eggs are laid. Fertilization takes place when the eggs are extruded. The fertilized eggs are cemented to the swimmerets under the abdomen where they may be carried for up to a year (see Figure 7a.3).

The larvae are planktonic. After a few molts, the fourth-stage (first postlarval stage) *Homarus* appear; these are similar to the adults, having large claws and a general lobster shape. They soon leave the plankton and join the benthos. In their natural environment, *Homarus* become sexually mature at about 6 years, when they reach approximately 500 g; lobsters may weigh up to 19 kg and live 50 to 100 years.

Traditionally, the **berried** females (those with the eggs attached to the pleo-

Figure 7a.3 A "berried" female lobster. (From Mead, 1908.)

pods) are caught for culture in the spring and brought into the laboratory. Alternatively, the culturist may

1. Use mated females that have not yet extruded their eggs.
2. Allow mating to take place in the laboratory by placing a mature male with a newly molted female.
3. Carry out an artificial fertilization by using a mild electrical current to stimulate the spermatophore's release from the male, and transferring it by hand to the female.

The eggs from fertilized females eventually are extruded in the culture facility. Under normal temperature conditions, most females spawn and molt on alternating years, so the farmer should select for artificial fertilization only those females that are known to have molted rather than spawned the prior season.

The regulating factors, controlling the production of yolk proteins and the extrusion of the eggs by the female, seem to be photoperiod and temperature. As in the case of other organisms, culturists would like to see the entire reproductive cycle under control so that strains of fast-growing, disease-resistant, energy-efficient animals can be developed. Thanks largely to research conducted in laboratories in Canada and California during the past decade, we now understand enough about the environmental influences over the lobster and its biology, that the cycle can be closed. Pilot-scale production of *Homarus* is technically feasible.

Rearing

The embryos remain attached to the female until they hatch into planktonic larvae. In the culture facility, multilevel "apartments" are constructed so the gravid animals are held in running seawater and the larvae, when they hatch, are swept into separate mesh baskets. The collected larvae are moved into a **kriesel** (a circular tank with a conical bottom) with moving seawater; if the larvae are allowed to collect at the kriesel bottom because the water is not moving, or if they are held in too high densities, they will attack and eat each other. Cannibalism can be avoided by keeping the water moving and feeding the larvae, being sure there is something to eat in the water at all times. The balance between enough food to inhibit aggressive behavior and so much that the water quality is significantly reduced is not always easy to achieve for the new culturist; if live feeds such as mysids or brine shrimp are used, the problem of water fouling is avoided. Other feeds that have been used include ground shellfish, liver, and frozen zooplankton or brine shrimp. The length of the larval stages is at least partially a function of the photoperiod to which the larvae are

exposed; it appears that lobster larvae reared in a short-daylight–long-nighttime photoperiod grow more quickly, but have an inferior survival rate, compared to lobster larvae exposed to long-daylight–short-nighttime photoperiods.

When the larvae begin the benthic stage of their life, they are transferred to individual compartments. These are about 100 cm² to begin with, but must be increased in size as the animals grow. If communal facilities are used, the young lobsters quickly cannibalize each other during ecdysis unless there are abundant empty mollusk shells for the animals to use as shelters; even if a communal tank is used initially, once the juveniles reach about 30 mm, they should be moved to individual apartments. This is one of the major limitations of lobster culture; individual compartments, which must be cleaned and receive food, means extensive labor.

Young lobsters grow best on live adult brine shrimp; larger juveniles require larger prey items. Considerable effort has been expended in recent years on the development of a prepared diet for lobsters; a few publications on this subject are summarized in Table 7a.1.

While no commercial growout of large numbers of lobsters exists, potential designs for mass culture facilities have been investigated on the laboratory scale, and suggestions for commercial operations have been given. The amount and quality of space that lobsters need for good growout, however, is not well understood; that is, growth rate is not just related to such questions as: How much area does each animal have? Is a communal or individual habitat being used? Is the diet acceptable? What is the volume of water replaced per animal per unit time? and What sort of shelter is available? Recent experiments suggest other considerations in the case of *Homarus*. More must be learned about the lobster's growth response before large-scale commercial culture becomes a reality. It is well estab-

TABLE 7a.1 Some Reports on Lobster Nutrition

Author(s)	Conclusions
1. Conklin, Devers, and Bordner, 1977	Increased growth of juvenile lobsters was observed when shrimp meal and water-soluble vitamins and minerals were added to the diet.
2. Capuzzo and Lancaster, 1979	The ratio (amount of protein/amount of carbohydrate) can be decreased as lobsters grow, reducing the cost of the feed ingredients.
3. Bartley, Carlberg, Van Olst, and Ford, 1980	In juvenile lobsters, the best growth efficiency was observed when they were fed 2% of their body weight per day and were kept at 20°C.
4. D'Abramo, Bordner, and Conklin, 1982	Soy phosphatidylcholine is an important part of the lobster diet because it acts to move cholesterol into the blood.
5. D'Abramo, Bordner, Conklin, and Baum, 1984	The sterol requirements for the juvenile lobster can be filled by cholesterol, but not by the other sterols tested.
6. Eagles, Aiken, and Waddy, 1984	Larval lobsters fed brine shrimp had a greater survival when fed in the dark and when not overfed.

lished that the growth rate of these lobsters can be greatly increased, at least for part of the life cycle, by using warmer waters (18°–22°C) than those *Homarus* would normally ever experience, but the cost of heating water is still prohibitive unless there is some sort of "waste heat" available. (Note: The lobster's ability to survive for extended periods of time in warm water has been questioned by some scientists.)

Some interesting suggestions have been made with reference to growout. If the eyestalks are removed from a crustacean, the rate of molting (and therefore growth) is increased because the source of molt-inhibiting hormone is eliminated. Rapid growth increases operating costs since the amount of food consumed per unit time increases, but that, in theory, could be easily offset by the increase in production of lobster per unit time (a slow growth rate is one of the problems faced by would-be lobster farmers). Under normal conditions, ablation of the eyestalks or hormone injections is not considered economically viable for mass culture of crustaceans because the value of each individual organism is not very great, but in the case of the lobster, where each animal is used as whole meal and the consumer is willing to pay a substantial price for it, this sort of treatment someday may be considered practical by aquaculturists. It has also been suggested that cultured lobsters be marketed at 200–250 g, rather than the traditional 450–500 g (roughly the legal capture size in most states). Smaller lobsters would reduce the space and feeding requirements and would mean that the crops could be turned over more often. Some evidence suggests that feeding efficiency is reduced and the growth rate slows down after the lobsters reach about 250 g.

APPENDIX 8

OTHER CULTURED FISH

RED DRUM

Red drum, *Sciaenops ocellatus,* sometimes also called the channel bass or **redfish,** along with the black drum, are members of the family Sciaenidae (the croakers). The red drum has been the subject of intensive aquaculture research during the past decade, primarily by scientists in the southeastern states and Texas. There are several reasons why the red drum has received this attention

1. Since 1981, commercial fishing of this species has been banned in Texas (although hatchery-reared fingerlings are now being released into coastal waters, so this ban, it is hoped, is a temporary measure).
2. The red drum is a very popular sport and food fish, so a market for it already exists. The demand for the "blackened redfish" dish during the 1980s, especially in New Orleans, has been well documented.
3. The red drum is a euryhaline and eurythermal species, making culture possible under a variety of conditions.

Red drum spawn in coastal waters during the late summer and early autumn. At this time, captured mature females can be induced to ovulate by injecting hCG. Males and females can be stripped, and fertilization may be carried out artificially. Outside of the spawning season, gonadal maturation in tanks can be induced by manipulating the temperature and photoperiod to simulate the early autumn conditions. Once the drums begin to spawn in the tanks, they will continue to do so until the environmental parameters change; drum have been held experimentally in this induced reproductive condition for over 3 years with approximately weekly spawnings. The females used are large (8 to 18 kg) and produce 10^5 to 2.5×10^6 eggs per spawn, but the quality of the eggs may decline if the fish stays in the reproductive state for too long.

The broodstock females are held with males in the tanks; the eggs are fertilized at spawning, float to the surface, and are easily collected. They are placed in gently aerated seawater tanks where, depending on the temperature, hatching takes place in 20 to 30 hours. After about 2 days, the yolk sac of the larval fish is absorbed, and it begins to search for food. In the hatchery, the larvae are started on a diet of rotifers, then switched to brine shrimp nauplii, and finally to adult brine shrimp and chopped fish or shrimp when the drums reach 10 to 12 mm. The fry can also be given prepared feeds at this point; the diets should contain 35% to 50% protein (the optimal amount of protein is probably a function of the temperature and salinity). Larvae can be raised to fingerlings in fertilized ponds; when the fry begin to feed, they may be stocked at 7.4 to 12.3×10^5/hectare and can be harvested in 30 to 40 days as fingerlings. Using this technique, survival rates are very high, typically 30% to 50%.

The amount of information on the growout of red drum is limited, since most efforts have been directed toward restocking natural waters with fingerlings. Drum kept in saltwater can tolerate temperatures of 2°–36°C, but are much less tolerant when reared in freshwater. Drum can be trained to take pellets if they are first offered with natural foods. Trout and catfish diets have been used for growout, but care should be taken because high-caloric diets may inhibit growth. The protein requirement for the small fingerlings is about 45%, which decreases to about 35% as the fish grow. Food conversion ratios of about 2 : 1 can be expected.

Drum have been raised largely in ponds. In an extensive system, fish of about 225 g were stocked at 500/hectare, fed forage fish and a prepared trout diet, and were harvested in five months at an average weight of 640 g. Drum have been raised intensively in ponds, raceways, and cages; while the data are still preliminary, initial reports are very promising.

BLACK DRUM

The **black drum,** *Pogonias cromis,* can be spawned with injections of carp pituitary extract or hCG, followed by stripping and dry fertilization. Mature black drum also spawn naturally if held in large, warm (21°C), outdoor saltwater (28 to 31 ppt) tanks. The eggs hatch best in full seawater. The fry can also be maintained in full seawater kept at 18°–20°C and will consume newly hatched brine shrimp.

Black drum have been cage cultured in thermal effluent (27°–30°C) with excellent results. When the fry were fed a prepared trout diet and kept in cages for 85 days, the survival ranged from 83% to 100%, with fish from the cages gaining over 1 g per day. The black drum has also been used in polyculture; when grown by biologists in brackish water 0.1 hectare ponds with striped mullet and pompano, the drum grew better in polyculture than in monoculture. When the drum and pompano were cultured together for 145 days, the pompano were harvested and the drum monocultured until day 278; the production of drum was 66% better than when the drum were in monoculture for the entire 278 days.

Red drum (male) × black drum (female) hybrids have been produced experimentally. The eggs may be fertilized by the dry method and hatched in 24 ppt water at 21°–22°C. When stocked in ponds, the hybrid takes prepared feed and grows more quickly than the black drum.

OTHER SCIAENIDS

The **spotted seatrout,** *Cynoscion nebulosus,* is an important commercial and recreational species in the Gulf of Mexico and along the Atlantic coast of the United States. Some government-funded culture of this species is being carried on.

Spawning can be easily induced with hCG, the fish stripped, and fertilization carried out in the hatchery. Alternatively, the female fish may be released into ponds or tanks with males after a hormone injection, and fertilized eggs can be collected in 1.5 to 2 days. Incubation of the eggs is done in tanks of clean, well-aerated seawater. Initial rearing can be conducted in the same tanks, or the fish can be transferred to ponds. Feeding the larval and juvenile seatrout has been a problem, for when the correct size zooplankton are unavailable, there may be significant losses due to cannibalism.

Culture in ponds starts with fertilization of the water with organic and/or inorganic materials. The results of fertilization have varied: Some programs have had good production and survival, while others have been less successful. An indoor rearing program that starts with feeding algae-packed rotifers to the seatrout, followed by the slow phasing in of brine shrimp nauplii and finally of prepared feeds, has been shown to be successful. The composition of the diet varies with the stage of the fish, younger seatrout needing proportionally more protein in feed than older individuals. A fish can be reared to 110 g in 5 months using this feeding regime; feed conversion ratios for the dry feeds as low as 0.78 have been reported.

Spawning of the **weakfish,** *C. regalis,* has recently been accomplished. Large adults kept in tanks were fed squid, shrimp, and calf liver; by manipulating the temperature and photoperiod, the fish were induced to spawn under "spring" conditions (14 hours light per day, 22°–24°C). The collected eggs hatched in 24 to 36 hours when held in aerated seawater (30 ppt and 23°C). The young weakfish initially eat rotifers, brine shrimp being added to the diet after about one week.

The **orangemouth corvina,** *C. xanthulus,* has been matured in closed systems and spawned with hormone injections. In preliminary laboratory experiments, they seem to grow quickly.

MARINE BAIT FISH AND BAIT SHRIMP

Marine bait fish are usually collected from brackish bodies of water by fishermen using small seines or traps. But collection is unreliable, so as the value of bait fish has increased, so has interest in their culture. Much of this, unlike freshwater bait fish culture, is still experimental in nature.

Fundulus grandis, the **gulf killifish** or **bull minnow,** has been cultured in Alabama and Texas by researchers and shows great promise. The killifish can be stocked as broodfish in brackish water ponds and fed a commercial trout diet. The fish spawn in the spring and summer on mats of Spanish moss like those used by freshwater bait fish culturists. The mats should be transferred to hatching/nursery ponds to separate the young from the cannibalistic adults. In one trial, about 1.2×10^6 eggs per hectare were placed in ponds in the spring; fish harvested one to three months later had an average weight of 500 mg and there was a 75% survival. During this time in the nursery ponds, the killifish were fed a commercial minnow feed and a conversion ratio of 1.9 was reported. The 500 mg fish were moved to growout ponds and fed a commercial trout ration two or three times per day; stocking densities ranged from 2.5 to 3.7×10^5 fish per hectare. Marketable fish (2.4 g) were harvested after seven weeks. The production during growout was about 660 kg per hectare, and there was 93% survival and a 2.4 feed conversion ratio. Less management (but lower production) is also possible by just keeping the ponds in bloom and adding no prepared diet for the fish. The production of *F. grandis* can be severely reduced when the sheepshead minnow, *Cyprinodon variegatus,* becomes established in the killifish ponds.

At the University of Hawaii, research has been conducted on the culture of a **topminnow,** *Poecilia vittata.* This fish's market is the commercial skipjack tuna industry rather than for recreational sports. Because of the success of the topminnow's culture in Hawaii, interest has begun to spread. These fish reproduce readily in ponds as long as the water quality remains high. As omnivores, they can be raised on the detritus and naturally growing organisms in the ponds, but these are supplemented with dry rations of tunafish meal and chicken mash. The newly hatched topminnows, 8 to 9 mm, feed on the same materials as the adults. This fish is hardy and withstands handling and crowding, such as would be experienced in live bait wells on board ships.

Shrimp are also used as bait organisms. Small shrimp of a few grams are collected in nets in coastal and estuarine waters and are sold to bait stores in several parts of the United States. Initial experiments using two species of *Penaeus* have suggested that bait-size shrimp could be raised from the postlarval stage very cheaply by enriching the pond water with a sewage effluent. Growth of large shrimp at high densities for human consumption requires additional feeding with specialized diets, but such supplements do not seem to affect growth significantly during the first few months of rearing. Ponds can be densely stocked, several crops can be generated each year, undesirable sewage water is utilized to generate income, there is no feed cost (one of, if not *the*, major expenses in shrimp culture), and the often-voiced problem concerning uptake of dangerous chemicals/pathogens is bypassed since the bait organisms are not digested by the fish they are used to catch. The major drawback is the availability of postlarval animals.

ORNAMENTALS

The most valuable fish, based on cost per unit weight, are the **ornamental fish.** Many of these are caught in their native tropical enrironment and are shipped to distributors around the world, but as the natural stocks decline, more of these fish are being cultured.

We are not sure how long people have kept ornamental fish, but records show that **goldfish** were kept in China during the Sung Dynasty (starting in the late tenth century). Goldfish of different colors and forms (see Figure 8a.1) were common pets in China by the end of the thirteenth century, spreading to Japan by 1500, and were brought by sailors to England and France in the seventeenth century. Goldfish were sent to the United States in 1878 and were put on display in Washington, D.C.; the public became excited over the new arrivals, and their popularity as pets quickly spread. Unlike many of the other ornamental fish, goldfish and their relatives **(koi)** are not tropical and may be raised in temperate climates without difficulty.

Commercial culture of tropical ornamental fish began in 1926 in Florida, which is still the center of the industry. In fact, tropical fish are the largest airfreight item shipped from that state.

Ornamental fish, which are common

Figure 8a.1   A ''shukin'' ornamental goldfish. (From Matsubara, 1908.)

pets in the United States, can be placed in four groups: the native species, saltwater fish, freshwater tropical livebearing fish, and freshwater tropical egg-laying fish. Hobbyist magazines and bulletins contain a good deal of specific literature on the breeding of many ornamentals.

Native Species

Relatively few **native species** of freshwater fish are presently sold in pet stores, although many hobbyists do collect them. Some of the species that have a potential for culture are the following:

1. The **golden shiner,** which is very important to the bait industry, has lively movements and a bright color. These characteristics, which make them excellent bait, also make them ideal ornamentals.

2. **Killifish** are popular with many aquarists because of the bright colors and hardiness of some strains. Although most of the killifish that are kept are exotics, the native species have taken part of the market in recent years.

3. **Sticklebacks,** which are also very hardy and easily kept, are popular with European hobbyists; they construct nests and breed readily in tanks and bowls (see Figure 8a.2).

4. **Sunfish** are very colorful and easy to maintain, and the aquaculturist can breed them with little trouble.

5. Others that might be considered are catfish, darters, small pickerel, and pupfish.

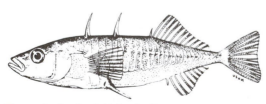

Figure 8a.2 A stickleback. (From Whitworth, Berrien, and Keller, 1968. Courtesy of Connecticut Department Environmental Protection.)

Saltwater Fish

The popularity of marine organisms is growing fast among aquarists. This is probably related to the bright colors of many reef fishes, and the great diversity of both fishes and invertebrates that potentially may be housed. Saltwater tanks are a bit more difficult to keep than freshwater tanks, but progress in tank and filter designs, the availability of salt mixtures, and some good popular literature on the subject have simplified their maintenance, so marine organisms can be kept by almost anyone interested in them.

Like many of the tropical freshwater fishes, marine species are generally caught rather than bred. This makes the supply less reliable, results in many injured fish being marketed, could potentially damage the environment, and naturally makes these fish rather expensive.

Breeding marine fishes is often more difficult than breeding freshwater fishes. Many species simply release small eggs into the water, but a few lay eggs that they attach to a substrate; the fish laying attached eggs were the subject of early culture work in the 1970s in Florida. In 1973 and 1974, the first cultured **percula clownfish,** *Amphiprion ocellaris* (see Figure 8a.3), and Atlantic **neon goby,** *Gobiosoma oceanops,* where produced by Martin Moe and Frank Hoff. More recently, Moe has cultured a number of fishes including the yellowhead jawfish, royal gramma, blackcap basslet, and several angelfish and gobys. These fishes were produced in some cases by broodstock held in tanks, while sometimes mature fishes were collected from the sea just prior to spawning; in general, hormones were not used to induce ovulation. The greatest difficulties in the culture of marine fishes for the aquarium trade is probably the lack of information available on the diets and diseases of the larval and juvenile fishes. Work on the culture of marine fish has been continued by several researchers, but it is still a relatively small industry.

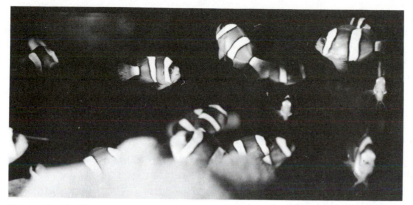

Figure 8a.3   Hatchery raised clownfish. (Photo courtesy of Dr. Robert Winfree, U.S. Fish and Wildlife Service.)

Livebearing Freshwater Fish

The first of the livebearing fish to be marketed in the United States was the **mosquito fish,** *Gambusia affinis,* in 1908, but these have been replaced by more colorful imports. Today, most of the live-bearers are cultured and belong to one of three groups: the guppies (*Lebistes*), the mollies (*Poecilia*), and swordtails and platies (*Xiphophorus*). There is some disagreement about the classification of these fishes—some authors combine all of them into the single genus *Poecilia*.

Livebearers are commercially bred in concrete vats or ponds. The ponds used to culture small ornamental fishes are much smaller than are those used to rear larger fish for human consumption. The first step in culture is to get a good stock of breeder fish. Overcrowding is prevented by constantly culling out the undesirable individuals; overcrowding inhibits breeding and also means increased feeding (this does not result in the same substantial increases in production costs for the ornamental fish farmer as it does for other fish culturists, but it may lead to water quality problems). In addition to feeding the fish, ponds may be fertilized because natural foods are often preferred and result in greater fecundity. Harvesting the ponds is done by traps, and the ponds are also drained periodically for maintenance.

Guppies are extremely prolific fish, mature females producing 150 to 250 young per brood. The females store sperm that can be used for five or more broods (which may be produced monthly). Under good conditions, the young begin to breed in a matter of six to eight weeks. These fish are popular because of their hardiness and the spectacular tail colors that have been selected for by breeders. However, the establishment of selected traits is more difficult than might be suspected. Since the fish begin to breed so early and the females store sperm, the young must be isolated nearly from birth until a mate is selected. In addition, many of the desirable traits are recessive, so the offspring of the first crosses must be back bred to the father, or the siblings of the initial cross must be bred.

Some **mollies** are collected wild and sold, but most are selected strains that have been developed by breeders over the last 50 years. The strains are easily crossed, and mollies will even hybridize with guppies. The most popular mollies are the **sailfins** produced by William Sternke in the 1930s. Like many other livebearers, they produce both fast-maturing and late-maturing males; the late-maturing males look like females during their first months and later develop into the large sailfins.

There are many strains of **swordtails** and **platies,** and hybrids are easily produced. The most important commercial swordtails are the bright red varieties, but several color patterns, as well as fin configurations, are available. The most important platies are red and have the "hifin" mutation.

Egg-Laying Fish

There are many egg-laying ornamentals in culture, although some are clearly more popular than others. The selection of a breeding stock is very important; the culturist should start with immature fish raised in conditions (especially water chemistry) similar to those that will be used for breeding. Culture conditions for egg layers vary with the species. Fish belonging to the same family have similar requirements.

Among the most important of the egg-laying fishes are those of the family Anabantidae: the gouramies, paradise fish, and bettas (or "Siamese fighting fish"). A few **anabantids** are most profitably raised in ponds, the majority being best reared in tanks. Female and unsexed bettas are mostly grown in Florida, while the large males are imported because of the high labor costs associated with their culture. Anabantids are bred in pairs (except kissing gouramies); they build bubble nests for their floating eggs that are tended by the male.

The fish of the family Cichlidae are also very important for aquarists. **Chichlids** include the discus, angelfish, kribensis, oscars, convicts, severum, chromides, and the deacon fish. The commonly cultured species are mouth breeders or substrate breeders. Some are cultured in tanks, but many can be bred in pools and ponds.

A few of the **tetras** (family Characinidae) are sometimes bred in ponds, but many species are kept exclusively in tanks. The male and female breeders are usually kept separately, paired only when breeding is desired, and then they are separated again. Water should be kept at 24°–28°C. Spanish moss mats are used as receptacles for the semiadhesive eggs.

Like the tetras, some of the ornamental fish in the family Cyprinidae (white clouds, zebra danios, giant danios, and barbs) can be pond bred, but tanks are more likely to give positive results for **cyprinids.** White clouds and barbs are bred like tetras. The danios should be bred in groups, and an egg trap must be used because they will eat their nonadhesive eggs.

The **killifishes,** which belong to the family Cyprinodontidae, will breed in ponds, as will some other aquarium fish, including the Japanese weather fish (family Cobitidae), glassfish (family Centropomidae), and the Australian purple striped gudgeon (family Eleotridae). Other exotics may require tanks for breeding.

SNOOK

Although there are no commercially cultured **snook (Centropomus),** great strides recently have been made in breeding and rearing this fish. Snook are commonly found only in tropical waters, from Florida south to Brazil; there is interest in establishing hatcheries that will release young snook for commercial and sport fishermen because natural stocks have been quickly declining. It has also been suggested in Brazil that they be grown out for food.

Mature fish may be captured and induced to ovulate by hCG injections. The fish may be strip-spawned and a high fertilization rate (95% or better) can be achieved. Newly hatched snook require live food (rotifers, copepods, brine shrimp) but can be weaned to take a salmon starter diet relatively quickly. The young fish grow best at about 27°C and in freshwater, although they can be reared in water at least up to 24 ppt. Snook reach 20 g in three to four months, at which time they can be stocked in ponds. Preliminary experiments suggest that growth is better in plastic-lined ponds than earthen ponds or tanks. The snook are fed salmon pellets during growout and are allowed to forage on small fishes (tilapias, minnows, platies, mosquitofish) in the ponds. They grow to a harvestable size (450 g) in about eight months. The feed conversions of snook are reported to be very good.

Several other members of the family Centropomidae are of interest. *Lates niloticus*, the **Nile "perch,"** is sometimes referred to as a "sea bass" (and mistakenly classified as belonging to the family Serranidae). This animal, which grows to 1.8 m, is reared in ponds in Africa, but it is sensitive to drops in DO and to turbidity. They reproduce easily in large ponds (but not small ponds) and grow to 0.5 kg in half a year. They are used effectively in controlling tilapia populations. In Thailand, *L. calcarifer*, the **"anadromous sea bass,"** is cultured. The young are collected in estuarine waters using traps and nets; these captured fish are usually at least 10 cm. They are stocked in cages or in ponds and are fed chopped fish. Ponds are stocked at 2500 to 5000 fish per hectare, and up to 2500 kg are produced in six months.

DOLPHIN FISH

The **dolphin fish** or **mahimahi,** *Coryphaena hippurus* (family Coryphaenidae), is found in the tropical and semitropical waters of the Pacific and Atlantic oceans. It is an active oceanic fish, and therefore for many years was not considered a viable aquaculture candidate. However, because it supports valuable fisheries in many parts of the world, initial work was done on the possibility of capturing small fish and rearing them in pens. Since that time, methods have been developed for keeping the adults so that spontaneous spawning takes place and for the rearing of the larvae. Several generations have

been kept in culture, but there are no large commercial dolphin culture operations yet.

Since these fish are active swimmers, they are best kept in round or oval tanks that do not require that the fish slow down or make quick changes in direction. In addition, the seawater used in the tanks should be 30 ppt to 36 ppt, of good quality, and warm. This means frequent water changes and/or extensive water treatment.

Newly hatched dolphin are reared on a diet of live foods (rotifers, copepods, and brine shrimp). After several weeks, the fry can be weaned away from live food and accept gels, dried feed, and animal tissue. Under good conditions, the fish grow very rapidly; juvenile animals of a few grams grow to over 1 kg in four to five months, with excellent survival (see Figure 8a.4). The biggest problem with the culture of these fish may be the expensive high-protein diet (chopped squid and fish).

SPARIDS

There is some confusion about the term **"bream."** The family Sparidae (the porgies) includes the "sea bream," such as *Sparus* and *Pargus*, although most fish given the name "bream" are freshwater cyprinids (e.g., *Abramis* and *Parabramis*).

The **gilthead sea bream,** *Sparus aurata,* has been almost incidentally cultured with mullet in brackish water ponds in Europe for many years. However, this species has recently been considered as a potentially important organism itself, although work is still largely experimental. These fish have been spawned and grown out to a marketable size.

The **sheepshead,** *Archosargus probatocephalus,* may have more of a potential for aquaculture than the gilthead sea bream (although at present it does not command a very high price in the United States). It can adapt to a wide range of

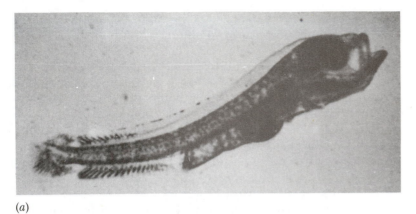

(a)

(b)

(c)

Figure 8a.4 (a) Larval dolphin, (b) dolphin fry, and (c) adult dolphin in a culture tank. (Photos courtesy of R. W. Hagood, NORAQUA, Florida.)

both temperatures and salinities, and may reach up to 14 kg. These fish mature at two years and spawn in estuaries and coastal waters. The gametes from ripe adults can be stripped, and fertilization in seawater is carried out by the culturist; the eggs are hatched in gently aerated seawater. The larvae may need live feed initially, such as rotifers and brine shrimp nauplii, but they can soon be switched to prepared feeds. After hatching and the initial feedings, the larvae can be transferred to larger tanks. The fish can be grown to a marketable size, about 0.5 kg, in one year.

The **red sea bream, *Pagrus (=Chrysophrys) major,*** is the major cultured sparid in Japan, although *Evynnis,* the crimson sea bream, *Taius,* the yellow back bream, and *Mylio,* the black sea bream are also grown commercially. The culture of *Pagrus* starts with the broodfish that are either collected by fishermen or produced by the culturist. The use of natural seed, once very important, is now declining. These fish spawn naturally in the spring, although they will spawn earlier if the water temperature is raised to 17°–20°C., Male and female fish are placed into spawning tanks containing 10^3–10^5 liters of seawater (1 : 1). The females begin to spawn at dusk each day for several weeks, each night producing 5×10^4 to 1.5×10^5 planktonic eggs. The fertilized eggs float out of the spawning tank and are caught in a collecting net (see Figure 8a.5).

Pargus eggs are incubated in aerated tanks of filtered seawater; at 16°C the eggs will hatch in about 3 days, at which point they may be transferred to larger tanks where they are raised for 45 to 60 days, growing to 20–35 mm. The larvae are fed rotifers packed with algae or yeast, and then the copepod, *Tigriopus,* when the fry reach about 7 mm (copepod nauplii may

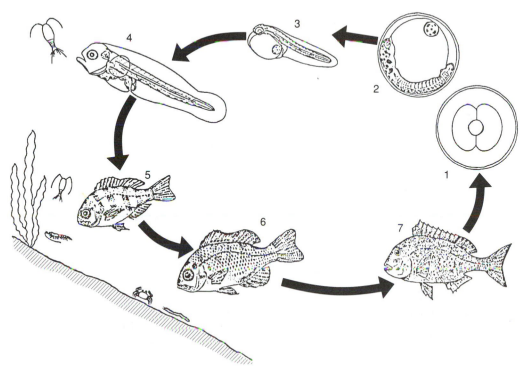

Figure 8a.5 Life cycle of the red sea bream. (From Kafuku and Ikenoue, 1983. Copyright © Elsevier Scientific Publishing Company, New York. Reprinted by permission.)

be fed to the bream when they are only 4 mm). Brine shrimp and wild zooplankton may also be added to the diet of the larvae. Much research has been conducted recently on larval nutrition and the development of artificial diets for the young bream. At 10 to 15 mm the fish begin to receive a slurry of fish, crustaceans, and a vitamin mixture.

After the larval growout, the red sea bream juveniles may be released to enhance the natural stocks of fish in the coastal waters, or they may be transferred to nylon net cages. The fish in the cages are fed a diet of fish (sand eel, mackerel, anchovy, sardine, and others) and euphasids if available (which prevent discoloration); pellets are also used. The growout to 0.5 kg takes 18 to 24 months.

SNAKEHEAD

The **snakeheads, *Channa*** and ***Ophicephalus*,** belong to a small family (Channidae) of freshwater African and Asian fishes with appreciable abilities as air breathers. These very hardy animals, which can live out of water for long periods of time, are carnivores with voracious appetites. There are two culture methods

1. Snakeheads are often used in polyculture pond situations where they may simply be used as predators to control the size of another fish's population (although some farmers may consider the snakehead a pest to be eradicated).
2. They have been grown in monoculture using floating cages, in Cambodia and Taiwan, where they have been trained to take a feed of 80% minced fish and 20% formulated eel feed or wheat flour.

Young snakeheads can be collected from natural waters or drainage canals, or brood stock kept in ponds can be used for a supply of eggs. Snakeheads build simple nests on the pond bottom where the females lay the eggs for male fertilization. Spawning can be induced before the normal spawning season begins by hormone injection.

The fertilized eggs float, and the males of some species build bubble nests in the vegetation on the pond surface; the nests, which are guarded by the males until the eggs hatch, can be easily collected with nets and transferred to nursery containers. Aeration is not needed; in fact, any change in the water must be done carefully so as not to disturb the eggs.

After the larvae have hatched and the yolk sac is absorbed (in about three days), the fish are moved to a nursery pond. The nursery pond should be first dried and then fertilized so there is a zooplankton bloom when the young snakeheads are added. *Daphnia* and rotifers are sometimes cultured separately and added to the nursery ponds. Over the next five to eight weeks the fish learn to accept tubifex worms and then chopped fish. When they reach 4–6 cm, they are transferred to larger rearing ponds or cages.

SEABASSES AND TEMPERATE BASSES

Of the **seabasses** (family Serranidae) the most important aquaculture species is *Dicentrarchus labrax.* This is a largely carnivorous fish with a valuable fillet. These seabass are stocked with eels in ponds in France and are grown in estuarine enclosures in Italy. In the Italian system, wild fry are caught and stocked in lagoons with reed fences or in ponds.

Future development of an intensive industry revolves around the fact that the growth of these fish can be sped up using warm water. Under such conditions, seabass will feed throughout the winter. *D. labrax* can also withstand low salinities; researchers therefore have mixed seawater with water from warm springs to increase the growth rate. Techniques for the mass production of larvae have also been developed.

Much research on the development of diets for *Dicentrarchus* has been carried out in Europe. Larvae are reared initially on rotifers and can be switched later to brine shrimp; the value of the brine shrimp to the young seabass is a function of what the *Artemia* consumes before being fed to the fish. As the larvae grow, they may be introduced to pelletized diets. The best growth and feed conversion ratios for larger fish receiving prepared feeds are seen with diets having protein levels of about 60%, although the amount of fat in the tissue is reduced when higher levels of protein are used.

The **gag, *Mycteroperca microlepis,*** is a commercially important grouper in the coastal waters of the southern United States. Females of this species are common in shallow water, but males are rare and often collected only at great depths, resulting in their being injured or severely stressed when brought to the surface. Scientists therefore took advantage of the fact that these fish are protogynous sex changers; sex inversion, females being changed to males, was induced by oral treatments of methyltestosterone. The normal females ovulate after hCG injection and can then be stripped of their eggs, and milt may be taken from the "new" males, for artificial fertilization.

Epinephelus akaara, the **red grouper,** is cultured in Hong Kong. There is some controlled production of the fingerlings, but most of the seed fry for farming come from the wild. The most important method of production entails the use of cages, taking two to three years to produce fish of 0.5 to 1 kg.

Two freshwater temperate basses (family Percichthyidae)—the **callop (*Plectroplites ambiguus*)** and the **Murray "cod" (*Maccullochella macquariensis*)**—may be important aquaculture organisms in Australia in the future. The callop can be stripped and fertilized by hand or may

spawn naturally in ponds when the temperature rises and the pond is flooded with freshwater. The Murray cod spawn when the ponds reach 20°C and there is simultaneous pond flooding.

BUFFALOFISH

Buffalofish belong to the family Catostomidae, the suckers. These are large (to 35 kg), benthic fish, getting a large portion of their nourishment from detritus generated by fertilization and agricultural wastes added to the ponds. When buffalofish are grown, the species of choice are *Ictiobus cyprinellus*, the bigmouth buffalofish, or a hybrid produced by crossing a female black buffalofish (*I. niger*) with the male bigmouth. The bigmouth differs from other suckers in being able to filter plankton from the water.

Ictiobus was once commonly cultured, but during the 1960s its popularity gave way to the channel catfish; buffalofish in Arkansas were cultured on over 1400 hectares in 1960, but by 1966 this declined to about 100 hectares. There are several reasons for this

1. During certain periods of their life, buffalofish are highly susceptible to parasites and disease.
2. The consumer has failed to purchase smaller fish (which, in turn, forces the farmer to extend the growout period to two years) and will pay much higher prices for catfish.

The commercial production of *Ictiobus* has all but ended; catfish cost more to raise but demand a much higher price than buffalofish. Most buffalofish sent to market today are either caught wild or polycultured with catfish.

Broodfish, about 1.3 kg, are placed in "wintering ponds" just before spring. Because these fish produce a substance that inhibits breeding when the density is too high, the farmer can stock them in such a way as to prevent unwanted spawns. As the temperature begins to rise, the amount of food given the fish in the wintering ponds is increased. Pelleted diets have been developed for these maturing fish. When the temperatures have risen to 18°–21°C, the fish are transferred to spawning ponds. The reduction in density and the "clean" water (without the **buffalofish inhibiting substance**) results in a spawn.

Hatching usually takes place within a week of spawning. The ponds are fertilized immediately after spawning so the fry can feed on plankton. The newly hatched fry are delicate; handling is not advised, but the parents should be removed from the pond to avoid the transmission of pathogens. After about two weeks, the farmer can begin feeding the fry. At 12 to 40 mm the buffalofish are transferred to nursery ponds where they are kept until being moved to a growout facility. The culture period may last up to two years for the production of large (2.5 kg) fish.

Like crawfish, buffalofish can be rotated with rice crops. Stocking at 25 to 250 fish per hectare is used; the fish are not fed, but the ponds are fertilized occasionally.

WALLEYE AND PERCH

The family Percidae consists of freshwater fish native to the northern hemisphere. Most of the fish commonly identified as **perch** are too small to be of interest to the culturist; exceptions are the yellow perch, *Perca flavescens* (see Figure 8a.6), and the European common perch, *P. fluviatilis*. (The white "perch" and its relatives in the genus *Morone* are not true perch but belong to the family Percichthyidae.) Some experimental work has been done on the culture of yellow perch, but there is no commercial production of this species. The European com-

Figure 8a.6 Yellow perch. (From Whitworth, Berrien, and Keller, 1968. Courtesy of Connecticut Department Environmental Protection.)

mon perch is not commercially grown out, but is sometimes spawned to repopulate a body of water; spawning takes place naturally or the eggs can be stripped and fertilized.

The North American **walleye,** *Stizostedion vitreum* (see Figure 8a.7), is the largest of the percid fish, and commonly cultured, as is its relative the European perch-pike, *S. lucioperca*. Walleye × yellow perch hybrids have been developed but generally are not used in culture. Walleye are not suitable for pond culture, but rather are released into rivers and lakes with sand or gravel bottoms and clean water, to support sport fishing. They are voracious predators that come near the surface in the evening to feed.

Walleye spawn on gravel beds in streams or lakes in the spring, when the water reaches about 10°C. To culture these fish, the brooders are captured at night with nets set near a spawning area, and the gametes from ripe adults are stripped. The newly stripped eggs are very fragile because the envelope of the egg has not yet hardened.

Figure 8a.7 Walleye. (From Whitworth, Berrien, and Keller, 1968. Courtesy of Connecticut Department Environmental Protection.)

Hardening the eggs is important because otherwise the sticky eggs will form clumps, which reduce the surface area and therefore the site of oxygen exchange. Hardening may be encouraged by one of several methods

1. Stirring the eggs with a large turkey feather one and a half to two hours after fertilization.
2. Scouring the surface of the eggs with particles by placing them in baths of starch or clay for a few hours, and then carefully washing them off.
3. Treating the eggs with tannic acid or a protease solution.

The hardened walleye eggs are put into incubation jars and gently suspended with a water current. Dead eggs float to the surface and can be removed. At 10°C, the eggs hatch in about 20 days. The fry are carried by the water current to a pond where they are kept for a few days until the yolk sac is absorbed. From these ponds, they may be released into natural waters or (more commonly) they may be kept and later released as fingerlings. If walleye fingerlings are to be produced, the fry are fed rotifers, planktonic crustaceans that are either collected or cultured, and insect larvae; the preferences of the young fish change as they develop. Some facilities use fertilized earthen ponds inoculated with copepods and *Daphnia* before the fry are added. The walleye reaches 7 to 8 g in just a few weeks, but during this period, cannibalism can cause significant losses.

There has been much research recently on the diets of walleye fry and fingerlings. Formulated diets are accepted by the newly hatched larvae, but survival is normally minimal, although researchers recently reported that the use of a special graded feed, the size of which was changed as the animals grew, yielded a 28% survival after 30 days with good growth. However, if there is a slow transition from planktonic crustaceans to a high-quality dry diet over the first two months after hatching, an even better survival probably will be observed.

Perch-pike are found in the warm, calm lakes of Europe. They begin their life as planktonivores and then become active piscivores. These fish do well in waters having a moderate turbidity as long as the DO levels are not too low. Spawning takes place in the spring when the water temperatures reach 12°–16°C; the fish form pairs to mate, and spawning takes place in an area selected by the male. The parents guard and fan the eggs, which hatch in 1 to 2 weeks. Like walleye, the perch-pike does not handle well, especially when young, and therefore is grown mostly to restock natural waters (although they are occasionally grown for food or to control overproduction of other pond fish).

Reproduction of perch-pike is allowed to occur naturally; stripping and hormone injections are not used. There are two spawning methods

1. The **semicontrolled method** of reproduction makes use of large spawning ponds into which brood fish are released.
2. **Controlled reproduction** involves the use of artificial nests made of roots, twigs, or branches wired to wood frames. The fish are released into a pond with no vegetation on the bottom and then lay their eggs on the nests. The nests are removed and the eggs are kept moist in a spray of water. When the eggs are about to hatch, they are transferred to nursery ponds, rich in plankton. If the perch-pike is to be raised past the first few weeks, a source of forage fish must be available.

PIKES

The **pikes** of the family Esocidae include only freshwater fishes of the northern hemisphere. The fish of interest for culturists are the **common** or **northern pike,** *Esox lucius,* the **chain pickerel, E. niger,** and the **muskellunge** or **"muskie," *E. masquinongy.*** In hatcheries a hybrid, the "tiger muskie" (female muskellunge × male common pike) can be produced (see Figure 8a.8). All these fish like calm, shallow water with plants such as reeds in which they can hide. They are stocked to support the natural populations, which may be heavily fished, and as piscivores they are useful in controlling the size of other fish populations.

Common pike spawn in the spring, the eggs adhering to vegetation. Broodfish may be collected in the natural spawning area or they can be grown in ponds specially constructed for this purpose. The gametes are taken from the mature fish and fertilization is carried out by the dry method (see Figure 8a.9). The eggs are incubated in flowing water in jars that are often kept in a hatching shed to prevent exposure to direct sunlight (although complete darkness is not necessary); the water must be clean and well oxygenated and may be heated if too cold. Hatching takes place after 120 degree days, but in some systems, just before hatching, the eggs are removed from the jars and trans-ferred to trays placed in a trough. The pike absorb their yolk sacs about 160 to 180 degree days after hatching.

The hatchlings attach themselves to the walls, periodically leaving for a short swim before returning. When they begin to swim without returning to the walls, they are transferred either to natural waters or to ponds built for the production of fingerlings. If the pike are to be released in natural waters, they should be handled very gently and distributed in calm water among the tall plants along the edge of a lake.

Liberation of fingerlings, rather than younger fish, is often a better way to repopulate a lake, since mortality is lessened and results are more predictable. Small pike are not cannibalistic and can be kept in relatively crowded conditions in the hatching trough until they are 3–5 cm, during which time they are fed zooplankton, then released. Alternatively, they can be moved to fertilized ponds after they begin to swim; stocking in these ponds should be $1.0-2.5 \times 10^4$ fry per hectare. The animals are released from the ponds or troughs after five to eight weeks in the summer as fingerlings. The fingerlings may also be used in polyculture ponds to control carp, roach, and tench.

Figure 8a.8 Tiger muskie. (Photo courtesy of the New Jersey Division of Fish, Game, and Wildlife.)

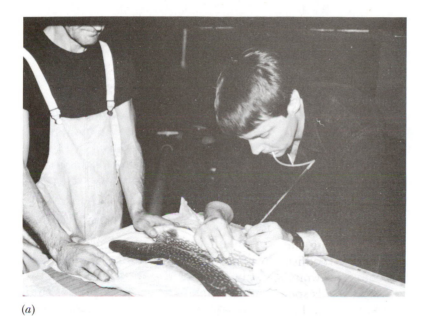

(a)

(c)

(b)

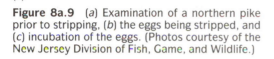

Figure 8a.9 (a) Examination of a northern pike prior to stripping, (b) the eggs being stripped, and (c) incubation of the eggs. (Photos courtesy of the New Jersey Division of Fish, Game, and Wildlife.)

SHAD

Shad, along with herring and menhadens, are members of the family Clupeidae. Several species are of interest to culturists, including *Alosa sapidissima* **(American shad),** *A. alosa* **(European** shad), *Hilsa ilisha* **(Indian shad),** *Dorosoma cepedianum* **(gizzard shad),** and *D. petenense* **(threadfin shad).** These are not intensely cultured, but are stocked in natural bodies of water as forage for other fish or to support an existing fishery.

The threadfin is a prey species used in

warm lake waters of the southern United States. They do not overwinter well in higher latitude lakes, but are sometimes stocked as adults in the spring because they spawn and their fry serves as forage. Brood shad are first spawned in ponds at a hatchery, and the offspring are transferred to an indoor facility for the winter; they accept prepared diets and may be released for stocking when the water in the northern lakes warms sufficiently.

The American shad, an anadromous species, has long been of interest to culturists because it is a tasty fish, and the roe is considered a great delicacy. Initial efforts were concerned with increasing dwindling Atlantic coast stocks by releasing young shad into the ocean; there is no evidence that these nineteenth-century programs significantly affected the shad stock, although modern programs with the same aims have recently begun and may have more success. However, cultured shad released on the Pacific coast about a century ago did establish themselves and now support a large fishery.

Ripe *Alosa* are collected, stripped, and fertilized using the dry method. After fertilization, the eggs are gently poured into a large container where, if shipping is required, they are water-hardened. The eggs are incubated in hatching jars with water at about 17°C flowing upward. As hatching approaches, the jars are moved to an area by the concrete or fiberglass rearing tanks, so that the larvae after hatching rise to the top of the jar, and are carried into the tanks with the overflow water. The fish at hatching are about 7 mm, and the yolk sac is absorbed in three to five days.

Formulated feeds have not been used with any success as starter diets for shad. The shad must be fed decapsulated brine shrimp cysts or newly hatched brine shrimp nauplii for at least 7 weeks; the brine shrimp may be supplemented with *Daphnia*. A starter diet may be introduced with the live food after 7 or 8 weeks, and at about 16 weeks the shad may be fed

only crumbles. The fish continue on a diet of crumbles until they are 8 to 9 cm long.

STURGEON AND PADDLEFISH

Sturgeon (family Acipenseridae) and **paddlefish** (family Polyodontidae) are large, primitive Chondrostean fish of the northern hemisphere with heterocercal caudal fins and skeletons that are largely cartilagenous. The sturgeons, *Acipenser* and *Huso*, have bony scutes along the sides of their body, barbels in the front of the mouth, and no teeth as adults; these are freshwater and anadromous organisms. The lake sturgeon of the United States, *A. fulvescens*, may reach nearly 3 m and weigh over 120 kg, while the giant Eurasian beluga, *H. huso*, may reach 5 m and weigh over 1.3 metric tons. The paddlefish, *Polyodon* and *Psephurus*, have a paddlelike snout, few scales, and minute teeth and barbels.

Sturgeon are disappearing from many of the waters that once supported them; culture of these fish, therefore, is largely for restocking. The reasons for the sturgeon's disappearance are overfishing, a relatively advanced age of sexual maturity (usually over 6 years depending on the location and population, but *H. huso* takes about 18 years), pollution, and dams that interfere with migration. Sturgeon swim upstream to spawn in gravel beds; with the exception of the freshwater species, the sturgeon leave freshwater after a year to go into brackish water where they remain for several more years before entering the sea. Culture was attempted in the United States during the late nineteenth and early twentieth centuries, but after many technical problems and very limited successes, this project was abandoned by governmental hatcheries. Later success was achieved in Europe, especially in the USSR.

Sturgeon of several types are used in the USSR, including one hybrid, the "**bester**" (beluga × sterlet). Culture of the sturgeon starts with the capture of ripe females. Securing enough ripe females at one time is still one of the culturist's major problems. Sturgeon can be brought to maturity in specially built ponds with a relatively strong current moving through them. Hormones also have been used for stimulating both male and female sturgeon. When the fish are ripe, the belly is soft, and gentle pressure causes the eggs or milt to run out. Because the fish are so large and therefore difficult to handle, they are killed for stripping.

Several million eggs may be obtained from a female, but because the eggs do not mature simultaneously, only $1-3 \times 10^5$ are in the correct developmental stage at any time. The eggs are collected in a pan under the fish and washed for a few minutes to remove the blood and mucus. A little water is added to the eggs after washing, and milt is spread over the eggs immediately; this is followed by hand mixing. After the eggs are mixed, they are allowed to stand for a few minutes, during which time fertilization is completed and hardening begins. The eggs are washed a second time to remove the excess sperm and the egg's sticky coating (which would cause clumping and adhering to the sides of the hatching vessel).

Hatching of the eggs takes place in incubators that are hung in lines in the river. The eggs must be inspected and cleaned of silt. Incubation is short, usually 90 degree days or less. The yolk sac is absorbed in 5 to 10 days. The sturgeon can be released at this point, but survival is so low that culture is often extended for a short period (2 to 3 weeks), a slightly longer period (4 to 6 weeks), or an extended period (2 years or more)

1. Short culture does not exceed 20 days after hatching and takes place in circular tanks or floating rectangular tanks in ponds. Fairly high densities are used, and after the yolk sac is absorbed, the fish are fed planktonic crustaceans, worms, and insect larvae.

2. The 4- to 6-week culture technique is carried out in ponds and begins with either sac fry or the 2- to 3-week-old fish from tanks. The ponds are fertilized and stocked with 4 to 6 fish per square meter. The sturgeon are 1.5 to 3.5 g at the end of the culture period and can be safely released.

3. When sturgeon are kept for 2 years or more they are sold as food, not used for restocking (see Figure 8a.10). Culture generally starts with the captured sterlets, *A. ruthenus*, which are caught as juveniles and transferred to carp ponds. This practice is uncommon in the USSR.

Culture of sturgeon in the United States, at least on an experimental basis, has been regaining momentum since the 1970s, thanks in part to techniques borrowed from the USSR. While the American technologies are still developing, there are now some basic differences in the techniques used by eastern and western culturists. Some of the American innovations include the use of hatching jars for the longer incubating species, and the use of prepared diets. Farms in California are currently marketing 3–5 kg sturgeons two to three years after hatching. The

Figure 8a.10 Cultured sturgeon. (Photo courtesy of Dr. Ferenc Muller, Fish Culture Research Institute, Szarvas, Hungary.)

species of interest are the lake sturgeon (*A. rubicundus*), white sturgeon (*A. transmontanus*), Atlantic sturgeon (*A. oxyrhyncus*), and the endangered shortnose sturgeon (*A. brevirostrum*).

A possible candidate for culture in the southern United States is the **"spoonhill catfish"** or **paddlefish,** *P. spathula,* which may weigh as much as 85 kg. Large-scale culture currently is essentially restricted to producing fingerlings for restocking, but this may change in the near future. There are several reasons for considering paddlefish farming

1. The fish is tasty and has recently supported a modestly large fishery, and the roe is a very acceptable caviar.
2. It feeds on zooplankton, which can be produced very simply by fertilizing the ponds.

In the past, the biggest problem with culturing paddlefish involved spawning. However, lately advances have been made in the collection of gravid females and the use of hormones to stimulate ovulation. In paddlefish, all the eggs are ovulated into the body cavity, and under normal conditions they would be released into the water slowly over a period of 8 hours to 1 day. Recently it has been shown that the eggs may be removed quickly by a Caesarean section operation, and the broodfish can be sutured closed and released in excellent condition. The eggs are then fertilized with sperm from a selected male; males release sperm within a day of hormone injection. Paddlefish milt is rather watery, and hand stripping produces large volumes but makes the males unable to reproduce for the rest of the day. Therefore, seminal fluid is collected by catheterization, which allows the culturist to collect only the necessary amount.

The fertilized eggs are hatched in jars. Hatching jars are generally supplied with a fairly vigorous flow of water, which not only aids in the hatching but also reduces fungus infection, a serious problem for these eggs which are sensitive to chemicals. The eggs hatch in 5 to 7 days when held at about 17°C. At hatching, the larvae are carried by the outflow from the jars into collection tanks. Culture is usually continued in one of three ways

1. The larvae are stocked in fertilized ponds at about 5×10^4 per hectare and are harvested by seining or draining. If the zooplankton population in the ponds is good, the fish will be about 30 cm when they are harvested after a period of 140 days.
2. Larvae can be placed in polyculture with catfish. Since the catfish wastes serve to stimulate production of plankton, the ponds need not be fertilized.
3. Paddlefish can be intensively cultured in tanks and troughs. These fish are started on a diet of *Daphnia* but can be trained to accept a pelletized prepared diet.

WHITEFISH

Whitefish are members of the family Salmonidae, and belong to one of three genera of the subfamily Coregoninae, *Coregonus, Prosopium,* and *Stenodus,* although in some areas the term "whitefish" refers only to *C. lavaretus,* a common fish in northern Europe reaching 0.5 m. Depending on the species of fish, the adults may be planktonivores, may feed on the benthos, or may be piscivores.

Coregonids live in the temperate lake waters of the northern hemisphere; like other salmonids, these fish require good water quality. They are cultured largely to repopulate and stock the waters that have supported these organisms in the past. Most of the work on whitefish is currently being conducted in the Soviet Union and other central and northern European countries. Coregonids are very important fishery products in Canada, so an increased Canadian interest in whitefish culture would not be surprising.

Several species of coregonids that are being grown in addition to the common whitefish (*C. lavaretus*) are

1. The vendace or European cisco (*C. albula*), which is found in lakes as well as marine environments.
2. The muksun (*C. muksun*), an anadromous fish found in coastal waters.
3. The peled (*C. peled*), found in rivers and lakes.
4. The broad whitefish (*C. nasus*).
5. The inconnu (*S. leucichthys*), which reaches 40 kg and is usually anadromous.
6. Several hybrids, including *C. nasus* or *C. muksun* × *S. leucichthys*.

Spawning takes place during the early winter, but spawning dates may differ significantly according to environmental conditions in the lake. For any particular lake, the spawning period is short (less than two weeks). Nets are laid out in the spawning areas in the evening, and the whitefish are harvested in the morning. The ripe eggs, which are transparent, are easily stripped from the female; if the female is not quite ready to spawn, she may be held for a few days until maturation is complete. The gametes are stripped from the fish in the same manner as they are for other salmonids, they are fertilized using the dry method, and incubation is carried out in jars with a water flow that allows a slow mixing. Dead and unfertilized eggs should be removed. The development of the whitefish embryos is rather slow, often requiring over 300 degree days (the water temperature should not exceed 10°C, and for some species must be even lower). When hatching takes place, the water current in the jar lifts out the fry, and they fall into rearing troughs below the jars.

The yolk sac is absorbed a few days after hatching, at which time the fish can be released into the lakes for restocking. Alternatively, the fry can be kept in the hatchery for a few weeks. Until recently it was believed that the fry must receive a diet of zooplankton, which restricted extended culture to those hatcheries having facilities for the culture or capture of plankton; however, scientists have developed dry diets that are acceptable to whitefish fry and support good growth.

OTHER FISH

Hundreds (at least) of additional species of fish have gone through some phase of culture: induced maturation, culture of the embryos through hatching, feeding of the larvae, growth of the adults under particular conditions, and so on. Some of those fish are now being grown or have been the subject of several pilot investigations. It is beyond the scope of this book to talk about *every* fish, but a few others can be introduced.

There are several fishes of the family Anabantidae that are important for cultivation. The **gouramy, *Osphronemus goramy*,** of Java and other parts of Southeast Asia, is a herbivore and planktivore that is cultured in freshwater ponds. They will reproduce in the ponds if the farmer supplies the correct nest-building materials (palm branches or bamboo sticks) to the fish. Two subspecies of *Trichogaster pectoralis,* the **blue gouramy** and **snakeskin gouramy,** are also found in Southeast Asia. These are planktivores and detritivores, which are harvested when they reach about 50 g. The male builds a bubble nest in the ponds and guards the eggs, but artificial fertilization is carried out during some parts of the year. The **kissing gouramy, *Helostoma temminicki*,** is also harvested for consumption although the fish do not get very large. They are planktivores and may be used in polyculture; kissing gouramies grow rapidly and are tolerant of low DO levels. They will reproduce in ponds, laying floating eggs. These fish are also grown in rice fields, reproducing after the rice has been harvested.

The **silverside** or **pejerrey, *Odonthestes (=Basilichthys) bonariensis,*** of South America belongs to the family Atherinidae. These fish have been raised in Argentina since the early twentieth century, growing well (up to 50 cm) in fresh and slightly brackish waters. The fish are generally captured when ripe, and artificial fertilization is carried out using the dry method. The eggs are incubated in jars. The fry are transferred to ponds to feed on plankton, although supplemental feeding may be carried out later. The fish are reared like this for several months and then released to restock natural waters. The adult pejerrey is considered a luxury food. Culture of a similar fish, *Chirostoma estor,* has been considered by some to be promising in Mexico.

The Characidae is a very large family of fishes, some of which have potential as cultured species in Latin America. A number of these organisms occupy different ecological niches, mostly low in the food chain. *Brycon* populations have been established in Peruvian lakes and rivers, and in experimental polyculture conditions grew at 2.3 g per day. Laterally compressed, disc-shaped fish such as *Colossoma macropomum, C. brachypomum,* and *C. mitrei* are cultured, either commercially or experimentally, in freshwater ponds in South America. *C. macropomum,* stocked at 5000 fish per hectare, grows at 1.5 to 4 g per day to yield 4.4 to 6.8 metric tons per hectare; this fish has also been raised in polyculture ponds, concrete tanks, and net cages. Less effort has been devoted to the other species, but a hybrid (*C. macropomum* × *C. brachypomum*) has been developed.

The cichlids include many ornamental fish as well as tilapias. The **pearl spot,** ***Etroplus suratensis,*** is a brackish water member of this family that is cultured on the Indian subcontinent. It is a herbivore, feeding on algae and decaying plants. Pearl spots are grown in ponds where they lay their eggs on structures placed in the water by the farmer.

A few freshwater **loaches** of the family Cobitidae, such as *Cobitis* and *Misgurnus,* are stocked in rice fields in Japan where they are later harvested and eaten. The mud loach, *M. anguillicaudatus,* can be spawned throughout the year using hormone injection.

The family Coracinidae, the **galjoens** of Africa, is a small group of coastal marine fishes that include *Coracinus capensis,* a greatly prized sport fish and a valuable food. This fish spawns in captivity when held in flowing seawater, and the larvae can be reared initially on a diet of rotifers and copepods.

Prochilodus is a member of the family Curimatidae. This South American detritivore has been the subject of research recently; in cattle watering tanks growth rates of 0.69 to 2.5 g per day have been reported.

The **lumpfish** and snailfish in the Cyclopteridae family may seem like unlikely candidates for culture. However, the lumpfish or "lumpsucker," *Cyclopterus lumpus,* produces a fine roe that is exported from Canada to Europe; in 1983 over 10^6 kg of this roe was marketed. The fish is also filleted and sold as food. Lumpfish can be spawned in aquaria, and the larvae can be reared on commercial dry feeds, although brine shrimp seem to support better growth.

Beside the carps, a number of other cyprinids are also cultured in Europe. The **tench (*Tinca tinca*)** is often in kept polyculture with the common carp, and the **rudd (*Scardinius erythrophthalmus*)** is sometimes used to restock canals and other bodies of water. The cyprinid **"bream"** (such as *Abramis* and *Parabramis*) have for many years played a minor role in European aquaculture. In lakes in Poland that have become eutrophic, "bream" are stocked and thrive; they are also raised in ponds in the USSR (mainly in the Ukrainian Republic, where they may be kept in polyculture with the common carp), and in China, where they are polycultured with Chinese carps.

The **sand goby, *Oxyeleotris marmoratus,*** is a member of the sleeper family, Eleotridae. These are grown in cages and ponds in Southeast Asia. In ponds they will eat benthic invertebrates and reach a weight of about 0.9 kg in one year.

Cod (family Gadidae), which were popular culture organisms years ago (Chapter 1), again are being grown in pilot studies in Norway and Canada. Rather than being released to support a fishery, however, they are being considered as candidates for cage culture.

The **mud skipper, *Boleophthalmus chinensis,*** is a member of the Gobiidae family. A small (15 cm long and 40 g in weight for fully grown adults), brackish water organism, it is among the most highly priced fish in Taiwan. The seed stock for the culturists are purchased from specialized collectors who gather small (1.5–3 cm) fish in the summer in estuaries. After fertilization of small, brackish ponds to encourage the growth of algae, the juveniles are stocked at about 3×10^4 fish per hectare. By the time the mud skippers reach about 5 cm, they have dug deep burrows and grazed down most of the algae; at this time the water can be drained and the bottom dried, refertilized, and refilled with brackish water while the animals are safe in their burrows. The mud skipper takes about two years to grow to marketable size (25–40 g). They are harvested with a dip net placed near the water inlet or by using small bamboo traps placed at the entrance of the burrows by professional mud skipper catchers.

A primitive fish of the family Osteoglossidae (bonytongues), *Heterotis* is commonly cultured in several parts of Africa. This fish spawns in large nests in swamps, and the eggs and young are guarded by the parents. They feed on plankton as well as on larger food particles, including peanut oil cakes or crushed cottonseed. They may be grown in polyculture. The **pirarucu, *Arapaima gigas,*** of South America may reach 3 m in length and weigh up to 200 kg. This fish has been spawned in ponds and used to stock lakes; in an intensive culture trial project, it grew from 15 g to 4 kg in 1 year and to 15 kg after 26 months. *Arapaima* feeds on live fishes, but can be trained to take artificial feeds.

The family Plecoglossidae, anadromous fish of the Orient, includes the **Ayu, *Plecoglossus altivelis*** (see Figure 8a.11). *Plecoglossus* has been reared from eggs stripped from the adult fish and is also often raised from fry that are collected from the wild. The fish are grown in raceways and ponds on prepared feeds. In

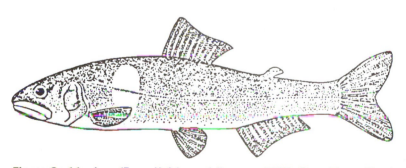

Figure 8a.11 Ayu. (From Kafuku and Ikenoue, 1983. Copyright © Elsevier Science Publishing Company, New York. Reprinted by permission.)

water of 21°–23.5°C they grow quickly if well fed, increasing from about 5 cm to a marketable 20 cm in about three months.

The Hawaiian **moi**, *Polydactylus sexfilis*, is a member of the Polynemidae family (threadfins). They are found in shallow coastal waters where they feed on small benthic crustaceans. Moi can be kept in net enclosures in lagoons and spawning (which is related to the lunar cycle) will take place spontaneously, or just prior to spawning they can be transferred to large tanks that allow better recovery of the eggs. The fry are started on a diet of rotifers followed by *Artemia* nauplii, and the larger fish take either sinking or floating artificial feeds. The fish grow quickly and have relatively efficient feed conversion ratios. Although moi are not produced commercially, they have great potential, especially in Hawaii, where an established market exists.

The **graylings** are salmonid fish found in freshwater streams and rivers in the northern hemisphere. Sexually mature *Thymallus* are captured in spawning areas. Spawning and fertilization are handled as they are for other salmonids, with the embryos being incubated in either trays or jars. Grayling fingerlings are produced the same as those of trout. The arctic char, *Salvelinus alpinus*, grows well in high densities (up to 90 kg/m^3). Char have been cultured in cold Canadian lakes, but recent research indicates that anadromous strains of this fish may be suitable for ocean growout. The principal obstacle to char culture is obtaining seed stock.

The **scats** (family Scatophagidae), including the **common spotted scat**, *Scatophagus argus,* are found on the coasts of South and Southeast Asia, and in the tropical Indo-Pacific. *S. argus* is a highly prized eating fish in many places (such as the Philippines) and is also in demand as an aquarium fish. Recently, researchers in Hawaii, Florida, and the Philippines have begun to work out culture methods for this species. It seems that hormone injections can be used with some success to stimulate maturation and spawning. Feeding newly hatched larvae is a problem, but older (9-day) larvae can eat rotifers and later can be switched to *Artemia* nauplii. The scats eventually pass through a larval **tholichthys** stage and become juveniles. Alternatively, the fry, which are very tolerant of changes in salinity, temperature, and pH, may be collected in large numbers during certain seasons. Much more work remains to be done, but the potential rewards are substantial.

Fugu is a highly prized gourmet food in Japan. The culture of these **puffers** (family Tetraodontidae) begins with the capture of broodfish during the spring spawning season. The eggs and milt are hand stripped and fertilized using the wet technique. The eggs are hatched in moving seawater. Larvae are fed first a diet of barnacle nauplii and are later switched to minced fish. The fish are finally stocked in floating cages or ponds and fed trash fish. Some culturists will start with wild 1-year fish instead of using the seedlings grown from eggs. *Fugu* are kept in culture for 1.5–2.5 years and are marketed when they reach 0.8 kg or more.

APPENDIX 9

CULTURED HIGHER VERTEBRATES

FROGS

The native frog populations of North America have been declining for several decades because of both the destruction of their habitat and the hunting of frogs for food or laboratory specimens. While there are few frog farms existing solely for the production of meat for human consumption, diversified operations that also produce frogs as laboratory animals and for other purposes are fairly common in the United States and other countries. The major exporters of frogs (both cultured and hunted) are Brazil, Indonesia, India, Bangladesh, Japan, Turkey, and Yugoslavia. Many of the frogs grown in culture belong to the genus *Rana*, the most important being the bullfrog, *R. catesbiana*. Bullfrogs are popular because they adapt to a variety of temperatures, and because they are larger than most other frogs (up to 0.9 kg).

Most frog farms are designed rather simply, with natural or slightly modified ponds. A bullfrog pond should have a small, relatively deep section that can serve as a refuge for the frogs when temperatures in the shallow areas get too high in the summer or when ice forms in the winter (when the frogs hibernate). However, to facilitate feeding, most of the pond area should be no deeper than 15 cm.

Modern frog farmers may divide their farms into sections including the spawning/hatching area, tadpole ponds, and frog ponds (there may be several of these for frogs of different sizes). Methods developed by researchers, especially D. D. Culley of Louisiana State University, call for more complex culture operations, including such components as breeding area, tadpole hatchery, tadpole and frog growout facilities, prepared feed preparation areas, a place to culture live feed animals, isolation area and disease diagnostic laboratory, and so on. These well-designed facilities should operate much more efficiently than their simple predecessors (see Figure 9a.1).

The bullfrog lays its eggs in the spring in warm regions, or in the summer in more temperate environments. Ponds should have some sort of vegetation to encourage reproductive behavior; the eggs are laid in large masses that adhere to aquatic weeds or other stationary objects. The length of the incubation period depends upon the temperature. The

389

Figure 9a.1 A modern frog culture facility. (Courtesy of Dr. D. D. Culley, Louisiana State University.)

swimming tadpoles that hatch from the eggs are omnivores, favoring soft foods like algae and decaying plants or animals. Metamorphosis of the bullfrog tadpoles to the adults may take as little as several months or up to two years, depending on the diet and the temperature (for this reason, most frog farms are in warm climates). Special diets for tadpoles have been developed by researchers to replace traditional feeds like vegetable wastes, boiled potatoes, or meat and fowl scraps; initially a fine mix of the prepared diet is spread on the surface of the water, and a semisolid food can be used after they have begun to adopt some benthic mannerisms.

The challenge to the frog farmer comes after metamorphosis. Several biological difficulties that must be overcome, in particular

1. Territoriality
2. Requirement of live food
3. Slow growth
4. Disease and predators

Bullfrogs will defend as a territory a section of the shoreline, up to 7.5 m. This obviously reduces the density of the culture. One possible solution is to build a pond with islands and peninsulas, increasing the shore area.

If the bullfrogs do not get enough food, they not only grow slowly and are susceptible to disease, but cannibalism becomes a problem. Adult frogs require live, moving food to stimulate their feeding response. They use their tongue to grab the active prey, and the movement of the captured organism stimulates the swallowing response. Several solutions to the problem of feeding frogs have been proposed. One solution is to culture bait fish with the frogs; the bait fish and their young serve as forage for the frogs, just as they do for piscivorous fish; also, when the bait fish are fed floating pellets, their feeding causes the pellet to bob on the surface of the water, thereby stimulating the frog to eat the pellet as well as the fish. In Japan, a tray with silkmoth pupae, dead silkworms, and other food is placed on a motorized shaker that slowly moves these items around to attract the frog's attention. The Japanese have developed a method of training the frogs to take silkworm pupae by presenting them to the frogs mixed with some form of moving prey; eventually the frogs get used to accepting just the pupae. Other live foods, such as fly larvae, have been cultured for frogs.

A method for rearing bullfrogs on nonliving food has been developed by Dozier Lester in Louisiana. The newly metamorphosed frogs are placed in front of a string that pulls along an extruded feed, mimicking the action of live prey. After the frogs learn how to recognize the food on the string, they will start taking feed that is dropped near them while they are feeding. Feeding takes place on a wire screen that allows the urine and feces produced by the feeding frogs to pass away from the food; it also allows the uneaten food to be collected and used again, and the movement of the frogs on the screen causes the food to "jump," thus increasing the feeding response (see Figure 9a.2). When frogs have learned to take dropped food rather than food pulled on a string, they are moved to a separate area. Older frogs are presented a diet of crawfish heads, a waste product of the local crawfish farming industry.

The problem of slow growth still exists, although this is less an impediment than many people believe. Frogs in culture generally grow more quickly than wild frogs, probably because of their diet, and under good conditions, a laboratory specimen is produced three to five months after metamorphosis, while edible-size frogs take two to three months more. Since frogs' metabolism slows down in cold weather, it has been suggested that indoor cultures could be maintained so that the frogs can be kept under the best possible conditions all year. The economics of this are questionable.

Common predators of frogs are some birds, alligators, snakes, and fish. Some insect larvae feed on tadpoles. Each of these predators has to be treated differently. The most common disease of cultured frogs is the bacterial infection, **"red leg,"** which is treated with antibiotics and salts. Environmental conditions also must be acceptable; for example, water with low calcium levels (less than 4 mg/liter)

Figure 9a.2 Breeder frogs feeding on a screen. (From Lester, 1988. Copyright © Archill River Corporation. Reprinted by permission.)

causes bone deformities unless the diet is supplemented, and pH levels above 8.1 lead to reduced survival (perhaps because there are greater numbers of pathogenic bacteria that flourish at this pH).

To start a bullfrog farm, eggs can be collected from the wild, but the farmer should be sure that the eggs belong to the correct species. The farmer also runs the risk that pathogenic microbes are already present in wild eggs. Clean eggs or brood frogs can be purchased to start the culture. Harvesting of frogs after they have grown depends on the growout conditions; they may be caught by spears, traps, or nets.

ALLIGATORS

There are about 30 operational **alligator** farms in the United States, most in Florida and Louisiana. During the early part of this century, when the human populations of the southern states started to grow very quickly, the numbers of wild alligators began to decline dramatically. In 1973 the American alligators was put on the list of "endangered" animals by the federal government. Since then, a number of management programs have been put into motion, and in 1978 alligators were reclassified as "threatened" but no longer endangered. In fact, the number of wild alligators has increased to the point where alligator hunting for meat, as well as for the valuable hides, is legal again (with the proper permits). However, the earlier decline in the alligator populations sparked considerable interest, in the 1960s, in the idea of farming these animals.

In farms, the alligators are fed five times per week a vitamin supplemented diet of red meat (diets of fish tend to reduce the nesting success). Animals are grown in pens, preferably made of concrete; the stocking density is high when the animals are small, about one alligator per square foot, but the density is reduced as the animals grow. Alligators grow at different rates depending on the culture conditions, especially the temperature and diet; ordinarily they are harvested at about 1.5 to 2 m, which is reached in about three years (see Figure 9a.3). Up to this size the feed conversion ratio is good, but quickly becomes unacceptable as the alligator grows larger.

Some large alligators are kept on the farms for breeding, which starts when the animals are about six years old. Eggs are carefully collected from nesting females, usually after five weeks of incubation; the tops of the eggs are marked since, unlike bird eggs, they must not be turned upside down before hatching. The females require a solitary habitat for nesting, but females that have been reared in pens require only about 10% of the space wild alligators need. One of the most critical steps in the culture operation is the incubation and hatching of the eggs. Aged hay and grass, which reduce the growth of fungi (copper sulfate may be used in addition), are used as the nesting medium in the incubators. The air in the incubator should be moist, and eggs must be kept at 28°–33°C. Temperature during incubation determines the sex of the alligators. Incubation of eggs at temperatures above 34.5°C yields all male hatchlings, while temperatures below 27°C yield all females.

TURTLES

The idea of raising turtles is not new. From the midnineteenth century until about 1915, turtle was found on the best restaurant tables, with Philadelphia being the center of turtle cookery. Diamondback terrapins raised in ponds cost $90 a dozen in 1920!

(a)

(b)

Figure 9a.3   (a) Cultured alligators (left to right: 2.5 years, 1.5 years, 0.5 years), and (b) environmentally controlled tanks on an alligator farm. (From Avault, 1985. Copyright © Archill River Corporation. Reprinted by permission.)

Modern culture of sea turtles began several decades ago in Florida and the Caribbean, largely as an effort by conservationists to maintain the dwindling natural stocks. Pilot farms are operating to determine the feasibility of producing **green sea turtles (*Chelonia mydas*)** (see Figure 9a.4). Animals, both wild stock and farm-reared stock, have been mated in beach/intertidal zone habitats, and fairly good hatching rates can be obtained from the turtles. These breeding animals can be fed a high-protein diet, and turtle grass (*Thalassia*) is sometimes added.

The **soft-shell turtle, *Trionyx sinensis*,** is cultured in Taiwan. This freshwater species is found in lakes, streams, and ponds, especially where there is a muddy bottom. An active feeder in warm waters, it digs a hole and hibernates in mud when temperatures drop below 12°C. These turtles are cultured in ponds with a mud bottom but with brick or concrete walls that have an overhanging lip to prevent escape; ponds should be in open sunlight and the water source should be warm (not from a cold spring). These turtles are cannibalistic, so they must be segregated by size into different ponds; this is particularly critical for newly hatched soft-shell turtles.

Larger *Trionyx* farms may have a special pond for spawners, with a small house having a gently sloping, clean sand bottom for egg laying and incubation. The house has a wooden plank that serves as a ladder for the brooders, leading from the water into the house. The turtles may start to spawn as early as the second year, but the eggs are irregular and the hatch is not very good; therefore, turtles that are eight or nine years old usually are used. After the turtles have laid their eggs in the sand, the farmer will examine them. The eggs are grayish brown in color; the unfertilized eggs develop white spots a few days after they are laid and can be discarded. The fertilized eggs are arranged in the sand to incubate at about 30°C and are kept slightly moist. Hatching begins in about 50 days and the hatching rate is often as high as 80%.

Young *Trionyx* are kept in nursery ponds where they are fed twice a day on a diet of tubifex worms and crushed shrimp, which later may be supplemented with an eel feed, minced fish, and boiled egg yolk. A large plastic sheet is some-

Figure 9a.4 Juvenile marine turtles (green turtle on left, loggerhead turtle on right). (From Stickney, 1979.)

times spread over the nursery pond to keep the ponds warm (but should be removed if the temperature exceeds 32°C). Within a year, the juvenile turtles reach 200 g and can be transferred to an adult turtle growout pond.

Sometimes walking catfish, *Clarius*, are added to the ponds to eat the unwanted food, thus maintaining the water quality. The turtles are fed a diet of trash fish mixed with either corn or soybean meal. The rate of growth is a function of the temperature, which ideally is 20° to 26°C. The turtles are harvested at about 600 g, which may be reached in one to three years. Harvesting is started by dragging a net along the bottom of the pond; after the initial collection of the turtles, the ponds are drained and the farmer wades in and removes the remaining turtles by hand.

APPENDIX 10

SUMMARY OF NEW JERSEY'S SHELLFISH BOTTOM LEASE POLICY*

I. Objectives

The objective of the leasing program is to provide bay bottom for commercial shellfishermen for the purpose of planting and cultivating shellfish including aquaculture, layout (wet storage) and cleansing. (specified in N.J.S.A. 50:1–23)

II. General requirements of a lease holder for both initial application and renewals

A. Must be a New Jersey resident. (specified in N.J.S.A. 50:1–25)

B. Must be a holder of a valid commercial shellfish license.

C. Staking requirements

1. There must be at least two (2) stakes at each corner.

2. Line stakes must be placed frequently enough so as to delineate a definite line between corners.

3. Stakes must be visible at high tide.

4. Lease must be staked to be worked.

5. Lease must be staked by May 1 of each year.

a. Failure to stake a lease by May 1 of each year will result in a warning and a 30-day grace period.

b. Failure to comply with the warning and the stake lease within 30-day grace period will result in termination of the lease.

6. Deliberate removal of a lessee's stakes will be considered as prima facie evidence of the intent to invade the lease and subject to a minimum penalty of $100 in addition to those prescribed in N.J.S.A. 23:2B–14.

III. General policies on leasing

A. Lease application procedure: will continue as has had been the practice recently.

1. Two acres maximum per application. The only exception to the 2-acre limitation would

* Typical of many coastal states.

be an old vacated lease within the interior of a block of lots that is greater than 2 acres but less than 3 acres.

2. Additional applications will be considered if the applicant can demonstrate the need.
3. A biological investigation will be conducted for all lease applications.
4. Leasing of areas classified as productive will be discouraged.
5. Application for leases in areas classified as condemned will not be accepted. Should the applicant desire a lease in condemned waters, he or she must first present the council with a written proposal demonstrating the need prior to the processing of an application.

B. Lease term
  1. Annual lease fee
    a. The fee is $2.00 per acre or $2.00 per 100 linear feet of shore line where appropriate.
    b. Special lease fees may be established for specific programs.

C. Lease transfers
  1. Assignee requirements
    a. Current New Jersey resident.
    b. Holder of a valid commercial shellfish license.
    c. Following the effective date of these regulations no transfers will be permitted on new leases acquired through the application process for a period of one year from the date of approval unless there are extenuating circumstances.
  2. All lease transfers must be resurveyed to verify their location.

D. Overstaking and staking of areas not leased
  1. Overstaking. If a resurvey of a lease reveals that the lessee is overstaked he or she shall be required to
    a. Pay the expense of the survey.
    b. Receive a monetary penalty in addition to the survey cost ($100).
    c. Relocate corners and line stakes to proper position immediately.
    d. Failure to relocate corners immediately will result in termination of lease.
  2. Staking of areas not leased. Illegally staking areas for which no valid lease or lease application exists so as to preclude other harvesters will be subject to a minimum penalty of $100.

IV. Disposition of leases
  A. Vacated leases. Any leases vacated or terminated by the lessee or department for any reason (nonpayment, failure to properly stake, leasing by nonresident, or lack of valid commercial license) will be considered to be public bottom and available for harvesting or lease application provided it is not prohibited by other regulations.
  B. Condemned leases. Following the effective date of these regulations any new leases granted through the application process and subsequently condemned for the harvest of shellfish will be terminated. However, the lessee will be given a period of two years during which time he may move any shellfish present to a lease in approved waters. Following the two-year period the lease will be terminated unless the lessee can

demonstrate the need to continue the lease.

V. Topics discussed but not regarded as necessary for inclusion into a lease policy at this time.

    1. No limit on the maximum amount of acreage leaseable by a single individual.

    2. No specific utilization requirements.

    3. No report of how the lease is being utilized.

APPENDIX 11

CONVERSION CHART

If You Take	And Multiply By	You Get	If You Take	And Multiply By	You Get
Acres	43,500	Square feet	Kilometers	3,280.8	Feet
Acres	0.00156	Square miles	Kilometers	0.62137	Miles
Acres	4,406.856	Square meters	Knots	1.1508	Miles per hour
Acres	0.40468	Hectares	Liters	33.814	Fluid ounces
Centimeters	0.3937	Inches	Liters	0.2642	Gallons
Centimeters	0.0328	Feet	Liters	0.03531	Cubic feet
Cubic centimeters	0.06102	Cubic inches	Liters	0.001	Cubic meters
			Meters	39.37	Inches
Cubic feet	1728	Cubic inches	Meters	3.2808	Feet
Cubic feet	0.028317	Cubic meters	Miles	0.869	Nautical miles
Cubic feet	7.48	Gallons	Miles	5280	Feet
Cubic feet	28.32	Liters	Miles	1609.344	Meters
Cubic inches	16.387064	Milliliters	Ounces (fluid)	0.00781	Gallons
Cubic meters	35.3145	Cubic feet	Ounces (fluid)	1.80469	Cubic inches
Feet	0.00019	Miles	Ounces (fluid)	0.00104	Cubic feet
Feet	0.0003	Kilometers	Ounces (fluid)	0.02957	Liters
Gallons	0.1337	Cubic feet	Square feet	0.0929	Square meters
Gallons	0.00379	Cubic meters	Square inches	6.452	Square centimeters
Gallons	3.7853	Liters	Square meters	10.765	Square feet
Grams	0.03527	Ounces	Square miles	640	Acres
Grams	0.0022	Pounds	Square miles	2.589998	Square kilometers
Hectares	10,000	Square meters	Tons	907.185	Kilograms
Hectares	2.471	Acres	Tons	0.907185	Metric tons
Inches	0.08333	Feet	Tons (metric)	2,204.62	Pounds
Inches	2.54	Centimeters	Tons (metric)	1000	Kilograms
Kilograms	2.20462	Pounds			

BIBLIOGRAPHY AND SUGGESTED READINGS

CHAPTER 1

Ackefors, H., 1988. Europe: The emerging force in aquaculture. *World Aquaculture* 19(3):5–9.

Anonymous, 1983. Freshwater aquaculture development in China. FAO Fisheries technical paper 215. Rome: Food and Agricultural Organization of the United Nations.

Anonymous, 1983. *Joint ADB/FAO (SCSP-INFOFISH) Market Studies,* Vol. 6: *The World Seaweed Industry and Trade,* SCS/DEV/83/26. Manila: South China Sea Fisheries Development and Coordinating Programme.

Anonymous, 1984. The status of aquaculture. *Aquaculture Magazine 1984 Buyer's Guide,* 4–10.

Anonymous, 1989. Status of world aquaculture. *Aquaculture Magazine 1989 Buyer's Guide,* 10–20.

Bardach, J. E., J. H. Ryther, and W. O. McLarney, 1972. *Aquaculture: The Farming and Husbandry of Freshwater and Marine Organisms.* New York: Wiley-Interscience.

Bell, F. W., and E. R. Canterbery, 1976. *Aquaculture for Developing Countries.* Cambridge, Mass: Ballinger Publishing Company.

Brown, E. E., 1983. *World Fish Farming: Cultivation and Economics.* Westport, Conn.: AVI Publishing Company, Inc.

Brummett, R. E., 1988. Aquaculture in Jordan. *World Aquaculture* 19(4):18–22.

Chang, W. Y. B., 1985. Pond fish culture in the Pearl River Delta, China. *Aquaculture Magazine* 11(4):45–46.

Coker, R. E., (1914). The Fairport fisheries biological station: Its equipment, organization, and functions. *Bulletin of the Bureau of Fisheries* 34:383–405.

Conrad, J., 1985. Mexico's cooperative oyster and shrimp farms. *Aquaculture Magazine* 11(5):46–49.

Conrad, J., 1986. Mussel production plant in Spanish Galicia. *Aquaculture Magazine* 12(4):57–58.

Deegan, R., 1988. Focus on British Columbia: The growth of an aquaculture industry. *Aquaculture Today* 1(1): 37–40.

Hanson, J. A., and H. L. Goodwin, 1977. *Shrimp and Prawn Farming in the Western Hemisphere.* Stroudsburg, Pa: Dowden, Hutchinson & Ross, Inc.

Iverson, E. S., 1968. *Farming the Edge of the Sea.* Farnham, Surrey, England: Fishing News Books Ltd.

Kirk, R., 1987. *A History of Marine Fish Culture in Europe and North America.* Farnham, Surrey, England: Fishing News Books Ltd.

Kochiss, J. M., 1974. *Oystering from New York to Boston.* Middletown, Conn.: Wesleyan University Press.

Kutty, M. N., 1980. Aquaculture in South East Asia: Some points of emphasis. *Aquaculture* 20:159–168.

Lusha, X., 1986. Jiading County's economic transformation. *China Reconstructs* 35 (July):19–23.

Mead, A. D., 1908. A method of lobster culture. *Bulletin of the Bureau of Fisheries* 28 (part 1):219–240.

Nash, C. E., 1988. A global view of aquaculture production. *Journal of the World Aquaculture Society* 19:51–58.

Palacio, F. J., 1979. Aquaculture policies in Latin America. *Marine Policy* 3 (April):99–105.

Provasoli, L., J. J. A. McLaughlin, and M. R. Droop, 1957. The development of artificial media for marine algae. *Archiv fur Mikrobiologie* 25:392–428.

Radcliffe, W., 1969. *Fishing from the Earliest Times*. New York: Burt Franklin.

Rhodes, R. J., 1985. Status of world aquaculture: 1984. *Aquaculture Magazine 1985 Buyer's Guide*, 4–14.

Rhodes, R. J., 1986. Status of world aquaculture: 1985. *Aquaculture Magazine 1986 Buyer's Guide*, 6–14.

Rhodes, R. J., 1987. Status of world aquaculture: 1986. *Aquaculture Magazine 1987 Buyer's Guide*, 4–16.

Rhodes, R. J., 1988. Status of world aquaculture: 1987. *Aquaculture Magazine 1988 Buyer's Guide*, 4–18.

Rhodes, R. J., 1989. Status of World Aquaculture: 1988. *Aquaculture Magazine 1989 Buyer's Guide*, 6–20.

Saint-Paul, U., 1986. Potential for aquaculture of South American freshwater fishes: A review. *Aquaculture* 54:205–240.

Sarig, S., 1988. The fish culture industry in Israel in 1987. *Bamidgeh* 40:43–49.

Shelbourne, J. E., 1971. *Artificial Propagation of Marine Fish*. London: Academic Press.

Stickney, R. R., 1979. *Principles of Warmwater Aquaculture*. New York: Wiley-Interscience.

Titcomb, M., 1952. Native use of fish in Hawaii. *Journal of the Polynesian Society*, Memoir no. 29.

Weidner, D. M., 1985a. A view of the Latin American shrimp culture industry, Part 1. *Aquaculture Magazine* 11(2):16–24.

Weidner, D. M., 1985b. A view of the Latin American shrimp culture industry, Part 2. *Aquaculture Magazine* 11(3):19–30.

Wildsmith, B. H., 1982. *Aquaculture: The Legal Framework*. Toronto: Edmond-Montgomery Limited.

CHAPTER 2

Aston, R. J., and A. G. P. Milner, 1981/82. Conditions required for the culture of *Branchiura sowerbyi* (Oligochaeta: tubificidae) in activated sludge. *Aquaculture* 26:155–160.

Bardach, J. E., J. H. Ryther, and W. O. McLarney, 1972. *Aquaculture*. New York: John Wiley & Sons, Inc.

Bower, C. E., 1978. Ionization of ammonia in seawater: Effects of temperature, pH, and salinity. *Journal of the Fisheries Research Board of Canada* 35:1012–1016.

Boyd, C. E., 1982. *Water Quality Management for Pond Fish Culture*. New York: Elsevier Science Publishing Company, Inc.

Brown, E. E., and J. B. Gratzek, 1980. *Fish Farming Handbook*. Westport, Conn.: AVI Publishing Company, Inc.

Davis, S. N., and R. J. M. DeWiest, 1966. *Hydrogeology*. New York: John Wiley & Sons, Inc.

Farkas, T., and S. Herodek, 1964. The effect of environmental temperature on the fatty acid composition of crustacean plankton. *Journal of Lipid Research* 5:369–377.

Freddi, A., L. Berg, and M. Bilio, 1981. Optimal salinity-temperature combinations for the early life stages of the gilthead bream, *Sparus auratus* (L.) *Journal of the World Mariculture Society* 12:130–136.

Gross, M. G., 1982. *Oceanography*, 3rd ed. Englewood Cliffs, N.J.: Prentice Hall.

Hanson, L. A., and J. M. Grizzle, 1985. Nitrite-induced predisposition of channel catfish to bacterial diseases. *The Progressive Fish-Culturist* 47:98–101.

Hernandorena, A., 1974. Effects of salinity on nutritional requirements of *Artemia salina*. *Biological Bulletin* 146:238–248.

Huguenin, J. E., and J. Colt, 1989. *Design and Operating Guide for Aquaculture Seawater Systems*. New York: Elsevier Science Publishing Company, Inc.

Huner, J. V., and E. E. Brown, 1985. *Crustacean and Mollusk Aquaculture in the United States*. Westport, Conn.: AVI Publishing Company, Inc.

Kissa, E., M. Moraitou-Apostolopoulou, and V. Kiortsis, 1984. Effects of four heavy metals on survival and hatching rate of *Artemia salina* (L.). *Archives fur Hydrobiologica* 102:255–264.

Knieper, L. H., and D. D. Culley, Jr., 1975. The effects of crude oil on the palatability of marine crustaceans. *The Progressive Fish-Culturist* 37:9–14.

Laing, I., and S. D. Utting, 1980. The influence of salinity on the production of two commercially important unicellular algae. *Aquaculture* 21:79–86.

Landau, M., and R. Pierce, 1986. Mercury content of shrimp (*Penaeus vannamei*) reared in a wastewater-seawater aquaculture system. *The Progressive Fish Culturist* 48:296–300.

Lannan, J. E., R. O. Smitherman, and G. Tchobanoglous, 1986. *Principles and Practices of Pond Aquaculture*. Corvallis: Oregon State University Press.

Lee, C.-S., and L. Krishnan, 1985. The influences of calcium and magnesium ions on embryo survival, percentage hatching and larval survival of the dolphin fish, *Coryphaena hippurus* (L.). *Journal of the World Mariculture Society* 16:95–100.

Losordo, T. M., R. H. Piedtahita, and J. M. Ebeling, 1988. An automated water quality data acquisition system for use in aquaculture ponds. *Aquacultural Engineering* 7:265–278.

Lovell, R. T., and D. Broce, 1985. Cause of musty flavor in pond-cultured penaeid shrimp. *Aquaculture* 50:169–174.

Marshall, G. A., R. L. Amborski, and D. D. Culley, Jr., 1980. Calcium and pH requirements in the culture of the bullfrog (*Rana catesbeiana*) larvae. *Proceeding of the World Mariculture Society* 11:445–453.

Neess, J. C., 1949. Development and status of pond fertilization in central Europe. *American Fisheries Society Transactions* 76:335–358.

McCarraher, D. E., 1962. Northern Pike *Esox lucius*, in alkaline lakes of Nebraska. *American Fisheries Society Transactions* 91:326–329.

Melancon, E. J., Jr., 1975. Design and use of a continuous-flow system for determining the oxygen tolerance of juvenile red swamp crawfish, *Procambarus clarkii*. M.S. thesis, Louisiana State University, Baton Rouge.

Moyle, P. B., and J. J. Cech, Jr., 1982. Fishes: An Introduction to Ichthyology. Englewood Cliffs, N.J.: Prentice Hall.

Riley, J. P., and R. Chester, 1971. *Introduction to Marine Chemistry*. New York: Academic Press.

Rosseland, B. O., and O. K. Skogheim, 1986. Neutralization of acidic brookwater using a shell-sand filter or seawater: Effects on eggs, alevins and smolts of salmonids. *Aquaculture* 58:99–110.

Schulte, E. H., 1975. Influence of algal concentration and temperature on the filtration rate of *Mytilus edulis*. *Marine Biology* 30:331–341.

Smith, C. E., and W. G. Williams, 1974. Experimental nitrite toxicity in rainbow trout and chinook salmon. *Transaction of the American Fisheries Society* 103:389–390.

Spotte, S., 1973. *Marine Aquarium Keeping: The Science, Animals, and Art*. New York: John Wiley & Sons, Inc.

Spotte, S., 1979a. *Fish and Invertebrate Culture: Water Management in Closed Systems*, 2nd ed. New York: John Wiley & Sons, Inc.

Spotte, S., 1979b. *Seawater Aquariums: The Captive Environment*. New York: John Wiley & Sons, Inc.

Taylor, A. C., 1977. The respiratory response of *Carcinus maenus* (L.) to changes in environmental salinity. *Journal of Experimental Marine Biology and Ecology* 29:197–210.

Thurston, R. V., and R. C. Russo, 1983. Acute toxicity of ammonia to rainbow trout. *Transactions of the American Fisheries Society* 112:696–704.

Thurston, R. V., C. Chakoumakos, and R. C. Russo, 1981. Effect of fluctuating exposures on the acute toxicity of ammonia to rainbow trout (*Salmo gairdneri*) and cutthroat trout (*S. clarki*). *Water Research* 15:911–917.

Thurston, R. V., G. R. Phillips, and R. C. Russo, 1981. Increased toxicity of ammonia to rainbow trout (*Salmo gairdneri*) resulting from reduced concentrations of dissolved oxygen. *Canadian Journal of Fisheries and Aquatic Science* 38: 983–988.

Wheaton, F. W., 1977. *Aquacultural Engineering*. New York: John Wiley & Sons, Inc.

Wilson, J., 1982. *Ground Water: A Non-Technical Guide*. Philadelphia: Academy of Natural Sciences.

CHAPTER 3

Adey, W. H., 1987. Food production in low-nutrient seas. *BioScience* 37:340–348.

Banerjee, R. K., and K. V. Srinivasan, 1983. Composted urban refuse and primary sewage sludge as a fish-pond manure. *Agricultural Wastes* 7:209–219.

Boyd, C. E., 1982. *Water Quality Management for Pond Fish Culture*. New York: Elsevier Science Publishing Company, Inc.

Brown, E. E., and J. B. Gratzek, 1980. *Fish Farming Handbook*. Westport, Conn.: AVI Publishing Company, Inc.

Chacon-Torres, A., L. G. Ross, and M. C. M. Beveridge, 1988. The effects of fish behavior on dye dispersion and water exchange in small net cages. *Aquaculture* 73:283–293.

Chamberlain, G. W., D. L. Hutchins, A. L. Lawrence, and J. C. Parker, 1980. Winter culture of *Penaeus vannamei* in

ponds receiving thermal effluent at different rates. *Proceedings of the World Mariculture Society* 11:30–43.

Clark, J. R., and R. L. Clark, 1964. *Sea-Water Systems for Experimental Aquariums*. Research Report 63, Bureau of Sport Fisheries and Wildlife, U.S. Department of the Interior, Washington, D.C.

Conrad, J., 1988. Growing caged catfish in Florida's rock pits. *Aquaculture Magazine* 14(2):34–37.

Dexter, S. C., 1986. Materials science in aquacultural engineering. *Aquacultural Engineering* 5:333–345.

Ford, R. F., J. R. Felix, R. L. Johnson, J. M. Carlberg, and J. C. Van Olst, 1979. Effects of fluctuating and constant temperatures and chemicals in thermal effluent on growth and survival of the American lobster (*Homarus americanus*). *Proceedings of the World Mariculture Society* 10:139–158.

Galtsoff, P. S., and V. L. Loosanoff, 1939. Natural history and methods of controlling the starfish (*Asterias forbesi*, Desor). *Bulletin of the Bureau of Fisheries* 49:75–132.

Goldman, J. C., K. R. Tenore, and H. I. Stanley, 1974. Inorganic nitrogen removal in a combined tertiary treatment— marine aquaculture system. I. Removal Efficiencies. *Water Research* 8:45–54.

Gordon, M. S., and R. A. Boolootian, 1964. *A Closed Circulating Sea-Water System*. Research Report 63, Bureau of Sport Fisheries and Wildlife, U.S. Department of the Interior, Washington, D.C.

Goyert, J. C., and J. W. Avault, Jr., 1977. Agricultural by-products as supplemental feed for crayfish, *Procambarus clarkii*. *Transactions of the American Fisheries Society* 106:629–633.

Hanebrink, E., and W. Byrd, 1989. Predatory birds in relation to aquaculture farming. *Aquaculture Magazine* 15(2): 47–51.

Huet, M., 1986. *Textbook of Fish Culture*, 2nd ed. Farnham, Surrey, England: Fishing News Books Ltd.

Huguenin, J. E., and J. Colt, 1989. *De-*

sign and Operating Guide for Aquaculture Seawater Systems. New York: Elsevier Science Publishing Company, Inc.

Huner, J. V., and O. V. Lindqvist, 1986. Nuclear power plant heat for culturing Atlantic salmon smolts. *Aquaculture Magazine* 12(6):22–24.

Inman, C. R. 1980. Construction hints and preliminary management practices for new ponds and lakes. Austin: Texas Parks and Wildlife Department booklet no. 3000-7.

Kraul, S., J. Szyper, and B. Burke, 1985. Practical formulas for computing water exchange rates. *Progressive Fish-Culturist* 47:69–70.

Lannan, J. E., R. O. Smitherman, and G. Tchobanoglous, 1986. *Principles and Practices of Pond Aquaculture.* Corvallis: Oregon State University Press.

Lapointe, B. E., L. D. Williams, J. C. Goldman, and J. H. Ryther, 1976. The mass outdoor culture of macroscopic marine algae. *Aquaculture* 8:9–21.

Lucchetti, G. L., and G. A. Gray, 1988. Prototype water reuse system. *The Progressive Fish-Culturist* 50:46–49.

Martin, R. M., and W. R. Heard, 1987. Floating vertical raceways to culture salmon (*Oncorhynchus* spp.). *Aquaculture* 61:295–302.

Moody, T. M., and R. N. McCleskey, 1978. Vertical raceways for production of rainbow trout. Bulletin no. 17, New Mexico Department of Game and Fish.

Nelson, P. V., 1985. *Greenhouse Operation and Managemeni,* 3rd ed. Reston, Va.: Reston Publishing Company, Inc.

Peterson, R. E., and K. K. Seo, 1977. Thermal aquaculture. *Proceedings of the World Mariculture Society* 8:491–503.

Pierce, B. A., 1980. Water reuse aquaculture systems in two solar greenhouses in northern Vermont. *Proceedings of the World Mariculture Society* 11:118–127.

Pillay, T. V. R., and W. A. Dill, 1979. *Advances in Aquaculture.* Farnham, Surrey, England: Fishing News Books Ltd.

Rice, T., D. H. Buck, R. W. Gorden, and P. P. Tazik, 1984. Microbial patho-gens and human parasites in an animal waste polyculture system. *Progressive Fish-Culturist* 46:230–238.

Rogers, G. L., and A. W. Fast, 1988. Potential benefits of low energy water circulation in Hawaiian prawn ponds. *Aquacultural Engineering* 7:155–165.

Ryther, J. H., W. M. Dunstan, K. R. Tenore, and J. E. Huguenin, 1972. Controlled eutrophication—increasing food production from the seas by recycling human wastes. *BioScience* 22:144–152.

Ryther, J. H., L. D. Williams, and D. C. Kneale, 1977. A freshwater waste recycling-aquaculture system. *Florida Scientist* 40:130–135.

Schwind, P., 1977. *Practical Shellfish Farming.* Camden, Me.: International Marine Publishing Company.

Singh, S. M., and P. N. Ferns, 1978. Accumulation of heavy metals in rainbow trout *Salmo gairdneri* (Richardson) maintained on a diet containing activated sewage sludge. *Journal of Fish Biology* 13:277–286.

Stickney, R. R., 1979. *Principles of Warmwater Aquaculture.* New York: John Wiley & Sons, Inc.

Svealv, T. L., 1988. Inshore versus offshore farming. *Aquacultural Engineering* 7:279–287.

Swift, D. R., 1985. *Aquaculture Training Manual.* Farnham, Surrey, England: Fishing News Books Ltd.

Tenore, K. R., and W. M. Dunstan, 1973. Growth comparisons of oysters, mussels and scallops cultivated on algae grown with artificial medium and treated sewage. *Chesapeake Science* 14:64–66.

Tobias, W. J., P. Sorgeloos, E. Bossuyt, and O. A. Roels, 1979. The technical feasibility of mass-culturing *Artemia salina* in the St. Croix "artificial upwelling" mariculture system. *Proceedings of the World Mariculture Society* 10:203–214.

Turner, J. W. D., R. R. Sibbald, and J. Hemens, 1986. Chlorinated secondary domestic sewage effluent as a fertilizer for marine aquaculture. III. Assessment of bacterial and viral quality and accumulation of heavy metals and chlorinated pesticides in culture fish and prawns. *Aquaculture* 53:157–168.

Uchida, R. N. and J. E. King, 1962. Tank culture of tilapia. *Fishery Bulletin of the Fish and Wildlife Service* 62:21–52.

Vaughn, J. M., and J. H. Ryther, 1974. Bacteriophage survival patterns in a tertiary sewage treatment-aquaculture model system. *Aquaculture* 4:399–406.

Wang, J.-K., and T. M. Losordo, 1985. The effect of earthen pond grass on prawn yield. *Journal of the World Mariculture Society* 16:217–224.

Wheaton, F. W., 1977. *Aquacultural Engineering*. New York: John Wiley & Sons, Inc.

Wheaton, F. W., T. B. Lawson, R. B. Brinsfield, and M. Yaramanoglu, 1983. Aquacultural ponds as energy storage and waste recycling systems. *Aquacultural Engineering* 2:233–245.

Wilson, C. B., 1924. Water beetles in relation to fishpond culture, with life histories of those found in fishponds at Fairport, Iowa. *Bulletin of the Bureau of Fisheries* 39:231–345.

CHAPTER 4

Baldwin, W. J., 1973. Results of tests to investigate the suitability of two fish pumps for moving live baitfish. *The Progressive Fish-Culturist* 35:39–43.

Carter, R., I. J. Karassik, and E. F. Wright, 1949. *Pump Questions and Answers*. New York: McGraw-Hill Book Company, Inc.

Holland, F. A., and F. S. Chapman, 1966. *Pumping of Liquids*. New York: Reinhold Publishing Corporation.

Huguenin, J. E., and J. Colt, 1989. *Design and Operating Guide for Aquaculture Seawater Systems*. New York: Elsevier Science Publishing Company, Inc.

Karassik, I. J., W. C. Krutzsch, W. H. Fraser, and J. P. Messina. 1986. *Pump Handbook*. New York: McGraw-Hill Book Company, Inc.

Kristal, F. A., and F. A. Annett, 1940. *Pumps*. New York: McGraw-Hill Book Company, Inc.

Lines, I. L., and V. L. Husa, 1979. Pumping live fish with a 10-inch volute pump. Paper no. 79-2115, presented at the Winnipeg summer meeting of the American Society of Agricultural Engineers and the Canadian Society of Agricultural Engineering.

Lobanoff, V. S., and R. R. Ross, 1985. *Centrifugal Pumps: Design & Application*. Houston: Gulf Publishing Company.

Parker, N. C., and M. A. Suttle, 1987. Design of airlift pumps for water circulation and aeration in aquaculture. *Aquacultural Engineering* 6:97–110.

Schwab, G. O., R. K. Frevert, T. W. Edminster, and K. K. Barnes, 1981. *Soil and Water Conservation Engineering*, 3rd. ed. New York: John Wiley & Sons, Inc.

Wheaton, F. W., 1977. *Aquacultural Engineering*. New York: John Wiley & Sons, Inc.

CHAPTER 5

Ahmad, T., and C. E. Boyd, 1988. Design and performance of paddle wheel aerators. *Aquacultural Engineering* 7:39–62.

Allen, H. E., and J. R. Kramer, 1972. *Nutrients in Natural Waters*. New York: John Wiley & Sons, Inc.

Bladetston, W. L., and J. M. Sieburth, 1976. Nitrate removal in closed-system aquaculture by columnar denitrification. *Applied Environmental Microbiology* 32: 808–818.

Bower, C. E., and J. P. Bidwell, 1978. Ionization of ammonia in seawater: Effects of temperature, pH, and salinity. *Journal of the Fisheries Research Board of Canada* 35:1012–1016.

Brune, D. E., and D. C. Gunther, 1981. The design of a new high rate nitrification filter for aquaculture water reuse. *Journal of the World Mariculture Society* 12(1):20–31.

Busch, C. D., C. A. Flood, Jr., and R. Allison, 1978. Multiple paddlewheels' influence on fish pond temperature and aeration. *Transactions of the American Society of Agricultural Engineers* 21:1222–1224.

Clark, J. R., and R. L. Clark, 1964. *Sea-Water Systems for Experimental Aquariums.* Research Report 63. Washington, D.C.: Bureau of Sport Fisheries and Wildlife, U.S. Department of the Interior.

Dryden, H. T., and L. R. Wesatherley, 1987. Aquaculture water treatment by ion-exchange: II. Selectivity studies with clinoptilolite. *Aquacultural Engineering* 6:51–68.

Fuss, J. T., 1986. Design and application of vacuum degassers. *The Progressive Fish-Culturist* 48:215–221.

Hassler, J. W., 1963. *Activated Carbon.* New York: Chemical Publishing Company Inc.

Huguenin, J. E., and J. Colt, 1989. *Design and Operating Guide for Aquaculture Seawater Systems.* New York: Elsevier Science Publishing Company, Inc.

Kaiser, G. E., and F. W. Wheaton, 1983. Nitrification filters for aquatic culture systems: State of the art. *Journal of the World Mariculture Society* 14:302–324.

LaBomascus, D. C., E. H. Robinson, and T. L. Linton, 1987. Use of water conditioners in water-recirculation systems. *The Progressive Fish-Culturist* 49:64–65.

Mahajan, O. P., A. Youssef, and P. L. Walker, Jr., 1978. Surface-treated activated carbon for removal of ammonia from water. *Separation Science and Technology* 13:487–499.

Marking, L. L., 1987. Evaluation of gass supersaturation treatment equipment at fish hatcheries in Michigan and Wisconsin. *The Progressive Fish-Culturist* 49:208–212.

Morales, J., J. de la Noue, and G. Picard, 1985. Harvesting marine microalgae species by chitosan flocculation. *Aquacultural Engineering* 4:257–270.

Oakes, D., P. Cooley, L. L. Edwards, R. W. Hirsch, and V. G. Miller, 1979. Ozone disinfection of fish hatchery waters: Pilot plant results, prototype design and control considerations. *Proceedings of the World Mariculture Society* 10:854–870.

Paller, M. H., and W. M. Lewis, 1988. Use of ozone and fluidized-bed biofilters for increased ammonia removal and fish loading rates. *The Progressive Fish-Culturist* 50:141–147.

Pond, W. G., and F. A. Mumpton. 1984. *Zeo-Agriculture.* Boulder, Colo.: Westview Press.

Rogers, G. L., and A. W. Fast, 1988. Potential benefits of low energy water circulation in Hawaiian prawn ponds. *Aquacultural Engineering* 7:155–165.

Salvato, J. A., Jr., 1972. *Environmental Engineering and Sanitation.* New York: John Wiley & Sons, Inc.

Schmid, G., 1988. Disinfection of water in aquaculture. *Aquaculture Today* 1(2): 43–45.

Scott, K. R., and L. Allard, 1984. A four-tank water recirculation system with a hydroclone prefilter and a single water reconditioning unit. *The Progressive Fish-Culturist* 46:254–261.

Spotte, S., 1979. *Seawater Aquariums.* New York: John Wiley & Sons, Inc.

Staff of Research and Education Association, 1978. *Modern Pollution Control Technology,* Vol. II. New York: Research and Education Association.

Sukenik, A., D. Bilanovic, and G. Shelef, 1988. Flocculation of microalgae in brackish and sea waters. *Biomass* 15:187–199.

Watten, B. J., and L. T. Beck, 1985. Model gas transfer in a U-tube oxygen absorption system: effects of off-gas recycling. *Aquacultural Engineering* 4:271–297.

Wheaton, F. W., 1977. *Aquacultural Engineering.* New York: John Wiley & Sons, Inc.

White, G. C., 1978. *Disinfection of Wastewater and Water for Reuse.* New York: Van Nostrand Reinhold Company.

CHAPTER 6

Ahne, W., I. Schwanz-Pfitzner, and I. Thomsen, 1987. Serological identification of 9 viral isolates from European eels

(*Anguilla anguilla*) with stomatopapilloma by means of neutralization tests. *Journal of Applied Ichthyology* 3:30–32.

Bardach, J E., J. H. Ryther, and W. O. McLarney, 1972. *Aquaculture.* New York: John Wiley & Sons, Inc.

Berg, C. J., Jr., 1983. *Culture of Marine Invertebrates.* Stroudsburg, Pa.: Hutchinson Ross Publishing Company.

Bierbaum, R. M., and S. Ferson, 1986. Do symbiotic pea crabs decrease growth rate in mussels? *Biological Bulletin* 170: 51–61.

Canzonier, W. J., 1972. *Cercaria tenuanus,* larval trematode parasite of *Mytilus* and its significance in mussel culture. *Aquaculture* 1:267–278.

Capuzzo, J. M., and B. A. Lancaster, 1979. The effects of dietary carbohydrate levels on protein utilization in the American lobster (*Homarus americanus*). *Proceeding of the World Mariculture Society* 10:689–700.

Chrisman, C. L., W. R. Wolters, and G. S. Libey, 1983. Triploidy in channel catfish. *Journal of the World Mariculture Society* 14:279–293.

Collins, C., 1988. Rearing channel catfish in cages—Part II. *Aquaculture Magazine* 14(2):56–58.

Colt, J., 1986. Gas supersaturation—impact on the design and operation of aquatic systems. *Aquaculture Engineering* 5:49–85.

Cowey, C. B., and J. R. M. Forster, 1971. The essential amino-acid requirements of the prawn *Palaemon serratus.* The growth of prawns on diets containing proteins of different amino-acid compositions. *Marine Biology* 10:77–81.

Davis, H. S., 1953. *Culture and Diseases of Game Fish.* Berkeley: University of California Press.

Deshimura, O., and Y. Yone, 1978. Requirement of prawn for dietary minerals. *Bulletin of the Japanese Society of Scientific Fisheries* 44:907–910.

Donaldson, L. R., 1971. Selective breeding of salmonid fishes. *Proceedings of the World Mariculture Society* 2:75–79.

Downing, S. L., and S. K. Allen, Jr., 1987. Induced triploidy in the Pacific oyster, *Crassostrea gigas:* Optimal treatments with cytochalasin B depend on temperature. *Aquaculture* 61:1–15.

Ellis, A. E., 1988. Current aspects of fish vaccinations. *Diseases of Aquatic Organisms* 4:159–164.

Griffin, B. R., 1987. Columnaris disease: Recent advances in research. *Aquaculture Magazine* 13(3):48–50.

Griffin, B. R., 1989. Screening of chemicals to control protozoan parasites of fish. *The Progressive Fish-Culturist* 51:127–132.

Griffin, B. R., 1990. Judgment and restraint are needed when fisheries chemicals are used. *Aquaculture Magazine* 16(1):70–72.

Halver, J. E., 1972. *Fish Nutrition.* New York: Academic Press, Inc.

Harrell, L. W., R. A. Elston, T. M. Scott, and M. T. Wilkinson, 1986. A significant new systematic disease of net-pen reared chinook salmon (*Oncorhynchus tshawytscha*) brood stock. *Aquaculture* 55:249–262.

Huet, M., 1986. *Textbook of Fish Culture.* Farnham, Surrey, England: Fishing News Books Ltd.

Holm, K. O., E. Strom, K. Stensvag, J. Raa, T. Jorgensen, 1985. Characteristics of a *Vibrio* sp. associated with the "Hitra disease" of Atlantic salmon in Norwegian fish farms. *Fish Pathology* 20:125–129.

Jana, B. B., and S. K. Roy, 1985. Spatial and temporal changes in nitrifying bacterial populations in fish ponds of differing management practices. *Journal of Applied Bacteriology* 59:195–204.

Johnson, S. K., 1978. *Handbook of Shrimp Diseases.* Texas A&M Sea Grant College Program publication TAMU-SG-75-603, College Station.

Jones, D., and D. H. Lewis, 1976. Gas bubble disease in fry of channel catfish (*Ictalurus punctatus*). *The Progressive Fish-Culturist* 38:41.

Kanazawa, A., S. Tokiwa, M. Kayama, and M. Hirata, 1977. Essential fatty acids in the diet of prawn—I. Effects of linoleic and linolenic acids on growth. *Bulletin of the Japanese Society of Scientific Fisheries* 43: 1111–1114.

Lee, J. S., 1981. Commercial Catfish Farming. Danville, Ill.: Interstate Printers & Publishers.

Leibovitz, L., E. F. Schott, and R. C. Karney, 1984. Diseases of wild, captive and cultured scallops. *Journal of the World Mariculture Society* 16:269–283.

Lewis, D. H. and J. K. Leong, 1979. *Proceedings of the 2nd Biennial Crustacean Health Workshop.* Sea Grant College Program, Texas A&M University, College Station.

Li, M. F., R. E. Drinnan, M. Drebot, Jr., and G. Newkirk, 1983. Studies on shell disease of the European flat oyster *Ostrea edulis* Linne in Nova Scotia. *Journal of Shellfish Research* 3:135–140.

Lightner, D. V., 1978. Possible toxic effects of the marine blue-green alga, *Spirulina subsalsa,* on the blue shrimp, *Penaeus stylirostris. Journal of Invertebrate Pathology* 32:139–150.

Lightner, D. V., R. P. Hedrick, J. L. Fryer, S. N. Chen, I. C. Liao, and G. H. Kou, 1987. A survey of cultured penaeid shrimp in Taiwan for viral and other important diseases. *Fish Pathology* 22:127–140.

Lightner, D. V., R. M. Redman, R. R. Williams, L. L. Mohney, J. P. M. Clerx, T. A. Bell, and J. A. Brock, 1985. Recent advances in penaeid disease investigations. *Journal of the World Mariculture Society* 16:267–274.

Lovell, T., 1989. Reevaluation of carbohydrates in fish feeds. *Aquaculture Magazine* 15(3):62–64.

Mackin, J. G., 1970. Oyster culture and disease. *Proceedings of the World Mariculture Society* 1:35–38.

Mellergaard, S., and I. Dalsgaard, 1987. Disease problems in Danish eel farms. *Aquaculture* 67:139–146.

Meyer, F. P., 1989. Solutions to the shortage of approved fish therapeutants. *Journal of Aquatic Animal Health* 1:78–80.

Meyers, S. P., and R. W. Hagood, 1984. Flake diets and larval crustacean culture. *The Progressive Fish-Culturist* 46:225–229.

Moriarty, D. J. W., and R. S. V. Pullin, 1987. *Detritus and Microbial Ecology in Aquaculture.* Manila, Phillippines: ICLARM.

Myers, J. M., 1986. Tetraploid induction in *Oreochromis* spp. *Aquaculture* 57:281–287.

New, M. B., 1976. A review of dietary studies with shrimp and prawns. *Aquaculture* 9:101–144.

Pillay, T. V. R., and W. A. Dill, 1979. *Advances in Aquaculture.* Farnham, Surrey, England: Fishing News Books Ltd.

Price, K. S., Jr., W. N. Shaw, and K. S. Danberg, 1976. *Proceedings of the First International Conference on Aquaculture Nutrition,* University of Delaware, Newark.

Pruder, G. D., C. J. Langdon, and D. E. Conklin, 1983. *Proceedings of the Conference on Aquaculture Nutrition.* Baton Rouge: Louisiana State University.

Purdom, C. E., 1986. Genetic techniques for control of sexuality in fish farming. *Fish Physiology and Biochemistry* 2:3–8.

Schnick, R. A., F. P. Meyer, and D. F. Walsh, 1986. Status of fishery chemicals in 1985. *The Progressive Fish-Culturist* 48: 1–17.

Schroeder, G. L., 1978. Autotrophic and heterotrophic production of microorganisms in intensely-manured fish ponds, and related fish yields. *Aquaculture* 14:303–325.

Sinderman, C. J., 1977. *Disease Diagnosis and Control in North American Marine Aquaculture.* New York: Elsevier Science Publishing Company, Inc.

Smith, T. I. J., and W. E. Wallace, 1984. Controlled spawning of F_1 hybrid striped bass (*Morone saxatilis* $\times$ *M. chrysops*) and rearing of F_2 progeny. *Journal of the World Mariculture Society* 15: 147–161.

Stewart, J. E., 1975. Gaffkemia, the fatal disease of lobsters (genus *Homarus*) caused by *Aerococcus viridans* (var.) *homari:*

A review. *Marine Fisheries Review* 37: 20–24.

Stickney, R. R., 1979. *Principles of Warmwater Aquaculture.* New York: John Wiley & Sons, Inc.

Stickney, R. R., 1986. *Culture of Nonsalmonid Freshwater Fishes.* Boca Raton, Fl.: CRC Press Inc.

Ribelin, W. E., and G. Migaki, 1975. *The Pathology of Fishes.* Madison: University of Wisconsin Press.

United States Fish and Wildlife Service, 1983. *Fish Hatchery Management.* Washington, D.C.: U.S. Department of Interior.

Valenti, R. J., 1975. Induced polyploidy in *Tilapia aurea* (Steindachner) by means of temperature shock treatment. *Journal of Fish Biology* 7:519–528.

Ward, H. B., and G. C. Whipple, 1918. *Fresh-Water Biology.* New York: John Wiley & Sons, Inc.

Williams, R. R., and D. V. Lightner, 1988. Regulatory status of therapeutants for penaeid shrimp culture in the United States. *Journal of the World Aquaculture Society* 19:188–196.

Wilson, C. B. 1903. North American parasitic copepods of the family Argulidae. *Proceedings of the U.S. National Museum* 25:635–742.

Wilson, C. B. 1911. North American parasitic copepods of the family Ergasilidae. *Proceedings of the U.S. National Museum* 39:263–400.

Wolf, K., and M. E. Markiw, 1985. *Salmonid Whirling Disease.* Washington, D.C.: U.S. Department of Interior, Fish and Wildlife Service, Fish Disease Leaflet 69.

CHAPTER 7

Asian Development Bank and Food & Agricultural Organization, 1983. *Joint ADB/FAO (SCSP-INFOFISH) Market Studies.* Vol. 6: *The World Seaweed Industry and Trade.* Manila, United Nations.

Bardach, J. E., J. H. Ryther, and W. O. McLarney, 1972. *Aquaculture.* New York: John Wiley & Sons, Inc.

Bird, K. T., and P. H. Benson, 1987. *Seaweed Cultivation for Renewable Resources.* New York: Elsevier Science Publishing Company, Inc.

Chapman, V. J., 1970. *Seaweeds and Their Uses,* 2nd ed. London: Methuen & Company Ltd.

Chen, T. P., 1976. *Aquaculture Practices in Taiwan.* Norwich, England: Page Brothers Limited.

Cheng, T. H., 1969. Production of kelp—a major aspect of China's exploitation of the sea. *Economic Botany* 23:215–236.

Dawson, E. Y., 1966. *Marine Botany.* New York: Holt, Rinehart and Winston, Inc.

Guist, G. G., C. J. Dawes, and J. R. Castle, 1982. Mariculture of the red seaweed, *Hypnea musciformis. Aquaculture* 28:375–384.

Hansen, J. E., 1983. A physiological approach to mariculture of red algae. *Journal of the World Mariculture Society* 14:380–391.

Huguenin, J. E., 1977. A subjective analysis of alternative approaches to the mass culturing of seaweeds. *Proceedings of the World Mariculture Society* 8:387–400.

Idyll, C. P., 1970. *The Sea Against Hunger.* New York: Thomas Y. Crowell Company.

Klausner, A., 1986. Algaculture: Food for thought. *Bio/Technology* 4:947–953.

Levring, T., H. A. Hoppe, O. J. Schmid, 1969. *Marine Algae.* Hamburg: Cram, De Gruyter & Co.

Limburg, P. R., 1980. *Farming the Waters.* New York: Beaufort Books, Inc.

Lobban, C. S., and M. J. Wynne, 1981. *The Biology of Seaweeds.* Los Angeles: University of California Press.

Madlener, J. C., 1977. *The Seavegetable Book.* New York: Clarkson N. Potter, Inc., Publishers.

McCoy, H. D., II, 1987. The commercial algaes: Prospects for one of the oldest industries. *Aquaculture Magazine* 13(4): 46–54.

McHugh, D. J., 1984. Marine phycoculture and its impact on the seaweed colloid industry. *Hydrobiologia* 116/117:351–354.

Mitchell, S. A., and A. Richmond, 1988. Optimization of a growth medium for *Spirulina* based on cattle waste. *Biological Wastes* 25:41–50.

Moreau, J., D. Pesando, and B. Caram, 1984. Antifungal and antibacterial screening of Dictyotales from the French Mediterranean coast. *Hydrobiologia* 116/117:521–524.

Niang, L. L., and X. Hung, 1984. Studies on the biologically active compounds of the algae from the Yellow Sea. *Hydrobiologia* 116/117:168–170.

Nisizawa, K., H. Noda, R. Kikuchi, and T. Watanabe, 1987. The main seaweed foods in Japan. *Hydrobiologia* 151/152:5–29.

Pillay, T. V. R., and W. A. Dill, 1976. *Advances in Aquaculture.* Farnham, Surrey, England: Fishing News Books Ltd.

Richmond, A., 1986. *CRC Handbook of Microalgal Mass Culture.* Boca Raton, Fl.: CRC Press, Inc.

Ryther, J. H., N. Corwin, T. A. DeBusk, and L. D. Williams, 1982. Nitrogen uptake and storage by the red algae *Gracilaria tikvahiae* (McLachlan, 1979). *Aquaculture* 26:107–115.

Saffo, M. B., 1987. New light on seaweeds. *BioScience* 37:654–664.

Stickney, R. R., 1988. The culture of macroscopic algae. *World Aquaculture* 19:54–58.

Sumich, J. L., 1984. *Biology of Marine Life.* Dubuque, Ia.: Wm. C. Brown, Publishers.

Switzer, L., 1982. *Spirulina: The Whole Food Revolution.* New York: Bantam Books.

Trainor, F. R., 1978. *Introductory Phycology.* New York: John Wiley & Sons, Inc.

Vonshak, A., and A. Richmond, 1988. Mass production of the blue-green alga

Spirulina: an overview. *Biomass* 15:233–247.

CHAPTER 8

Angell, C. L., J. Tetelepta, and L. S. Smith, 1984. Culturing the spiny oyster, *Saccostrea echinata,* in Ambon, Indonesia. *Journal of the World Mariculture Society* 15:433–441.

Anonymous, 1986. Abalone on shore. *Aquaculture Magazine* 12(4):6–10.

Avault, J. W., 1987. Species profile—mussels and clams. *Aquaculture Magazine* 13(5):56–58.

Bardach, J. E., J. H. Ryther, and W. O. McLarney, 1972. *Aquaculture.* New York: John Wiley & Sons, Inc.

Blogoslawski, W., and M. E. Stewart, 1983. Depuration and public health. *Journal of the World Mariculture Society* 14:535–545.

Castagna, M., 1983. Review of recent bivalve culture methods. *Journal of the World Mariculture Society* 14:567–575.

Caturano, S., L. S. Glanz, D. C. Smith, L. Tsomides, and J. R. Moring, 1988. Shellfish mariculture: The status of mussel power in Maine. *Fisheries* 13(3):18–21.

Chaves, L. A., and K. K. Chew, 1975. Mussel culture studies in Puget Sound, Washington. *Proceedings of the World Mariculture Society* 6:185–191.

Chen, T. P., 1976. Aquaculture Practices in Taiwan. Norwich, England: Page Brothers Limited.

Chew, K. K., 1989. Recent trends in bivalve culture in the orient. *Aquaculture Magazine* 15(3):41–54.

Coker, R. E., 1908. The fisheries and the guano industry of Peru. *Bulletin of the Bureau of Fisheries* 28:333–365.

Conrad, J., 1986. Mussel production plant in Spanish Galicia. *Aquaculture Magazine* 12(4):57–58.

Coon, S. L., D. B. Bonar, and R. M. Weiner, 1986. Chemical production of cultchless oyster spat using epinephrine and norepinephrine. *Aquaculture* 58:255–262.

Cross, M. L., G. F. Newkirk, and G. Johnson, 1987. Strategies for the cultivation of European oysters (*Ostrea edulis*) in the Maritime provinces of Canada: Preliminary analyses. *Canadian Journal of Fisheries and Aquatic Sciences* 44:674–679.

Davy, F. B. and M. Graham, 1982. *Bivalve Culture in Asia and the Pacific*. Ottawa, Canada: International Development Research Centre.

Ebert, T. B., and E. E. Ebert, 1988. An innovative technique for seeding abalone and preliminary results of laboratory and field trials. *California Fish and Game* 74:68–81.

Epifanio, C. E., C. C. Valenti, and C. L. Turk, 1981. A comparison of *Phaeodactylum tricornutum* and *Thalassiosira pseudonana* as food for the oyster *Crassostrea virginica*. *Aquaculture* 23:347–353.

Field, I. A. 1922. Biology and economic value of the sea mussel *Mytilus edulis*. *Bulletin of the Bureau of Fisheries* 38:127–260.

Freudenthal, A. R., and J. L. Jijina, 1988. Potential hazards of *Dinophysis* to consumers and shellfisheries. *Journal of Shellfish Research* 7:695–701.

Genade, A. B., A. L. Hirst, and C. J. Smit, 1988. Observations on the spawning, development and rearing of the South African abalone *Haliotis midae* Linn. *South African Journal of Marine Science* 6:3–12.

Gibbons, M. C., and M. Castagna, 1985. Responses of the hard clam *Mercenaria mercenaria* (Linne) to induction of spawning by serotonin. *Journal of Shellfish Research* 5:65–67.

Goldsmith, J. B., 1989. Massachusetts quahogs flourish with excellent management. *Aquaculture Magazine* 15(2):52–57.

Hadley, N., 1988. Improving growth rates of hard clams through genetic manipulation. *World Aquaculture* 19(3):65–66.

Hahn, K. O., 1988. *Handbook of Culture of Abalone and Other Marine Gastropods*. Boca Raton, Fl.: CRC Press, Inc.

Haley, L. E., and G. F. Newkirk, 1977. Selecting oysters for faster growth. *Proceedings of the World Mariculture Society* 8:557–565.

Huner, J. V., and E. E. Brown, 1985. *Crustacean and Mollusk Aquaculture in the United States*. Westport, Conn.: AVI Publishing Company, Inc.

Jory, D. E., and E. S. Iversen, 1985. Molluscan mariculture in the greater Caribbean: An overview. *Marine Fisheries Review* 47(4):1–10.

Kafuku, T., and H. Ikenoue, 1983. *Modern Methods of Aquaculture in Japan*. New York: Elsevier Science Publishing Company, Inc.

Korringa, P., 1976. *Farming Marine Organisms Low in the Food Chain*. New York: Elsevier Science Publishing Company, Inc.

Krantz, G. E., 1984. *Oyster Hatchery Technology Series*, publication no. UM-SG-MAP-82-01. University of Maryland Sea Grant.

Kraeuter, J. N., and M. Castagna, 1985. The effects of seed size, shell bags, crab traps, and netting on the survival of the northern hard clam *Mercenaria mercenaria* (Linne). *Journal of Shellfish Research* 5:69–72.

Laing, I., 1987. The use of artificial diets in rearing bivalve spat. *Aquaculture* 65:243–249.

Laing, I., and P. F. Millican, 1986. Relative growth and growth efficiency of *Ostrea edulis* L. spat fed various algal diets. *Aquaculture* 54:245–262.

Lannan, J. E., A. Robinson, and W. P. Breese, 1980. Broodstock management of *Crassostrea gigas*. II. Broodstock conditioning to maximize larval survival. *Aquaculture* 21:337–345.

Lutz, R. A., 1979. Bivalve molluscan mariculture: A *Mytilus* perspective. *Proceedings of the World Mariculture Society* 10:596–608.

Manzi, J. J., N. H. Hadley, and M. B. Maddox, 1986. Seed clam, *Mercenaria mercenaria*, culture in an experimental-scale upflow nursery system. *Aquaculture* 54:301–311.

Matthiessen, G. C., and L. J. Smith, 1979. Analysis of methods for the culture of *Crassostrea virginica* in New England. *Proceedings of the World Mariculture Society* 10:609–623.

Menzel, R. W., 1971. Quahog clams and their possible mariculture. *Proceedings of the World Mariculture Society* 2: 23–36.

Menzel, W., 1977. Selection and hybridization in quahog clams (*Mercenaria* spp.). *Proceedings of the World Mariculture Society* 8:507–521.

Morse, A. N. C., C. A. Froyd, and D. E. Morse, 1984. Molecules from cyanobacteria and red algae that induce larval settlement and metamorphosis in the mollusc *Haliotis rufescens*. *Marine Biology* 81:293–298.

Newkirk, G. F., K. R. Freeman, and L. M. Dickie, 1980. Genetic studies of the blue mussel, *Mytilus edulis,* and their implications for commercial culture. *Proceedings of the World Mariculture Society* 11:596–604.

Ogle, J., S. M. Ray, and W. J. Wardle, 1977. A summary of oyster mariculture utilizing an offshore petroleum platform in the Gulf of Mexico. *Proceedings of the World Mariculture Society* 8:447–455.

Owen, B., L. H. DiSalvo, E. E. Ebert, and E. Fonck, 1984. Culture of the California red abalone *Haliotis rufescens* Swainson (1822) in Chile. *The Veliger* 27:101–105.

Rhodes, R. J., 1988. Mariculture in Maine: The future is now. *Aquaculture Magazine* 14(3):42–48.

Schulte, E. H., 1975. Influence of algal concentration and temperature on the filtration rate of *Mytilus edulis*. *Marine Biology* 30:331–341.

Schwind, P., 1977. *Practical Shellfish Farming.* Camden Me.: International Marine Publishing Company.

Stephano, J. L., and M. Gould, 1988. Avoiding polyspermy in the oyster (*Crassostrea gigas*). *Aquaculture* 73:295–307.

Walker, R. L., 1983. Feasibility of mariculture of the hard clam, *Mercenaria mercenaria* Linne in coastal Georgia. *Journal of Shellfish Research* 3:169–174.

Walne, P. R., 1963. Observations on the food value of seven species of algae to the larvae of *Ostrea edulis*. I. Feeding experiments. *Journal of the Marine Biological Association of the United Kingdom* 43:767–784.

Wikfors, G. H., J. W. Twarog, Jr., and R. Ukeles, 1984. Influence of chemical composition of algal food sources on growth of juvenile oysters, *Crassostrea virginica*. *Biological Bulletin* 167:251–263.

CHAPTER 9

AQUACOP, 1979. Penaeid reared brood stock: Closing the life cycle of *P. monodon, P. stylirostris,* and *P. vannamei. Proceedings of the World Mariculture Society* 10:445–452.

AQUACOP, 1984. Review of ten years of experimental penaeid shrimp culture in Tahiti and New Caledonia (South Pacific). *Journal of the World Mariculture Society* 15:73–91.

Arieli, Y., and U. Rappaport, 1982. Experimental cultivation of the freshwater prawn *Macrobrachium rosenbergii. Bamidgeh* 34:140–143.

Bardach, J. E., J. H. Ryther, and W. O. McLarney, 1972. *Aquaculture.* New York: John Wiley & Sons, Inc.

Bliss, D. E., and L. H. Mantel, 1985. *Biology of the Crustacea,* Vol. 9. *Integument, Pigments, and Hormonal Processes.* New York: Academic Press, Inc.

Brown, A., Jr., J. McVey, B. S. Middleditch, and A. L. Lawrence, 1979. Maturation of white shrimp (*Penaeus setiferus*) in captivity. *Proceedings of the World Mariculture Society* 10:435–444.

Brunson, M. W. 1987. Pre-flood evaluation of seven grasses (*Graminae* L.) as planted forage for crawfish. *Journal of the World Aquaculture Society* 18:186–189.

Chen, H.-Y., Z. P. Zein-Eldin, and

D. V. Aldrich, 1985. Combined effects of shrimp size and dietary protein source on growth of *Penaeus setiferus* and *P. vannamei*. *Journal of the World Mariculture Society* 16:288–296.

Cohen, D., A. Sagi, Z. Ra'anan, and G. Zohar, 1988. The production of *Macrobrachium rosenbergii* in monosex populations. III. Yield characteristics under intensive monoculture conditions in earthen ponds. *Bamidgeh* 40:57–63.

Cruz-Suarez, L. E., J. Guillaume, and A. Van Wormhoudt, 1987. Effect of various levels of squid protein on growth and some biochemical parameters of *Penaeus japonicus* juveniles. *Nippon Suisan Gakkaishi* 53:2083–2088.

Culley, D. D., and L. Doubinis-Gray, 1987. Molting, mortality, and the effects of density in a soft-shell crawfish culture system. *Journal of the World Aquaculture Society* 18:242–246.

Deshimaru, O., and K. Shigeno, 1972. Introduction to the artificial diet for prawn *Penaeus japonicus*. *Aquaculture* 1:115–133.

Deshimaru, O., and Y. Yone, 1978. Requirement of prawn for dietary minerals. *Bulletin of the Japanese Society of Scientific Fisheries* 44:907–910.

Dobkin, S., 1961. Early developmental stages of the pink shrimp, *Penaeus duorarum* from Florida waters. *Fisher Bulletin* 190, United States Fish and Wildlife Service.

Goyert, J. C., and J. W. Avault, Jr., 1977. Agriculture by-products as supplemental feed for crayfish, *Procambarus clarkii*. *Transactions of the American Fisheries Society* 106:629–633.

Hanson, J. A., and H. L. Goodwin, 1977. *Shrimp and Prawn Farming in the Western Hemisphere*. Stroudsburg, Pa.: Dowden, Hutchinson & Ross, Inc.

Huner, J. V., 1985. An update on crawfish aquaculture. *Aquaculture Magazine* 11(4):33–40.

Huner, J. V., 1988. Crayfish culture in Europe. *Aquaculture Magazine* 14(2):48–52.

Huner, J. V., and E. E. Brown, 1985. *Crustacean and Mollusk Aquaculture in the United States*. Westport, Conn.: AVI Publishing Company, Inc.

Hunter, B., G. Pruder, and J. Wyban, 1987. Biochemical composition of pond biota, shrimp ingesta, and relative growth of *Penaeus vannamei* in earthen ponds. *Journal of the World Aquaculture Society* 18:162–174.

Hysmith, B. T., and R. L. Colura, 1976. Effect of salinity on the growth and survival of penaeid shrimp in ponds. *Proceedings of the World Mariculture Society* 7:289–304.

Issar, G., E. R. Seidman, and Z. Samocha, 1987. Preliminary results of nursery and pond culture of *P. semisulcatus* in Israel. *Bamidgeh* 39:63–74.

Jones, D. A., K. Kurmaly, and A. Arshard, 1987. Penaeid shrimp hatchery trials using microencapsulated diets. *Aquaculture* 64:133–146.

Kanazawa, A., M. Shimaya, M. Kawasaki, and K. Kashiwada, 1970. Nutritional requirements of prawn—I. Feeding on artificial diet. *Bulletin of the Japanese Society of Scientific Fisheries* 36:949–954.

Kafuku, T., and H. Ikenoue, 1983. *Modern Methods of Aquaculture in Japan*. New York: Elsevier Science Publishing Company, Inc.

Karplus, I., G. Hulata, G. W. Wohlfarth, and A. Halevy, 1986. The effect of size-grading juvenile *Macrobrachium rosenbergii* prior to stocking on their population structure and production in polyculture. I. Dividing the population into two fractions. *Aquaculture* 56:257–270.

King, J. E., 1948. A study of the reproductive organs of the common marine shrimp, *Penaeus setiferus* (Linnaeus). *Biological Bulletin* 94:244–262.

Landau, M., and R. T. Eifert, 1985. The suitability of two strains of *Artemia* (Phyllopoda, Anostraca) as food for developing *Penaeus vannamei* Boone (Decapoda Natantia). *Crustaceana* 48:106–110.

Laufer, H., D. Borst, F. C. Baker, C. Carrasco, M. Sinkus, C. C. Reuter, L. W.

Tsai, and D. A. Schooley, 1987. Identification of a juvenile hormone-like compound in a crustacean. *Science* 235:202–205.

Lawson, T. B., and F. W. Wheaton, 1983. Crawfish culture systems and their management. *Journal of the World Mariculture Society* 14:325–335.

Lawrence, A. L., Y. Akamine, B. S. Middleditch, G. W. Chamberlain, and D. L. Hutchins, 1980. Maturation and reproduction of *Penaeus setiferus* in captivity. *Proceedings of the World Mariculture Society* 11:481–487.

Leung-Trujillo, J. R., and A. L. Lawrence, 1985. The effect of eyestalk ablation on spermatophore and sperm quality in *Penaeus vannamei*. *Journal of the World Mariculture Society* 16:258–266.

Liu, P.-C., Y.-T. Lin, and C.-K. Lee, 1988. Super intensive culture of the red-tailed shrimp, *Penaeus penicillatus*. *Journal of the World Aquaculture Society* 19:127–131.

Lumare, F., 1981. Artificial reproduction of *Penaeus japonicus* Bate as a basis for the mass production of eggs and larvae. *Journal of the World Mariculture Society* 12(2):335–344.

Manzi, J. J., and M. B. Maddox, 1976. Algal supplement enhancement of static and recirculating system culture of *Macrobrachium rosenbergii* larvae. *Helgolander wiss. Meeresunters* 28:447–455.

Mathavan, S., and S. Murugadass, 1988. An improved design for *in vitro* hatching of *Macrobrachium* eggs. *Asian Fisheries Science* 1:197–201.

McVey, J. P., 1983. *Handbook of Mariculture*. Vol. I, *Crustacean Aquaculture*. Boca Raton, Fla.: CRC Press, Inc.

Menz, A., and B. F. Blake, 1980. Experiments on the growth of *Penaeus vannamei* Boone. *Journal of Experimental Marine Biology and Ecology* 48:99–111.

Meltzoff, S. K., and E. LiPuma, 1986. The social and political economy of coastal zone management: shrimp mariculture in Ecuador. *Coastal Zone Management Journal* 14:349–380.

Middleditch, B. S., S. R. Missler, H. B. Hines, E. S. Chang, J. P. McVey, A. Brown, and A. L. Lawrence, 1980. Maturation of penaeid shrimp: Lipids in the marine food chain. *Proceedings of the World Mariculture Society* 11:463–470.

Miltner, M. R., A. Granados, R. Romaire, J. W. Avault, Jr., Z. Ra'anan, and D. Cohen, 1983. Polyculture of the prawn, *Macrobrachium rosenbergii*, with fingerlings and adult catfish, *Ictalurus punctatus*, and Chinese craps, *Hypophthalmichthys molitrix* and *Ctenopharyngodon idella*, in earthen ponds in south Louisiana. *Journal of the World Mariculture Society* 14:127–134.

Moller, T. H., 1978. Feeding behavior of larvae and postlarvae *Macrobrachium rosenbergii* (de Man) (Crustacea: Palaemonidae). *Journal of Experimental Marine Biology and Ecology* 35:251–258.

New, M. B., 1980. A bibliography of shrimp and prawn nutrition. *Aquaculture* 21:101–128.

New, M. B., and S. Singholka, 1985. *Freshwater Prawn Farming*, FAO Fisheries Technical paper 225. Rome: United Nations.

Pavel, D. L., S. W. Cange, and J. W. Avault, Jr., 1985. Polyculture of the channel catfish, *Ictalurus pinctatus*, with postlarval and juvenile prawns, *Macrobracium rosenbergii*. *Journal of the World Mariculture Society* 16:464–470.

Perry, W. G., and J. Tarver, 1984. Production trials of prawns comparing a marine ration, catfish diet and agricultural range pellet. *Journal of the World Mariculture Society* 15:120–128.

Primavera, J. H., F. Apud, and C. Usigan, 1978. Effect of different stocking densities on survival and growth of sugpo (*Penaeus monodon* Fabricius) in a milkfish rearing pond. *The Philippine Journal of Science* 105:193–203.

Pruder, G. D., C. Landgon, and D. Conklin, 1983. *Proceedings of the Second International Conference on Aquaculture Nutrition: Biochemical and Physiological Approaches to Shellfish Nutrition*. Baton Rouge: Louisiana State University.

Ra'anan, Z., D. Cohen, U. Rappaport, and G. Zohar, 1984. The production of the freshwater prawn *Macrobrachium rosenbergii* in Israel: The effect of added substrates on yields in a monoculture system. *Bamidgeh* 36:35–40.

Ra'anan Z., and A. Sagi, 1985. Alternative mating strategies in male morphotypes of the freshwater prawn *Macrobrachium rosenbergii* (deMan). *Biological Bulletin* 169:592–601.

Sanchez M., R. 1986. Rearing of mysid stages of *Penaeus vannamei* fed cultured algae of three species. *Aquaculture* 58:139–144.

Sandifer, P. A., J. S. Hopkins, and A. D. Stokes, 1987. Intensive culture potential of *Penaeus vannamei. Journal of the World Aquaculture Society* 18:94–100.

Sandifer, P. A., and T. I. J. Smith, 1977. Intensive rearing of postlarval Malaysian prawns, *Macrobrachium rosenbergii*, in a closed cycle nursery system. *Proceedings of the World Mariculture Society* 8:225–235.

Schmitt, W. L., 1973. *Crustaceans.* Newton Abbot, England: David & Charles Limited.

Simon, C. M., 1978. Culture of the diatom *Chaetoceros gracilis* and its use as a food for penaeid protozoel larvae. *Aquaculture* 14:105–113.

Simon, C. M., 1981. Design and operation of a large-scale, commercial penaeid shrimp hatchery. *Journal of the World Mariculture Society* 12(2):322–334.

Smith, T. I. G., P. A. Sandifer, and W. C. Trimble, 1976. Pond culture of the Malaysian prawn, *Macrobrachium rosenbergii* (de Man) in South Carolina, 1974–1975. *Proceedings of the World Mariculture Society* 7:625–645.

Subramoniam, T., and S. Varadarajan, 1982. *Progress in Invertebrate Reproduction and Aquaculture.* Madras, India: New Century Printers.

Tobias-Quinitio, E., and C. T. Villegas, 1982. Growth, survival and macronutrient composition of *Penaeus monodon* Fabricus larvae fed with *Chaetoceros cal-*

citrans and *Tetraselmis chuii. Aquaculture* 29:253–260.

Trimble, W. C., and A. P. Gaude III, 1988. Production of red swamp crawfish in a low-maintenance hatchery. *The Progressive Fish-Culturist* 50:170–173.

Weidner, D. M., 1985. A view of the Latin American shrimp culture industry, Part II. *Aquaculture Magazine* 11(3):19–31.

Wickins, J. F., 1976. Prawn biology and culture. *Oceanography and Marine Biology Annual Review* 14:435–507.

Wickins, J. F., and T. W. Beard, 1974. Observations on the breeding and growth of the giant freshwater prawn *Macrobrachium rosenbergii* (de Man) in the laboratory. *Aquaculture* 3:159–174.

Wilkenfeld, J. S., A. L. Lawrence, and F. D. Kuban, 1984. Survival, metamorphosis, and growth of penaeid shrimp larvae reared on a variety of algal and animal foods. *Journal of the World Mariculture Society* 15:31–49.

Wyban, J. A., J. M. Sweeney, and R. A. Kanna, 1988. Shrimp yields and economic potential of intensive round pond systems. *Journal of the World Aquaculture Society* 19:210–217.

Young, J. H. 1959. Morphology of the white shrimp *Penaeus setiferus* (Linnaeus 1758), *Fishery Bulletin* 145. Washington, D.C., United States Fish and Wildlife Service.

CHAPTER 10

Adams, M. A., P. B. Johnsen, and Z. Hong-Qi, 1988. Chemical enhancement of feeding for the herbivorous fish *Tilapia zillii. Aquaculture* 72:95–107.

Anonymous, 1986. *Stocking and Management Recommendations for Texas Farm Ponds,* Special Publication No. 1. Texas Chapter of the American Fisheries Society.

Avault, J. W., Jr., 1986. New advances

in catfish farming research. *Aquaculture Magazine* 12(3):41–43.

Avault, J. W., Jr., 1987. Species profile—*Tilapia*. *Aquaculture Magazine* 13(1):47–49.

Avault, J. W., Jr., 1987. Species profile—minnows and other baitfish. *Aquaculture Magazine* 13(2):45–47.

Balon, E. K., 1975. Reproductive guilds in fishes: A proposal and definition. *Journal of the Fisheries Research Board of Canada* 32:821–864.

Bardach, J. E., J. H. Ryther, and W. O. McLarney, 1972. *Aquaculture.* New York: John Wiley & Sons, Inc.

Behrends, L. L., and R. O. Smitherman, 1984. Development of a cold-tolerant population or red tilapia through introgressive hybridization. *Journal of the World Mariculture Society* 15:172–178.

Benz, G. W., and R. P. Jacobs, 1986. Practical field methods of sexing largemouth bass. *The Progressive Fish-Culturist* 48:221–225.

Brandt, T. M., R. M. Jones, Jr., and R. J. Anderson, 1987. Evaluation of prepared feeds and attractants for largemouth bass fry. *The Progressive Fish-Culturist* 49:198–203.

Brown, E. E. and J. B. Gratzek, 1980. *Fish Farming Handbook.* Westport, Conn.: AVI Publishing Company, Inc.

Bye, V. J., and R. F. Lincoln, 1986. Commercial methods for the control of sexual maturation in rainbow trout (*Salmo gairdneri* R.). *Aquaculture* 57:299–309.

Carmichael, G. J., J. H. Williamson, C. A. C. Woodward, and J. R. Tomasso, 1988. Responses of northern, Florida, and hybrid largemouth bass to low temperature and low dissolved oxygen. *The Progressive Fish-Culturist* 50:225–231.

Carro-Anzalotta, A. E., and A. McGinty, 1986. Effects of stocking density on growth of *Tilapia nilotica* cultured in cages in ponds. *Journal of the World Aquaculture Society* 17:52–57.

Chen, T. P., 1976. *Aquaculture Practices in Taiwan.* Norwich, England: Page Brothers Limited.

Collins, C., 1988. Rearing channel catfish in cages. *Aquaculture Magazine* 14(1):53–55.

Collins, C., 1989. Some of the old and new in channel catfish farming. *Aquaculture Magazine* 15(4):47–54.

Conrad, J., 1985. Trout farming in Idaho. *Aquaculture Magazine* 11(1):32–36.

Conrad, J., 1985. Eels in catfish ponds? Danish research suggests possibility. *Aquaculture Magazine* 11(6):46–48.

Conrad, J., 1987. America's only exclusive eel farm. *Aquaculture Magazine* 13(4):24–27.

Davis, A. T., and R. S. Stickney, 1978. Growth response of *Tilapia aurea* to dietary protein quality and quantity. *Transactions of the American Fisheries Society* 107:479–483.

Degani, G., and D. Levanon, 1984. Influence of shapes of indoor and outdoor containers on adaptation to artificial food, growth, and survival of elvers. *The Progressive Fish-Culturist* 46:191–194.

Degani, G., and D. Levanon, 1986. The influence of different feeds on growth and body composition of glass eels (*Anguilla anguilla* L.). *Bamidgeh* 38:13–21.

Dobie, J., O. L. Meehean, S. F. Snieszko, and G. N. Washburn, 1956. *Raising Bait Fish,* Circular 35, Washington, D.C.: Fish and Wildlife Service, U.S. Department of the Interior.

Flickinger, S. A., 1971. Pond culture of bait fish, Bulletin 478A, Cooperative Extension Service. Fort Collins: Colorado State University.

Foltz, J. W., 1982. A feeding guide for single-cropped channel catfish. *Journal of the World Mariculture Society* 13, 274–281.

Gall, G. A. E., J. Baltodano, and N. Huang, 1988. Heritability of age at spawning for rainbow trout. *Aquaculture* 68:93–102.

Giudice, J. J., D. L. Gray, and J. M. Martin, undated. Manual for bait fish culture in the South, Publication EC 550. U.S. Fish and Wildlife Service and the University of Arkansas Cooperative Extension Service.

Halver, J. E., 1972. *Fish Nutrition*. New York: Academic Press, Inc.

Halvorsen, L., 1984. Improving the tilapia strain. *Aquaculture Magazine* 10(3): 26–30.

Hepher, B., and Y. Pruginin, 1981. *Commercial Fish Farming*. New York: John Wiley & Sons, Inc.

Hida, T. S., J. R. Harada, and J. E. King, 1962. *Rearing Tilapia for Tuna Bait*, Fishery Bulletin 198, Fish and Wildlife Service. Washington, D.C.: U.S. Department of the Interior.

Hoar, W. S., and D. J. Randall, 1969. *Fish Physiology*. Vol. 3, *Reproduction and Growth, Bioluminescence, Pigments, and Poisons*. New York: Academic Press, Inc.

Hudon, B., and J. de la Noue, 1985. Amino acid digestibility in rainbow trout: Influence of temperature, meal size and type of food. *Journal of the World Mariculture Society* 16:101–103.

Huet, M., 1986. *Textbook of Fish Culture*, 2nd ed. Farnham, Surrey, England: Fishing News Books Ltd.

Hulata, G., S. Rothbard, J. Itzkovich, G. Wohlfarth, and A. Halevy, 1985. Differences in hybrid fry production between two strains of the Nile tilapia. *The Progressive Fish-Culturist* 47:42–49.

Jhingran, V. G., and R. S. V. Pullin, 1988. *A Hatchery Manual for the Common, Chinese and Indian Major Carps*. Metro Manila, Philippines: International Center for Living Aquatic Resources Management.

Kafuku, T., and H. Ikenoue, 1983. *Modern Methods of Aquaculture in Japan*. New York: Elsevier Science Publishing Company, Inc.

Kingsley, J. B., 1987. Legal constraints to tilapia culture in the United States. *Journal of the World Aquaculture Society* 18:201–203.

Laird, L., and T. Needham, 1988. *Salmon and Trout Farming*. Chichester, England: Ellis Horwood Limited.

Lannan, J. E., R. O. Smitherman, and G. Tchobanoglous, 1986. *Principles and Practices of Pond Aquaculture*. Corvallis: Oregon State University Press.

Lee, J. S., 1981. *Commercial Catfish Farming*, 2nd ed. Danville, Ill.: The Interstate Printers & Publishers, Inc.

Lovell, T., 1987. Growing popularity of tilapia culture increases importance of nutrition. *Aquaculture Magazine* 13(1):45–46.

Marshall, N. B., 1972. *The Life of Fishes*. New York: Universe Books.

Martin, M., 1988. Black and hybrid crappie culture and crappie management. *Aquaculture Magazine* 14(3):35–41.

McGinty, A. S., 1985. *Tilapia Production in Ponds*, Mayaguez, Sea Grant Publication No. 25. University of Puerto Rico.

McGinty, A. S., 1987. Efficacy of mixed-species communal rearing as a method for performance testing of tilapias. *The Progressive Fish-Culturist* 49:17–20.

McLarney, W., 1984. *The Freshwater Aquaculture Book*. Point Roberts, Wash.: Hartley & Marks.

Meske, C., 1985. *Fish Aquaculture*. New York: Pergamon Press.

Mirza, J. A., and W. L. Shelton, 1988. Induction of gynogenesis and sex reversal in silver carp. *Aquaculture* 68:1–14.

Moody, T. M., and R. N. McCleskey, 1978. Vertical raceways for production of rainbow trout, Bulletin No. 17. New Mexico Department of Game and Fish.

Morrison, J. K., and C. E. Smith, 1986. Altering the spawning cycle of rainbow trout by manipulating water temperature. *The Progressive Fish-Culturist* 48:52–54.

Moyle, P. B., and J. J. Cech, Jr., 1982. *Fishes: An Introduction to Ichthyology*. Englewood Cliffs, N.J.: Prentice Hall.

Nagel, T., 1976. Rearing largemouth bass yearlings on artificial diets, In-Service Note 335. Ohio Department of Natural Resources, London Division of Wildlife.

Newman, M. W., H. E. Huezo, and D. G. Hughes, 1979. The response of all-male tilapia hybrids to four levels of protein in isocaloric diets. *Proceedings of the World Mariculture Society* 10:788–792.

Ostrowski, A. C., and D. L. Garling, Jr.,

1987. Changes in dietary protein to metabolizable energy needs of 17-alpha-methyltestosterone treated juvenile rainbow trout. *Journal of the World Aquaculture Society* 18:61–70.

Piper, R. G., I. B. McElwain, L. E. Orme, J. P. McCraren, L. G. Fowler, and J. R. Leonard, 1982. *Fish Hatchery Management.* Washington, D.C.: U.S. Department of the Interior, Fish and Wildlife Service.

Roell, M. J., G. D. Schuler, and C. G. Scalet, 1986. Cage-reared rainbow trout in dugout ponds in eastern South Dakota. *The Progressive Fish-Culturist* 48:273–278.

Saeed, M. O., and C. D. Ziebell, 1986. Effects of dietary nonpreferred aquatic plants on the growth of redbelly tilapia (*Tilapia zilli*). *The Progressive Fish-Culturist* 48:110–112.

Santiago, C. B., M. B. Aldaba, and O. S. Reyes, 1987. Influence of feeding rate and diet form on growth and survival of Nile tilapia (*Oreochromis niloticus*) fry. *Aquaculture* 64:277–282.

Smith, D. W., 1988. Phytoplankton and catfish culture: A review. *Aquaculture* 74:167–189.

Stickney, R. R., 1979. *Principles of Warmwater Aquaculture.* New York: John Wiley & Sons, Inc.

Stickney, R. R., 1986. *Culture of Nonsalmonid Freshwater Fish.* Boca Raton, Fla.: CRC Press, Inc.

Stickney, R. R., 1986. Tilapia tolerance of saline waters: A review. *The Progressive Fish-Culturist* 48:161–167.

Sumich, J. L., 1984. *An Introduction to the Biology of Marine Life,* 3rd ed. Dubuque, Ia.: Wm. C. Brown Publishers.

Swift, D. R., 1985. *Aquaculture Training Manual.* Farnham, Surrey, England: Fishing News Books Ltd.

Swingle, H. S., 1960. Comparative evaluation of two tilapias as pondfishes in Alabama. *Transactions of the American Fisheries Society* 89:142–148.

Tave, D., 1987. Improving productivity in catfish farming by selection. *Aquaculture Magazine* 13(5):53–55.

Tave, D., 1989. All-male catfish could improve yield. *Aquaculture Magazine* 15(4):67–69.

Thompson, B. Z., R. J. Wattendorf, R. S. Hestand, and J. L. Underwood, 1987. Triploid grass carp production. *The Progressive Fish-Culturist* 49:213–217.

Usui, A., 1974. *Eel Culture.* Surrey, Farnham, England: Fishing News Books Ltd.

Whitworth, W. R., P. L. Berrien, and W. T. Keller, 1968. *Freshwater Fishes of Connecticut,* Bulletin 101. State Geological and Natural History Survey of Connecticut.

Yant, D. R., R. O. Smitherman, and O. L. Green, 1975. Production of hybrid (blue × channel) catfish and channel catfish in ponds. *Proceedings of the Southeastern Association of Game & Fish Commissioners* 29:82–86.

Young, M. J. A., A. W. Fast, and P. G. Olin, 1989. Induced maturation of the Chinese catfish *Clarias fuscus. Journal of the World Aquaculture Society* 20:7–11.

CHAPTER 11

Aiken, D., 1989. Salmon farming in Canada. *World Aquaculture* 20(2):11–18.

Avault, J. W., Jr., 1987. Species profile—salmon. *Aquaculture Magazine* 13(4):65–66.

Bardach, J. E., J. H. Ryther, and W. O. McLarney, 1972. *Aquaculture.* New York: John Wiley & Sons, Inc.

Bonn, E. W., and W. M. Bailey, J. D. Bayless, K. E. Erickson, and R. E. Stevens, 1976. *Guidelines for Striped Bass Culture.* Bethesda, Md: American Fisheries Society.

Bromley, P. J., P. A. Sykes, and B. R. Howell, 1986. Egg production of turbot (*Scophthalmus maximus* L.) spawning in tank conditions. *Aquaculture* 53:287–293.

Brown, E. E., 1977. Production and culture of yellowtail (*Seriola quinqueradiata*) in Japan. *Proceedings of the World Mariculture Society* 8:765–771.

Brown, E. E., 1983. *World Fish Farming. Cultivation and Economics,* 2nd ed. Westport, Conn.: AVI Publishing Company, Inc.

Chen, T. P., 1976. *Aquaculture Practices in Taiwan.* Norwich, England: Page Brothers Limited.

Conrad, J., 1988. A seatrout farm in Denmark. *Aquaculture Magazine* 14(5): 35–40.

Donaldson, E. M., 1986. The integrated development and application of controlled reproduction techniques in Pacific salmonid aquaculture. *Fish Physiology and Biochemistry* 2:9–24.

Finucane, J. H., 1970. Progress in pompano mariculture in the United States. *Proceedings of the World Mariculture Society* 1:69–72.

Fitzmayer, K. M., J. I. Broach, and R. D. Estes, 1986. Effects of supplemental feeding on growth, production, and feeding habits of striped bass in ponds. *The Progressive Fish-Culturist* 48:18–24.

Fukuhara, O., T. Nakagawa, and T. Fukunaga, 1986. Larval and juvenile development of yellowtail reared in the laboratory. *Bulletin of the Japanese Society of Scientific Fisheries* 52:2091–2098.

Geiger, J. G., 1983. A review of pond zooplankton production and fertilization for the culture of larval and fingerling striped bass. *Aquaculture* 35:353–369.

Geiger, J. G., and N. C. Parker, 1985. Survey of striped bass hatchery management in the southeastern United States. *The Progressive Fish-Culturist* 47:1–13.

Hardy, R., 1986. Nutritional considerations for salmon culture in North America. *Aquaculture Magazine* 12(3): 44–46.

Henderson-Arzapalo, A., and R. L. Colura, 1987. Laboratory maturation and induced spawning of striped bass. *The Progressive Fish-Culturist* 49:60–63.

Hepher, B., and Y. Pruginin, 1981. *Commercial Fish Farming.* New York: John Wiley & Sons, Inc.

Huet, M., 1986. *Textbook of Fish Culture,* 2nd ed. Farnham, Surrey, England: Fishing News Books Ltd.

Ingram, M., 1987. The flatfish are coming. *Aquaculture Magazine* 13(3):44–47.

Iversen, E. S., and F. H. Berry, 1968. Fish mariculture: Progress and Potential. *Proceedings of the Gulf and Caribbean Fisheries Institute* 21:163–176.

Johannesson, B., 1987. Observations related to the homing instinct of Atlantic salmon (*Salmo salar* L.). *Aquaculture* 64:339–341.

Jory, D. E., E. S. Iversen, and R. H. Lewis, 1985. Culture of fishes of the genus *Trachinotus* (Carangidae) in the western Atlantic: Prospects and problems. *Journal of the World Mariculture Society* 16:87–94.

Kafuku, T., and H. Ikenoue, 1983. *Modern Methods of Aquaculture in Japan.* Amsterdam: Elsevier Science Publishing Company, Inc.

Kerby, J. H., J. M. Hinshaw, and M. T. Huish, 1987. Increased growth and production of striped bass × white bass hybrids in earthen ponds. *Journal of the World Mariculture Society* 18:35–43.

Korringa, P., 1976. *Farming Marine Fishes and Shrimps.* Amsterdam: Elsevier Science Publishing Company, Inc.

Laird, L., and T. Needham, 1988. *Salmon and Trout Farming.* Chicester, England: Ellis Horwood Limited.

Lee, C.-S., C. S. Tamaru, J. E. Banno, C. D. Kelley, A. Bocek, and J. A. Wyban, 1986. Induced maturation and spawning of milkfish, *Chanos chanos* Forsskal, by hormone implantation. *Aquaculture* 52:199–205.

Lee, C.-S., C. S. Tamaru, and G. M. Weber, 1987. Studies on the maturation and spawning of milkfish *Chanos chanos* Forsskal in a photoperiod-controlled room. *Journal of the World Aquaculture Society* 18:253–259.

Marte, C. L., and F. Lacanilao, 1986. Spontaneous maturation and spawning of milkfish in floating net cages. *Aquaculture* 53:115–132.

McMaster, M. F., 1988. Pompano

aquaculture: Past success and present opportunities. *Aquaculture Magazine* 14(3): 28–34.

McNeil, W. J., 1988. *Salmon Production, Management, and Allocation.* Corvallis: Oregon State University Press.

Meske, C., 1985. *Fish Aquaculture: Technology and Experiments.* New York: Pergamon Press.

Murray, C. B., and T. D. Beacham, 1986. Effect of incubation density and substrate on the development of chum salmon eggs and alevins. *The Progressive Fish-Culturist* 48:242–249.

Pantastico, J. B., J. P. Baldia, and D. M. Reyes, Jr., 1986. Feed preference of the milkfish (*Chanos chanos* Forsskal) fry given different algal species as natural feed. *Aquaculture* 56:169–178.

Piper, R. G., I. B. McElwain, L. E. Orme, J. P. McCraren, L. G. Fowler, and J. R. Leonard, 1982. *Fish Hatchery Management.* Washington, D.C.: U.S. Department of the Interior, Fish and Wildlife Service.

Roa, M. C., C. Huelvan, Y. Le Borgne, and R. Metailler, 1982. Use of rehydratable extruded pellets and attractive substances for the weaning of the sole (*Solea vulgaris*). *Journal of the World Mariculture Society* 13:246–253.

Ruyet, J. P.-L., D. L'Elchat, and G. Nedelec, 1981. Research on rearing turbot (*Scophthalmus maximus*): Results and perspectives. *Journal of the World Mariculture Society* 12:143–152.

Shelbourne, J. E., 1971. *Artificial Propagation of Marine Fish.* Jersey City, N.J.: T.F.H. Publications, Inc.

Smith, T. I. J., 1988. Aquaculture of striped bass and its hybrids in North America. *Aquaculture Magazine* 14(1):40–49.

Smith, T. I. J. 1989. Culture potential of striped bass and its hybrids. *World Aquaculture* 20(1):32–38.

Stickney, R. R., 1979. *Principles of Warmwater Aquaculture.* New York: John Wiley & Sons, Inc.

Tang, J., M. D. Bryant, and E. L. Brannon, 1987. Effect of temperature extremes on the mortality and development rates of coho salmon embryos and alevins. *The Progressive Fish-Culturist* 49:167–174.

Thompson, K. S., W. H. Weed III, and A. G. Taruski, 1971. Saltwater fishes of Connecticut, Bulletin 105. State Geological and Natural History Survey of Connecticut.

Valenti, R. J., J. Aldred, and J. Liebell, 1976. Experimental marine cage culture of striped bass in northern waters. *Proceeding of the World Mariculture Society* 7:99–108.

Wainwright, G., 1985. Scottish salmon aquaculture now three times wild catch. *Aquaculture Magazine* 11(1):26–29.

Williams, S., R. T. Lovell, and J. P. Hawke, 1985. Value of menhaden oil in diets of Florida pompano. *The Progressive Fish-Culturist* 47:159–165.

CHAPTER 12

Anonymous, 1989. U.S. states enacting aquaculture legislation. *Aquaculture Magazine* 15(2):20–25.

Colson, D. A., 1984/1985. Transboundary fishery stocks in the EEZ. *Oceanus* 27:48–51.

Davies, C. B., and H. W. Shields, 1970. Mariculture and the law. *Proceeding of the World Mariculture Society* 1:55–59.

Gaucher, T. A., 1971. *Aquaculture: A New England Perspective.* Narragansett, R.I.: New England Marine Resources Information Program.

Gordon, W. G., and R. E. Gutting, Jr., 1984/1985. The coastal fishing industry and the EEZ. *Oceanus* 27:35–40.

Hanson, J. A., 1974. *Open Sea Mariculture.* Stroudsburg, Pa.: Dowden, Hutchinson, & Ross Inc.

Joint Subcommittee on Aquaculture, 1987. *Aquaculture: A Guide to Federal Government Programs.* Washington, D.C.: National Agricultural Library, United States Department of Agriculture.

Kendall, R. L., 1985. *Role of the U.S.*

Government in Aquaculture: Nonfederal Perspectives. Bethesda, Md.: American Fisheries Society.

McLarney, W., 1984. *The Freshwater Aquaculture Book.* Point Roberts, Wash.: Hartley & Marks.

McNeil, W. J., 1988. *Salmon Production, Management, and Allocation.* Corvallis: Oregon State University Press.

Milne, P. H., 1979. *Fish and Shellfish Farming in Coastal Waters.* Farnham, Surrey, England: Fishing News Books Ltd.

National Research Council, 1978. *Aquaculture in the United States: Constraints and Opportunities.* Washington, D.C.: Printing and Publishing Office, National Academy of Science.

New York Sea Grant Institute, 1985. *Aquaculture Development in New York.* Albany: New York Sea Grant Institute.

Owen, S., 1978. The response of the legal system to technical innovations in aquaculture: A comparative study of the mariculture legislation in California, Florida, and Maine. *Coastal Zone Management Journal* 4:269–297.

Parker, H. S., 1986. *Strategies for Aquaculture Development in Massachusetts.* Proceedings of a conference held at Southeastern Massachusetts University, North Dartmouth, March 12–13, 1986.

Rhodes, R. J., 1988. Mariculture in Maine: the future is now. *Aquaculture Magazine* 14(3):42–48.

Royce, W. F., 1984. *Introduction to the Practice of Fishery Science.* New York: Academic Press, Inc.

Wildsmith, B. H., 1982. *Aquaculture: The Legal Framework.* Toronto, Canada: Edmond-Montgomery Limited.

Wypyszinski, A. W., 1988. Legal considerations in aquaculture production. Paper presented at New Jersey Aquaculture III, Atlantic City.

CHAPTER 13

Anderson, J. L., 1985. Private aquaculture and commercial fisheries: bioeconomics of salmon ranching. *Journal of Environmental Economics and Management* 12:353–370.

Anderson, L. G., and D. C. Tabb, 1971. Some economic aspects of pink shrimp farming in Florida. *Proceeding of the Gulf & Caribbean Fisheries Institute* 23:113–124.

Anonymous, 1988. Financing for aquaculture. *Aquaculture Today* 1(2):40.

Chaston, I., 1984. *Business Management in Fisheries and Aquaculture.* Farnham, Surrey, England: Fishing News Books Ltd.

Diana, J. S., S. L. Kohler, and D. R. Ottey, 1988. A yield model for walking catfish production in aquaculture systems. *Aquaculture* 71:23–35.

Hatch, U., S. Sindelar, D. Rouse, and H. Perez, 1987. Demonstrating the use of risk programming for aquacultural farm management: The case of a *Penaeid* in Panama. *Journal of the World Aquaculture Society* 18:260–269.

Hed, A., 1985. National agriculture export council building new markets. *Aquaculture Magazine* 11(4):28–32.

Klemetson, S., and G. L. Rogers, 1984. A realistic approach to small aquaculture facilities. *Aquacultural Engineering* 3:153–157.

Laird, L., and T. Needham, 1988. *Salmon and Trout Farming.* Chicester, England: Ellis Horwood Limited.

Lee, J. S., 1981. *Commercial Catfish Farming.* Danville, Ill.: Interstate Printers and Publishers, Inc.

Leeds, R., 1986. Financing aquaculture projects. *Aquacultural Engineering* 5:109–113.

Meade, J. W., 1985. Determine combinations that will maximize fish farm income. *Aquaculture Magazine* 11(2):27–34.

McLarney, W., 1984. *The Freshwater Aquaculture Book.* Point Roberts, Wash.: Hartley & Marks.

National Research Council, 1978. *Aquaculture in the United States: Constraints and Opportunities.* Washington, D.C.: Printing and Publishing Office, National Academy of Science.

Pillay, T. V. R., and W. A. Dill, 1979. *Advances in Aquaculture*. Farnham, Surrey, England: Fishing News Books Ltd.

Pomeroy, R. S., D. B. Luke, and J. M. Whetstone, 1986. The economics of crawfish production in South Carolina. *Aquaculture Magazine* 12(2):36–39.

Shang, Y. C., 1981. *Aquaculture Economics*. Boulder, Color.: Westview Press.

Shepherd, C. J., 1974. The economics of aquaculture—a review. *Oceanography and Marine Biology Annual Review* 13:413–420.

Skurla, J. A., and M. E. McDonald, 1988. Predicting the economics of a walleye aquaculture operation using bioenergetics models. *World Aquaculture* 19(4):74–75.

APPENDIX 1

Bower, C. E., and T. Holm-Hansen, 1980. A salicylate-hypochlorite method for determining ammonia in seawater. *Canadian Journal of Fisheries and Aquatic Science* 27:794–798.

Franson, M. A., 1976. *Standard Methods for the Examination of Water and Wastewater*, 14th ed. Washington, D.C.: American Public Health Association.

Grasshoff, K., 1976. *Methods of Seawater Analysis*. New York: Verlag Chemie.

Landau, M., and J. W. Tucker, Jr., 1984. Acute toxicity of EDB and aldicarb to young of two estuarine fish species. *Bulletin of Environmental Contamination and Toxicology* 33:127–132.

Strickland, J. D. H., and T. R. Parsons, 1972. *A Practical Handbook of Seawater Analysis*, Bulletin 167, 2nd ed. Ottawa: Fisheries Research Board of Canada.

Van Son, M., R. C. Schothorst, G. den-Boef, 1983. Determination of total ammoniacal nitrogen in water by flow injection analysis and a gas diffusion membrane. *Analytica Chim. Acta* 153:271–275.

APPENDIX 2

Eshbach, O. W., 1952. *Handbook of Engineering Fundamentals*. New York: John Wiley & Sons, Inc.

Hansen, A. G., 1967. *Fluid Mechanics*. New York: John Wiley & Sons, Inc.

Moody, L. F., 1944. Friction factors for pipe flow. *Transactions of the American Society of Mechanical Engineers* 66:671–678.

Souders, M., 1966. *The Engineer's Companion*. New York: John Wiley & Sons, Inc.

Wheaton, F. W., 1977. *Aquacultural Engineering*. New York: John Wiley & Sons, Inc.

APPENDIX 3

Adams, A., N. Auchinachie, A. Bundy, M. F. Tatner, and M. T. Horne, 1988. The potency of adjuvanted injected vaccines in rainbow trout (*Salmo gairdneri* Richardson) and bath vaccines in the Atlantic salmon (*Salmo salar* L.) against furunculosis. *Aquaculture* 69:15–26.

Ainsworth, A. J., G. Capley, P. Waterstreet, and D. Munson, 1986. Use of monoclonal antibodies in the indirect fluorescent antibody technique (IFA) for the diagnosis of *Edwardsiella ictaluri*. *Journal of Fish Diseases* 9:439–444.

Austin, B., I. Bishop, C. Gray, B. Watt, and J. Dawes, 1986. Monoclonal antibody-based enzyme-linked immunosorbent assays for the rapid diagnosis of clinical cases of enteric redmouth and furunculosis in fish farms. *Journal of Fish Diseases* 9:469–474.

Colwell, R. R., A. J. Sinskey, and E. R. Pariser, 1984. *Biotechnology in the Marine Sciences*. New York: John Wiley & Sons, Inc.

Brem, G., B. Brenig, G. Horstgen-Schwark, and E.-L. Winnacker, 1988. Gene transfer in tilapia (*Oreochromis niloticus*). *Aquaculture* 68:209–219.

Davis, L. G., M. D. Dibner, and J. F.

Battey, 1986. *Methods in Molecular Biology.* New York: Elsevier Science Publishing Company, Inc.

Down, N. E., E. M. Donaldson, H. M. Dye, K. Langley, and L. M. Souza, 1988. Recombinant bovine somatotropin more than doubles the growth rate of coho salmon (*Oncorhynchus kisutch*) acclimated to seawater and ambient winter conditions. *Aquaculture* 68:141–155.

Ellis, A. E., 1988. Current aspects of fish vaccines. *Diseases of Aquatic Organisms* 4:159–164.

Freifelder, D., 1985. *Essentials of Molecular Biology.* Boston: Jones and Barlett Publishers, Inc.

Hood, L. E., I. L. Weissman, and W. B. Wood, 1978. *Immunology.* London: The Benjamin/Cummings Publishing Company, Inc.

Laufer, H., D. Borst, F. C. Baker, C. Carrasco, M. Sinkus, C. C. Reuter, L. W. Tsai, D. A. Schooley, 1987. Identification of a juvenile hormone-like compound in a crustacean. *Science* 235:202–205.

Morse, D. E., 1984. Biochemical and genetic engineering for improved production of abalones and other valuable molluscs. *Aquaculture* 39:263–282.

Pascho, R. J., and D. Mulcahy, 1987. Enzyme-linked immunosorbent assay for a soluble antigen of *Renibacterium salmoninarum*, the causative agent of salmonid bacterial kidney disease. *Canadian Journal of Fisheries and Aquatic Sciences* 44:183–191.

Poline-Fuller, M., and A. Gibor, 1986. Calluses, cells, and protoplasts in studies toward genetic improvement of seaweeds. *Aquaculture* 57:117–123.

Saeed, M. O., and J. A. Plumb, 1986. Immune response of channel catfish to lipopolysaccharide and whole cell *Edwardsiella ictaluri* vaccines. *Diseases of Aquatic Organisms* 2:21–25.

Sato, N., K. Murata, K. Watanabe, T. Hayami, Y. Kariya, M. Sakaguchi, S. Kimura, M. Nonaka, and A. Kimura, 1988. Growth-promoting activity of tuna growth hormone and expression of tuna growth hormone cDNA in *Escherichia coli*. *Biotechnology and Applied Biochemistry* 10:385–393.

Secombes, C. J., A. Van Winkoop, J. G. M. Van den Boogaart, L. P. M. Timmermans, and I. G. Priede, 1986. Immunological approaches to control maturation in fish. I. Cytotoxic reactions against germ cells using monoclonal antibodies. *Aquaculture* 52:125–135.

Tave, D., 1988. Genetic engineering. *Aquaculture Magazine* 14(2):63–65.

APPENDIX 5

Bennett, W. N., and M. E. Boraas, 1988. Isolation of a fast-growing strain of the rotifer *Brachionus calyciflorus* Pallas using turbidostat culture. *Aquaculture* 73, 27–36.

Berg, C. J., Jr., 1983. *Culture of Marine Invertebrates.* Stroudsburg, Pa.: Hutchinson Ross Publishing Company.

Betouhim-El, T., and D. Kahan, 1972. *Tisbe pori* n. sp. (Copepoda: Harpacicoida) from the Mediterranean coast of Israel and its cultivation in the laboratory. *Marine Biology* 16:201–209.

Brand, L. E., 1984. The salinity tolerance of forty-six marine phytoplankton isolates. *Estuarine, Coastal and Shelf Science* 18:543–556.

Brisset, P., D. Versichele, E. Bossuyt, L. De Ruyck, and P. Sorgeloos, 1982. High density flow-through culturing of brine shrimp *Artemia* on inert feeds—preliminary results with a modified culture system. *Aquacultural Engineering* 1:115–119.

Brune, D. E., 1982. Design and development of a flowing bed reactor for brine shrimp culture. *Aquacultural Engineering* 1:63–70.

DePaw, N., P. Laureys, and J. Morales, 1981. Mass cultivation of *Daphnia magma* Straus on ricebran. *Aquaculture* 25:141–152.

Fogg, G. E., 1975. *Algal Cultures and Phytoplankton Ecology.* Madison: University of Wisconsin Press.

Fontaine, C. T., and D. B. Revera, 1980. The mass culture of the rotifer *Brachionus plicatus* for use as a foodstuff in aquaculture. *Proceedings of the World Mariculture Society* 11:211–218.

Gatesoupe, F.-J., and J. H. Robin, 1981. Commercial single cell proteins either as sole food source or in formulated diets for intensive and continuous production of rotifers (*Brachionus plicatilis*). *Aquaculture* 25:1–15.

Giliberto, S., and A. Mazzola, 1981. Mass culture of *Brachionus plicatilis* with an integrated system of *Tetraselmis suecica* and *Saccharomyces cerevisiae*. *Journal of the World Mariculture Society* 12:61–62.

Kahan, D., 1979. Vegetables as food for marine harpacticoid copepods. *Aquaculture* 16:345–350.

Kahan, D., G. Uhlig, D. Schwenzer, and L. Horowitz, 1981/82. A simple method for cultivating harpacticoid copepods and offering them to fish larvae. *Aquaculture* 26:303–310.

Kosiorek, D., 1974. Development cycle of *Tubifex tubifex* Mull. in experimental culture. *Polskie Archiwum Hydrobiologii* 21:411–422.

Laws, E. A., S. Taguchi, J. Hirata, and L. Pang, 1988. Optimization of microalgal production in a shallow outdoor flume. *Biotechnology and Bioengineering* 32:140–147.

Needham, J. G., F. E. Lutz, P. S. Welch, and P. S. Galtsoff, 1937. *Culture Methods for Invertebrate Animals*. New York: Dover Publications, Inc.

Persoone, G., P. Sorgeloos, O. Roels, and E. Jaspers, 1980. *The Brine Shrimp Artemia* (3 vols). Wetteren, Belgium: Universa Press.

Rees, J. T., and J. M. Oldfather, 1980. Small-scale mass culture of *Daphnia magna* Straus. *Proceedings of the World Mariculture Society* 11:202–210.

Richmond, A., 1986. *CRC Handbook of Microalgal Mass Culture*. Boca Raton, Fla.: CRC Press, Inc.

Sarma, S. S. S., and T. R. Rao, 1987. Effect of food level on body size and egg size in a growing population of the rotifer *Brachionus patulus* Muller. *Arch. fur Hydrobiol.* 111:245–253.

Scott, A. P., and S. M. Baynes, 1978. Effect of algal diet and temperature on the biochemical composition of the rotifer *Brachionus plicatilis*. *Aquaculture* 14:247–260.

Snell, T. W., and F. H. Hoff, 1988. Recent advances in rotifer culture. *Aquaculture Magazine* 14(5):41–45.

Snell, T. W., M. Childress, E. M. Boyer, and F. H. Hoff, 1987. Assessing the status of rotifer mass culture. *Journal of the World Aquaculture Society* 18:270–277.

Sommer, T., 1988. Commercial microalgal production: The state of the art. *Aquaculture Magazine* 14(2):44–47.

Sorgeloos, P., E. Bossuyt, E. Lavina, M. Baeza-Mesa, and G. Persoone, 1977. Decapsulation of *Artemia* cysts: A simple technique for the improvement of the use of brine shrimp in aquaculture. *Aquaculture* 12:311–315.

Stottrup, J. G., K. Richardson, E. Kirkegaard, and N. J. Pihl, 1986. The cultivation of *Acartia tonsa* Dana for use as a live food source for marine fish larvae. *Aquaculture* 52:87–96.

Theilacker, G. H., and M. F. McMaster, 1971. Mass culture of the rotifer *Brachionu plicatis* and its evaluation as food for larval anchovies. *Marine Biology* 10:183–188.

Trotta, P., 1981. A simple and inexpensive system for continuous monoxenic mass culture of marine microalgae. *Aquaculture* 22:383–387.

Ward, S. H., 1984. A system for laboratory rearing of the mysid, *Mysidopsis bahia* Molenock. *Progressive Fish-Culturist* 46:170–175.

Wikfors, G. H., J. W. Twarog, Jr., and R. Ukeles, 1984. Influence of chemical composition of algal food sources on growth of juvenile oysters, *Crassostrea virginica*. *Biological Bulletin* 167:251–263.

Yufera, M., 1982. Morphometric characterization of a small-sized strain of *Bra-*

chionus plicatilis in culture. *Aquaculture* 27:55–61.

APPENDIX 6

Adey, W. H., 1987. Food production in low-nutrient seas. *BioScience* 37:340–348.

Anonymous, 1984. Research program on giant clams initiated. *Aquaculture Magazine* 10(5):26–29.

Bardach, J. E., J. H. Ryther, and W. O. McLarney, 1972. *Aquaculture*. New York: John Wiley & Sons, Inc.

Berg, C. J., Jr., 1983. *Culture of Marine Invertebrates*. Stroudsburg, Pa.: Hutchinson Ross Publishing Company.

Berg, C. J., Jr., and P. Alatalo, 1982. Mariculture potential of shallow-water Bahamian bivalves. *Journal of the World Mariculture Society* 13:294–300.

Berry, S. S., 1912. A review of cephalopods of the western Atlantic. *Bulletin of the Bureau of Fisheries* 30:267–336.

Breber, P., 1981. The controlled reproduction of the carpet-shell clam (*Venerupis decussata* [L]: preliminary results. *Journal of the World Mariculture Society* 12(2):172–179.

Braley, R. D., 1988. Farming the giant clam. *World Aquaculture* 20(1):6–17.

Broom, M. J., 1985. *The Biology and Culture of Marine Bivalve Molluscs of the Genus* Anadara. Metro Manila, Philippines: International Center for Living Aquatic Resources Management.

Chen, T. P., 1976. *Aquaculture Practices in Taiwan*. Norwich, England: Page Brothers Limited.

Conrad, J., 1985. Producing clams for the U.S. market. *Aquaculture Magazine* 11(3):38–40.

Costello, T., 1985. Peru has a thriving scallop industry. *Aquaculture Magazine* 11(3):32–36.

Creswell, L., 1984. Ingestion, assimilation, and growth of juveniles of the queen conch *Strombus gigas* Linne fed experimental diets. *Journal of Shellfish Research* 4:23–30.

Davis, M., B. A. Mitchell, and J. L. Brown, 1984. Breeding behavior of the queen conch *Strombus gigas* Linne held in a natural enclosed habitat. *Journal of Shellfish Research* 4:17–21.

Davy, F. B., and M. Graham, 1982. *Bivalve Culture in Asia and the Pacific*. Ottawa, Canada: International Development Research Centre.

Forsythe, J. W., and R. T. Hanlon, 1988. Effect of temperature on laboratory growth, reproduction and life span of *Octopus bimaculoides*. *Marine Biology* 98:369–379.

Hahn, K. O., 1988. *Handbook of Culture of Abalone and Other Marine Gastropods*. Boca Raton, Fla.: CRC Press, Inc.

Hanlon, R. T., 1977. Laboratory rearing of the Atlantic reef octopus, *Octopus briareus* Robson, and its potential for mariculture. *Proceedings of the World Mariculture Society* 8:471–482.

Hanlon, R. T., and J. W. Forsythe, 1985. Advances in the laboratory culture of octopuses for biomedical research. *Laboratory Animal Science* 35:33–40.

Hefferman, P. G., R. L. Walker, and D. M. Gillespie, 1988. Biological feasibility of growing the northern bay scallop, *Argopecten irradians irradians* (Lamarck, 1819), in coastal water of Georgia. *Journal of Shellfish Research* 7:83–88.

Heslinga, G. A., and W. K. Fitt, 1987. The domestication of reef-dwelling clams. *BioScience* 37:332–339.

Huner, J. V., and E. E. Brown, 1985. *Crustacean and Mollusk Aquaculture in the United States*. Westport, Conn.: AVI Publishing Company, Inc.

Kafuku, T., and H. Ikenoue, 1983. *Modern Methods of Aquaculture in Japan*. New York: Elsevier Science Publishing Company, Inc.

Korringa, P., 1976. *Farming Marine Organisms Low in the Food Chain*. New York: Elsevier Science Publishing Company, Inc.

Leighton, D. L., and C. F. Phleger, 1977. The purple-hinge rock scallop: a new candidate for marine aquaculture.

Proceedings of the World Mariculture 8:457–469.

Morgan, D. E., J. Goodsell, G. C. Matthiessen, J. Garey, and P. Jacobson, 1980. Release of hatchery-reared bay scallop (*Argropecten irradians*) onto a shallow coastal bottom in Waterford, Connecticut. *Proceedings of the World Mariculture Society* 11:247–261.

Muir, J. E., and R. J. Roberts, 1982. *Recent Advances in Aquaculture.* Boulder, Colo.: Westview Press.

Pillay, T. V. R., and W. A. Dill, 1979. *Advances in Aquaculture.* Farnham, Surrey, England: Fishing News Books Ltd.

Shaw, W. N., 1985. The purple-hinge rock scallop: A mariculture species of the future? *Aquaculture Magazine* 11(1):43–44.

Spotts, D. G., 1987. Farming queen conch in the Netherland Antilles and Aruba. *Aquaculture Magazine* 13(2):32–35.

Stickney, R. R., 1979. *Principles of Warmwater Aquaculture.* New York: John Wiley & Sons, Inc.

Toll, R. B., and C. H. Strain, 1988. Freshwater and terrestrial food organisms as an alternative diet for laboratory culture of cephalopods. *Malacologia* 29:195–200.

Wada, K. T., A. Komaru, and Y. Uchimura, 1989. Triploid production in the Japanese pearl oyster, *Pinctada fucata martensii. Aquaculture* 76:11–19.

Weymouth, F. W., H. C. McMillin, and H. B. Holmes, 1925. Growth and age at maturity of the Pacific razor clam, *Siliqua patula. Bulletin of the Bureau of Fisheries* 41:201–236.

APPENDIX 7

Aiken, D., 1988. Marron farming. *World Aquaculture* 19(4):14–17.

Aiken, D. E., D. J. Martin, J. D. Meisner, and J. B. Sochasky, 1981. Influence of photoperiod on survival and growth of larval American lobsters (*Homarus ameri-*

canus). *Journal of the World Mariculture Society* 12(1):225–230.

Avault, J. W., Jr., 1986. Soft shell crabs and crawfish. *Aquaculture Magazine* 12(2):45–48.

Bardach, J. E., J. H. Ryther, and W. O. McLarney, 1972. *Aquaculture.* New York: John Wiley & Sons, Inc.

Bartley, D. M., J. M. Carlberg, J. C. Van Olst, and R. F. Ford, 1980. Growth and conversion efficiency of juvenile American lobsters (*Homarus americanus*) in relation to temperature and feeding level. *Journal of the World Aquaculture Society* 11:355–368.

Brownell, W. N., A. J. Provenzano, Jr., and M. Martinez, 1977. Culture of the West Indian spider crab (*Mithrax spinosissimus*) at Los Roques, Venezuela. *Proceeding of the World Mariculture Society* 8:157–168.

Capuzzo, J. M., and B. A. Lancaster, 1979. Effect of dietary carbohydrate levels on protein utilization in the American lobster (*Homarus americanus*). *Proceedings of the World Mariculture Society* 10:689–700.

Chen, T. P., 1976. *Aquaculture Practices in Taiwan.* Norwich, England: Page Brothers Limited.

Churchill, E. P., Jr., 1919. Life history of the blue crab. *Bulletin of the Bureau of Fisheries* 36:91–128.

Conklin, D. E., K. Devers, and C. Bordner, 1977. Development of artificial diets for the lobster, *Homarus americanus. Proceedings of the World Mariculture Society* 8:841–852.

D'Abramo, L. R., C. E. Bordner, and D. E. Conklin, 1982. Relationship between dietary phosphatidylcholine and serum cholesterol in the lobster *Homarus* sp. *Marine Biology* 67:231–235.

D'Abramo, L. R., C. E., Bordner, D. E. Conklin, and N. A. Baum, 1984. Sterol requirement of juvenile lobsters, *Homarus* sp. *Aquaculture* 42:13–25.

Eagles, M. D., D. E. Aiken, and S. L. Waddy, 1984. Effect of food quality on survival, growth and development of larval American lobsters fed frozen adult brine shrimp. *Journal of the World Mariculture Society* 15:142–143.

Freshwater Australian Crayfish Traders, 1981–1984. *Newsletters 1 to 12.* Stafford, Australia: FACT.

Hartman, M. C., 1977. A mass rearing system for the culture of brachyuran crab larvae. *Proceeding of the World Mariculture Society* 8:147–155.

Huner, J. V., and E. E. Brown, 1985. *Crustacean and Mollusk Aquaculture in the United States.* Westport, Conn.: AVI Publishing Company, Inc.

Inyang, N. M., 1977/78. Effects of some environmental factors on the growth and food consumption of the Baltic palaemonid shrimp, *Palaemon adspersus* var. *bavricii* (Rathke). *Meeresforschung* 26:30–41.

Kafuku, T. and H. Ikenoue, 1983. *Modern Methods of Aquaculture in Japan.* New York: Elsevier Science Publishing Company, Inc.

Korringa, P., 1976. *Farming Marine Organisms Low in the Food Chain.* New York: Elsevier Science Publishing Company, Inc.

McConaugha, J. R., K. McNally, J. W. Goy, and J. D. Costlow, 1980. Winter induced mating in the stone crab, *Menippe mercenaria. Proceeding of the World Mariculture Society* 11:544–547.

McVey, J. P., 1983. *Handbook of Mariculture,* Vol. I. Boca Raton, Fla.: CRC Press, Inc.

Mead, A. D., 1908. A method of lobster culture. *Bulletin of the Bureau of Fisheries* 28:219–240.

Morrissy, N. M., 1974. The ecology of the marron *Cherax tenuimannus* (Smith) introduced into some farm dams near Boscabel in the great southern area of the wheatbelt region of Western Australia. *Fisheries Research Bulletin of Western Australia* 12:1–55.

Morrissy, N. M., 1979. Experimental pond production of marron, *Cherax tenuimanus* (Smith) (Decapoda: Parastacidae). *Aquaculture* 16:319–344.

Muir, J. E., and R. J. Roberts, 1982.

Recent Advances in Aquaculture. Boulder, Color.: Westview Press.

Nelson, K., D. Hedgecock, and W. Borgeson, 1988. Factors influencing egg extrusion in the American lobster (*Homarus americanus*). *Canadian Journal of Fisheries and Aquatic Sciences* 45:797–804.

O'Sullivan, D., 1988. Queensland cray farmers opt for local species. *Aquaculture Magazine* 14(5):46–49.

Pillay, T. V. R., and W. A. Dill, 1979. *Advances in Aquaculture.* Farnham, Surrey, England: Fishing New Books Ltd.

Provenzano, A. J., and W. N. Brownell, 1977. Larval and early post-larval stages of the West Indian spider crab, *Mithrax spinosissimus* (Lamark) (Decapoda: Majidae). *Proceedings of the Biological Society of Washington* 90:735–752.

Richard, P., 1978. Effect of temperature on growth and molting of *Palaemon serratus* in relation to their size. *Aquaculture* 14:13–22.

Stickney, R. R., 1979. *Principles of Warmwater Aquaculture.* New York: John Wiley & Sons, Inc.

Tunberg, B. G., and R. L. Creswell, 1988. Early growth and mortality of the Caribbean king crab *Mithrax spinosissimus* reared in the laboratory. *Marine Biology* 98:337–343.

Waddy, S. L., 1988. Farming the homarid lobsters: state of the art. *World Aquaculture* 19(4):63–71.

Yang, W. T., 1971. Preliminary report on the culture of the stone crab. *Proceedings of the World Mariculture Society* 2:53–54.

APPENDIX 8

Baldwin, W. J., and M. J. McGrenra, 1979. Problems with the culture of topminnows (family Poeciliidae) and their use as live baitfish. *Proceeding of the World Mariculture Society* 10:249–259.

Bardach, J. E., J. H. Ryther, and W. O. McLarney, 1972. *Aquaculture.* New York: John Wiley & Sons, Inc.

Boyd, V., 1984. High intensity indoor aquaculture. *Farm Pond Harvest* 18(2): 10–11.

Brown, E. E., and J. B. Gratzek, 1980. *Fish Farming Handbook.* Westport, Conn.: AVI Publishing Company, Inc.

Buttner, J. K., 1989. Culture of fingerling walleye in earthen ponds, state of the art 1989. *Aquaculture Magazine* 15(2): 37–43.

Chamberlain, G. W., and G. McCarty, 1985. Why chose redfish? *Aquaculture Magazine* 11(2):35–42.

Chen, H.-C., and C. Y. Su, 1980. Studies on the induced spawning of the mud loach, *Misgurnus anguillicaudatus,* by hormone injection throughout the year. *Journal of the Fisheries Society of Taiwan* 7: 13–20.

Chen, T. P., 1976. *Aquaculture Practices in Taiwan.* Norwich, England: Page Brothers Limited.

Colesante, R. T., N. B. Youmans, and B. Ziolkoski, 1986. Intensive culture of walleye fry with live food and formulated diets. *The Progressive Fish-Culturist* 48: 33–37.

Daniels, W. H., and E. H. Robinson, 1986. Protein and energy requirements of juvenile red drum *(Sciaenops ocellatus)*. *Aquaculture* 53:243–252.

Davy, F. B., and A. Chouinard, 1981. *Induced Fish Breeding in Southeast Asia.* Ottawa, Canada: International Development Research Centre.

Dillard, J. G., L. K. Graham, and T. R. Russell, 1986. *Paddlefish: Status, Management and Propagation,* North Central Division of the American Fisheries Society Special Publication No. 7. Jefferson City, Mo.: Modern Litho-Print Company.

Ellison, D. G., J. A. Gleim, and D. Kapke, 1983. Overwintering threadfin shad through intensive culture. *The Progressive Fish-Culturist* 45:90–93.

Epifanio, C. E., D. Gorshorn, and T. E. Targett, 1988. Induction of spawning in the weakfish, *Cynoscion regalis. Fishery Bulletin* 86:168–171.

Fast, A. W., 1988. *Spawning Induction and Pond Culture of the Spotted Scat (Scatophagus argus Linnaeus) in the Philippines.* Manoa: University of Hawaii. Mariculture Research Training Center.

Foscarini, R., 1988. A review: Intensive farming procedures for red sea bream *(Pagrus major)* in Japan. *Aquaculture* 72:191–246.

Garza, G., W. H. Bailey, and J. L. Lasswell, 1978. Rearing of black drum in freshwater. *The Progressive Fish-Culturist* 40:170.

Girin, M., 1979. Prospects for commercial culture of the European sea bass *(Dicentrarchus labrax)* and other marine finfish in France and neighboring countries. *Proceedings of the World Mariculture Society* 10:272–279.

Hagood, R. W., G. N. Rothwell, M. Swafford, and M. Tosaki, 1981. Preliminary report on the aquacultural development of the dolphin fish *Coryphaena hippurus* (Linnaeus). *Journal of the World Mariculture Society* 12:135–139.

Hamilton, J., and K. Graham, 1988. Paddlefish by C-section. *The Missouri Conservationist* 49:18–21.

Henderson-Arzapalo, A., and R. L. Colura, 1984. Black drum × red drum hybridization and growth. *Journal of the World Mariculture Society* 15:412–420.

Howey, R. G., 1985. Intensive culture of juvenile American shad. *The Progressive Fish-Culturist* 47:203–212.

Huet, M., 1986. *Textbook of Fish Culture,* 2nd ed. Farnham, Surrey, England: Fishing News Books Ltd.

Hung, S. S. O., 1989. Practical feeding of white sturgeon. *Aquaculture Magazine* 15(1):60–62.

Jones, F. V., and K. Strawn, 1985. The effects of feeding rates on the dynamics of growth of black drum and spot cage cultured in a heated water lake. *Journal of the World Mariculture Society* 16:19–31.

Kafuku, T., and H. Ikenoue, 1983. *Modern Methods of Aquaculture in Japan.* New York: Elsevier Science Publishing Company, Inc.

Kanazawa, A., S. Koshiom, and S.-I. Teshima, 1989. Growth and survival of larval red sea bream *Pagrus major* and Japanese flounder *Paralichthys olivaceus* fed microbound diets. *Journal of the World Aquaculture Society* 20:31–37.

Krise, W. F., and J. W. Meade, 1986. Review of the intensive culture of walleye fry. *The Progressive Fish-Culturist*, 48: 81–89.

Krise, W. F., L. Bulkowski-Cummings, A. D. Shellman, K. A. Kraus, and R. W. Gould, 1986. Increased walleye egg hatch and larval survival after protease treatment of eggs. *The Progressive Fish-Culturist* 48:95–100.

Landau, M., and J. H. Ryther, 1985. Culture of the shrimp *Penaeus vannamei* Boone using feed and treated wastewater. *Journal of Shellfish Research* 5:26–27.

Loadman, N. L., J. A. Mathias, and G. E. E. Moodie, 1989. Method for the intensive culture of walleye. *The Progressive Fish-Culturist* 51:1–9.

Luczynski, M., P. Majkowski, and K. Dabrowski, 1986. Rearing of larvae of four coregonid species using dry and live food. *Aquaculture* 56:179–185.

Matsubara, S., 1908. Goldfish and their culture in Japan. *Bulletin of the Bureau of Fisheries* 28:381–397.

May, R. C., and M. Santerre, 1977. Some effects of temperature and salinity on laboratory reared eggs and larvae of *Polydactylus sexfilis* (Pisces: Polynemidae). *Aquaculture* 10:341–351.

Mazzola, A., and B. Rallo, 1981. Further experiences in the intensive culture of the seabream (*Sparus aurata* L.). *Journal of the World Mariculture Society* 12:137–142.

McLarney, B., 1984. Native aquarium fish—a new aquaculture crop, part II. *Aquaculture Magazine* 10(3):18–21.

McLarney, B., 1985. Pioneers in saltwater aquarium fish, part I. *Aquaculture Magazine* 11(6):38–41.

McLarney, B., 1986. Pioneers in saltwater aquarium fish, part II. *Aquaculture Magazine* 12(1):31–33.

Metailler, R., J. F. Aldrin, J. L. Messager, G. Mevel, and G. Stephan, 1981. Feeding of European sea bass *Dicentrarchus labrax:* Role of protein level and energy source. *Journal of the World Mariculture Society* 12:117–118.

Monaco, G., R. K. Buddington, and S. I. Doroshov, 1981. Growth of the white sturgeon (*Acipenser transmontanus*) under hatchery conditions. *Journal of the World Mariculture Society* 12:113–121.

Muir, J. E., and R. J. Roberts, 1982. *Recent Advances in Aquaculture*. Boulder, Colo.: Westview Press.

Perschbacher, P. W., and K. Strawn, 1988. The sheepshead minnow *Cyprinodon variegatus*—a major pest in experimental mariculture of the gulf killifish *Fundulus grandis*. *Journal of the World Mariculture Society* 19:113–117.

Pillay, T. V. R., and W. A. Dill, 1979. *Advances in Aquaculture*. Farnham, Surrey, England: Fishing News Books Ltd.

Porter, C. W., and A. F. Maciorowski, 1984. Spotted seatrout fingerling production in saltwater ponds. *Journal of the World Mariculture Society* 15:222–232.

Prentice, J. A., and R. L. Colura, 1984. Preliminary observations of orangemouth corvina spawn inducement using photoperiod, temperature and salinity cycles. *Journal of the World Mariculture Society* 15:162–171.

Roberts, D. E., and R. A. Schlieder, 1983. Induced sex inversion, maturation, spawning and embryogeny of the protogynous grouper, *Mycteroperca microlepis*. *Journal of the World Mariculture Society* 14:639–649.

Robin, J. H., F. J. Gatesoupe, and R. Ricardez, 1981. Production of brine shrimp (*Artemia salina*), using mixed diet: consequences on rearing of sea bass larvae (*Dicentrarchus labrax*). *Journal of the World Mariculture Society* 12:119–120.

Rossberg, K. S., and R. K. Strawn, 1980. Induced feeding and growth enhancement of black drum when cultured with Florida pompano and striped mullet. *Proceedings of the World Mariculture Society* 11:226–234.

Saint-Paul, U., 1986. Potential for aquaculture of South American freshwater fishes: A review. *Aquaculture* 54:205–240.

Smith, T. I. J., and E. K. Dingley, 1984. Review of biology and culture of Atlantic (*Acipenser oxyrhynchus*) and shortnose sturgeon (*A. brevirostrum*). *Journal of the World Mariculture Society* 15:210–218.

Smith, T. I. J., E. K. Dingley, and D. E. Marchette, 1981. Culture trials with Atlantic sturgeon, *Acipenser oxyrhynchus. Journal of the World Mariculture Society,* 12:78–87.

Stickney, R. R., 1979. *Principles of Warmwater Aquaculture.* New York: John Wiley & Sons, Inc.

Stickney, R. R., and S. P. Meyers, 1983. *Proceedings of the Warmwater Fish Culture Workshop.* Baton Rouge: Louisiana State University.

Szyper, J. P., R. Bourke, and L. D. Conquest, 1984. Growth of juvenile dolphin fish, *Coryphaena hippurus,* on test diets differing in fresh and prepared components. *Journal of the World Mariculture Society* 15:219–221.

Trimble, W. C., W. M. Tatum, and S. A. Styron, 1981. Pond studies on gulf killifish (*Fundulus grandis*) mariculture. *Journal of the World Mariculture Society* 12(2):50–60.

Tseng, W.-Y., 1983. Prospects for commercial netcage culture of red grouper (*Epinephelus akaara* T. & S.) in Hong Kong. *Journal of the World Mariculture Society* 14:650–660.

Tucker, J. W., Jr., 1987a. Snook and tarpon snook culture and preliminary evaluation for commercial farming. *The Progressive Fish-Culturist* 49:49–57.

Tucker, J. W., Jr., 1987b. Sheepshead culture and preliminary evaluation for farming. *The Progressive Fish-Culturist* 49:224–228.

Tucker, J. W., Jr., 1988. Growth of the spotted seatrout on dry feeds. *The Progressive Fish-Culturist* 50:39–41.

Whitworth, W. R., P. L. Berrien, and W. T. Keller, 1968. *Freshwater Fishes of Connecticut,* Bulletin 101. State Geological and Natural History Survey of Connecticut.

Wiggins, T. A., T. R. Bender, Jr., V. A. Mudrak, and M. A. Takacs, 1983. Hybridization of yellow perch and walleye. *The Progressive Fish-Culturist* 45:131–132.

APPENDIX 9

Adams, I. K., and A. C. Bruinsma, 1987. Intensive commercial bullfrog culture: the Brazilian experience. *Aquaculture Magazine* 13(4):28–44.

Anonymous, 1988. Alligators' steamy comfort boosts growth rate. *Aquaculture Magazine* 14(6):26–29.

Avault, J. W., Jr., 1985. The alligator story. *Aquaculture Magazine* 11(4):41–44.

Bardach, J. E., J. H. Ryther, and W. O. McLarney, 1972. *Aquaculture.* New York: John Wiley & Sons, Inc.

Chen, T. P., 1976. *Aquaculture Practices in Taiwan.* Norwich, England: Page Brothers Limited.

Culley, D. D., 1986. Bullfrog culture still a high risk venture. *Aquaculture Magazine* 12(5):28–35.

Joanen, T., and L. McNease, 1977. Artificial incubation of alligator eggs and post hatching culture in controlled environmental chambers. *Proceeding of the World Mariculture Society* 8:483–490.

Lester, D., 1988. Raising bullfrogs on non-living food. *Aquaculture Magazine* 14(2):20–27.

Marshall, G. A., R. L. Amborski, and D. D. Culley, Jr., 1980. Calcium and pH requirements in the culture of bullfrog (*Rana catesbeiana*) larvae. *Proceedings of the World Mariculture Society* 11:445–453.

Riedman, S. R., and R. Witham, 1974. *Turtles: Extinction or Survival.* New York: Abelard-Schuman.

Wood, J. R., and F. E. Wood, 1977. Captive breeding of the green sea turtle (*Chelonia mydas*). *Proceeding of the World Mariculture Society* 8:533–541.

INDEX

431